RECENT INSIGHTS IN PETROLEUM SCIENCE AND ENGINEERING

Edited by **Mansoor Zoveidavianpoor**

Recent Insights in Petroleum Science and Engineering
http://dx.doi.org/10.5772/65538
Edited by Mansoor Zoveidavianpoor

Contributors

Naga Raju Maddela, Laura Scalvenzi, Tatjana Paulauskiene, Galba Maria Campos-Takaki, Marcos Antonio Babosa Lima, Luciana De Oliveira Franco, Manfred Schwartz, Kaoru Okada, Aline Elesbão Do Nascimento, Carlos Alberto Alves Da Silva, Patricia Mendes Souza, Thayse Alves De Lima E Silva, Ian Garrard, Elias B Tambourgi, Nabil Mohammed Al-Areeq, Nohra Violeta Gallardo Rivas, Ernestina Elizabeth Banda Cruz, Ulises Paramo-Garcia, Ana María Mendoza-Martínez, José Aarón Melo Banda, Oscar E. Vázquez-Noriega, Javier Guzmán-Pantoja, Maria Yolanda Chavez-Cinco, Reinaldo David Martínez-Orozco, Luciano Aguilera-Vázquez, Luis A. Alcazar-Vara, Ignacio R. Cortés-Monroy, Reda Abdelazim, Sara Faiz, Nour Al Obaidi, Ahmed Al Bagir, Bing Wei, Lezorgia Nekabari Nwidee, Ahmed Barifcani, Stefan Iglauer, Mohammad Sarmadivaleh, Mouna Mahjoubi, Simone Cappello, Yasmine Souissi, Ameur Cherif, Yongtu Liang, Haoran Zhang, Jing Ma, Pengwei Di, Xiaohan Yan, Zhongliang Huang, Adolfo Pires, Rafael Scardini, Viatcheslav Priimenko, David Lacerda, Andre Vinhal, Igor Evdokimov, Aleksey Fesan, Aleksandr Losev, Abiodun Olagoke Adeniji, Omobola Okoh, Anthony Okoh, Muftah El-Naas, Taghreed Al-Khalid, Pradip Mandal, Mitsuru Sasaki, Dmitry Borisov, Dmitry Milordov, Svetlana Yakubova, Makhmut Yakubov

Published by InTech
Janeza Trdine 9, 51000 Rijeka, Croatia

Publishing Process Manager Iva Simcic
Technical Editor SPi Global
Cover InTech Design team

First published February, 2018
Printed in Croatia
Legal deposit, Croatia: National and University Library in Zagreb

Additional hard copies can be obtained from orders@intechopen.com

Recent Insights in Petroleum Science and Engineering, Edited by Mansoor Zoveidavianpoor
p. cm.
Print ISBN 978-953-51-3809-9
Online ISBN 978-953-51-3810-5

For EU product safety concerns:
IN TECH d.o.o., Prolaz Marije Krucifikse Kozulić 3, 51000 Rijeka, Croatia,
info@intechopen.com or visit our website at intechopen.com.

Meet the editor

Dr. Mansoor Zoveidavianpoor has over 18 years of multidisciplinary oil and gas experience, built upon his technical, operational, and management roles in the industry and academia. He received his BSc degree in Geology and both his MSc and PhD degrees in Petroleum Engineering. He was involved in different disciplines such as geology, flow assurance, piping construction, artificial intelligence, environmental engineering, petroleum engineering, and project management. Dr. Zoveidavianpoor has lectured several courses at the University of Technology Malaysia (UTM), the Petroleum University of Technology (PUT), and the Islamic Azad University (IAU). He is a member of Society of Petroleum Engineers (SPEs) and registered as a chartered petroleum engineer. He has published more than 50 publications in international peer-reviewed journals and conferences, contributed to 5 textbooks, and served in many scientific committees. Previously, Dr. Zoveidavianpoor was working as an assistant professor at UTM, and currently, he is working as a senior petroleum engineer at the National Iranian Oil Company (NIOC). He is actively involved in multidisciplinary studies, and currently, his main focus is on integrated reservoir management.

Contents

Preface IX

Section 1 Exploration and Production 1

Chapter 1 **Petroleum Source Rocks Characterization and Hydrocarbon Generation 3**
Nabil Mohammed Al-Areeq

Chapter 2 **Petroleum Extraction Engineering 31**
Tatjana Paulauskiene

Chapter 3 **Optimal Planning for Deepwater Oilfield Development Under Uncertainties of Crude Oil Price and Reservoir 55**
Zhang Haoran, Liang Yongtu, Ma Jing, Di Pengwei, Yan Xiaohan and Huang Zhongliang

Chapter 4 **Drilling Fluids for Deepwater Fields: An Overview 71**
Luis Alberto Alcázar-Vara and Ignacio Ramón Cortés-Monroy

Chapter 5 **Evaluation of Different Correlation Performance for the Calculation of the Critical Properties and Acentric Factor of Petroleum Heavy Fractions 99**
Dacid B. Lacerda, Rafael B. Scardini, André P. C. M. Vinhal, Adolfo P. Pires and Viatcheslav I. Priimenko

Chapter 6 **Characterization of Crude Oils and the Precipitated Asphaltenes Fraction using UV Spectroscopy, Dynamic Light Scattering and Microscopy 117**
Ernestina Elizabeth Banda Cruz, Nohra Violeta Gallardo Rivas, Ulises Páramo García, Ana Maria Mendoza Martinez and José Aarón Melo Banda

Chapter 7 **Density Anomalies in Crude Oil Blends Reflect Multiple Equilibrium States of Asphaltene Colloidal Aggregates 137**
Igor N. Evdokimov, Aleksey A. Fesan and Aleksandr P. Losev

Chapter 8 **A Realistic Look at Nanostructured Material as an Innovative Approach for Enhanced Oil Recovery Process Upgrading 155**
Lezorgia Nekabari Nwidee, Ahmed Barifcani, Maxim Lebedev, Mohammad Sarmadivaleh and Stefan Iglauer

Chapter 9 **Lessons Learned from Our Recent Research in Chemical Enhanced Oil Recovery (C-EOR) Methods 189**
Bing Wei, Peng Wei, Shuai Zhao and Wanfen Pu

Chapter 10 **Numerical Study of Low Salinity Water Flooding in Naturally Fractured Oil Reservoirs 211**
Reda Abdel Azim, Sara Faiz, Shaik Rahman, Ahmed Elbagir and Nour Al Obaidi

Chapter 11 **Experimental Study of the Effect of Composite Solvent and Asphaltenes Contents on Efficiency of Heavy Oil Recovery Processes at Injection of Light Hydrocarbons 229**
Dmitry N. Borisov, Dmitry V. Milordov, Svetlana G. Yakubova and Makhmut R. Yakubov

Section 2 Environmental Solutions 249

Chapter 12 **Total Acid Number Reduction of Naphthenic Acids Using Supercritical Fluid and Ionic Liquids 251**
Pradip Chandra Mandal and Mitsuru Sasaki

Chapter 13 **Conducting Polymers Films Deposited on Carbon Steel and Their Interaction with Crude Oil 273**
Oscar E. Vázquez-Noriega, Javier Guzmán, Nohra V. Gallardo-Rivas, Reinaldo David Martínez Orozco, Ana M. Mendoza-Martínez, María Yolanda Chávez Cinco, Luciano Aguilera Vázquez and Ulises Páramo-García

Chapter 14 **Desulfurization of Dibenzothiophene by Pseudomonas fluorescens (UCP 1514) Leading to the Production of Biphenyl 293**
Thayse A.L. Silva, Manfred Schwartz, Patrícia M. Souza, Ian Garrard, Galba M. Campos-Takaki and Elias B. Tambourgi

Chapter 15 **Reduction in the Sulfur Content of Fossil Fuels by Cunninghamella elegans (UCP 0596) to Dibenzothiophene Compound 309**
Patrícia Mendes de Souza, Thayse Alves de Lima e Silva, Marcos Antonio Barbosa Lima, Luciana de Oliveira Franco, Manfred Schwartz, Paulo Henrique da Silva, Lúcia Roberta Barbosa, Aline Elesbão do Nascimento, Kaoru Okada and Galba Maria de Campos-Takaki

Chapter 16 **Microbial Bioremediation of Petroleum Hydrocarbon–Contaminated Marine Environments 325**
Mouna Mahjoubi, Simone Cappello, Yasmine Souissi, Atef Jaouani and Ameur Cherif

Chapter 17 **Petroleum Degradation: Promising Biotechnological Tools for Bioremediation 351**
Maddela Naga Raju and Laura Scalvenzi

Chapter 18 **Organic Contaminants in Refinery Wastewater: Characterization and Novel Approaches for Biotreatment 371**
Taghreed Al-Khalid and Muftah H. El-Naas

Chapter 19 **Analytical Methods for Polycyclic Aromatic Hydrocarbons and their Global Trend of Distribution in Water and Sediment: A Review 393**
Abiodun Olagoke Adeniji, Omobola Oluranti Okoh and Anthony Ifeanyi Okoh

Preface

This book presents new insights into the development of different aspects of petroleum science and engineering and covers the different topics from upstream to downstream sectors. The book contains 19 chapters divided into two main sections: (i) Exploration Production and (ii) Environmental Solutions. There are 11 chapters in the first section, and the focus is on the topics related to exploration and production of oil and gas, such as characterization of petroleum source rocks, drilling technology, characterization of reservoir fluids, and enhanced oil recovery (EOR). The first three chapters of the present book are about the characterization of petroleum source rocks, the purpose and principles of rotary drilling rig, and a new technique in planning for deep-water oilfield development; the next threechapters deal with the characterization of reservoir fluids, and the remaining five chapters are about the new approaches to an enhanced oil recovery (EOR) using nanoparticles, solvent injection novel chemical EOR methods, water flooding in naturally fractured reservoirs, and techniques for EOR. There are 8 chapters in the second section, and the special emphasis is on waste technologies and environmental cleanup in the downstream sector. The first twochapters present new approaches to dealing with corrosion problems in the petroleum industry; the next two chapters cover the biodegradation of wastes, and the remaining three chapters deal with the different bioremediation techniques. The reduction in polycyclic aromatic hydrocarbons is discussed in the last chapter. Intended for readers wishing to acquire understanding of the current trends in petroleum science and engineering and comprehension of the issues, this book addresses exciting topics in this field.

Mansoor Zoveidavianpoor, PhD
Sr. Petroleum Engineer
National Iranian Oil Company (NIOC)—Arvandan Oil & Gas Company (AOGC)
Khorramshahr, Khuzestan, Iran

Section 1

Exploration and Production

Chapter 1

Petroleum Source Rocks Characterization and Hydrocarbon Generation

Nabil Mohammed Al-Areeq

Additional information is available at the end of the chapter

http://dx.doi.org/10.5772/intechopen.70092

Abstract

This chapter is proposed to give the principal learning on the application of the formation of petroleum source rocks and hydrocarbon generation to exploration activities. The evaluation of petroleum source rocks and hydrocarbon generation is a very important skill for explorationists to define the location and type of petroleum prospects in a region. In this chapter, subsurface samples from case study (Sayun-Masilah basin) were used to determine the source rock characteristics and petroleum generative potentials of prospective source rocks. Qualitative and quantitative evaluation of the source rock in this basin was done by means of geochemical and geophysical approaches for four rock units. It is clear that Madbi Formation is considered the main source, in which the organic carbon content reached up to more than 5.2 wt%. The types of organic matter from rock-eval pyrolysis data indicated that type I kerogen is the main type, in association with type II, and a mixture of types II and III kerogens. The study of the different maturation parameters obtained from rock-eval pyrolysis, such as T_{max} and vitrinite reflectance, reflects that the considered rock units are occurred in different maturation stages, ranging from immature to mature sources. One-dimensional basin modeling was performed to analyze the hydrocarbon generation and expulsion history of the source rocks in the study area based on the reconstruction of the burial and thermal maturity histories in order to improve our understanding of the hydrocarbon generation potential. Calibration of the model with measured vitrinite reflectance (%Ro) and borehole temperature (BHT) data indicates that the paleo-heat flow was high at Late Jurassic. The models also indicate that the early hydrocarbon generation in the Madbi source rock occurred during late Cretaceous and the main hydrocarbon generation has been reached approximately at Early Eocene. Therefore, the Madbi source rock can be considered as generative potentials of prospective source rock horizons in the Sayun-Masilah basin.

Keywords: source rocks, thermal maturity, basin modeling, hydrocarbon generation modeling, Sayun-Masilah basin, Yemen

1. Introduction

Petroleum source rock is defined as the fine-grained sediment with sufficient amount of organic matter, which can generate and release enough hydrocarbons to form a commercial accumulation of oil or gas [1]. Source rocks are commonly shales and lime mudstones, which contain significant amount of organic matter [2]. A petroleum source rock is defined as any rock that has the capability to generate and expel enough hydrocarbons to form an accumulation of oil or gas. Source rocks are classified according to oil generation into three classes [1], as follows:

1. Immature source rocks that have not yet generated hydrocarbons.
2. Mature source rocks that are in generation phase.
3. Post mature source rocks are those which have already generated all crude oil type hydrocarbons.

Waples [3] distinguished the petroleum source rocks into potential, possible, and effective, as follows:

A. Potential source rocks are immature sedimentary rocks capable of generating and expelling hydrocarbons, if their level of maturity were higher.

B. Possible source rocks are sedimentary rocks whose source potential has not yet been evaluated, but which may have generated and expelled hydrocarbons.

C. Effective source rocks are sedimentary rocks, which have already generated and expelled hydrocarbons.

The hydrocarbon source evaluation is generally based on the organic matter quantity (organic richness), quality (kerogen type), and the thermal maturation generation capability and of the organic matter disseminated in the rock [2–4]. Organic matter content can be determined directly from laboratory analyses of the source rock samples (shale, limestone, or marl), and through indirect methods based on wireline data offer the advantages of economic, ready availability of data, and continuity of sampling of vertically heterogeneous shale section.

This chapter displays how to evaluate source rocks and hydrocarbon generation using geochemical data, with an easy method and evaluate source rocks from gas chromatography for crude oil and extract bitumen. This chapter will be presented with case study and some examples for understanding petroleum source rocks and hydrocarbon generation. Also, in this chapter, quantitative one-dimensional basin modeling is performed for evaluating the thermal histories and timing of hydrocarbon generation and expulsion of the source rocks in the sedimentary basin.

2. Evaluation of petroleum source rocks using geochemical data

The source rock evaluation within any study area involved the recognition of petroleum source, which depends on the determination of its proportion of organic matter (organic matter quantity), which is usually expressed as total organic carbon (TOC wt%). It also depends on the type (or quality) of organic matter (kerogen) preserved in the petroleum source. The geochemical data such as total organic carbon (TOC wt%), rock-eval pyrolysis data, bitumen extraction, and vitrinite reflectance are presented and discussed for the proposed Upper Jurassic and Lower Cretaceous rock units in the Sayun-Masilah basin in Yemen. The total organic carbon, S2 and genetic potential from rock-eval pyrolysis and extractable organic matter (bitumen) from selected rock samples were used to identify the source-richness in terms of quantity and generation potential. Plots of T_{max} (C°) against hydrogen index (HI mgHC/gm of TOC) and hydrogen index (HI) against oxygen index (OI) from rock-eval pyrolysis are used to identify the kerogen type (quality) and depositional environment. Rock-eval T_{max} (C°) was used to evaluate source rock maturity stage, in conjunction with vitrinite reflectance pattern as a maturity tool.

Pyrolysis is almost the best routine tool for determining the kerogen type [5]. The rock-eval pyrolysis data are considered to be the most valuable geochemical exploration tool used to evaluate the type of organic matter, thermal maturity, and the generation capability of source rocks. The generated thermo-vaporized free hydrocarbons already present in the rock "S1" are released at temperatures lower than those needed to break down the kerogen, hence monitoring the hydrocarbons released by steadily increasing temperature, providing a way for obtaining the amount of generated hydrocarbons relative to the total potential. The "S2" peak represents the genetic potential of the sample, which is the hydrocarbon that would generate at optimum maturity. The "S1" and "S2" are expressed in milligrams of hydrocarbon per gram of rock (mg/g). The "S3" peak represents the quantity of evolved CO_2 expressed in milligrams of CO_2 per gram of rock (mg/g). The temperature (T_{max}) at which the pyrolysis peak S2 occurs has been used as a measure of maturity; it increases with increasing levels of maturity. Two useful parameters are obtained from rock-eval pyrolysis data: the hydrogen index (HI = S2/TOC wt%, equivalent to H/C atomic ratio in van Krevelen diagram) and the oxygen index (OI = S3/TOC wt%, equivalent to O/C atomic ratio in the kerogen).

2.1. Source rock generative potential

The organic matter richness and hydrocarbon generative potential of the source rocks in the Sayun-Masilah basin can be evaluated by bulk geochemical data such as TOC content and pyrolysis S1 and S2 yields (**Table 1**). The organic richness of a rock is usually expressed as the total organic carbon content (wt% TOC). The minimum acceptable TOC value for clastic type rocks indicating good source potential is 1.0% [1, 6]. The Upper Jurassic sediments samples have moderate to high TOC content (0.85–33.3 wt%), revealing organic-rich intervals within stratigraphic levels (**Figure 1**). The L. Qishn Member (Lower Cretaceous age) consists of sandstone and shale and with small intercalations of carbonates. The L. Qishn shale samples contain rich organic matter and have TOC content of 1.4–3% (**Figure 1**). Based on the classification proposed by [7], L. Qishn shale sample is considered to be a fair to good source rock (**Figure 1**). The amount of hydrocarbon yield (S2) expelled during pyrolysis is a useful measurement to evaluate the generative potential of source rocks [7, 8]. Most of the analyzed samples have more than 1.0 mg HC/g rock (**Figure 1**). Thus, pyrolysis S2 yields indicate that the L. Qishn shale samples are poor to fair generative potential (**Figure 1**).

Age	Rock Units	Depth (m)	TOC (wt%)	Rock-eval pyrolysis								Ro (%)
				S_1 (mg/g)	S_2 (mg/g)	S_3 (mg/g)	T_{max} (C)	HI (mg/g)	OI (mg/g)	PI (mg/g)	PY ($S_1 + S_2$)	
Lower Cretaceous	Harshiat Fm	900	3.47	0.0694	6.8706	0.3817	433	198	11	0.01	6.94	0.36
		1000	33.3	0.6963	68.931	2.664	428	207	8	0.01	69.6	0.36
		1100	34	0.5564	55.08	2.72	430	162	8	0.01	55.6	0.34
		1500	0.85	0.0201	0.3825	0.476	427	45	56	0.05	0.4	0.4
	L. Qishn Member	1860	1.87	2.1442	3.0855	3.927		165	210	0.41	5.2	0.39
		1875	17	23.803	37.23	24.99		219	147	0.39	61.0	
		1890	2.99	3.8757	5.3521	3.9767		179	133	0.42	9.2	
		1920	1.4	1.0478	1.946	3.024		139	216	0.35	2.9	0.41
Upper Jurassic	Nayfa Fm	2310	0.82	0.0306	0.5822	0.164	436	71	20	0.05	0.6	0.43
		2400	2.6	1.7967	6.37	4.264		245	164	0.22	8.1	0.5
		2430	0.83	0.0743	0.9877	0.1826	442	119	22	0.07	1.0	
		2457	1.01	0.114	1.515	0.2828	444	150	28	0.07	1.6	0.5
		2490	1.57	0.3072	3.5325	0.3454	444	225	22	0.08	3.8	
		2510	1.65	0.2359	3.696	0.396	442	224	24	0.06	3.9	0.5
		2540	1.62	0.2456	4.6656	0.2754	441	288	17	0.05	4.9	
		2555	1.29	0.3668	4.2183	0.4257	435	327	33	0.08	4.5	
		2570	2.23	0.5082	9.6559	0.4014	440	433	18	0.05	10.1	
	Madbi Fm	2585	8.37	5.5056	49.55	0.7533	443	592	9	0.1	55.0	0.74
		2586.4	6.1	6.9342	33.855	0.61	442	555	10	0.17	40.7	
		2587	4.07	4.4224	20.147	0.4477	439	495	11	0.18	24.5	
		2588	8.16	6.5111	43.574	0.5712	441	534	7	0.13	50.0	
		2589	11.8	7.3684	66.316	0.472	442	562	4	0.1	73.6	
		2590	6.85	4.9697	40.21	0.3425	441	587	5	0.11	45.1	
		2591	7.9	5.1359	41.554	0.869	443	526	11	0.11	46.6	0.87
		2591.3	8.82	5.3472	54.067	0.3528	438	613	4	0.09	59.4	
		2592.7	2.75	3.1625	12.65	0.55	437	460	20	0.2	15.8	
		2593.4	3.48	3.9917	17.017	0.3828	432	489	11	0.19	21.0	
		2595	7.37	5.265	42.599	0.6633	442	578	9	0.11	47.8	
		2597	7.41	5.507	40.385	0.8151	441	545	11	0.12	45.8	
		2598.5	5.92	5.9989	31.494	0.6512	442	532	11	0.16	37.4	
		2602	4.63	4.1098	23.289	0.5093	444	503	11	0.15	27.3	
		2605	4.08	3.3276	20.441	0.4896	440	501	12	0.14	23.7	0.78
		2606	5.96	4.3237	31.707	0.2384	442	532	4	0.12	36.0	0.81

Age	Rock Units	Depth (m)	TOC (wt%)	Rock-eval pyrolysis								Ro (%)
				S_1 (mg/g)	S_2 (mg/g)	S_3 (mg/g)	T_{max} (C)	HI (mg/g)	OI (mg/g)	PI (mg/g)	PY (S_1+S_2)	
		2606.5	4	4.5828	17.24	0.64	439	431	16	0.21	21.8	
		2610.7	4.05	4.2737	22.437	0.5265	440	554	13	0.16	26.7	0.73
		2615	1.24	0.7307	4.4888	0.2232	436	362	18	0.14	5.21	
	Shuqra Fm	2615.5	5.09	3.3802	24.788	0.7126	442	487	14	0.12	28.1	
		2625	5.94	3.9718	32.135	0.594	440	541	10	0.11	36.1	
		2665	1.47	0.3286	4.3659	0.4557	440	297	31	0.07	4.6	
		2680	1.13	0.163	3.0962	0.3729	442	274	33	0.05	3.2	0.82

S1: volatile hydrocarbon (HC) content, mg HC/g rock; TOC: total organic carbon, wt.%.
S2: remaining HC generative potential, mg HC/g rock; S3: Volatile carbon dioxide (CO2) content, mg HC/g rock; PI: production index = S1/(S1 + S2).
T_{max}: temperature at maximum of S2 peak; PY: potential yield = S1 + S2 (mg/g).
HI: hydrogen index = S2 × 100/TOC, mg HC/g TOC; OI: Oxygen Index = S3 × 100/TOC; Ro: Vitrinite reflectance.

Table 1. Results of pyrolysis and TOC content analyses with calculated parameters with measured vitrinite reflectance of the source rocks in Sayun-Masilah basin.

L. Qishn shale samples have low hydrogen index (HI) values in the range of 139–219 mg HC/g TOC (**Figure 1**). The shale samples in the Nayfa Formation (Upper Jurassic age) contains a total organic carbon content ranging between 0.82 and 2.2 wt%; thus indicating a fair to good source rock (**Figure 1**). The pyrolysis yield S2 and petroleum potential yield ranging from 0.5 to 9.6 and 0.07 to 1.7 mg HC/g rock, respectively; **Figure 1**, consequently, is considered to be poor to good source generative potential. Total organic carbon (TOC) analysis showed high TOC values of the samples from the Madbi Formation (Upper Jurassic age) and ranging from 1.2 to 8.8 wt% (**Figure 1**). The TOC contents meet the accepted standards of a source rock with good to excellent hydrocarbon generative potential as suggested by [1]. This is confirmed by the pyrolysis S2 yield, petroleum potential yield S1, and extract of organic matter (EOM) (**Figure 1**) as a useful parameter to evaluate the generation potential of source rocks [7, 8].

The hydrocarbon yields (S2) are in agreement with TOC content, indicating that the shales of Madbi Formation are good to excellent source rock generative potential based on the classification by Peters and Cassa [6] (**Figure 2**). The shale samples could become the most promising source rock for hydrocarbon generation as reflected by high pyrolysis yield (S2) and total organic carbon (TOC wt%) content (**Figure 2**).

Overall, the relation between genetic petroleum potential yield (PY; S1 + S2) and TOC of the studied units in the Sayun-Masilah basin confirms the above results, where it suggests that most of the samples from Madbi source rocks locate in the zone of the potential source rocks for hydrocarbon generation (**Figure 3**).

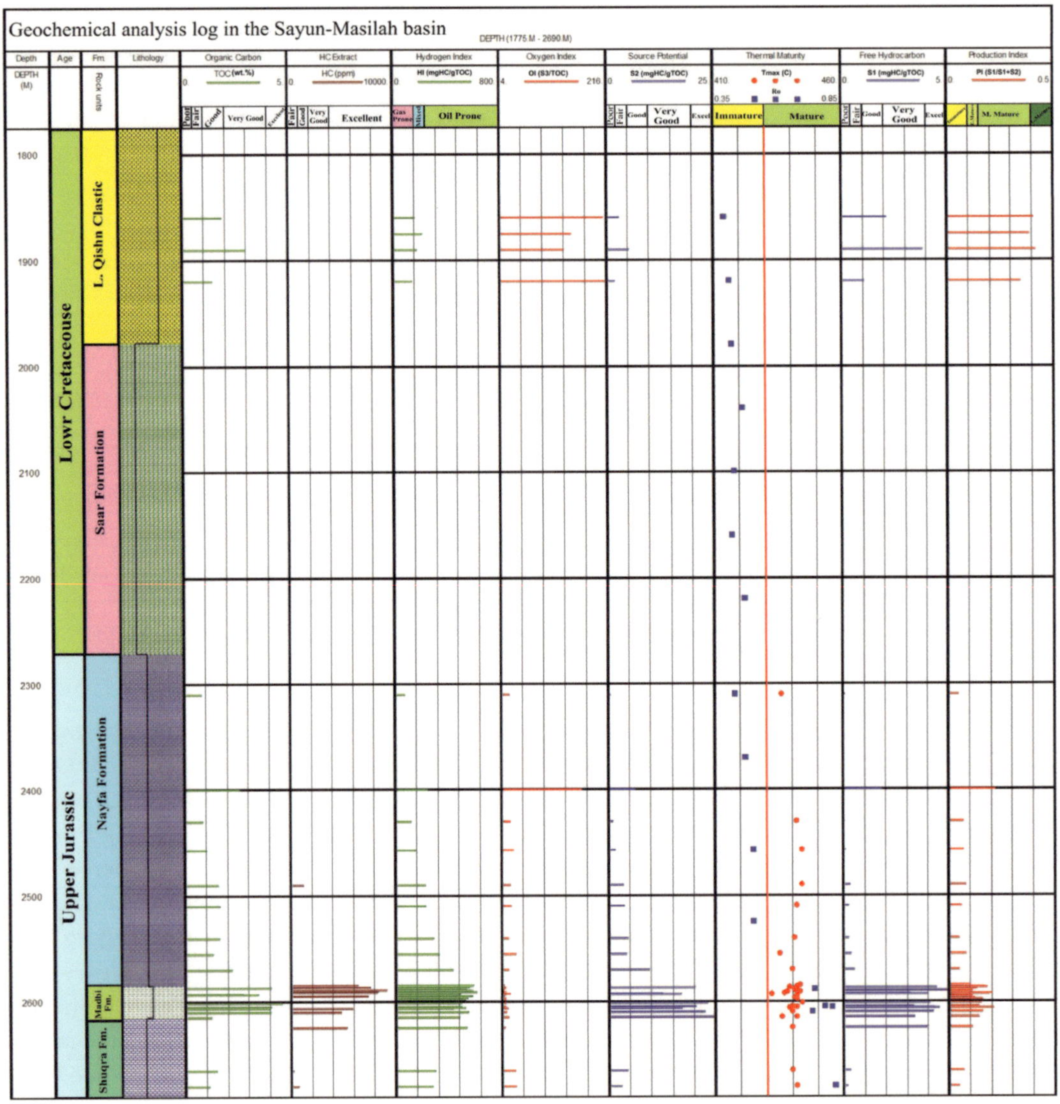

Figure 1. Organic geochemical log of the Upper Jurassic and Lower Cetaceous source rock samples according to Rock-eval pyrolysis and TOC content results.

2.2. Types of source rocks and depositional mechanisms of source rocks

One of the requirements needed for source rocks to generate commercial amounts of oil is that they must contain sufficient quantity of organic matter enough to generate and expel hydrocarbons [9]. Ronov (op.cit) also reported that the organic matter content in the open marine argillaceous sediments reaches about 1.1%, whereas that of the continental and lagoonal sediments reaches about 0.43%. Thomas [10] classified the potentiality of source rocks on the basis of their weight percentage of organic carbon; into poor source (<0.5 wt%), fair source

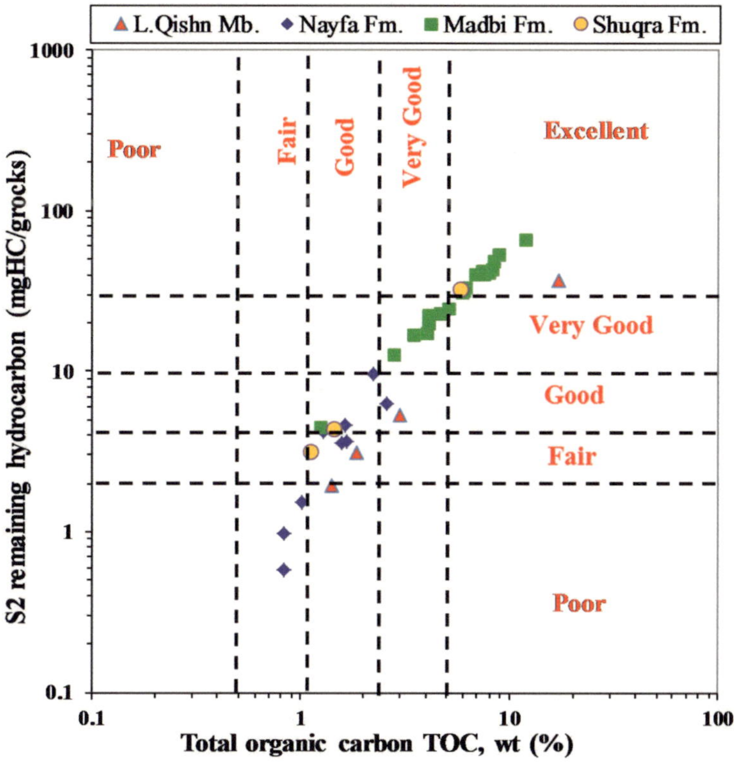

Figure 2. Pyrolysis S2 versus total organic carbon (TOC) plot showing generative source rock potential for the rock units in the study area.

(0.5–1.0 wt%), good source (1.0–2.0 wt%), and very good source (>2.0 wt%). Tissot and Welte [2] stated that, "clastic rocks, which are considered as source for petroleum contain a minimum of 0.5 wt% of the total organic carbon (TOC wt%), while good source rocks contain an average of about 2.0 wt% of TOC." The type of organic matter has important influence on the nature of generated hydrocarbons. Espitalie et al. [11] found that organic richness alone may not suffice to evaluate source rocks, where the organic matter is mainly inertinite, i.e., oxidized or biodegraded, is not capable of generating hydrocarbons, even if present at high concentrations. Peters and Cassa [6] presented a scale for the assessment of source rocks used in a wide scale and is applied in this work; a content of 0.5 wt% TOC as a poor source, 0.5–1.0 wt% as a fair source, 1.0–2.0 wt% as a good source, and more than 2.0 wt% TOC as a very good source rock, and also based on the rock-eval pyrolysis data, such S1 and S2, as shown in **Table 2**.

Pristane (pr) and phytane (ph) are usually the most important acyclic isoprenoid hydrocarbons in terms of concentration [12] and present in sediments and oil. Both are assumed to be diagenetic products of the phytyl side chain of chlorophyll 11 [13]. In certain restricted environments, for example, hypersaline, phytane can be derived from archae bacteria [14]. The pristane to phytane ratio (pr/ph) is similar for petroleum, which has resulted from material deposited under similar conditions [15]. Reducing conditions preferentially would lead to the formation of phytane.

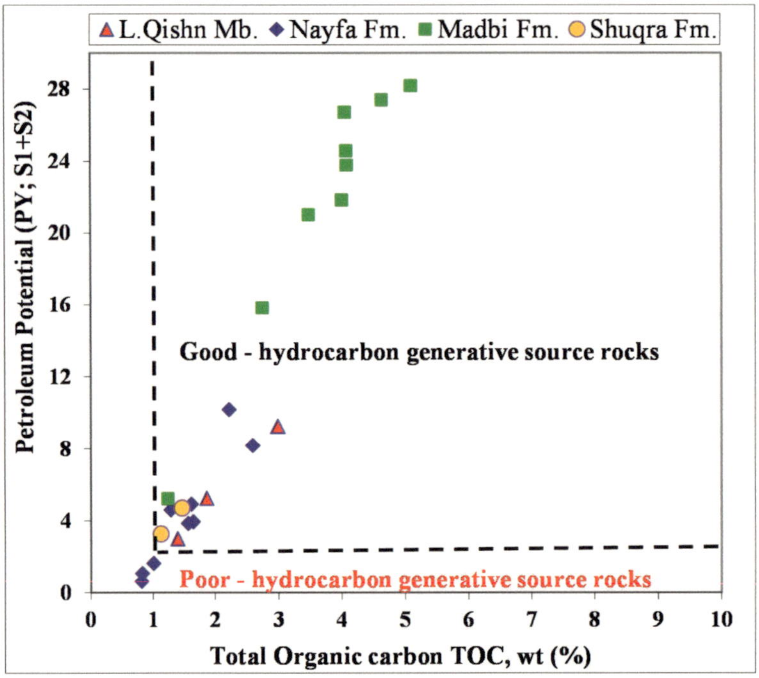

Figure 3. Source rock rating and hydrocarbon generative potential based on the plot of petroleum potential yield (PY) versus TOC for the analyzed rock samples.

Rowlands et al. [16] stated that, "the phytane carbon skeleton is extensively preserved under the condition, where H2S is available." Therefore, the pristane to phytane ratios of ancient sediments and oil reflect the paleoenvironmental conditions of source rock sedimentation [17] and are considered as potential indicators of the redox conditions during sedimentation and diagenesis [18]. Pristane/phytane ratio of less than 1 is ascribed to anoxic depositional environments, whereas ratios greater than 1 are ascribed to oxic depositional environments [17]. The use of the pr/ph ratio as an indicator, however, is not recommended to describe the paleoenvironment at low maturity levels [19]. Within the oil-generative window, the high pr/ph ratios (>3) indicate terrestrial organic matter input under oxic condition and low values typify anoxic, commonly under hypersaline environments. Hughes et al. [14] suggested that specific depositional environments and lithologies are associated with specific values for pr/ph ratios. Values less than 1 have associated with marine carbonates, between 1 and 3 with marine shales, and larger than 3 with

Quality	TOC (wt. %)	S_1 (mg HC/gm rock)	S_2 (mg HC/gm rock)
Poor	0.0–0.5	0.0–0.5	0.0–2.5
Fair	0.5–1.0	0.5–1.0	2.5–5.0
Good	1.0–2.0	1.0–2.0	5.0–10.0 >10
Very good	> 2	> 2	

Table 2. Source rock generative potential (after [7]).

http://dx.doi.org/10.5772/intechopen.70092

nonmarine shales and coals [20]. Lijmbach [21] proved that the ph/n-C17 ratio is less than 0.5 in environment with abundant aerobic bacterial activity and more than 1 in low aerobic bacterial activity environment. Shanmugam [22] made a combination between the isoprenoids and normal alkanes, which provides valuable information about the source of organic matter, organic facies, biodegradation, and maturation levels. These information are obtained from plotting the pristane /n-C17 versus phytane/n-C17. Abdullah [23], on the other hand, documented high pr/n-C17 and Ph/n-C18 values (1.5–1.6 and 1.2–1.6, respectively) in shallow marine shale and lower values (0.4–1.1 and 0.4–0.9, respectively) in deep marine shale conditions. Gas chromatography of the saturated hydrocarbons obtained from the extracted bitumen in the source rocks at Sayun-Masilah basin are used to identify the nature of organic matter present in these formations, preservation conditions, and depositional environments of the present organic matter. The isoprenoid pr/ph ratios of the studied rock units in the Sayun-Masilah basin (**Figure 4**) vary between 1.92 and 3.36, which point that the present organic matters were preserved under oxic to anoxic conditions in a deep marine environment as reflected from the lower values of pr/nC17 and ph/n-C18. The higher values of pr/ph ratio of >3 in some samples indicate a terrestrial input organic matter, i.e., presence of vitrinite macerals.

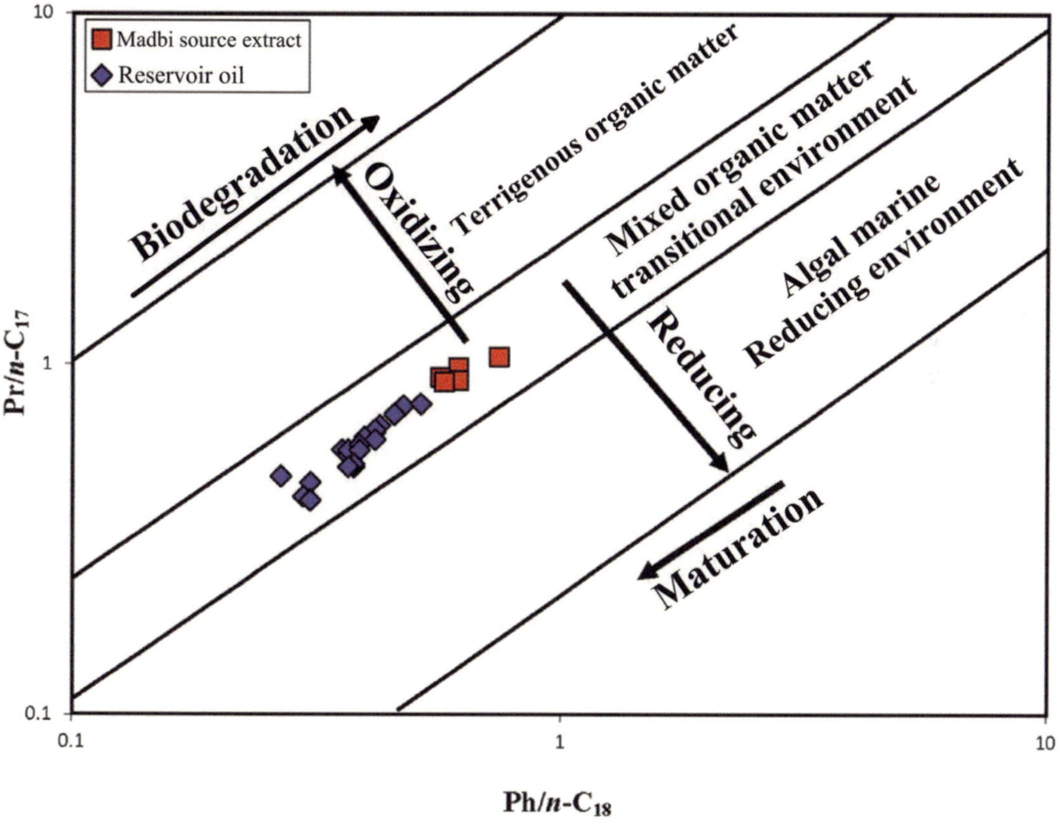

Figure 4. Phytane to n-C18 alkane (Ph/n-C18) versus pristane to n-C17 alkane (Pr/n-C17) ratios for reservoir crude oils and source rock extracts.

2.3. Types of organic matter (kerogen types)

The type of organic matter (kerogen) is considered the second most important parameter in evaluating the source rock. The kerogen type can be differentiated by optical microscopic or by physicochemical methods. The differences among them are related to the nature of the original organic matter. The organic matter in potential source rocks must be of the type that is capable of generating petroleum [2, 3]. It has been established that the organic matter is classified into three classes [24]:

1. (types I and II) equivalent Sapropelic type.
2. (type III) equivalent Humic type.
3. Mixed type from the two other types equivalent (II/III or III/II).

Espitalie et al. [5] used the pyrolysis yield to differentiate between the types of organic matter by plotting the hydrogen index versus the oxygen index on a modified Van Krevelen diagram, as follows:

1. Type I: mainly oil-prone organic matter with minor gas.
2. Type II: mixed oil and gas-prone organic matter.
3. Type III: mainly gas-prone organic matter.

The organic matter type is an important factor for evaluating the source rock potentiality and has important influence on the nature of the hydrocarbon products [2, 4]. Peters and Cassa [6] proposed that, for mature source rock, HI for gas-prone organic matter is less than 150, gas-oil-prone organic matter is between 150 and 300, whereas oil-prone organic matter is more than 300 HI. So, it is very important to determine the kerogen types, in order to detect the hydrocarbon products. The pyrolysis results can be used for the determination of the organic matter types. This can be achieved by drawing the relation between the hydrogen index (HI) and the oxygen index (OI).

In this study, the kerogen types present in the source rocks of the Sayun-Masilah basin identified from the modified Van Krevelen diagram (**Figure 5**) show that the Madbi shales interval contain kerogen of type I (algenite) oil prone, whereas type II (exinite) could be capable to generate mixed oil and gas. The hydrogen index (HI) value of Madbi Formation ranges from 362 to 613 mg/g with low oxygen index, indicating a capability of this formation to generate oil and mixed oil and gas hydrocarbons (**Figure 5**). The plots of the Naifa and L. Qishn Formations are characterized by the type II kerogen of lower hydrogen index (**Figure 5**), which could be capable to generate gases and oil, because they have higher (OI) than the Madbi Formation. Generally, most of the studied samples of Madbi source rocks of Sayun-Masilah basin are characterized by low OI, which reflect the capability of these sources to produce more oil than gases in type II kerogen.

2.4. Bitumen bulk geochemical parameters

The amounts of extractable organic matter (EOM), total hydrocarbon yield as well as the relative proportions of saturated, aromatic fractions, and nitrogen/sulfur/oxygen (NSO) compounds. The saturated and aromatic fractions together create the petroleum-like hydrocarbon fraction; thus,

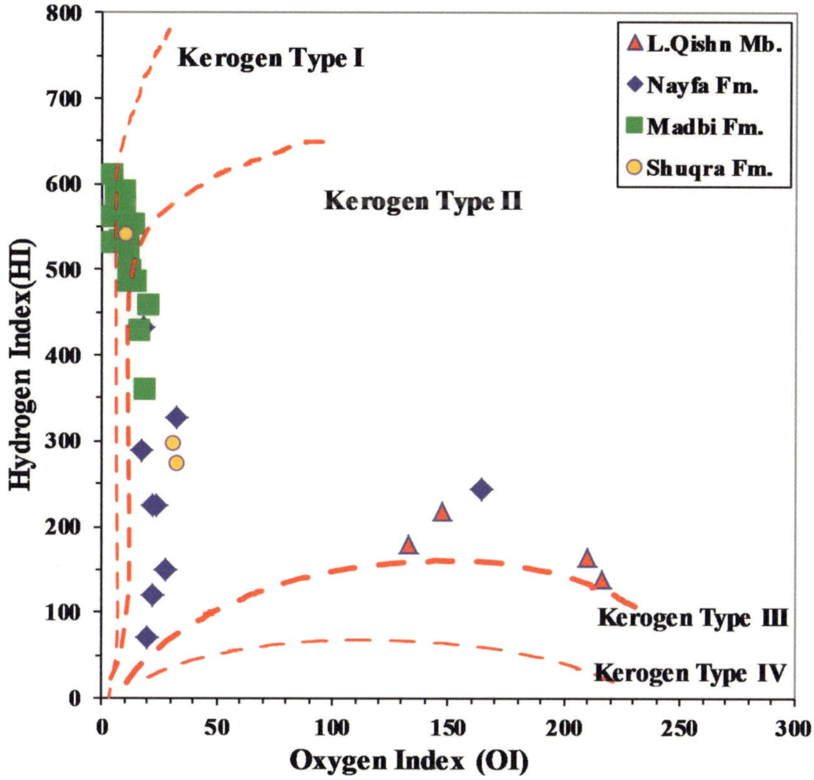

Figure 5. Plots of Hydrogen index (HI) versus Oxygen index (OI), showing kerogen quality of the Upper Jurassic samples in the Sayun-Masila basin, Yemen.

the sum of these two fractions is referred as hydrocarbons (HCs) [25]. Since the hydrocarbon portion of the bitumen extracted from sediment is the petroleum-like portion, it is used as an important parameter in the source-rock evaluation [26, 27]. These parameters are very important in petroleum source-rock evaluation (e.g., [26–28]). In this respect, most of the Upper Jurassic samples in Sayun-Masilah basin appear to be prolific petroleum sources where abundant naphthenic oils might be expected to be generated. The plots of TOC content versus extractable organic matter (EOM) and hydrocarbon yields (**Figure 6**) show the Upper Jurassic sediments in the study area as good to very good source rocks with good to very good potential for oil generation potential (e.g., [6]).

2.5. Thermal maturity of organic matter

As a rock containing kerogen and is progressively buried in a subsiding basin, it is subjected to increasing temperature and pressure. A source rock is defined as mature when it is reached to generate hydrocarbons. A rock that does not reach to the level of generation of hydrocarbons is defined as an immature source, and that which passed the time of significant generation and expulsion, it is considered as over-mature source rock. Generally, various parameters have been used for estimating source rock maturation. These parameters include vitrinite reflectance (Ro) and rock-eval pyrolysis data such as T_{max} and

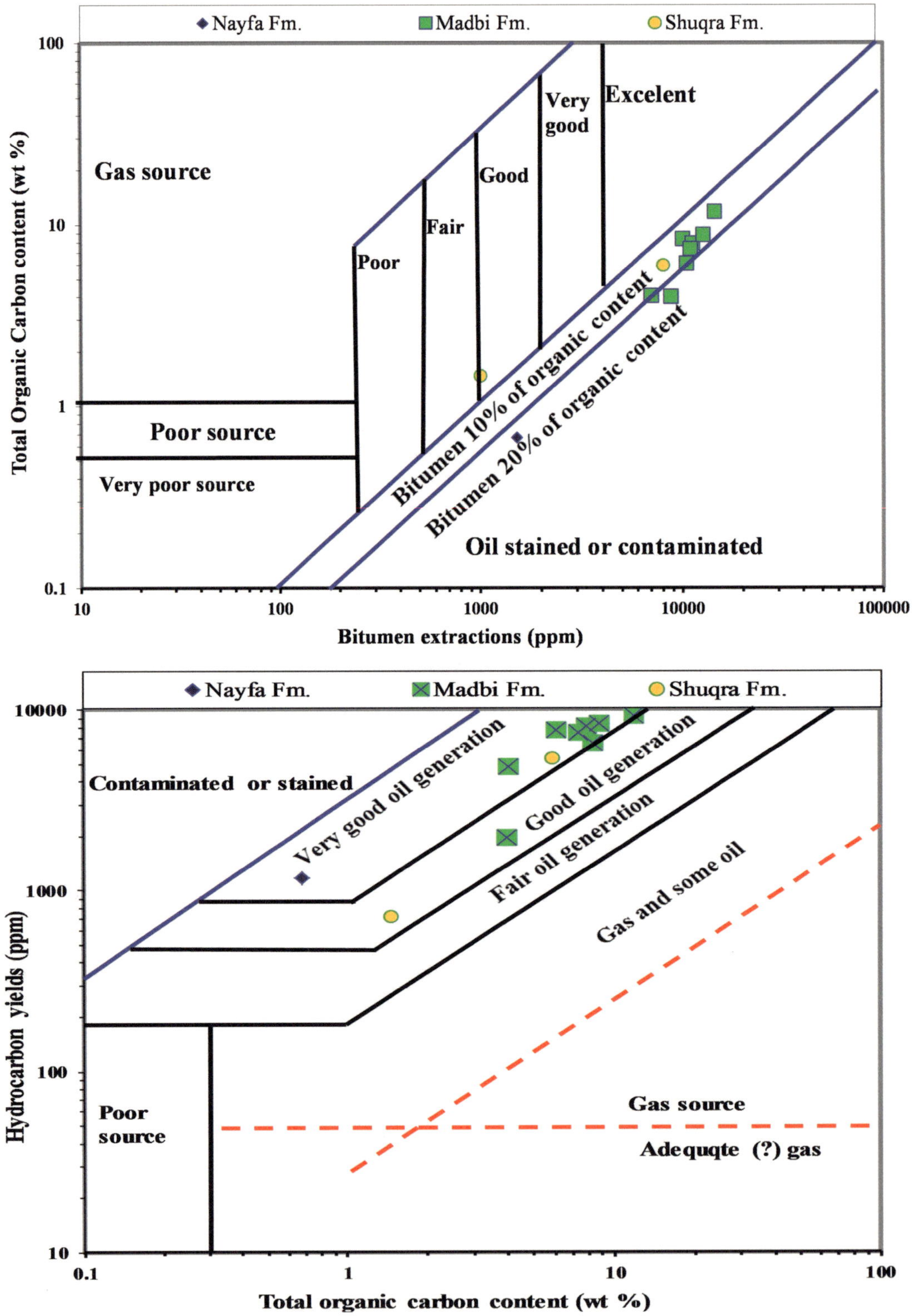

Figure 6. Plots of TOC content versus bitumen extractions and hydrocarbon yields, showing source potential rating and hydrocarbon source-rock richness for the selected samples.

production index (PI). The study of thermal maturation of source rocks is one of the main steps in the source rocks evaluation in the study area. This is because it is possible from the maturation stage to determine the position of the sediments with the respect to the oil generation. It can also help in oil exploration from knowing the relation between hydrocarbon generation, migration, and accumulation with the tectonics, which lead to the development of the structural traps in the study area.

A. Vitrinite reflectance (Ro %)

The most common method used for determining the stage of maturation is the vitrinite reflectance (Ro), which was discussed by several workers. Hood et al. [29] noted that one of the most useful measures of organic metamorphism is the reflectance of vitrinite. Tissot and Welte [2] considered the vitrinite reflectance as the most powerful tool for detecting maturation of organic matter. Waples [30] considered that a vitrinite reflectance (Ro%) of 0.6% mark the early stage of oil generation, while the peak of oil generation is at Ro ≈ 0.8%, and the late stage or the end of oil generation is marked at Ro ≈ 1.35%. The Ro% is considered as the most powerful maturation measure tool. It measures the ability of tiny vitrinite particles (called macerals) in kerogen to reflect incident light. This method depends on the separation of the organic macerals and measuring its vitrinite reflectance in oil immersion lens using a reflecting polarizing microscope connected with a photometer. The vitrinite macerals are increased in its reflectivity, as the maturation of their host rocks increases. Tissot and Welte [2] detected the onset, peak, and end of oil generation for the different types of kerogen according to Ro% (**Table 3**). In the Sayun-Masilah basin, the vitrinite reflectance values ranges from 0.32 to 0.87 Ro%. These reflect that Madbi and Shuqra Formations are mature stage, whereas the samples in L. Qishn Member and Nayfa Formation lie mainly in the mature stage (**Figure 7**). The lowering of the values of vitrinite reflectance in the studied samples from Sunah field, in spite of their occurrence at greater depths, may be related to the presence of high content of unstructured lipids of the type II kerogen [31]

B. Rock-eval pyrolysis

The relationship of the T_{max} with depth for the studied intervals in the Sayun-Masilah basin (**Figure 1**) shows values ranging from 432 to 443C°; this reflects that Madbi Formation is in the mature stage. The values in the Nayfa and Shuqra Formations reflect immature to mature stages. The production index (PI) data plotted against depth in (**Figure 1**) indicate that the phases of maturation of kerogen of these rock units are in the immature to mature stages. Most of the studied samples in the Madbi Formation at Sayun-Masilah basin lie in the mature stage (**Figure 1**). Reversely, the L. Qishn and Nayfa samples reveal a marginal mature stage. The types of organic matter ranges from oil- and gas-prone (HI ranged from >200 to 625) to oil-prone organic matter in Madbi Formation, and gas-prone kerogen in L. Qishn and Nayfa formations.

Generation	Type I	Type II	Type III
On set of oil generation	0.65	0.5	0.55 Ro%
Peak of oil generation	1.1	0.8	0.9 Ro%
End of oil generation	>1.4	>1.4	>1.4 Ro%

Table 3. The generation for the different types of kerogen with Ro%.

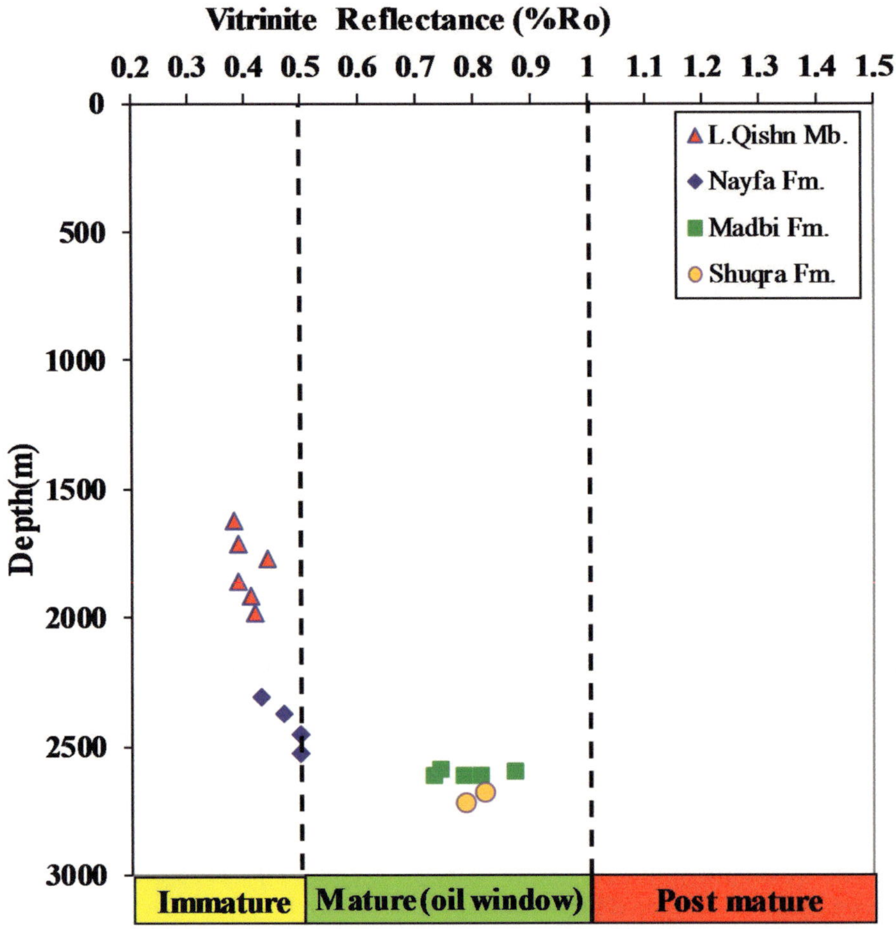

Figure 7. Plot of vitrinite reflectance data (Ro) versus depths showing thermal maturity stages of the Source rocks samples in the study area.

3. Basin modeling procedure

Geohistory diagrams [32] and similar diagrams have been widely used in geology, particular in hydrocarbon exploration. These diagrams were adapted to perform numerical modeling of burial, erosion, and thermal histories in sedimentary basins, e.g., [33, 30, 34, 35]. This method has become an important tool and successfully applied to search for new petroleum plays or for the evaluation of exploitable oil and gas accumulations around the world (e.g., [31, 36, 37]). In this chapter, quantitative one-dimensional basin modeling is performed for evaluating the thermal histories and timing of hydrocarbon generation and expulsion of the Nayfa, Madbi, and Shuqra source rocks in the Sayun-Masilah basin. The reconstruction of the burial, thermal, and maturity histories were modeled in order to evaluate the remaining hydrocarbon potential using Schlumberger's PetroMod (1D) modeling software. Sunah exploration well was created as a result of geochemical, well log, and further geologic data were used. The geologic model consisting of the depositional, nondepositional, and erosional events in absolute ages was compiled using stratigraphic data that were provided from well reports and previous stratigraphic studies, e.g., [38]. Hydrocarbon generation modeling was based on TOC and HI of the Nayfa, Madbi, and Shuqra source rocks in the Sayun-Masilah basin as example, and the maturity modeling was calculated using the EASY% Ro model of Sweeney and Burnham [39].

3.1. Subsidence and burial history

The tectonic evolution of the region has significantly influenced burial and thermal history of the study area. The burial (subsidence) and thermal histories are necessary in order to predict the timing of hydrocarbon generation and expulsion. To describe the resulting models clearly, we review first the results of our reconstruction of the subsidence curves [40]. Based on well profile, subsidence curves (**Figure 8**) were first constructed for the studied well by decompacting the sedimentary section using formation thicknesses (present day thickness) and lithologies assigned from mud logs and composite well log. The subsidence curves and basin history filling of one representative well is shown in **Figure 8**, it illustrate that the Upper Jurassic section have a long burial history although it has thin sedimentary cover (400 m), due to thick sedimentary sections (2300 m) precipitated during the Cretaceous and Tertiary epoch (**Figure 8**),which help oil generation in this area. However, the Madbi source rocks during that time were buried deeply, and the petroleum generation can be generated in this time.

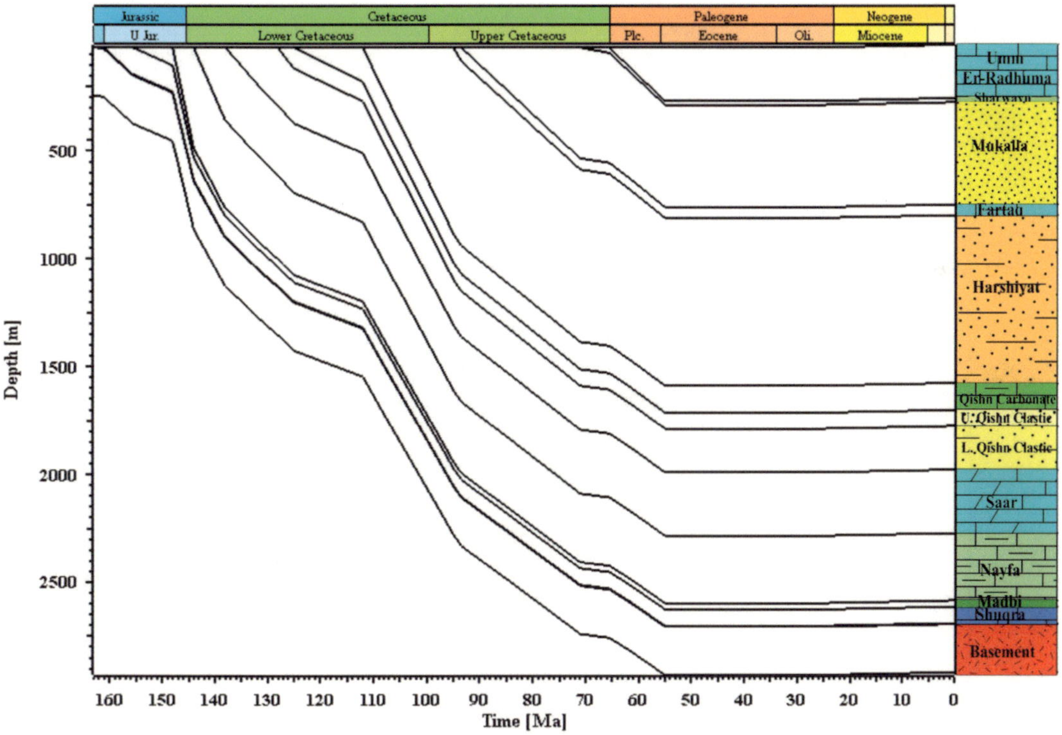

Figure 8. Burial history modeling for investigated well in the Sayun-Masilah basin.

3.2. Thermal history and paleo-heat flow

The thermal history of the source rocks in the sedimentary basins can be evaluated based not only on the deposition and erosion history but also on the heat-flow evolution [41, 42]. The borehole temperatures increase systematically with depth in the Earth and were used to calibrate the present day heat-flow regime. The increase of temperatures, indicating that heat is transferred through sediment layers to the surface. The transfer of heat is mainly controlled by thermal conductivity of the formations and geothermal gradient. Therefore, the thermal conductivity and geothermal gradient need to be determined to estimate the heat-flow history [43, 44, 45]. The present day geothermal gradient of borehole location was calculated using BHTs that were corrected for the circulation of drilling fluids. The maximum temperatures were reached at Upper Jurassic and Oligocene and Miocene time (**Figure 9**). The heat-flow is an important value in the input of the basin modeling, but needs to be determined for the geological past [45, 46]. Therefore, the reconstruction of the thermal history of the basin is simplified and calibrated with thermal maturation measurements (e.g., temperatures and vitrinite reflectance) (**Figure 10a**). Vitrinite reflectance was measured from maturity measurements of three stratigraphic units (Upper Jurassic), including Naifa, Madbi, and Shuqra formations (**Table 1**), and used to predict paleo-heat flow. Heat-flow model (**Figure 10b**) is used to calculate maturity, which is generally calibrated with a thermal maturity parameter such as vitrinite reflectance, e.g. [31, 47–49]. In the Sayun-Masilah basin, paleo-heat flow was affected

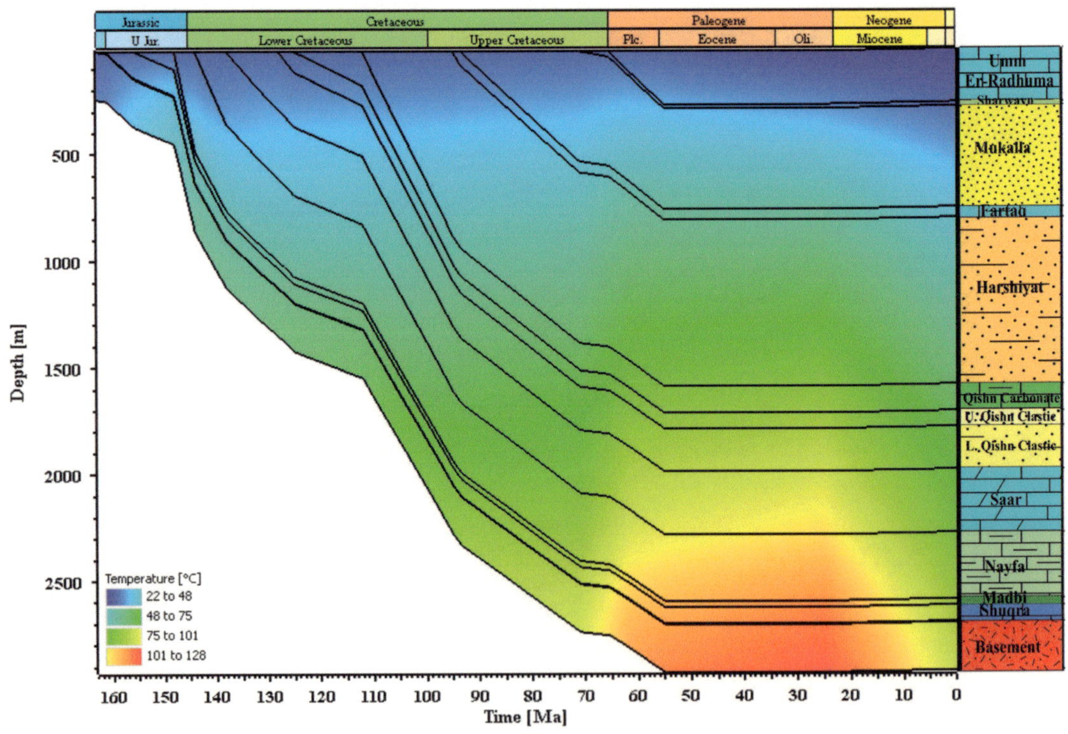

Figure 9. Paleo-temperature modeling in well calibrated using borehole temperature.

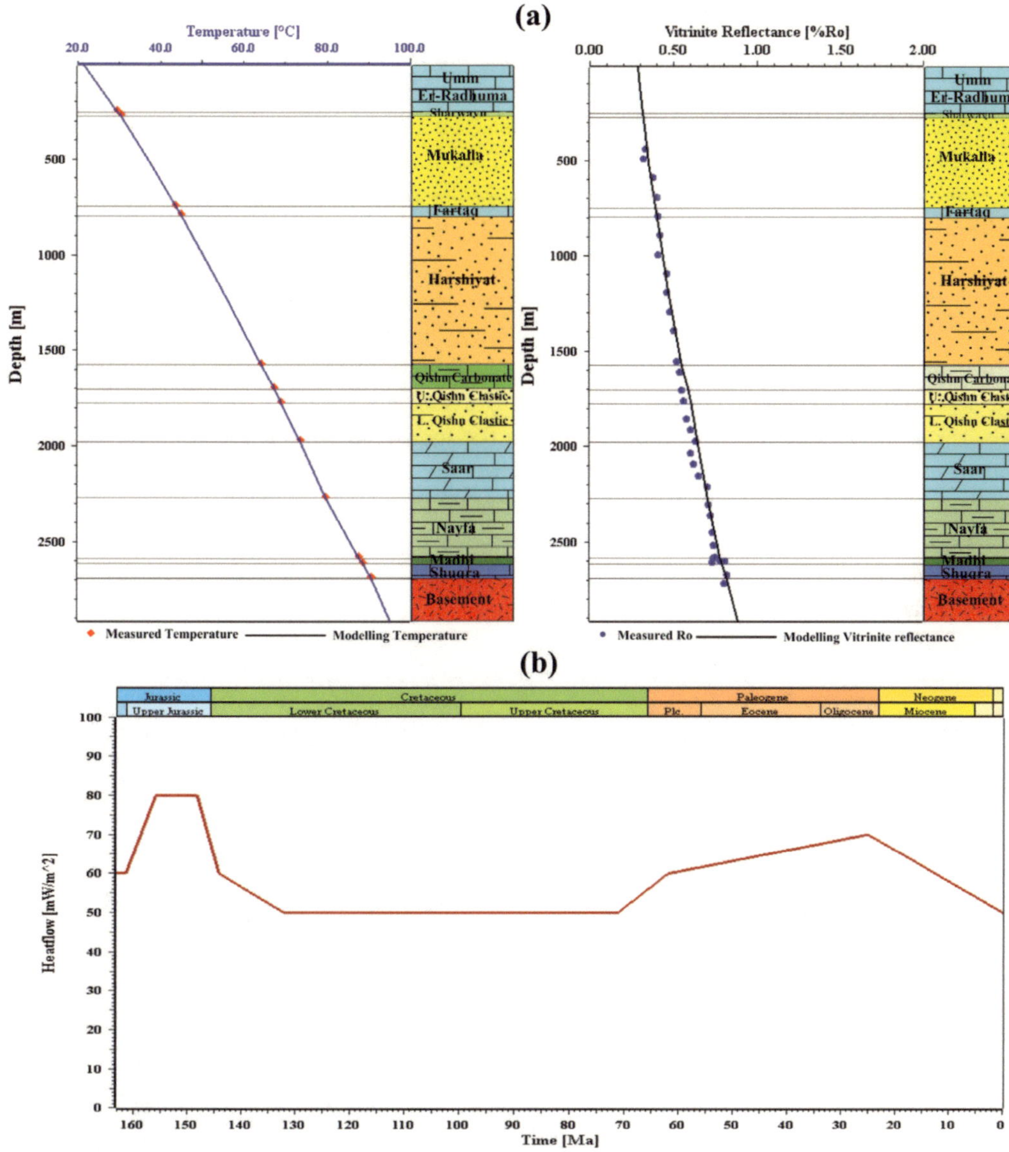

Figure 10. (a) Calibration of the thermal and maturity modeling in the studied in the Sayun-Masilah basin. Notice that there is a good correlation between measured data and calculated curves of temperature and measured vitrinite reflectance. (b) Heat flow through time in the investigated well, which were used to model the most probable scenario for hydrocarbon generation and expulsion in the Upper Jurassic source rocks.

by the tectonic evolution and rifting phase. The rift influenced heat-flow model, which incorporates a higher heat flow episode during the rift phase and an exponential reduction during the post-rift phase [50]. Based on the geological evolution of the Sayun-Masilah basin, the two rifting phases were incorporated in the heat flow model by peaks of heat flow during the periods of rifting (**Figure 10b**).

3.3. Source rocks maturity history model

In thermal history reconstructions of the study area, the influence of the tectonic evolution on the heat-flow distribution through time was applied. Thermal maturity levels of the Upper Jurassic source rocks are calculated based on the Easy% Ro routine [39] using one-dimensional modeling of single well. The detailed maturity history model of source rocks was used to determine the time when source rocks passed through the oil window. The detailed maturity history of source rocks in the Upper Jurassic source rocks is modeled for the representative well in the Sayun-Masilah basin (**Figure 11**). Based on the thermal maturity model, the hydrocarbon generation history of the source rocks in the model are different because of the variation in thermal and buried history (**Figure 11**). Assigning a heat-flow value of 80 mW/m^2 during 155.7 Ma gives the best fit between measured and calculated vitrinite reflectance and bottom hole temperatures (**Figure 10**). The Madbi Formation has reached the required levels of maturity in the model probably due to the temperatures (**Figure 11**). The model also shows that the source rock in this unit has reached the required levels of thermal maturity to onset of the oil window (0.64–0.87% Ro) from about 77 Ma at a depth 2315 m (**Figure 11**).

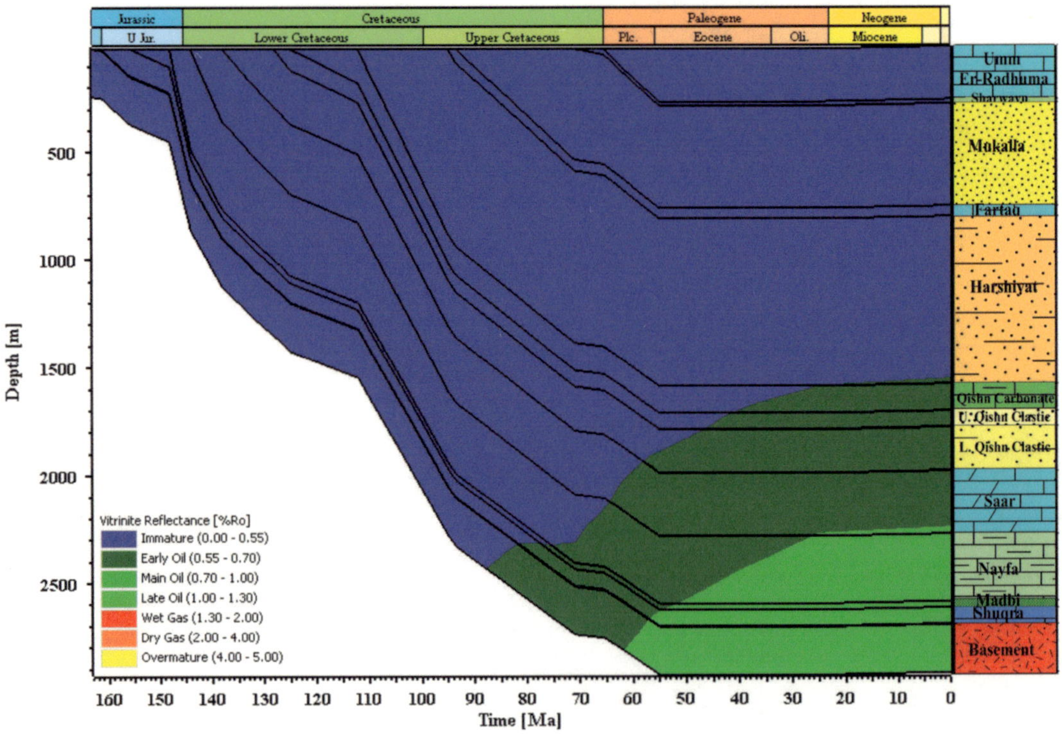

Figure 11. Burial and thermal maturity histories of the Upper Jurassic source rocks for the studied well showing the positions of the oil window.

3.4. Hydrocarbon generation and expulsion modeling

The timing of petroleum generated and expelled from the Upper Jurassic source rocks were modeled (**Figure 12**). Oil generation is defined in this model by transformation ratios between 10 and 50%. Immature source rocks have transformation ratios less than 10% (no generation). Peak oil generation occur at a transformation ratio of 50% when the main phase of oil generation is reached [49]. The modeled hydrocarbon generation and expulsion of the studied well shows that the Madbi source rocks were generating hydrocarbon with oil as the main product (**Figure 12**). In general, the hydrocarbon generation and expulsion history of the Madbi source rock in the studied model was represented by only two stages (**Figure 12**). The first stage of hydrocarbon generation of the Madbi source rock was occurred during Late Cretaceous-Early Eocene time at 70–54.6 Ma (**Figure 12**). This stage is the early phase of oil generation without any expulsion. The transformation ratio of the source rock varied from 10 to 25% during this stage, with computed VR of 0.55–0.65% Ro. The second stage (approximately 25.12-0 Ma) is the main phase of the oil generation and no gas generation has been detected. The transformation ratio of the source rock in this stage varies from 25 to 36%, with calculated VR of 0.65–0.87% Ro (**Figure 12**).

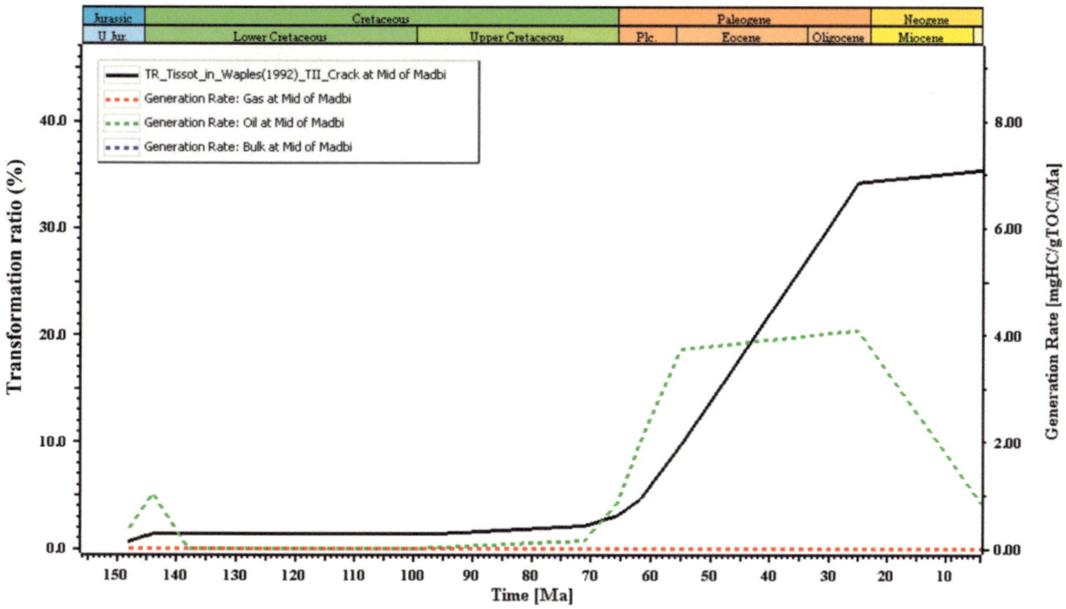

Figure 12. Plots of evolution of the transformation ratio and rate of hydrocarbon generation with age from the Madbi source rocks in the studied well.

4. Crude oil characterization

The petroleum was defined as a liquid substance referred to as "crude oil" or simply "oil" occurred in underground natural reservoirs, but the definition has been broadened to include hydrocarbon gases referred to as "natural gases" occurring in similar reservoirs. Oil is a complex mixture containing a large number of closely related compounds [2]. The compounds present and their relative amounts are controlled initially by the nature of the organic matter in the source rock. With more specific words, the relative amounts of normal alkanes, isoprenoids, aromatics, and sulfur compounds are characteristic of the source and should be essentially the same for all oil derived from a particular source rock. The fact that variations in crude oil composition are to a certain extent inherited from different source rocks. For instance, coaly material in general yields more gaseous compounds, whereas high wax crude oils are commonly associated with source material containing high proportions of lipids of terrestrial higher plants and of microbial organisms [2]. High-sulfur crude oils are frequently related to carbonate-type source rock. A side from the influence of source rock facies, the state of maturity of the source material is also of

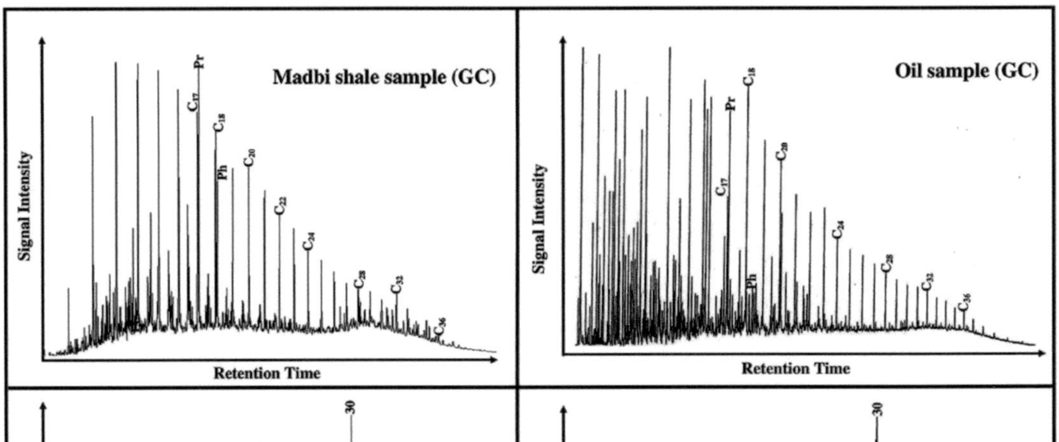

Figure 13. Gas chromatography traces and m/z 191, m/z 217 mass fragmentograms for the representative two oil samples.

importance. However, much larger variations in composition can cause processes operating in the reservoir. In other words, crude oil alteration processes (thermal alteration, deasphalting, biodegradation, and water washing) tend to obscure the original character of the oil, and therefore affects crude oil correlation, furthermore influence the quality and economic value of petroleum [2]. Therefore, the careful studying of the chemical compositions of the rock extracts, seeps, and produced oil can minimize the risk associated with finding petroleum accumulations.

The crude oils from the Sayun-Masilah basin have API gravity values in the range of 24.4–35.6 [51]. The crude oils have high saturated and aromatic fractions and ranging from 40.0 to 65.9% and 28.0 to 46.5%, respectively [51]. The high saturated and aromatic fractions with low amount of asphaltene and resin components indicate that these oils are naphthenic oils and have no sign of biodegradation. The similar bulk property and composition of the analyzed crude oils indicate that only one oil type is present. Biodegradation process may occur in an oil reservoir, and the process dramatically affects the fluid properties of the hydrocarbons, e.g., [52]. The early stages of oil biodegradation are characterized by the loss of n-alkanes or normal alkanes followed by the loss of acyclic isoprenoids (e.g., pristane and phytane). Compared with those compound groups, other compound classes (e.g., highly branched and cyclic saturated hydrocarbons as well as aromatic compounds) are more resistant to biodegradation [53]. In this respect, there is no sign of biodegradation among the studied oil samples, where the analyzed oils contain a complete suite of n-alkanes in the low-molecular weight region and acyclic isoprenoids (e.g., pristane and phytane); (**Figure 13**). This is also indicated by the analyzed oil samples generally contain more saturated hydrocarbons than aromatic hydrocarbons with generally saturate/aromatic ratios >1.

5. Oil-source correlation

The correlation of crude oils with one another and with extracts from their source rocks provide valuable tools for helping the exploration geologist to answer production and exploration trends [2]. Are there one or more families of oils in a particular rock sequence? Each family of oils represents one element of distinct petroleum system. Oil-source rock correlations are more difficult than oil-oil correlation; this is because many problems are involved in both sampling and interpreting the data. Tissot and Welte [2] showed that source rock oil is not usually similar in composition to its corresponding reservoir oil for several reasons. First, there is an evidence for the oil fractionates during the process of leaving the source and migrating to the reservoir accumulation. Second, source rocks do not yield oils of the same composition throughout their generation history. Third, degradation processes can affect the reservoir oil. All these problems require that the correlation be made by parameters (e.g., gross composition of oil and source rock extracts, biomarker analyses… etc.) that are possibly unchanged by the preceding factors. These parameters solve most of problems in oil-source correlations, because the differences in the chemical composition of the oil and the organic matters retained in the source rock are a function of migration fractionation and post-migration alteration.

Various parameters have been used for oil-source correlation purposes. These parameters depend mostly on the pre-burial environments of living organisms, the depositional environments of the organic matter, and the diagenetic processes in the source rocks. In this respective study, the applied parameters used for oil-source correlation are the steranes ternary diagrams of oils and source rock extracts by gas chromatographic (GC) analysis of C_{27}, C_{28}, and C_{29} regular steranes distribution. The distribution of C_{27}, C_{28}, and C_{29} homologous sterols on a ternary diagram was first suggested by Tissot and Welte [2] as a source indicator.

The objective of this part in this study is to investigate the genetic link between the oils recovered from Sayun-Masilah oilfield and Upper Jurassic source rocks. In an attempt to develop an oil-source rock correlation, we extracted soluble bitumens from four samples of the Madbi shale and analyzed their biomarkers using GC and GC–MS analyses. Overall, the oil data closely match the Upper Jurassic source rock data. Key factors include biomarker parameters and the similar positions on the cross-plots (**Figures 14** and **15**) [51].

The results of the steranes ternary diagrams of the oil and source rock extract samples are illustrated in **Figure 14**. The steranes distribution shows composed of C_{27}–C_{29} regular steranes and relatively low C_{27}/C_{29} regular steranes ratios (**Figures 14** and **15**), suggest a combination of marine and terrestrial organic matter input [54, 55]. **Figure 14** shows that the C_{27}, C_{28}, and C_{29}

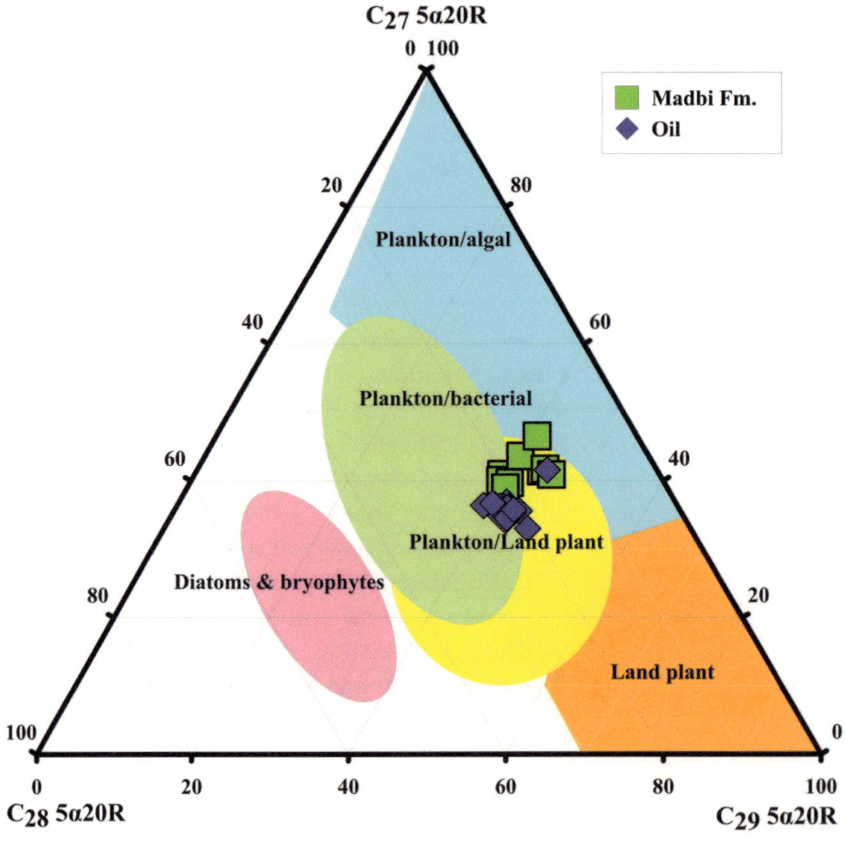

Figure 14. Ternary diagram of regular steranes (C_{27}–C_{29}) showing the relationship between sterane compositions, source organic matter input (modified after [56]).

http://dx.doi.org/10.5772/intechopen.70092

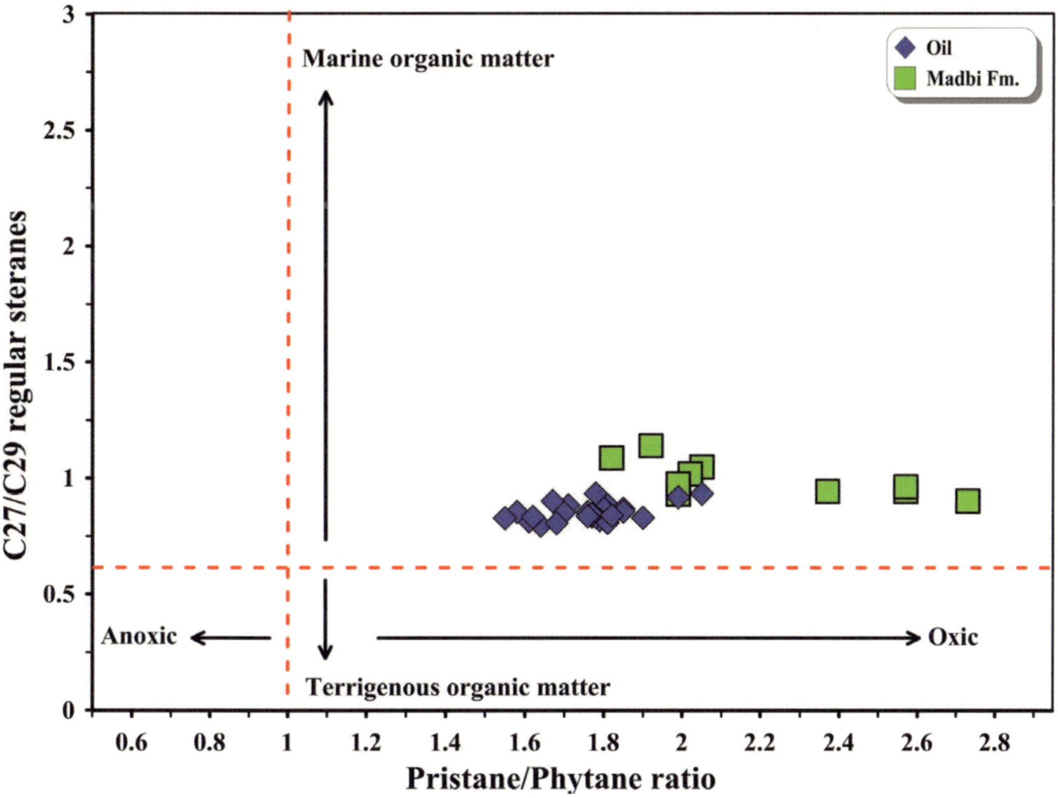

Figure 15. A plot of pristane/phytane versus C_{27}/C_{29} regular steranes, indicating organic matter input and depositional conditions [51].

are plotted in ternary diagram for oil extracts from the Madbi source rock, and the crude oils from the reservoir rocks, as shown in **Figures 14** and **15**, display the plotted points in the same area that mean the source oil in the Sayun-Masilah basin reflects that, the oils extracted from the source rock, and reservoir rocks are genetically have one family derived from the same basin.

6. Summary and conclusions

This chapter overviews the petroleum source rocks characterization and hydrocarbon generation, based on organic geochemical characteristics (e.g., total organic carbon content (TOC), rock-eval pyrolysis and bitumen extraction); in addition, burial and thermal histories and timing of petroleum generation/expulsion for petroleum source rock intervals using one-dimensional basin modeling software. The results obtained in this study give a strong indication as follows:

1. The organic geochemical data show that the Upper Jurassic sequence is the main source for hydrocarbon generation due to the high content of organic matter, which reached up

to more than 5.0 wt%, indicating a fair to very good source rock generative potential. The samples of Lower Cretaceous units vary between poor and fair source rocks.

2. The types of organic matter (kerogen) in these formations are of the type I and mixed types II and III, which were originally deposited in anoxic to suboxic depositional environment, as indicated from the low oxygen index OI, thus considered to be mainly oil- and gas-prone.

3. Maturity data such as vitrinite reflectance and pyrolysis T_{max} show that the Lower Cretaceous samples are thermally immature for hydrocarbon generation, whereas the Upper Jurassic samples are early-mature to peak-mature oil window mature stages for hydrocarbon generation.

4. The basin modeling study indicates that the source rocks in the Madbi Formation have entered the mature to peak-oil window mature stages for hydrocarbon generation and the mean-oil generation has been reached during Early Eocene (54.6 Ma), generating significant amount of oils with TR in the range of 25–36%.

5. In summary, results from the case study reveal that the Madbi Formation (Upper Jurassic) act as effective source rock and significant amount of hydrocarbons can be expected to generate in the study area.

6. From geochemical analysis, conclude the oils extracted from the source rock and reservoir rocks are genetically one family derived from the same basin.

Acknowledgements

The author would like to thank to Petroleum Exploration and Production Authority (PEPA), Ministry of oil and Minerals in Yemen for supplying the data and samples for this study. Schlumberger is acknowledged for providing the PetroMod Basin Modeling software.

Author details

Nabil Mohammed Al-Areeq

Address all correspondence to: nabilalareeq@yahoo.com

Department of Geology and Environment, Faculty of Applied Sciences, Thamar University, Yemen

References

[1] Hunt JM. Petroleum Geochemistry and Geology. 2nd ed. San Francisco: Freeman; 1995

[2] Tissot BP, Welte DH. Petroleum Formation and Occurrence. 2nd ed. Berlin: Springer; 1984. p. 699

[3] Waples DW. Maturity modelling: Thermal indicators, hydrocarbon generation, and oil cracking. American Association of Petroleum Geologists Bulletin. 1994;**60**:285-306

[4] Hunt JM. Petroleum Geochemistry and Geology. 1st ed. San Francisco: Freeman; 1979. p. 617

[5] Espitalie J, Laporte JL, Madec M, Marquis F, Leplat P, Pauletand J, Boutefeu A. Methode rapide de caracterisation des roches meres, de leur potential petrolier et de leur degre d'evolution. Revue de l'Institut Francais du Petrole. 1977;**32**:23-42

[6] Peters KE, Cassa MR. Applied source rock geochemistry. In: Magoon LB, Dow WG, editors. The Petroleum System—From Source to Trap. 60th ed. USA: American Association of Petroleum Geologists; 1994. pp. 93–120

[7] Peters KE. Guidelines for evaluating petroleum source rock using programmed pyrolysis. American Association of Petroleum Geologists Bulletin. 1986;**70**:318-329

[8] Bordenave ML. Applied petroleum geochemistry. In: Buker C, Littke R, Welte DH, editors. 2D–Modelling of Thermal Evolution of Carboniferous and Devonian Sedimentary Rocks of the Eastern Ruhr Basin and Northern Rheinish Massif. 146th ed. Germany: Zeitschrift der deutschen Geologischen Gesellschaft; 1993. pp. 321–339

[9] Ronov DA. Organic carbon in sedimentary rocks. Geochemistry. 1985;**5**:510-536

[10] Thomas BM. Geochemical analysis of hydrocarbon occurrence, northern Perth basin Australia. AAPG Bulletin. 1979;**63**:1092-1107

[11] Espitalie J, Deroo G, Marquis F. Rock–Eval Pyrolysis and Its Application. 1st ed. Paris: Inst. Fr. Preprint; 1985. p. 72

[12] Powel TG, MCRirdly DM. Relationship between ratio of pristane to phytane, crude oil composition and geological environment. Nature. 1973;**243**:37-39

[13] Brooks JD, Gould K, Smith JW. Isoprenoid hydrocarbons in coal and petroleum. Nature. 1969;**222**:257-259

[14] Hughes WB, Holba AG, Dzou LIP. The ratios of dibenzothiophene to phenanthrene and pristane to phytane as indicators of depositional environment and lithology of petroleum source rocks. Geochimica et Cosmochimica Acta. 1995;**59**:3581-3598

[15] Alexander R, Kagi RI, Woodhouse GW. Geochemical correlation of Windalia oil and extracts of winning group (cretaceous) potential source rocks, Berrow Subbasin, Western Australia. AAPG Bulletin. 1981;**65**:235-249

[16] Rowlands S, Rockeg C, Al-Lihaibi S, Wolff GA. Incorporation of sulphur into phytol derivatives during simulated early diagenesis. Organic Geochemistry. 1993;**20**:1-5

[17] Didyk BM, Simoneit BRT, Brassell SC, Eglinton G. Organic geochemical indicators of palaeoenvironmental conditions of sedimentation. Nature. 1978;**272**:216-222

[18] Largen DJ, Gize AP. Pristane/phytane ratio in the mineralized Kup ferschiefer of the Fore-Sudetic Monocliney, southwest Poland. Ore Geology Reviews. 1996;**11**:89-103

[19] Volkman JK, Maxwell JR. Acyclic isoprenoids as biological makers. In: Johns RB, editor. Biological Markers in the Sedimentary Record. 1st ed. New York: Elsevier; 1984. p. 1-42

[20] Koopmans MP, Rijpstra WIC, Klapwijk MM, Leenw d, Lewan MD, Damste JSS. A thermal and chemical degradation approach to decipher pristine and phytane precursors in sedimentary organic matter. Organic Geochmistry. 1999;**30**:1089-1104

[21] Lijmbach GW. Proceedings of the 9th World Petroleum Congress 2, Applied Science Publ., London; 1975. pp. 357-369

[22] Shanmugam G. Significance of coniferous rainforests and related organic matter in generating commercial quantities of oil, Gipps-land Basin. American Association of Petroleum Geologists Bulletin. 1985;**69**:1241-1254

[23] Abdullah WA. Organic facies variations in the Triassic shallow marine and deep marine shales of central Spitsbergen. Marine and Petroleum Geology. 1999;**16**:467-481

[24] Tissot B, Durand B, Espitalie J, Compaz A. Influence of nature and diagenesis of organic matter in formation of petroleum. AAPG Bulletin. 1974;**58**(3):499-506

[25] Hakimi MH, Abdullah WH, Say-Gee S, Makeen YM. Organic geochemical and petrographic characteristics of tertiary coals in the northwest Sarawak, Malaysia: Implications for palaeoenvironmental conditions and hydrocarbon generation potential. Marine and Petroleum Geology Journal. 2013;**48**:31-46

[26] Philippi GT. Identification of oil source beds by chemical means. International Geological Congress. 1957;**20**:25-38

[27] Baker DR. Organic geochemistry and geological interpretations. Journal of Geological Education. 1972;**20**:221-234

[28] Mustapha KA, Abdullah WH. Petroleum source rock evaluation of the Sebahat and Ganduman formations, Dent Peninsula, Eastern Sabah, Malaysia. Journal of Asian Earth Sciences. 2013;**76**:346-355

[29] Hood A, Gutjahar CM, Heacock RL. Organic metamorphism and the generation of petroleum. AAPG Bulletin. 1975;**59**:986-996

[30] Waples DW. Time and temperature in petroleum formation: Application of Lopatin's method to petroleum exploration. American Association of Petroleum Geologists Bulletin. 1988;**64**:916-926

[31] Al-Areeq NM, Abu El Ata AS, Maky AF, Omran AA. Hydrocarbon potentialities of some upper Jurassic rock units in Masila block, Sayun-Masila Basin, Yemen. Journal of Applied Geophysics. 2011;**10**(2):147-168

[32] Van Hinte JE. Geohistory analysis-application of micropaleontology in exploration geology. American Association of Petroleum Geologists Bulletin. 1978;**62**:201-222

[33] Lopatin NV. Temperature and geologic time as factors in coalification. Altai Izvestiya Akademiya Nauk Kazakh SSR, Seriya Geologicheskaya. 1971;**3**:95-106

[34] Welte DH, Yukler A. Petroleum origin and accumulation in basin evolution—A quantitative model. American Association of Petroleum Geologists Bulletin. 1981;**65**:1387-1396

[35] Littke R, Buker C, Luckge A, Sachsenhofer RF, Welte DH. A new evaluation of palaeoheat flows and eroded thicknesses for the Carboniferous Ruhr Basin, western Germany. International Journal of Coal Geology. 1994;**26**:155-183

[36] Shalaby MR, Hakimi MH, Abdullah WH. Modeling of gas generation from the AlamEl-Bueib formation in the Shoushan Basin, northern western desert of Egypt. International Journal of Earth Sciences. 2013;**102**:319-332

[37] Baur F, Di Primio R, Lampe C, Littke R. Mass balance calculations for different models of hydrocarbon migration in the Jeanne D'Arc Basin, offshore Newfoundland. Journal of Petroleum Geology. 2011;**34**:181-198

[38] Beydoun ZR, AL As-Saruri M, El-Nakhal H, Al-Ganad IN, Baraba RS, Nani ASO, Al-Aawah MH. International Lexicon of Stratigraphy. 2nd ed. Republic of Yemen: International Union of Geological Sciences and Ministry of Oil and Mineral Resources; 1998. p. 245

[39] Sweeney JJ, Burnham AK. Evaluation of a simple model of vitrinite reflectance based on chemical kinetics. American Association of Petroleum Geologists Bulletin. 1990;**74**:1559-1570

[40] El Nady MM, Hakimi MH. The petroleum generation modeling of prospective affinities of Jurassic–Cretaceous source rocks inTut oilfield, northWestern desert, Egypt: An integrated bulk pyrolysis and 1D-basin modelling. Arabian Journal of Geosciences. 2016;**9**:430-441

[41] Allen PA, Allen JR. Basin Analysis Principles and Applications. 1st ed. Oxford: Blackwell Scientific; 1990

[42] Lachenbruch A. Crustal temperature and heat productivity: Implications of the linear heat flow relation. Journal of Geophysical Research. 1970;**75**:3291-3300

[43] Frielingsdorfa J, Islamb SA, Blockc M, Rahmanb MM, Rabbanid MG. Tectonic subsidence modelling and Gondwana source rock hydrocarbon potential, Northwest Bangladesh Modelling of Kuchma, Singra and Hazipur wells. Marine and Petroleum Geology. 2008;**25**:553-564

[44] Xiaowen G, Sheng H, Keyu L, Zhongsheng S, Sani B. Modelling the petroleum generation and migration of the third member of the Shahejie formation (Es3) in the Banqiao depression of Bohai Bay Basin, eastern China. Journal of Asian Earth Sciences. 2011;**40**:287-302

[45] Al-Areeq NM, Maky AF, Ramdan MA, Salman A. Thermal conductivity, radiogenic heat production and heat flow of some upper Jurassic rock units, Sabatayn Basin, Yemen. Journal of Applied Sciences Research. 2013;**9**(1):498-518

[46] Hakimi MH, Abdullah WH. Thermal maturity history and petroleum generation modelling for the upper Jurassic Madbi source rocks in the Marib-Shabowah Basin western Yemen. Marine and Petroleum Geology. 2015;**59**:202-216

[47] He S, Middleton M. Heat flow and thermal maturity modelling in the northern Carnarvon Basin, north west shelf, Australia. Marine and Petroleum Geology. 2002;**19**:1073-1088

[48] Li MJ, Wang TG, Chen JF, He FQ, Yun L, Akbar S, Zhang WB. Paleo heat flow evolution of the Tabei uplift in Tarim Basin, northwest China. Journal of Asian Earth Sciences. 2010;**37**:52-66

[49] Makeen YM, Abdullah WH, Pearson MJ, Hakimi MH, Elhassan OMA, Yousif TAH. Thermal maturity history and petroleum generation modelling for the lower cretaceous Abu Gabra formation in the Fula sub-basin, MugladMuglad Basin, Sudan. Marine and Petroleum Geology. 2016;**75**:310-324

[50] McKenzie D. Some remarks on the development of sedimentary basins. Earth and Planetary Science Letters. 1978;**40**:25-32

[51] Al-Areeq NM, Maky AF. Organic geochemical characteristics of crude oils and oil-source rock correlation in the Sunah oilfield, Masila region, eastern Yemen. Marine and Petroleum Geology Journal. 2015;**63**:17-27

[52] Miiller DE, Holba AG, Huges WB. Effects of biodegradation on crude oils. In: Meyer RF, editor. Exploration for Heavy Crude Oil and Natural Bitumen. American Association of Petroleum Geologists Studies, USA; 1987. pp. 233-241

[53] Larter SR, Head IM, Huang H, Bennett B, Jones M, Aplin AC, Murray A, Erdmann M, Wilhelms A, di Primio R. Biodegradation, gas destruction and methane generation in deep subsurface petroleum reservoirs: An overview. In: Dore AG, Vining B, editors. Petroleum Geology: Northwest Europe and Global Perspectives. 6th ed. London: Geological Society; 2005. pp. 633–640

[54] Moldowan JM, Seifert WK, Gallegos EJ. Relationship between petroleum composition and depositional environment of petroleum source rocks. American Association of Petroleum Geologists Bulletin. 1985;**69**:1255-1268

[55] Peters KE, Moldowan JM. The biomarker guide: Interpreting molecular fossils. In: Petroleum and Ancient Sediments. 1st ed. Prentice Hall: Upper Saddle; 1993. p. 363

[56] Huang WY, Meinschein WG. Sterols as ecological indicators. Geochimica et Cosmochimica Acta. 1979;**43**:739-745

Chapter 2

Petroleum Extraction Engineering

Tatjana Paulauskiene

Additional information is available at the end of the chapter

http://dx.doi.org/10.5772/intechopen.70360

Abstract

In this chapter, the information about rotary drilling rig components, their purpose and principles of operation is presented through the in-depth analysis of hoisting, rotating and circulating equipment. Detailed classification of drilling fluids and its content is followed by the thorough investigation of the phenomenon of drilling fluid losses. The effects of drilling mud additives and loss circulation materials on rheology and the rate of penetration of drilling mud are supported by the studies of comparing the rate of penetration of drilling mud with various loss circulation materials. Finally, the fluid capability to form filter cake on the borehole walls is presented through the physical simulation of flow.

Keywords: petroleum extraction, oil well drilling, drilling fluid, drilling mud loss, filter cake

1. Introduction

For more than a century, oil is well known as a good primary energy source competing coal, natural gas, nuclear energy and renewables in various regions and fields of the energy sector. According to the last statistical reports, oil is dominant fuel in America and Africa, whereas natural gas dominates in Europe and Eurasia and coal in the Asia Pacific. The use of oil and gas in the Middle East reach 98% of total energy consumption in this region.

Oil is the world's leading fuel (accounting for 32.9% of global energy consumption) with the 10-year average rate of growth of 1.9%. However, the rate of growth recorded in 2015 (1.0%) is slightly lower and similar to the rate recorded in 2014 (+1.1%) (**Figure 1**) [1].

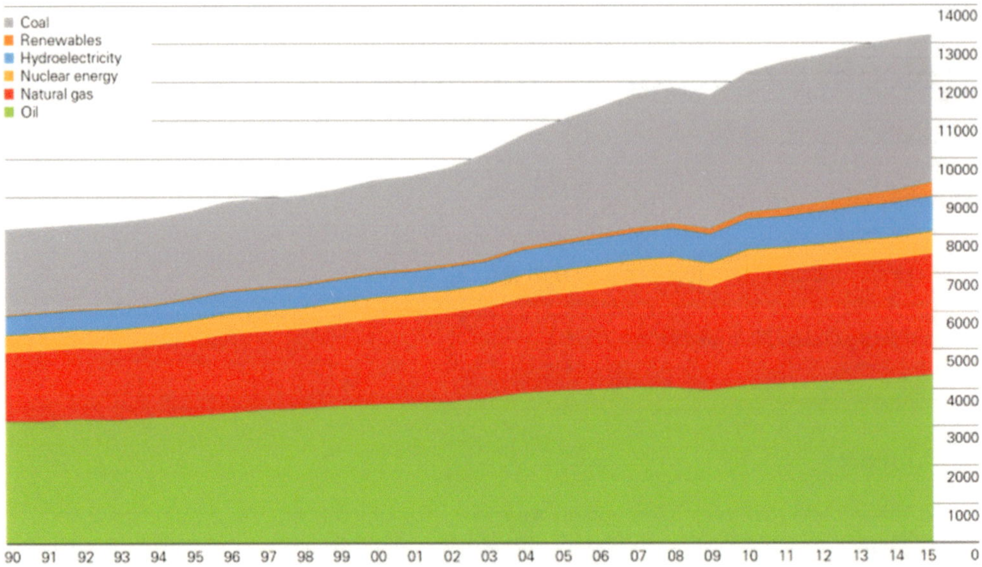

Figure 1. Primary world energy consumption, million tonnes of oil equivalent [1].

Oil, originated from ancient fossilized organic materials, is considered as nonrenewable primary energy source with limited amounts. There are two indicators used to represent remaining oil reserves—proved oil reserves and reserves-to-production ratio. Proved oil reserves is amount of oil that geological information indicates with reasonable certainty can be recovered to the future under existing economic and operating conditions, whereas reserves-to-production ratio represents the length of time that those remaining reserves would last if production were to continue at the previous year's rate [1].

Constant growth of proved oil reserves from 1126.2 thousand million barrels in 1995 until 1697.6 thousand million barrels in 2015 is presented in **Figure 2**. Nearly half of proven oil reserves are located in the Middle East.

According to the last statistical overview, oil reserves increased by 24% over the past decade and meet 50.7 years of global production. On a regional basis, South and Central American reserves have the highest oil reserves-to-production ratios—117 years and Asia Pacific have the lowest reserves-to-production ratios—14.05 years.

In various regions all over the world, oil is found in the geological structures that form oil reservoirs. According to the depth of the oil reservoir, they are classified as follows: shallow, 30–800 m; medium, 800–2000 m; deep, 2000–5000 m and over deep, more than 5000 m. This classification is constantly changing as advances in drilling equipment with opportunity to achieve greater depth. However, irrespective of the depth of the oil reservoir, the main principle of oil extraction stays the same and is based on the life cycle of the oil field (**Figure 3**).

There are five stages of oil and gas fields' life cycle: exploration, appraisal, development, production and abandonment.

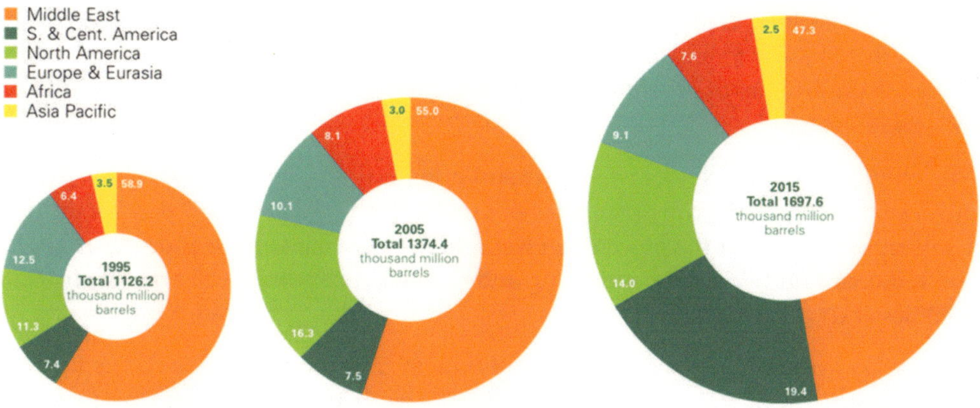

Figure 2. Distribution of proved oil reserves: 1995, 2005 and 2015, percentage [1].

Exploration is a method used for searching potentially viable oil and gas sources through geological surveys and drilling exploration wells to identify areas of potential interest. During the drilling process, general information and samples are collected to know about the rocks, fluids to find out how much oil and gas may be available at the explored area and what is the depth of the oil and gas window.

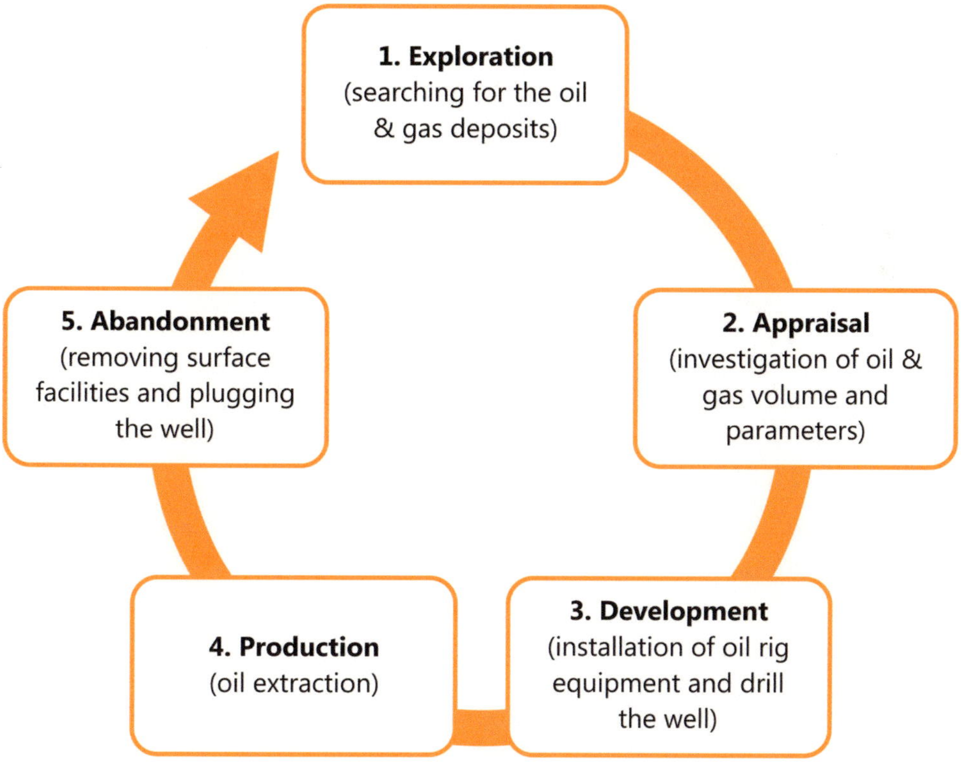

Figure 3. Life cycle of the oil and gas field.

After successful drilling exploration wells, the **appraisal** stage of the lifecycle starts. The main purpose of this phase is to improve the field description through further data acquisition and to reduce the uncertainty or possibility of losses about the size, shape and marketability of the oil and gas reservoir.

The **development** stage occurs after successful appraisal and before production. The main activities are formation of a conceptual development plan (in order to develop the oil and gas field, to prepare design for the production wells, to decide what surface and subsurface facilities are required and to describe operating and maintenance principles) and construction of the facilities and production units.

The **production** phase starts with the first oil flow in the wellhead. Oil and gas fields have a lifespan ranging from 15 to 30 years (from first oil to abandonment) and may be extended up to 50 years or more for the largest deposits. After extraction, oil and gas transported for processing and distribution.

When the oil and gas production is no longer cost-effective, wells are plugged and **abandoned**, production facilities are removed and this is the last stage of oil and gas fields' life cycle.

Thereinafter, we will be focusing on the third step of the life cycle of the oil field—development of the well.

2. Rotary drilling rig components

During the first phase of the development of the well, a rotary drilling rig is installed to bore a hole in the ground and reach the oil reservoir. The main rotary drilling rig components are derrick or mast, power and prime movers, hoisting equipment, rotating component, circulating system, tubular and tubular handling equipment and bit.

Derrick is mainly used offshore and is a large load-bearing vertical structure, usually of bolted construction and pyramidal in shape, for the equipment used to lower and raise the drill string into and out of the wellbore. The height of the derrick does not affect its load-bearing capacity, but it shows the maximum length of the drill pipe section. The standard derrick has square-shaped rig floor with four legs standing at the corners of the substructure. It provides work space for the necessary equipment on the rig floor.

Mast is mainly used with onshore rigs and is a portable derrick that can be raised as unit but for the transporting can be divided into two or more sections. It is usually rectangular or trapezoidal in shape.

Power and prime movers. The power generated by the power system is used for five main operations such as rotating, hosting, drilling fluid circulation, rig lighting and hydraulic systems. It is important to note that the most of the generated power is consumed by the hoisting and drilling fluid circulation systems. Internal combustion engine (mostly diesel) connected to electric generators or turbine is the source of power on the rig. Some rotary rigs can use electricity directly from power lines.

Hoisting component is used to perform all lifting activities on the rig and helps in lowering or raising equipment into or out of the well. It consists of drawworks, crown block, traveling block, deadline anchor, supply reel and drilling line (**Figure 4**).

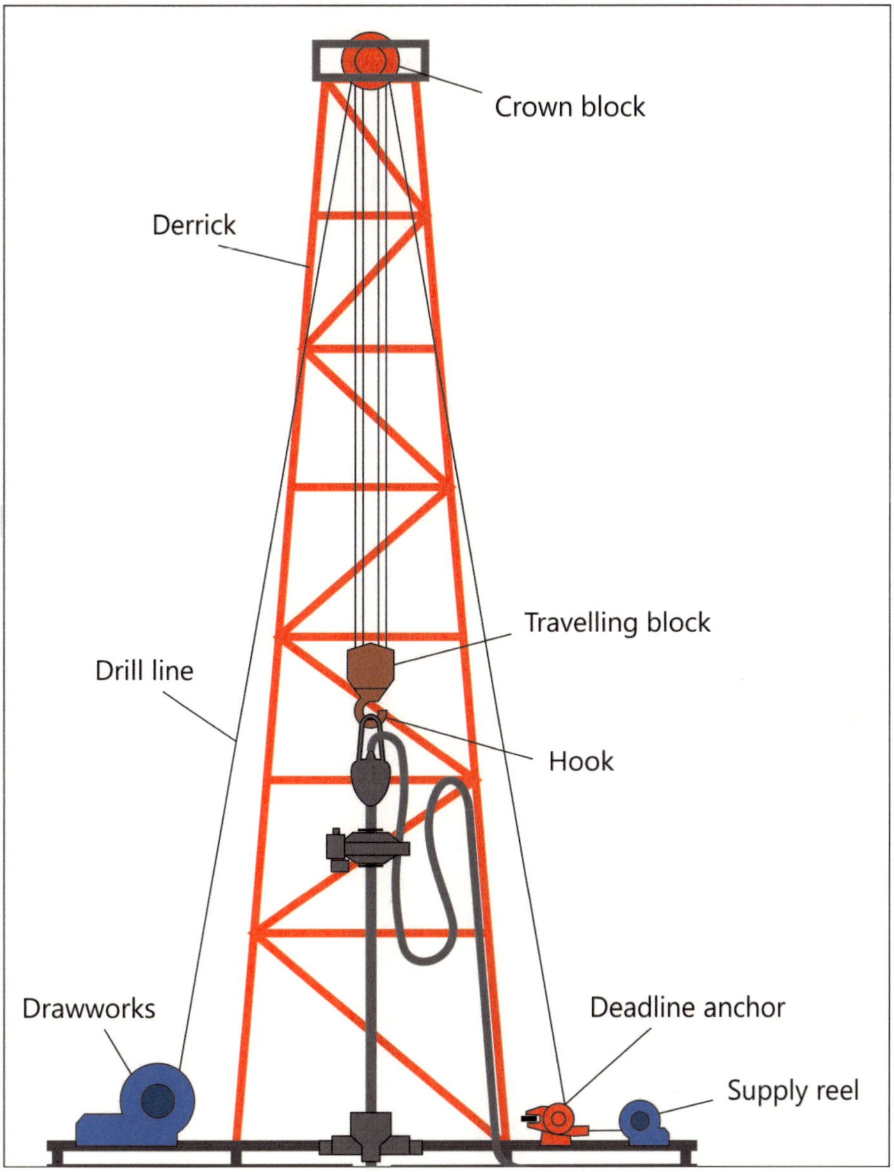

Figure 4. Hoisting equipment of the drilling rig.

- **Drawworks** is the main operating component of the hoisting system and is used to transmit power from prime movers to the hoisting drum that lifts drill string, casing or tubing string out of and to lower it back into the borehole. They consist of a large diameter steel spool, brakes, a power source and assorted auxiliary devices. The primary function of the drawworks is to reel out and reel in the drill line, a large diameter wire rope, in a controlled manner. The speeds for hoisting the drill string could be changes by driller via integrated gear system.
- **Crown block** is fixed assembly of sheaves (single or double) with a wire rope drilling line running between it and is located at the top of the derrick or mast and over which the drilling line is threaded. It is used to change the direction of pull from the drawworks to the traveling block.
- **Traveling block and hook** combination is used to safely and efficiently raise or lower tools and equipment in the well. It is the set of sheaves or pulleys through which the drill line (wire rope) is threaded or reeved, is opposite the crown block and enabling heavy loads to be lifted out of or lowered into the wellbore. Hook is located beneath the traveling block and is used to pick up and secure the swivel and Kelly.
- **Deadline anchor** is usually bolted on to the substructure and is the equipment that holds down the deadline part of the wire rope. It provides weight measurements and secure deadlines.
- **Supply reel** is a spool that stores the unused portion of the drill line.
- **Drill line** is the wire rope used to support the drilling tools. It is threaded or reeved through the traveling block and crown block to facilitate the lowering and lifting of the drill string into and out of the borehole. Drill line then clamped to the rig floor by the deadline anchor.

Rotating component is the equipment responsible for rotating the drill string. It consists of the swivel, Kelly spinner, Kelly or top drive, Kelly bushing, master bushing and rotary table (**Figure 5**).

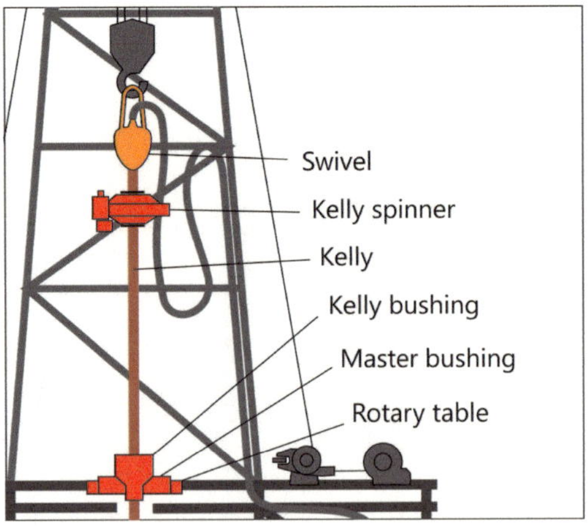

Figure 5. Rotating equipment of the drilling rig.

- **Swivel** is a mechanical device that is hung from the hook of the traveling block to support the weight of the drill string and allows it to rotate freely. It provides connection for the rotary hose as well as passageway for the flow of drilling fluid into the drill stem.
- **Kelly spinner** is a pneumatically controlled device mounted below the swivel to spin the Kelly to make up tool joints when making connections.
- **Kelly** is the heavy steel square or hexagonal member that is suspended from the swivel through the rotary table and connected to the topmost joint of drill pipe to turn the drill stem as the rotary table turns. It has a hole drilled through the middle that permits fluid to be circulated into the drill stem and up the annulus or vice versa. The Kelly goes through the Kelly bushing, which is driven by the rotary table.
- **Top drive** is a hydraulically powered device on the drilling rig and is located at the swivel place. It allows the drill stem to spin and facilitate the process of drilling a borehole. Top drive means a power swivel, which directly turns the drill string without need for a Kelly and rotary table.
- **Kelly bushing** is a device that fits into a part of rotary table called master bushing, transmits torque to the Kelly and simultaneously permits vertical movement of the Kelly to make hole. The Kelly bushing as Kelly is square or hexagonal and has an inside profile matching the Kelly's outside profile with slightly larger dimensions so that the Kelly can freely move up and down inside.
- **Master bushing** is a tool that fits into the rotary table of a drilling rig to accommodate the slips and drive the Kelly bushing so that the rotating motion of the rotary table can be transmitted to the Kelly.
- **Rotary table** is section of the drill floor used to turn the drill stem. It has a beveled gear arrangement to create the rotational motion and opening into which bushings are fitted to drive and support the drilling assembly.

Circulating component is the rig equipment responsible for the movement of drilling fluid within the well as well as solids removal incurred by the drilling fluid (**Figure 6**). Normally, the circulation would start from the mud pits or tanks that are located besides the rig. Powerful pumps force the drilling through the surface high-pressure connections to a set of valves called pump manifold, located at the derrick floor. Then, the fluid goes up the rig within a pipe called standpipe to approximately 1/3 of the height of the mast. From there, the drilling fluid flows through a flexible high-pressure rotary hose to the top of the drill string. The flexible hose allows the fluid to flow continuously as the drill string moves up and down during normal drilling operations. The fluid enters in the drill string through a special piece of equipment called swivel located at the top of the Kelly. The swivel permits rotating the drill string while the fluid is pumped through the drill string. In wellbore, the drilling fluid then floes down the rotating string and jets out through the nozzles in the drill bit at the bottom of the hole. Drilling fluid carrying the drilled cuttings and flows out the center of the drill bit and is forced back up the outside of the drill pipe between the drill string and walls of the well (annular) onto the surface of the ground where it is cleaned and circulated back to the well. The cleaning process starts from the shale shaker, which is basically a vibrating screen.

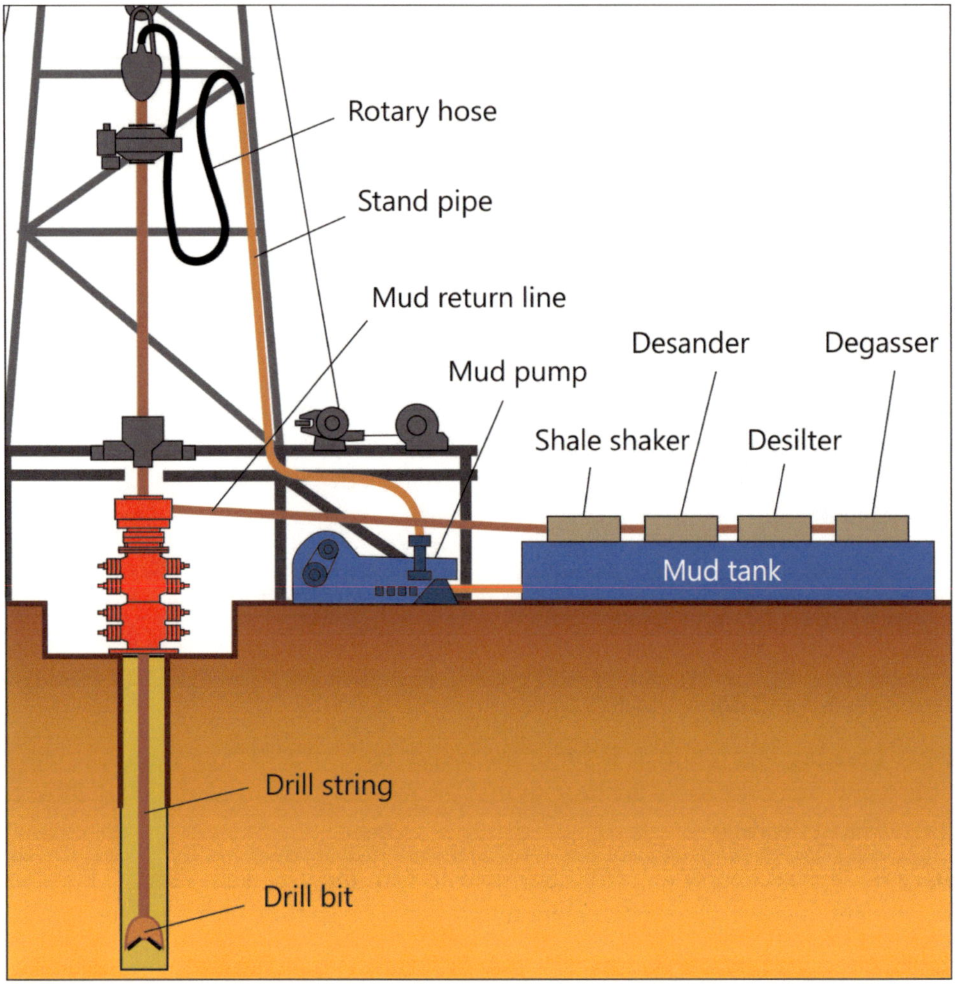

Figure 6. Circulation system of the drilling rig.

This will remove the larger particles, while allowing the residue to pass into settling tanks. The finer material can be removed using other solids removal equipments such as desander and desilter. If the mud contains gas from the formation, it will be passed through a degasser that separates the gas from the liquid mud. Having passed through all the mud processing equipment, the mud is returned to the mud pits or tanks for recycling.

The principal components of the drilling fluid circulation system are as follows:

- **Mud pump** is a large, high-pressure and high-volume pump used to circulate the drilling fluid down the drill pipe and out of the annulus on an oil rig. It could be double acting duplex (2 cylinder) pump, which has four pumping actions per pump cycle or single acting triplex (3 cylinder) with three pumping actions per pump cycle whose pistons or plungers travel in replaceable liners and are driven by a crankshaft actuated by an engine or motor.

- **Pump manifold** is an arrangement of piping and valves that receives drilling fluid from mud pumps and transmits the drilling fluid to the succeeding circulating component. It is designed to control, distribute and monitor drilling fluid flow.
- **Stand pipe** is the vertical rigid pipe rising along the side of the derrick or mast, which joins mud pump manifold to the rotary hose.
- **Drill string** is the mechanical assemblage connecting the rotary drive on the surface to the drilling bit on bottom of the wellbore.
- **Mud return line or flow line** is the large diameter metal pipe and is the passageway of the drilling fluid as it comes out of the well.
- **Shale shaker** is the primary solids-removing device with one or more vibrating screens, which is used to remove cuttings from the circulating fluid for reuse. Screens vibrate while the mud flows on top of it. The liquid phase with solids which are smaller than the wire mesh pass through the screen as well as larger solids are retained on the screen and eventually fall to the special container and can be disposed in an environmentally friendly manner.
- **Desander** is a centrifugal device for removing sand-size particles from the drilling fluid to prevent abrasion of the pumps. There are no moving parts of a desander, and the removal of particles is done by gravity and pressure. As the drilling fluid flows around and gradually down the inside of the cone shape, particles are separated from the liquid by centrifugal forces.
- **Desilter** is also a centrifugal device for removing free particles of silt from the drilling fluid. Comparing with desander, its design incorporates a greater number of smaller cones, which allow removing smaller diameter particles than a desander does.
- **Degasser** is device designed to remove air, methane, hydrogen sulfide (H_2S), carbon dioxide (CO_2) and other gases from drilling fluids and allow it to be reused continuously. It helps to reduce the risk of explosions and other dangers during the drilling process.
- **Mud pit** is an excavated earthen-walled pit and is used only to store used or waste drilling fluid and cuttings.
- **Mud tank** is an open-top steel container with possibility to observe the consistency of drilling fluid and monitor it level in the tanks. It is used as a reserve store for the drilling fluid.

Tubular and tubular handling equipment. Tubular consists of the following equipments:

- **Drill pipe** is the longest section of the drill string and is heavy hot-rolled, pierced and seamlessly tubing. It connects the surface equipment with the bottom hole assembly and the bit is used to rotate the bit and for drilling fluid circulation.
- **Drill collar** is thick-walled, heavy and large diameter steel tube placed between the drill pipe and the bit in the drill stem to provide weight on a bit. It can be cylindrical or spiral shape and is threaded at both ends (male and female) to allow multiple drill collars to be joined above the bit assembly.
- **Heavy weight drill pipe** is thick-walled tube and is used as transition pipe between drill collar and drill pipe. In high-angled and horizontal wellbore, it is used in lieu of drill collars.

- **Subs** are short component of the drill string, threaded piece of pipe used to adapt parts of the drilling string that cannot otherwise be screwed together because of difference in thread size or design.

Tubular handling equipment is made of the following equipments:

- **Elevator** clamps that grip a stand of casing, tubing, drill pipe or drill collars so that the stand or joint can be lifted and lowered into the wellbore opening of the rotary table. The elevators are connected to the traveling block by means of bails, which are solid steel bars with eyes at both sides. Elevator could be side door, center latch or single joint types.
- **Elevator links** is device designed to support the elevators and attach them to the hook.
- **Slips** are a wedge-shaped piece of metal with teeth or other gripping elements that supports and transmits the weight of the drill string to the rotary table and are used to hold the pipe in place as well as to prevent pipe from slipping down into borehole. Different types of slips are used during oil well drilling such as drill pipe, drill collar or casing slips.
- **Safety clamp** is a mechanical device used on tubulars above the slips and is used to keep parts of the tool string from falling down the wellbore if other safety measures fail.
- **Tongs** are large wrenches used to make or break out tubular. It must be used in opposing pairs—make up or breakout tongs to make or break connection.
- **Drill pipe spinner** is a pneumatically operated device usually suspended on the rig floor used to make fast connections and spin off of drill pipe.
- **Iron roughneck** is a pneumatically operated machine that replaces the functions performed by the Kelly spinner, drill pipe spinner and tongs and is used to connect and disconnect tubular.

Drill bits are cutting tools used to create cylindrical holes. Bits are located at the bottom of the drill string and are suited for particular conditions, such as formation, which is to be drilled. There are three different types of bit designs, such as:

- Roller cone bits with milled tooth or tungsten carbide insert (TCI) could have 2–6 cone-shaped steel devices that are free to turn as the bit rotates.
- Fixed cutter bits could be drill bit or core bit. The first one could be polycrystalline diamond compact bit (PDC-bit), surface set diamond bit and impregnated diamond bit. It consists of bit bodies and cutting elements integrated with the bit bodies and do not have moving parts.
- Hybrid bits combine both rolling cutter and fixed cutter elements.

If the drill bit needs to be changed, the whole string of pipe must be raised to the surface.

3. Classification of the drilling fluids

Modern drilling fluids (muds) are complex heterogeneous fluids (water based, oil based) and are complex mixtures of more than 200 minerals and chemicals. It is used in a drilling operation and circulates from the surface, down the drill string, through the bit and back to the surface via the annulus. The original use of the drilling fluids was to remove cuttings continuously. Progress in drilling engineering demanded more sophistication from the drilling mud. In order to enhance the usage of drilling fluids, numerous additives were introduced and a simple fluid became a complicated mixture of liquids, solids and chemicals. As the drilling fluids evolved, their design changed to have common characteristic features that aid in safe, economic and satisfactory completion of a well. In addition, drilling fluids are also now required to perform following functions:

- Clean the rock formation beneath the bit for rock cuttings.
- Remove cutting from the well.
- Control formation pressures while drilling and maintain wellbore stability.
- Suspend and release cuttings.
- Seal permeable formations to prevent excessive mud loss.
- Minimize reservoir damage by using reservoir drill-in fluid.
- Cool, lubricate and clean the bit and drilling assembly.
- Transmit hydraulic energy to downhole assembly.
- Ensure adequate formation evaluation.
- Control corrosion.
- Facilitate downhole measurement (measurement while drilling, logging while drilling).
- Facilitate cementing and completion.
- Minimize impact on the environment.

However, excessive use of oil-based drilling fluids may harm the environment and it is important to develop more environmentally friendly drilling fluids. In this respect, water-based drilling fluids are more acceptable. As well known, bentonite is widely applied in the water-based drilling fluids, which could enhance the clean properties and form a thin filter with low permeability. The functions of bentonite are to make the fluids more viscous and reduce the loss of fluids.

There are four types of drilling fluids **(Figure 7)**:

1. **Water-based drilling fluid (WBM)** is the mud in which water is continuous phase. The water could be fresh, brackish or seawater. The most basic WBM system begins with water, then clays and other chemical and is incorporated into the water to create a homogenous blend. The clay (called "shale" in its rock form or bentonite) is frequently referred to in the oilfield as "gel." Many other chemicals (e.g. potassium formate, $KHCO_2$) are added to a WBM system to achieve various effects, including velocity control, shale stability, enhance drilling rate of penetration, cooling and lubricating of equipment [2–4].

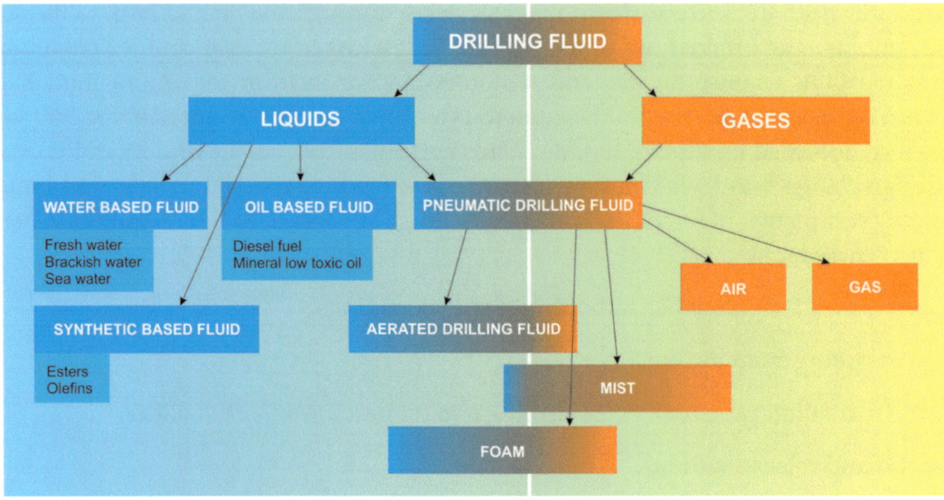

Figure 7. Classification of the drilling fluids.

Advantages

- Low cost
- High rate of penetration
- Good cuttings removal
- Good geoscientific investigations
- The pressure in the cutting area increases with increasing hydrostatic pressure of drilling fluid.

Disadvantages

- Low borehole stability [5]
- Insufficient cutting transport efficiency
- Insufficient lubricating properties
- Drilling fluid loss.

2. **Oil-based drilling fluid** has best technical properties such as stability, lubricity and temperature stability. Oil-based mud can be a mud where the base fluid is a petroleum product such as diesel fuel or mineral low toxic oil. The authorities do not permit the discharge of oil-based drilling fluid and cuttings drilled with oil-based drilling fluids because of their special nature of being a mixture of two immiscible liquids (oil and water). In that case special treatment and testing are required. The terms oil-based mud and inverted or invert-emulsion mud used to distinguish among the different types of oil-based drilling fluids. Traditionally, an oil-based mud is a fluid with 0–5% by volume of water, whereas an invert-emulsion mud contains more than 5% by volume of water.

http://dx.doi.org/10.5772/intechopen.70360

Advantages

- Excellent lubricating properties (reduce drilling torque and drag)
- Good temperature stability
- Favorable to borehole stability
- High rate of penetration
- Will not hydrate clays
- Long bit life
- Low reservoir damage
- Low drilling fluid loss
- Salt not dissolved
- Corrosion resistance
- Can be reused.

Disadvantages

- High initial cost
- Electric log difficulty
- Viscosity varies with temperature
- Environmental issue
- Difficult to keep the rig clean while drilling
- Difficult to identify gas kick
- Messy working environment
- Fire hazards.

3. **Synthetic drilling fluids** are based on ether, ester or olefin. They have technical properties that are similar to oil-based drilling fluids and are most often used on offshore rigs or in environmentally sensitive areas, because it has the properties of an oil-based mud, but the toxicity of the fluid fumes is much less than an oil-based fluid. This is often used on offshore rigs.

Advantages

- Favorable to borehole stability
- High rate of penetration
- Good wellbore stability
- Good control of drilling fluid properties

- Good cutting transport efficiency and removal
- Good filtration properties.

Disadvantages

- Complex system with high solid content
- Geoscientific investigations difficulty.

4. **Pneumatic drilling fluids**—Fluids, which are based on air/gas, mist, aerated fluid or foam. Air drilling is used primarily in hard rock areas and in special cases to prevent formation damage while drilling into production zones or to circumvent severe lost-circulation problems. Air drilling includes dry air drilling, mist or foam drilling and aerated mud drilling. In dry air drilling, dry air/gas is injected into the standpipe at a volume and rate sufficient to achieve the annular velocities needed to clean the hole of cuttings. Mist drilling is used when water or oil sands are encountered that produce more fluid than can be dried up using dry air drilling. A mixture of foaming agent and water injected into the air stream, producing foam that separates the cuttings and helps remove fluid from the borehole. In aerated mud drilling, both mud and air pumped into the standpipe at the same time. Aerated mud is used when it is impossible to drill with air alone because of water sands and/or lost-circulation situations.

Advantages

- High rate of penetration
- Low reservoir damage
- Good bit performance
- Low drilling fluid loss
- Low water consumption
- Low air quality requirements for foam drilling
- Low hydrostatic pressure
- Good cleaning of the borehole.

Disadvantages

- There are restrictions on the possible lithological structures
- Drilling could be limited by the length of the horizontal section of the well
- Possibility of fire
- Possible additional costs to rent equipment
- Gas costs
- Gas and foam utilization issues

- Aerated fluids require specialized equipment for the injections
- Aerated fluids and foam have potential corrosion problems and the need to use additional inhibitor
- The quality of the foam changes in exchange pressure
- The foam is a complicated system and may require computer modeling of foam movement in the borehole.

3.1. Mud ingredients

Various materials may be added at the surface to change or modify the characteristics of the mud:

1. **Weighting materials** (usually barite) are added to increase the density of the mud, which helps to control subsurface pressures and build the filter cake. Salts are sometimes added to protect downhole formations or to protect the mud against future contamination, as well as to increase density. Dispersants or deflocculants may be added to thin the mud, which helps to reduce surge, swab and circulating pressure problems.
2. **Viscosifying materials** (clays, polymers and emulsified liquids) are added to thicken the mud and increase its hole cleaning ability. [6]
3. **Filtration control materials**. Clays, polymers, starches, dispersants and asphaltic materials may be added to reduce filtration of the mud through the borehole wall. This reduces formation damage, differential sticking and problems in log interpretation.
4. **pH control and lubricating materials**. Mud additives may include lubricants, corrosion inhibitors, chemicals that tie up calcium ions and flocculants to aid in the removal of cuttings at the surface. Caustic soda is often added to increase the pH of the mud, which improves the performance of dispersants and reduces corrosion.
5. **Other additives**. Preservatives, bactericides, emulsifiers and temperature extenders may all be added to make other additives work better.

Most of these additives have distinct properties that help in countering specific challenges encountered during the drilling process as well as in accomplishing the drilling work with efficiency and precision [7]. However, to select the proper fluid, it is necessary to calculate the cost of the fluid, understand the environmental impact of using the fluid and to know the impact of the fluid on production from the pay zone.

4. Drilling mud losses and its prevention

The complex drilling fluids represent 15–18% of the total cost of petroleum well drilling. Lost circulation is major problem in the drilling operations and is defined as the loss of drilling fluid through the pores or fissures in the rock formations to be drilled, sometimes referred to as "thief zones." It occurs when hydrostatic pressure of fluid column in the wellbore is higher than the formation pressure and is defined as the loss of drilling fluid into the formation. Lost circulation influences directly effect the non-productive time, a drilling operation that includes the cost of time and all services that support the drilling operation. It is usually accompanied by wellbore stability problems, which can result in stuck pipe and even the loss of the well [8–10].

The fluid loss of circulation is most commonly responsible for 10–20% of the total cost of a productive or an exploration well. Well bore costs, in turn, represent 35–50% of the total capital costs of a geothermal typical project; therefore, about 3.5–10% of the total costs can be attributed to the loss of circulation [11].

4.1. Physical simulator of flow in the formation

The physical simulator of flow in the formation (SFF) device allows determining the mud loss to the formation. It consists of a fluid storage tank with mixer, well-simulated pipe with formation packing system, pump, temperature and pressure-measuring device and so on.

An experimental procedure was developed with the purpose of studying effects of additives and loss circulation material on mud loss to the formation. The mud sample that was prepared and mixed in the separate storage tank transferred into the stand's storage tank (1) (**Figure 8**). Formation packing system was filled with the formation that was tested against the drilling mud. Then, the formation packing system connected to the well-simulated pipe (5). The hollow cable from the pump (3) is connected to the compressor to make the pump run

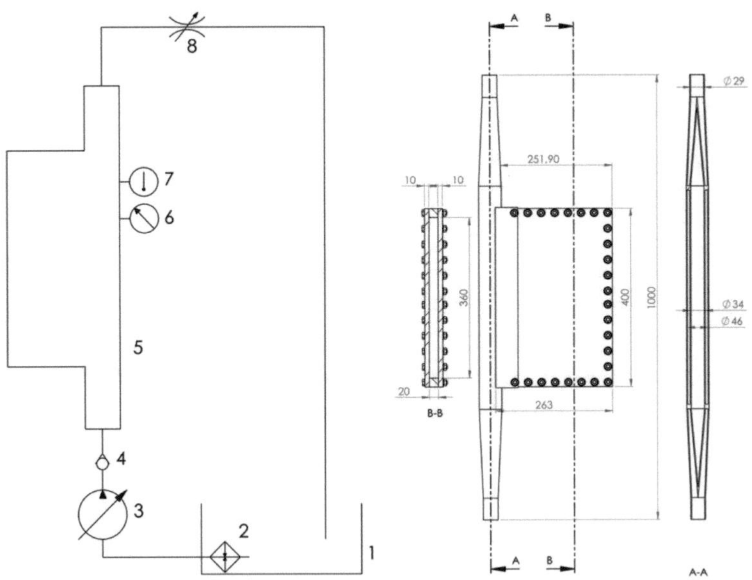

Figure 8. Kinematic scheme of the physical simulator of flow in the formation: 1—fluid storage tank with mixer; 2—heater; 3—pump; 4—valve; 5—well-simulated pipe with formation packing system; 6—pressure measuring device; 7—temperature measuring device and 8—pressure regulator.

http://dx.doi.org/10.5772/intechopen.70360

at desired pressure. After the pump has been turned on, the drilling mud started to circulate from the storage tank through the well-simulated pipe. The process runs for 30 minutes and when it is finished it is possible to measure the fluid penetration rate.

The visualization of the physical simulator of flow in the formation is shown in **Figure 9**.

Combating loss by the proper use of reinforcement materials of wells, well strengthening and loss circulation materials is fundamental for a successful drilling [12–14]. In **Figure 10**, simulation results (Flow 3D) show the influence of the loss circulation materials on the drilling process. In case of the water drilling, the drilling fluid losses are significant, whereas in case of the loss circulation materials, rate of penetration considerably decreases.

Industries use coke, attapulgite, nutshells, mica flakes, cellulose nanoparticles and other materials to mitigate the loss of circulation [15]. The use of such materials increases the cost of drilling, but by using the materials such as cotton, sawdust and used oil would employ the same purpose in most cost-effective and eco-friendly way.

Various materials such as cotton waste, used oil, saw dust etc. are commonly employed as fluids loss control agent. The evaluation of the rate of penetration of various mud samples to the formation proposes an effective way to minimize the mud loss by forming a static filter cake on the walls by changing the components of the water-based drilling fluid.

It is important to evaluate the amount of drilling fluid loss to the formation and to overcome it by forming a static filter cake on the borehole walls by changing the components of the water-based drilling fluid [16, 17].

Figure 9. Visualization of physical simulator of flow in the formation.

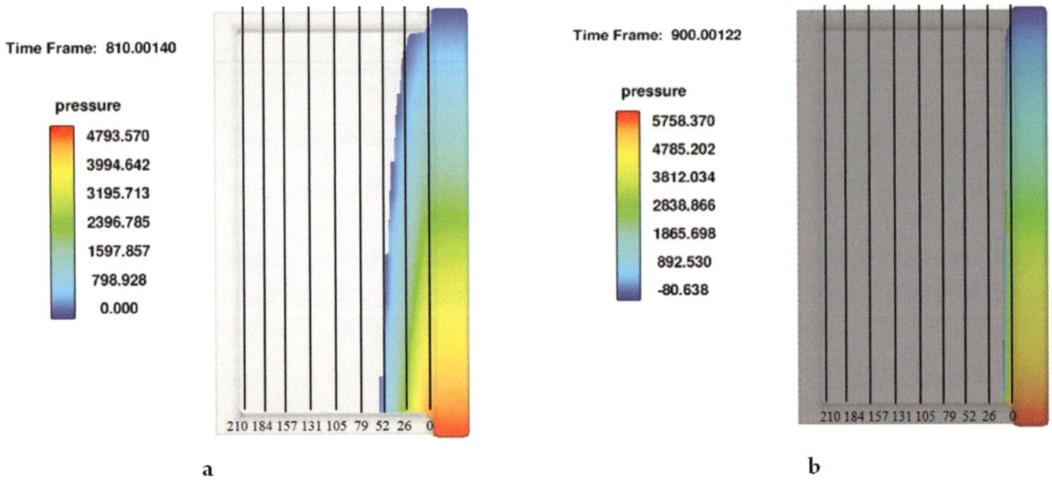

Figure 10. Comparison of the fluid loss and rate of the penetration by pure water and the water-based fluid drilling: a—water; b—drilling fluid.

4.2. Preparation of mud samples and properties of the formation

Formation porosity directly affects the mud loss, if the pore size of the formation is high, it means that the formation pore size do have much space to retain any fluid or small particle which passes through it [18]. During the experiment, the density of the formation was 1.606 g·cm^{-3}. The pH of the formation was 8.73 and is alkaline, so it will not play a vital on altering any significant property of the drilling mud. Humidity does not play a major role in mud loss, but it has to be measured to determine the filtration property of the sand. Humidity of formation was 4.26%.

Base mud sample, containing only water and bentonite clay, was prepared by adding 720 g of bentonite to 12 liters of water to obtain a bentonite mass fraction of 5.66% and a bentonite-to-water ratio of 6% (**Figure 11**).

The bentonite-to-water ratio was maintained constant for all subsequent mud samples used in this research.

All mud samples were prepared at ambient conditions (at 17°C or 62.6°F). Respectively, their density and rheological properties were measured. Sodium hydroxide (NaOH) was used to adjust the pH of mud samples to ensure that each sample has same pH value of 10.85. Potassium chloride (KCl) was used as a clayish rock swelling inhibitor because the loam formation used during the research has clay content and was constant for all mud samples. Sodium carbonate (NaOH) was used to regulate the calcium concentration in the drilling mud.

Mud samples with varying additive concentrations and loss circulation materials such as saw dust, waste cotton and used oil were prepared as it is shown in **Table 1**.

The water-based mud samples presented in **Table 1** are named according to the loss circulation materials added to it. For instance, the mud sample with sawdust is named from S1 to S3

Figure 11. Prepared mud sample.

according to the weight of the material present in it. The sawdust was added from 232 to 522 g in three mud samples and cotton is added from 25 to 75 g in C1–C3. The used oil was added in milliliter and their relative weight of oil was calculated and presented in g. The used oil was added from 135 to 404 g to O1 to O3 mud samples.

In this experiment, rheological properties of drilling mud additives were studied (**Table 2**). Mud samples with a varying concentration of additives were prepared; their properties were studied and compared.

Mud samples		Mass of water	Mass of bentonite	Mass of Na_2CO_3	Mass of NaOH	Mass of saw dust	Mass of waste cotton	Mass of used oil
		Kg	g	g	g	g	g	g
Bentonite	B1	12	720	60	14.4	0	0	0
Sawdust	S1			68	17.2	232	0	0
	S2			70.5	19	397	0	0
	S3			74.2	20.5	522	0	0
Cotton	C1			61	17.8	0	25	0
	C2			63.2	18.05	0	50	0
	C3			66.6	18.60	0	75	0
Used oil	O1			60	23.4	0	0	135
	O2			63.7	24.2	0	0	269
	O3			66.6	24.8	0	0	404

Table 1. The composition of water-based mud samples.

Mud samples		Mud density	Plastic viscosity	Yield point	10-sec gel strength
		lb/gal	cP	lb/100 ft^2	lb/100 ft^2
Bentonite	B1	9.260	18	18	4
Sawdust	S1	9.290	23	25	6
	S2	9.290	24	28	7
	S3	9.296	26	31	9
Cotton	C1	10.26	29	33	16
	C2	10.43	35	35	24
	C3	10.68	37	43	31
Used oil	O1	9.280	20	21	5
	O2	9.280	21	24	6
	O3	9.296	22	26	8

Table 2. Variation of the muds' rheological properties by using additives.

The **mud density** comparison of all three mud indicated that a cotton-based mud will give a higher mud density than the other two (10.26–10.68 lb/gal). This is because of the higher specific gravity of cotton as it greater than of saw dust and oil. In addition to its use as loss circulation materials, cotton-based mud can also act as weighting agent. To prevent the flow of formation fluids into the hole, the drilling mud must exert a greater pressure than that of the fluids in the porous rocks that are penetrated by the bit. In that case, the cotton-based mud with slightly high density can act as an effective loss circulation materials.

Plastic viscosity (PV). The experiment results show that the mud samples with sawdust and cotton have high viscosity ranging from 23 to 26 cP in the case of saw dust and 29 to 37 cP in the case of cotton. This is due to the fact that the mud with considerable suspended particle will always have high plastic viscosity as well as the force existing between the particles and the force between the particles and the liquid. So, this result indicates that the mud samples with cotton and sawdust have much high solid content as 75 g in cotton and 522 g in sawdust and always depend upon the concentration of mud solids.

Yield point (YP) is used to evaluate the ability of mud to lift cuttings out of the annulus. When the yield point is higher, the mud loss inside the fracture is less. The shear stress required to initiate the flow of mud also increases as the yield point increases. Over time, this yield point helps in preventing the mud from flowing further into the fracture, which leads eventually to it becoming plugged [19]. A higher YP implies that drilling fluid has ability to carry cuttings better than a fluid of similar density but lower YP. From the above result, compare the mud samples S3 and O3, in both mud samples, the density is quite the same, it is 9.296 lb/gal, whereas the YP of S3 is 5% greater than O3. In this case, the S3 mud sample will work better in carrying cuttings than O3. The bentonite (B1) sample is taken as a reference sample.

10-sec gel strength results. The gel strength is one of the important drilling fluid properties because it demonstrates the ability of the drilling mud to suspend drill solid and weighting material when circulation is ceased. The results were achieved based on standard API procedure. It is investigated that the mud sample with cotton as a loss circulation materials has gel strength twice higher than the other two. It means that it will work well in the case of suspending drill cutting when the circulation is halt for 1 to 2 days.

The rate of penetration of all mud samples is represented in **Figure 12**.

Figure 12 consolidates all the result obtained from the experimental work. In the case of sawdust mud sample, the weight percentage of sawdust added were ranging from 1 to 3%, and in the case of used oil, it is 1, 1.5, 2%, but in the case of cotton, it is just 0.2, 0.4, 0.6% because the cotton make the mud more viscous and heavily dense, which makes it hard for the pump to deliver the same pump rate as it was done with sawdust and used oil mud samples. From **Figure 12**, it is evident that the mud samples with additives can be used as loss circulation material during oil well drilling.

In this work, it is evident that the prepared and tested mud samples work well with the unconsolidated coarse-grained formation in terms of mud loss.

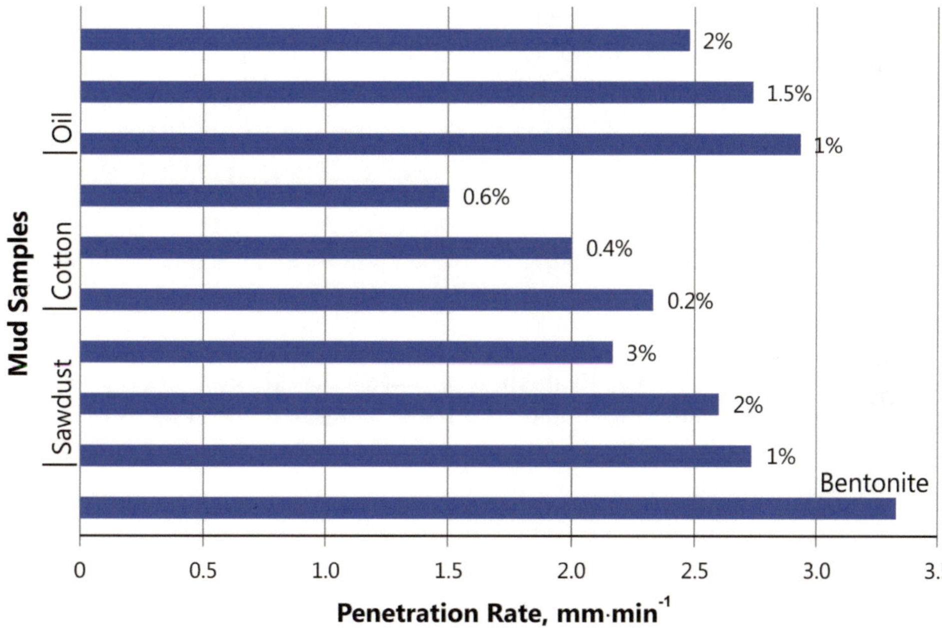

Figure 12. Rate of penetration of all mud samples in formation.

5. Conclusions

The concentration of loss circulation material is vital to control the rheological properties of drilling mud. Significant changes in mud density, plastic viscosity, yield point and gel strength were noted to correspond to changes in the concentration of mud loss circulation material.

Waster-based mud with cotton as loss circulation material gave a remarkably higher value of density, yield point, gel strength and plastic viscosity when used at lesser concentration than sawdust and used oil. Moreover, water-based mud samples with cotton having the least penetration rate. The lack of loss circulation material could result in significant mud loss.

Author details

Tatjana Paulauskiene

Address all correspondence to: tatjana.paulauskiene@ku.lt

Klaipeda University, Klaipeda, Lithuania

References

[1] BP. Statistical Review of World Energy 2016 [Internet]. June 2016.Available from: http://www.bp.com/content/dam/bp/pdf/energy-economics/statistical-review-2016/bp-statistical-review-of-world-energy-2016-full-report.pdf [Accessed: September 2016]

[2] Khodja M et al. Shale problems and water-based drilling fluid optimisation in the Hassi Messaoud Algerian oil field. Applied Clay Science. 2010;**49**(4):383-393. DOI: 10.1016/j.clay.2010.06.008

[3] Khodja M et al. Products and Services; from R&D to Final Solutions. Drilling Fluid Technology: Performances and Environmental considerations. Intech, Croatia; 2010. 434 p. DOI: 10.5772/55742

[4] Jain R, Mahto V, Sharma VP. Evaluation of polyacrylamide-grafted-polyethylene glycol/silica nanocomposite as potential additive in water based drilling mud for reactive shale formation. Journal of Natural Gas Science and Engineering. 2015;**26**:526-537. DOI: 10.1016/j.jngse.2015.06.051

[5] Mahto V et al. Development of non-damaging and inhibitive water based oil well drilling fluids. Petroleum Science and Technology. 2013;**31**(7):721-726. DOI: 10.1080/ 10916466.2010.531353

[6] Ay A, Gucuyener IH, Kok MV. An experimental study of silicate—polymer gel systems to seal shallow water flow and lost circulation zones in top hole drilling. Journal of Petroleum Science and Engineering. 2014;**122**:690-699. DOI: 10.1016/j.petrol.2014.09.011

[7] Yang X et al. A biomimetic drilling fluid for wellbore strengthening. Petroleum Exploration and Development. 2013;**40**(4):531-536. DOI: 10.1016/S1876-3804(13)60069-5

[8] Akhtarmanesh S, Shahrabi MJA, Atashnezhad A. Improvement of wellbore stability in shale using nanoparticles. Journal of Petroleum Science and Engineering. 2013;**112**: 290-295. DOI: 10.1016/j.petrol.2013.11.017

[9] Liang C et al. Wellbore stability model for shale gas reservoir considering the coupling of multi-weakness planes and porous flow. Journal of Natural Gas Science and Engineering. 2014;**21**:364-378. DOI: 10.1016/j.jngse.2014.08.025

[10] Kang Y et al. Constructing a tough shield around the wellbore: Theory and method for lost-circulation control. Petroleum Exploration and Development. 2014;**41**(4):520-527. DOI: 10.1016/S1876-3804(14)60061-6

[11] Calçada LA et al. Evaluation of suspension flow and particulate materials for control of fluid losses in drilling operation. Journal of Petroleum Science and Engineering. 2015;**131**:1-10. DOI: 10.1016/j.petrol.2015.04.007

[12] Safi B et al. Physico-chemical and rheological characterization of water-based mud in the presence of polymers. Journal of Petroleum Exploration and Production Technology. 2016;**6**(2):185-190. DOI: 10.1007/s13202-015-0182-x

[13] Abduo MI et al. Comparative study of using water-based mud containing multiwall carbon nanotubes versus oil-based mud in HPHT fields. Egyptian Journal of Petroleum. 2016;**25**(4):459-464. DOI: 10.1016/j.ejpe.2015.10.008

[14] Samavati R et al. Rheological and fluid loss properties of water based drilling mud containing HCl-modified fufu as a fluid loss control agent. International Journal of Chemical Engineering and Applications. 2014;**5**(6):446-450. DOI: 10.7763/IJCEA.2014.V5.426

[15] Song K et al. Water-based bentonite drilling fluids modified by novel biopolymer for minimizing fluid loss and formation damage. Colloids and Surfaces A: Physicochemical and Engineering Aspects. 2016;**507**:58-66. DOI: 10.1016/j.colsurfa.2016.07.092

[16] Rugang Y et al. Effect of water-based drilling fluid components on filter cake structure. Powder Technology. 2014;**262**:51-61. DOI: 10.1016/j.powtec.2014.04.060

[17] Caenn R, Darley HCH, Gray GR. Composition and Properties of Drilling and Completion Fluids. 6th ed. Gulf professional publishing, United States; 2011. 720 p. DOI: 10.1016/B978-0-12-383858-2.00026-3

[18] Zhiyong H et al. Establishment and application of drilling sealing model in the spherical grouting mode based on the loosing-circle theory. International Journal of Mining Science and Technology. 2012;**22**(6):895-898. DOI: 10.1016/j.ijmst.2012.12.004

[19] Abdo J, Haneef MD. Clay nanoparticles modified drilling fluids for drilling of deep hydrocarbon wells. Applied Clay Science. 2013;**86**:76-82. DOI: 10.1016/j.clay.2013.10.017

Chapter 3

Optimal Planning for Deepwater Oilfield Development Under Uncertainties of Crude Oil Price and Reservoir

Zhang Haoran, Liang Yongtu, Ma Jing, Di Pengwei,
Yan Xiaohan and Huang Zhongliang

Additional information is available at the end of the chapter

http://dx.doi.org/10.5772/intechopen.71078

Abstract

The development planning of deepwater oilfield directly influences production costs and benefits. However, the uncertainties of crude oil price and reservoir and the special production requirements make it difficult to optimize development planning of deepwater oilfield. Although there have been a number of scholars researching on this issue, previous models just focused on several special working conditions and few have considered energy supply of floating production storage and offloading (FPSO). In light of the normal deepwater production development cycles, in this paper, a multiscenario mixed integer linear programming (MS-MILP) method is proposed based on reservoir numerical simulation, considering the uncertainties of reservoir and crude oil price and the constraint of energy consumption of FPSO, to obtain the globally optimal development planning of deepwater oilfield. Finally, a real example is taken as the study objective. Compared with previous researches, the method proposed in this paper is testified to be practical and reliable.

Keywords: uncertainty, deepwater oilfield, development planning, multiscenario mixed integer linear programming, optimization

1. Introduction

The construction of deepwater oilfield development (DWOD) costs much, which is the highest proportion of oilfield development investment [1]. The DWOD facilities and connection modes will be fixed in the working cycles once they are determined and put into use [2]. Hence, it is important to work out totally optimal oilfield development planning, considering the future production during planning stages. Deepwater oilfield exploration and development generally consists of four stages: exploration, evaluation, and deliverability construction and production [3], each of which contains a number of uncertain factors [4, 5] that exert influence on the total operation. In the evaluation stage of the early period of oilfield development, there are a great number of uncertain factors and some have strong uncertainties [6, 7]. The essence of optimal planning for DWOD is to calculate the totally optimal oilfield development plan including the type of floating production storage and offloading (FPSO), commissioning plan, drilling plan, connection mode between productive well and FPSO, production plan of each production well, and FPSO energy supply plan. In this way, the optimal planning of DWOD is complex, since it requires various optimization decisions under the premise of strong uncertainty.

Recently, the issue of field development planning has attracted many scholars. Midthun et al. [8] established an optimal model for natural gas field development, which considers the construction rule of processing facilities and pipeline infrastructure. Arredondo-Ramrez et al. [9] focused on the optimal planning for nonconventional shale gas development and proposed a model applicable for multistage development. As for offshore field infrastructure planning, Gupta and Grossmann [10] put forward a multiperiod mixed integer linear programming (MILP) model which involves oil-gas-water, three phases in reservoir. Based on the previous work, Gupta and Grossmann [11] took the production-sharing agreements and the endogenous uncertainties into consideration, making the proposed multistage model more accordant with practical situations. When building up the oilfield development planning model, it is inevitable to consider uncertain factors such as reservoir behavior and crude oil price. Tarhan et al. [12] thought the reservoir uncertainty was related to initial production output of each well, reservoir scale, and water breakthrough time. These variable factors were exhibited as eight scenarios through enumeration. However, it is possible that the reality is too more sophisticated to describe the field's uncertainty based on the eight scenarios. And when it comes to the uncertainty of crude oil price, Jonsbraten [13] build up an MILP model and employed the scenario and policy aggregation technique to solve the construction planning of oilfield development under crude oil price uncertainty. Aseeri et al. [14] addressed the financial risk management of offshore field development planning and scheduling and paid attention to the uncertainties of crude oil price and oil well productivity index. Kang et al. [15] proposed an optimal model of oilfield development programming under stochastic oil price.

As to the model resolution, Dawson and Fuller [16] established a multistage nonconvex MILP model taking the highest net present value (NPV) as the objective function for the offshore oilfield development. During the model resolution, the continuous variables should be discretized, and then, heuristic algorithm was adopted. Heever and Grossmann [17] put forward a multistage MILP model which was solved by an iterative aggregation/disaggregation algorithm. Chen and Feng [18] established a model of oilfield measure program which is predicted by the BP network. Carvalho et al. [19] established an MIP model to work out offshore oilfield infrastructure planning and applied decomposition method to solve the model. Ge et al. [20] investigated the drilling of cluster horizontal wells and set up a platform program ming model. Zhang et al. [21, 22] built an MILP model for optimal offshore oilfield gathering system and a unified MILP model for topological structure of production well gathering pipeline network, and both were solved by the branch-and-bound method. Li et al. [23] proposed a nonlinear programming model for the integrated development of multiple gas fields, using the improved genetic algorithm (GA) to solve the model.

Although domestic and foreign scholars have researched on the design for offshore field engineering system, reservoir uncertainty, and output fluctuation are considered by simply enumerating several special conditions and few have taken the FPSO energy supply constraint into account. In light of the normal deepwater production development cycles, this paper proposes a multiscenario MILP (MS-MILP) model based on reservoir numerical simulation, considering the uncertainties of reservoir and crude oil price, to obtain the globally optimal planning of DWOD.

2. Issue description

In this paper, the study issue is to design and draw up the commissioning plan of infrastructure required for DWOD in the given development cycle. As shown in **Figure 1**, the reservoir includes three parts and there is one FPSO working. On FPSO, oil-gas-water separation, as well as storage and transportation to shuttle tankers, can be carried out for the produced liquid. And there are two types of FPSO: one is the small FPSO of smaller throughput, coming from old oil tanker's conversion, and the other is the large new FPSO of larger throughput. Furthermore, production wells are connected to FPSO by means of drilling vessels or semi-submerged platforms (SSP). One production well can only correspond to one FPSO, while one FPSO can correspond to multiple wells.

The optimization is to draw up the investment and operation decisions during the development cycle. The investment decisions include the type, number, and corresponding processing capacity of development facilities, the manufacture, installation, and service time of these facilities, as well as the wells to be drilled, drilling sequence, and drilling facility. The operation decisions

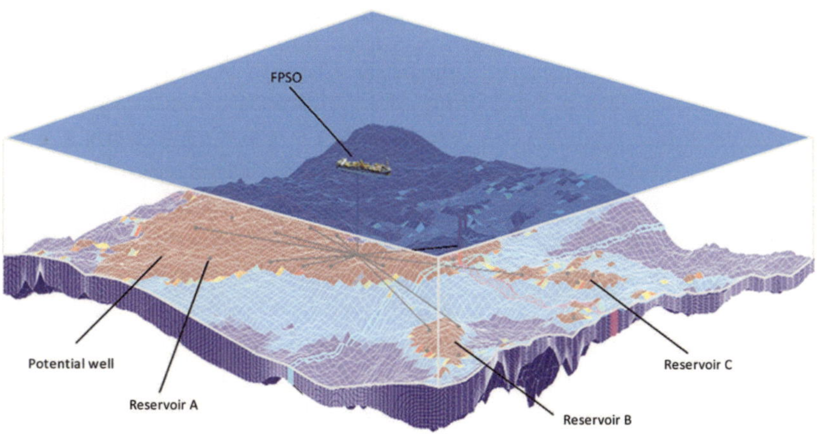

Figure 1. Diagram of DWOD.

need to consider the uncertainties of reservoir and crude oil price and simultaneously provide the recovery rates of reservoirs in each period. The general target is to balance the complex investment and operation decisions, in order to maximize the expected NPV of the project.

The predrilling well number of each reservoir, the construction cycles, and the available FPSO type and corresponding throughput should be known to solve the model. The model decisions include annual drilling site and production output of reservoirs, FPSO commissioning, connection relationship between production well and FPSO, FPSO oil-gas-water throughput, and FPSO energy supply plan.

3. Model establishment

3.1. Objective function

The discrete-time representation is adopted in this model. Suppose $s \in S$ stands for the set that considers uncertainties; $r \in R$ stands for the set of reservoirs; $t \in T$ stands for the set of years; $i \in I_r$ stands for the set of predrilling of reservoir r; $l \in L$ stands for the set of available FPSO; and $a \in A$ stands for the set of accumulated output ranges. Since multiple uncertainties are involved in the model, the maximum mean NPV of various conditions is taken as the objective function. The income is related to the annual crude oil price and output in this year. The expenditure is related to the construction and purchase cost of FPSO, daily maintenance cost, drilling cost, and diesel consumption cost.

$$\max f = \left[\sum_s \sum_t I_t (INC_{s,t} - COS_{s,t})\right] / N_S \quad s \in S \tag{1}$$

$$INC_{s,t} = C_{PERs,t} \sum_r \sum_i Q_{OPs,r,t,i} \quad s \in S, r \in R, t \in T, i \in I_r \tag{2}$$

$$\begin{aligned} COS_{s,t} = {} & \sum_l C_{BTLP\,l} B_{BTLP\,t+N_{Ll},l} + \sum_l C_{FMO\,l} \sum_{t'=1}^{t} B_{BTLP\,t',l} + C_{DIT} \sum_r \sum_i B_{DIT\,r,t+N_D,i} \\ & + C_{Ds,t} \sum_l Q_{Ds,t,l} \quad s \in S, r \in R, t \in T, i \in I_r, l \in L \end{aligned} \tag{3}$$

where N_S represents the number of the considered uncertainties; N_{Ll} represents the construc tion time for $FPSO_l$; N_D represents the drilling time; I_t represents the discount rate in the year t; $INC_{s,t}$ represents the total income in the year t under the specified scenario s, \$/y; $COS_{s,t}$ represents the total expenditure in the year t under the specified scenario s, \$/y; $C_{PERs,t}$ represents the crude oil price per unit volume in the year t under the specified scenario s, \$/m^3; C_{BTLPl} represents the construction cost for $FPSO_l$, \$; C_{FMOl} represents the daily maintenance cost for $FPSO_l$, \$/y; C_{DIT} represents the unit price of drilling cost, \$; $C_{Ds,t}$ represents the diesel price unit under the scenario s, \$/m^3; $Q_{OPs,r,t,i}$ represents the crude oil output from the predrilling well i of the reservoir r in the year t under the specified scenario s, m^3/y; $Q_{Ds,t,l}$ represents the con sumption diesel for $FPSO_l$ electricity generation in the year t under the specified scenario s, m^3/y; $B_{BTLPt,l}$ represents a binary variable, if $FPSO_l$ needs to be put into production in the year t, $B_{BTLPt,l} = 1$, otherwise $B_{BTLPt,l} = 0$; $B_{DITr,t,i}$ represents a binary variable, if the predrilling well i of the reservoir r is drilled in the year t, $B_{DITr,t,i} = 1$, otherwise $B_{DITr,t,i} = 0$.

3.2. Drilling constraint

There are two states for all the predrilling wells: development or not.

$$H_{\mathrm{DIT}r,i} + \sum_t B_{\mathrm{DIT}r,t,i} \le 1 \quad r \in R, t \in T, i \in I_r \tag{4}$$

where $H_{\mathrm{DIT}r,i}$ represents a binary variable, if the drilling operation for the predrilling well of the reservoir r has been finished before the start time of the study, $H_{\mathrm{DIT}r,i} = 1$, otherwise $H_{\mathrm{DIT}r,i} = 0$; other variables are defined the same as before.

The drilling number of drilling vessels should be less than the maximum drilling capacity per year.

$$\sum_r \sum_i B_{\mathrm{DIT}r,t,i} \le N_{\mathrm{CFmax}t} \quad r \in R, t \in T, i \in I_r \tag{5}$$

where $N_{\mathrm{CF\,max}\,t}$ represents the maximum drilling number of the year t; other variables are defined the same as before.

3.3. Output constraint

The accumulated output of each reservoir equals to that of the last year plus all the output of this reservoir in this year.

$$V_{\mathrm{AP}s,r,t} = V_{\mathrm{AP}s,r,t-1} + \sum_i Q_{\mathrm{OP}s,r,t,i} \quad s \in S, r \in R, t \in T \tag{6}$$

where $V_{\mathrm{AP}s,r,t}$ represents the accumulated crude output of the reservoir r in the year t under the specified scenario s, m^3; other variables are defined the same as before.

Binary variables can be divided by ranges to determine the range that accumulated output belongs to. When $B_{\mathrm{AP}s,r,t,a} = 1$, $V_{\mathrm{AP\,min}\,a} < V_{\mathrm{AP}s,r,t} \le V_{\mathrm{AP\,max}\,a}$ should be met.

$$V_{\mathrm{APmin}a} + (B_{\mathrm{AP}s,r,t,a} - 1)M < V_{\mathrm{AP}s,r,t} \le V_{\mathrm{APmax}a} + (1 - B_{\mathrm{AP}s,r,t,a})M \quad s \in S, r \in R, t \in T, a \in A \tag{7}$$

where $V_{\mathrm{APmin}s,r,a}$ represents the minimum value of the accumulated crude output range a in the reservoir r under the specified scenario s, m^3; $V_{\mathrm{APmax}s,r,a}$ represents the maximum value of the accumulated crude output range a in reservoir r under the specified scenario s, m^3; M represents a positive maxima; $B_{\mathrm{AP}s,r,t,a}$ represents a binary variable, if the accumulated crude output of the reservoir r in the year t belongs to the range a under the specified scenario s, $B_{\mathrm{AP}s,r,t,a} = 1$, otherwise $B_{\mathrm{AP}s,r,t,a} = 0$; other variables are defined the same as before.

The accumulated output must only exist in one set of range.

$$\sum_a B_{\mathrm{AP}s,r,t,a} = 1 \quad s \in S, r \in R, t \in T, a \in A \tag{8}$$

After the accumulated output range is determined, the maximum output of single well can be obtained by the linear fitting formula between the maximum output of single well and the accumulated output in the range. If the accumulated output locates in range a, $B_{APs,r,t,a} = 1$, and then, $Q_{UWPMs,r,t} = \omega_{s,r,a} V_{APs,r,t} + \mu_{s,r,a}$.

$$Q_{UWPMs,r,t,i} \leq \omega_{s,r,a,i} V_{APs,r,t} + \mu_{s,r,a,i} + (1 - B_{APs,r,t,a})M \quad s \in S, r \in R, t \in T, i \in I_r, a \in A \quad (9a)$$

$$Q_{UWPMs,r,t,i} \geq \omega_{s,r,a,i} V_{APs,r,t} + \mu_{s,r,a,i} + (B_{APs,r,t,a} - 1)M \quad s \in S, r \in R, t \in T, i \in I_r, a \in A \quad (9b)$$

where $\omega_{s,r,a,i}$ and $\mu_{s,r,a,i}$ represent the coefficients of the linear fitting formula for the maximum output of single well and the accumulated crude output; $Q_{UWPMs,r,t,i}$ represents the maximum output of the predrilling i of the reservoir r in the year t under the scenario s, m^3/y; other variables are defined the same as before.

The single well output needs to be less than the maximum output of the single well.

$$Q_{OPs,r,t,i} \leq Q_{UWPMs,r,t,i} \quad s \in S, r \in R, t \in T, i \in I_r \quad (10)$$

If drilling operation does not begin, the well output should be zero.

$$Q_{OPs,r,t,i} \leq \left(H_{DITr,i} + \sum_{t'=1}^{t} B_{DITr,t,i}\right)M \quad s \in S, r \in R, t \in T, i \in I_r \quad (11)$$

3.4. Production facility constraint

If one predrilling well is to be developed, one FPSO should be determined to be connected.

$$\sum_t B_{DITr,t,i} = \sum_l B_{OITr,i,l} \quad r \in R, t \in T, i \in I_r, l \in L \quad (12)$$

where $B_{OITr,i,l}$ is a binary variable, if the predrilling well i of the reservoir r is connected to FPSO$_l$, $B_{OITr,i,l} = 1$, otherwise $B_{OITr,i,l} = 0$; other variables are defined the same as before.

If predrilling wells need to be connected to one FPSO, the FPSO should be put into production before being connected.

$$\frac{B_{DITr,t,i} + B_{OITr,i,l} - 1}{2} \leq H_{HTLPl} + \sum_{t'=1}^{t} B_{BTLPt,l} \quad r \in R, t \in T, i \in I_r, l \in L \quad (13)$$

where H_{HTLPl} is a binary variable, if the FPSO$_l$ has been put into production before the start time of the study, $H_{HTLPl} = 1$, otherwise $H_{HTLPl} = 0$; other variables are defined the same as before.

All the available FPSO can be put into production or not.

$$H_{HTLPl} + \sum_t B_{BTLPr,t,l} \leq 1 \quad r \in R, t \in T, l \in L \quad (14)$$

The predrilling wells to be developed can only be connected to one FPSO, and the transportation flow to the FPSO should be equal to the well output. If the predrilling wells to be developed do not connect to an FPSO, the transportation flow must be zero.

$$Q_{OPs,r,t,i} = \sum_l Q_{OPLs,r,t,i,l} \quad s \in S, r \in R, t \in T, i \in I_r, l \in L \tag{15}$$

$$Q_{OPLs,r,t,i,l} \le B_{OITr,i,l} M \quad s \in S, r \in R, t \in T, i \in I_r, l \in L \tag{16}$$

where $Q_{OPLs,r,t,i,l}$ is the transportation flow from the predrilling well i of the reservoir r to the in FPSO$_l$ the year t under the specified scenario s, m^3/y; other variables are defined the same as before.

The total oil flow of one reservoir received by an FPSO should equal to the total output of all the predrilling wells in the reservoir that are connected to the FPSO.

$$Q_{LPRs,r,t,l} = \sum_i Q_{OPLs,r,t,i,l} \quad s \in S, r \in R, t \in T, i \in I_r, l \in L \tag{17}$$

where $Q_{LPRs,r,t,l}$ is the total flow of the reservoir r received by FPSO$_l$ in the year t under the specified scenario s, m^3/y; other variables are defined the same as before.

When the accumulated output range is determined, the total flow of water and gas of one reservoir received by an FPSO can be obtained from the range.

$$Q_{LGRs,r,t,l} \le Q_{LPRs,r,t,l} R_{PGs,r,a} + (1 - B_{APs,r,t,a}) M \quad s \in S, r \in R, t \in T, a \in A \tag{18a}$$

$$Q_{LGRs,r,t,l} \ge Q_{LPRs,r,t,l} R_{PGs,r,a} + (B_{APs,r,t,a} - 1) M \quad s \in S, r \in R, t \in T, a \in A \tag{18b}$$

$$Q_{LWRs,r,t,l} \le Q_{LPRs,r,t,l} R_{PWs,r,a} + (1 - B_{APs,r,t,a}) M \quad s \in S, r \in R, t \in T, a \in A \tag{19a}$$

$$Q_{LWRs,r,t,l} \ge Q_{LPRs,r,t,l} R_{PWs,r,a} + (B_{APs,r,t,a} - 1) M \quad s \in S, r \in R, t \in T, a \in A \tag{19b}$$

where $R_{PGs,r,a}$ is the gas/oil ratio of the accumulated output range a in the reservoir r under the specified scenario s; $R_{PWs,r,a}$ is the water/oil ratio of the accumulated output range a in the reservoir r under the specified scenario s; $Q_{LWRs,r,t,l}$ is the transportation water flow from the predrilling well i of the reservoir r to the FPSO$_l$ in the year t under the specified scenario s, m^3/y; $Q_{LGRs,r,t,l}$ is the transportation gas flow from the predrilling well i of the reservoir r to the FPSO$_l$ in the year t under the specified scenario s, m^3/y; other variables are defined the same as before.

The water and gas flow received by each FPSO should be less than the throughput.

$$\sum_r Q_{LPRs,r,t,l} \le Q_{LPmaxl} \quad s \in S, r \in R, t \in T, l \in L \tag{20}$$

$$\sum_r Q_{LWRs,r,t,l} \le Q_{LWmaxl} \quad s \in S, r \in R, t \in T, l \in L \tag{21}$$

where Q_{LPmaxl} is the maximum oil throughput of the FPSO$_l$, m^3; Q_{LWmaxl} is the maximum water throughput of the FPSO$_l$, m^3; other variables are defined the same as before.

The FPSO energy consumption in production is related to the received oil and water flow, which can be supplied by natural gas or diesel.

$$\alpha_{PE} \sum_r Q_{LP\ s,r,t,l} + \alpha_{WE} \sum_r Q_{LWR\ s,r,t,l} = \beta_{GE}\left[\sum_r \left(Q_{LGR\ s,r,t,l}\right) - Q_{LGP\ s,t,l}\right] + \beta_{DE}\, Q_{D\ s,t,l}, \quad s \in S,\ r \in R,\ t \in T,\ l \in L \tag{22}$$

where α_{PE} is the energy consumption required for processing the oil of a unit volume, MJ/m^3; α_{WE} is the energy consumption when the water of a unit volume is processed, MJ/m^3; β_{GE} is the available energy produced by the gas of a unit volume, MJ/m^3; β_{DE} is the available energy produced by the diesel of a unit volume, MJ/m^3; $Q_{LGPs,t,l}$ is the gas emitted from the FPSO$_l$ in the year t under the specified scenario s, m^3; other variables are defined the same as before.

If production wells are connected to FPSO before the start time of the study, it will be unnecessary to identify the connection relationship. In other words, when $H_{OITr,i,l} = 1$, $B_{OITr,i,l} = 1$.

$$B_{OITr,i,l} \geq H_{OITr,i,l} \quad r \in R, i \in I_r, l \in L \tag{23}$$

where $H_{OITr,i,l}$ is a binary variable, if the predrilling well i of the reservoir r has been connected to the FPSO$_l$ before the start time of the study, $H_{OITr,i,l} = 1$, otherwise $H_{OITr,i,l} = 0$; other variables are defined the same as before.

4. Model solution

The crude oil price and the reservoir parameters (i.e., porosity, permeability, and thickness of reservoir structure) play an important role in the construction planning of deepwater oilfield infrastructure. The crude oil price volatility is full of randomness and thereby hard to be characterized by statistical probability functions [6]; thus, its uncertainty is defined as the range uncertainty. The reservoir parameters can be measured according to the data from exploration or production wells, and the measurement accuracy will be further improved along with oilfield developing and historical data increasing. Therefore, the reservoir parameters can be roughly characterized by statistical probability functions and their uncertainty is defined as stochastic uncertainty whose variance will decrease accordingly with oilfield developing [24].

5. Case study

5.1. Initial data

In this paper, an oilfield is presented as a case study. The water depth in the field comes up to 1350–1525 m, and the area is 10.5 km^2 or so. There are no other neighboring oilfields, and the field consists of A and B reservoir. It is evaluated that the geologic reserve of reservoirs A and B is about 1.9 × 107 m^3 and 3.7 × 107 m^3, respectively, and rich in the natural water. The mean

http://dx.doi.org/10.5772/intechopen.71078

and standard deviation of reservoir parameters are shown in **Table 1**. Considering the top-priority economic benefit and quick cost recovery, the oilfield is to be developed depending on natural energy. The recovery cycle is 10 year. The predrilling wells of each reservoir are 15, and the drilling cycle is 1 year. At the beginning of development, there are 5 kinds of FPSO optional, as shown in **Table 2**. FPSO lease preparation and construction cycle both are one year. The crude oil price forecast over the next decade is shown in **Figure 2**, and the uncertain fluctuation range of the oil price is given as 2%. Several series of reservoir parameters that are selected stochastically are incorporated into the reservoir numerical model, to carry out the relationship between the recovery degree of each reserve and the maximum output of single well, as shown in **Figure 3**.

5.2. Calculation result

The scenario number exerts great influence on the model solution since the smaller scenario number leads to poor convergence, while the bigger leads to low calculation speed. To explore the influence of scenario number on the model solution, the uncertain scenario number is increased successively and the MILP solver, GUROBI, is applied in MATLAB R2014a to solve the model. The implementation result is shown in **Figure 4**. It can be seen that the model tends to be convergent when the number comes up to 40 and finally the net profit is 1.551 billion dollars for 10-year development of the oilfield. The NPV variation with years is shown in **Figure 5**.

The final result shows 18 drilling wells and 1 FPSO converted from old oil tanks are required, and all the drilling and construction can be finished in the first three years. The detailed construction and drilling plans are shown in **Table 3**. The annual oil, gas, and water production outputs of the oilfield are shown in **Figure 6**. The annual diesel consumption of FPSO is shown in **Figure 7**. In the first year, 10 wells are to be developed and one oil tank should be turned into the FPSO. In the next two years, six wells and two wells are required to be developed, respectively, in order to stabilize the production. Since there is higher crude output in the third and fourth years, resulting in the produced natural gas insufficient for FPSO energy supply, addictive diesel is necessary for FPSO.

To verify the solving effect of the proposed MS-MILP method, three different methods, namely the MILP method that does not involve uncertainty, the improved GA [21], and the multistage goal programming (MGP) method in literature, are determined to solve this case. In this paper, 45 groups of reservoir parameters and crude oil price are generated

Reservoir	Permeability (mD)		Porosity (%)		Thickness of reservoir structure (m)	
	Mean	Standard deviation	Mean	Standard deviation	Mean	Standard deviation
A	180	10	15.9	1	38	1
B	180	10	14.0	1	38	1

Table 1. Uncertainty parameters of reservoir.

FPSO type	Conversion of old oil tank	Lease	New construction of small size	New construction of medium size	New construction of large size
Construction cost (10^8\$)	0.5	0	1	1.6	2.4
Assistant production cost (10^4\$/y)	200	1200	200	250	300
Oil throughput (10^4m^3/y)	140	200	210	280	330
Water throughput (10^4m^3/y)	240	265	273	420	570

Table 2. Optional FPSO cost and throughput.

stochastically as the test group. Based on the field planning by each method, the field NPV of 45 groups is calculated and the result is shown in **Figure 8** [25].

As seen from the **Figure 8**, the improved GA has the lowest mean NPV because its self-limitation causes converging to locally optimal solution. Compare with the MILP method, the MGP method is better because it considers the situation changing with the field development and models are established corresponding to different periods. However, the involved factors of MGP are less than MS-MILP; thus, the mean NPV of the former is lower than the latter. Particularly, when the oil price is lower than the expected and the reservoir scale is smaller than the expected, the NPV by MS-MILP is far higher than the other. In this way, considering

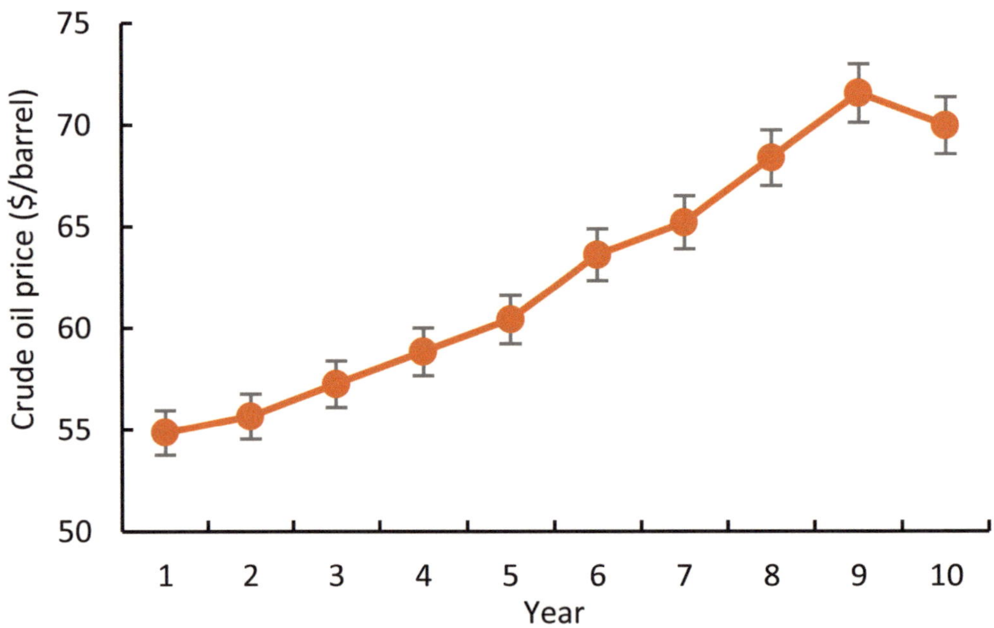

Figure 2. Crude oil price forecast.

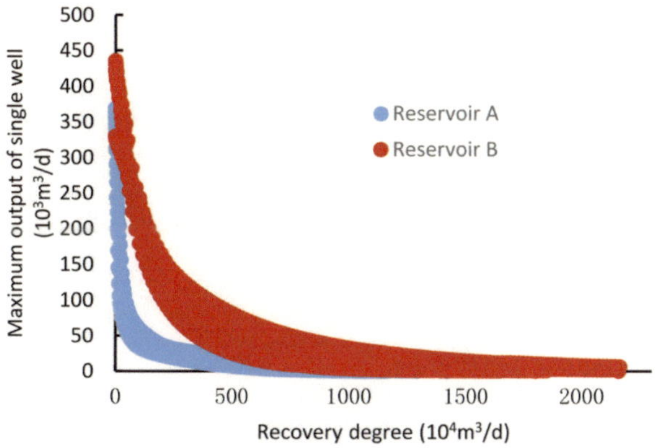

Figure 3. Result of reservoir numerical simulation.

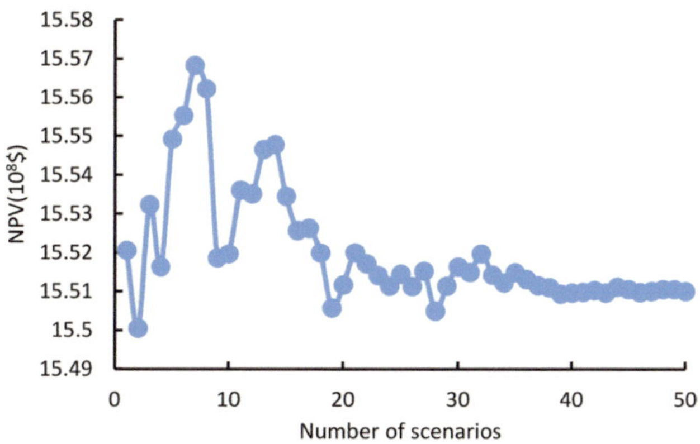

Figure 4. NPV for different numbers of scenarios.

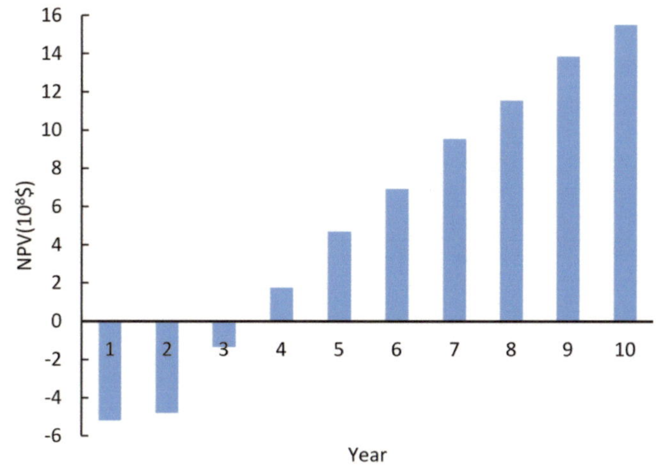

Figure 5. Variation of NPV.

Year	Well number		Construction
	Reservoir A	Reservoir B	
1	4	6	One FPSO converted from old oil tank
2	4	2	-
3	-	2	-

Table 3. Construction and drilling plans.

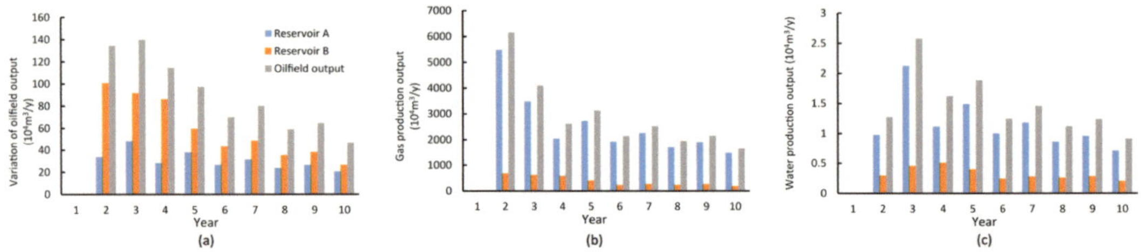

Figure 6. Variation of oilfield output. (a) Crude oil production output. (b) Gas production output. (c) Water production output.

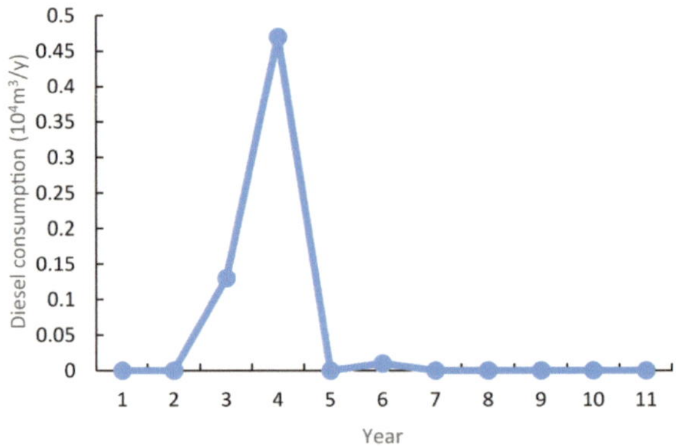

Figure 7. Variation of diesel consumption.

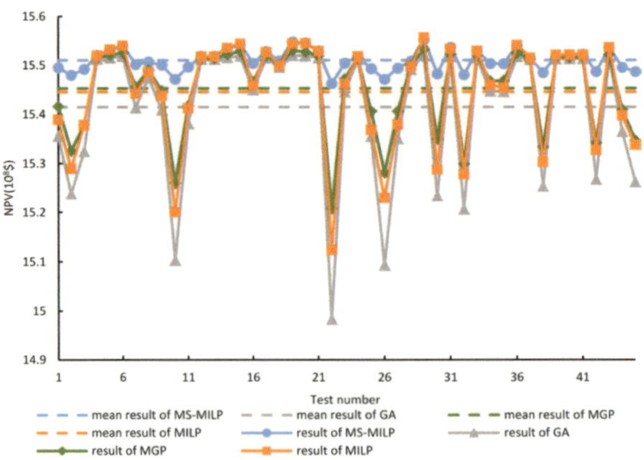

Figure 8. Comparison of different methods.

the complex uncertainties of reservoir and oil price, the oilfield planning by MS-MILP has higher rate of return and its anti-risk ability is superior to the other.

6. Conclusion

This paper put forward an optimal planning method for DWOD under the uncertainties of reservoir and crude oil price. The method takes the maximum total NPV as the objective function. The MS-MILP model is established, coupling with reservoir numerical simulation model and taking the constraints including drilling, output, production facilities, and energy consumption into account. The GUROBI solver is used to solve out the globally optimal planning of DWOD.

Finally, a study case based on a deepwater oilfield is given to work out an optimal development planning and evaluate the model's practicality. The proposed method is compared with the previous, illustrating that the oilfield development planning calculated by this paper's takes the advantage of high rate of return and strong anti-risk ability.

Acknowledgements

This work was part of the program of "Study on the mechanism of complex heat and mass transfer during batch transport process in products pipelines" funded under the National Natural Science Foundation of China, grant number 51474228. The authors are grateful to all study participants.

Author details

Zhang Haoran[1]*, Liang Yongtu[1], Ma Jing[1], Di Pengwei[2], Yan Xiaohan[1] and Huang Zhongliang[3]

*Address all correspondence to: 656673379@qq.com

1 Beijing Key Laboratory of Urban Oil and Gas Distribution Technology, China University of Petroleum, Beijing, China

2 CMOE Key Laboratory of Petroleum Engineering, China University of Petroleum, Beijing, China

3 CNPC Offshore Engineering Company Ltd, Beijing, China

References

[1] Wang D, Li X, Zhang H, et al. A model for estimating the drilling and completion investment in offshore oilfields in West Africa and the Asia-Pacific region. Petroleum Exploration & Development Online. 2012;**39**(4):534-538

[2] Gupta V, Grossmann IE. Development Planning of Offshore Oilfield Infrastructure. Springer International Publishing Switzerland, M. Martín (ed.), Alternative Energy Sources and Technologies. 2016

[3] Zhang Y, Sun XP, Zheng XD. Application of seismic reservoir prediction techniques in different exploration and development stages. Petroleum Exploration & Development, 2006;**33**(1):59-63

[4] Huo CL, Liu S, Gu L, et al. Quantitative method for appraising reservoir geological model uncertainty. Petroleum Exploration & Development. 2007;**34**(5):574-579

[5] Wang J, Tao L, Chen F, et al. Reduction of probabilistic gas reserves uncertainty by geological constraints. Petroleum Exploration & Development. 2011;**38**(6):764-768

[6] Soleimani M. Simulation of petroleum exploration based on a conceptual decision model: Taking the Dezful embayment in southwestern Iran as an example. Petroleum Exploration & Development. 2013;**40**(4):476-480

[7] Luo DK, Hua W. Real option valuation model for petroleum development projects. Petroleum Exploration & Development. 2007;**34**(4):493-496

[8] Midthun KT, Fodstad M, Hellemo L. Optimization model to analyse optimal development of natural gas fields and infrastructure ☆. Energy Procedia. 2015;**64**:111-119

[9] Arredondo-RamRez K, Ponce-Ortega JM, El-Halwagi MM. Optimal planning and infrastructure development for shale gas production. Energy Conversion & Management. 2016;**119**:91-100

[10] Gupta V, Grossmann IE. An efficient multiperiod MINLP model for optimal planning of offshore oil and gas field infrastructure. Industrial & Engineering Chemistry Research. 2012;**51**(19):6823-6840

[11] Gupta V, Grossmann IE. Multistage stochastic programming approach for offshore oilfield infrastructure planning under production sharing agreements and endogenous uncertainties. Journal of Petroleum Science & Engineering. 2014;**124**:180-197

[12] Tarhan B, Grossmann IE, Goel V. Stochastic programming approach for the planning of offshore oil or gas field infrastructure under decision-dependent uncertainty. Industrial & Engineering Chemistry Research. 2009;**48**(6):3078-3097

[13] Jonsbraten TW. Oil field optimization under price uncertainty. Journal of the Operational Research Society. 1998;**49**(8):811-818

[14] Aseeri A, Patrick Gorman A, Bagajewicz MJ. Financial risk management in offshore oil infrastructure planning and scheduling. Industrial & Engineering Chemistry Research, 2004;**43**(12):3063-3072

[15] Kang XJ, Zhao-Min LI, Liu ZB. Optimal model of oilfield development programming under stochastic oil price. Petroleum Exploration & Development. 2007;**34**(6):765-768

[16] Dawson RG, Fuller JD. Mixed integer nonlinear program for oilfield production planning. Information Systems & Operational Research. 1999;**37**(2):121-140

[17] Heever SAVD, Grossmann IE. An iterative aggregation/disaggregation approach for the solution of a mixed-integer nonlinear oilfield infrastructure planning model. Industrial & Engineering Chemistry Research. 2000;**39**(6):1955-1971

[18] Ling C, Feng QH, Hong-Jun YU. The application of BP network to oil field measure program. Petroleum Exploration & Development. 2002;**29**(3):78-80

[19] Carvalho MCA, Pinto JM, Carvalho MCA, et al. An MILP model and solution technique for the planning of infrastructure in offshore oilfields. Journal of Petroleum Science & Engineering. 2006;**51**(1):97-110

[20] Ge YH, Yan AM, Gao YR, et al. Drilling pad optimization for oilfield development by cluster horizontal wells. Petroleum Exploration & Development. 2005;**32**(5):94-100

[21] Zhang H, Liang Y, Ma J, et al. An MILP method for optimal offshore oilfield gathering system. Ocean Engineering. 2017;**141**:25-34

[22] Zhang H, Liang Y, Zhang W, et al. A unified MILP model for topological structure of production well gathering pipeline network. Journal of Petroleum Science & Engineering. 2017;**152**:284-293

[23] Qiang LI, Zhong H, Wang Y, et al. Integrated development optimization model and its solving method of multiple gas fields. Petroleum Exploration & Development. 2016;**43**(2):293-300

[24] Dong W, Jiao J, Xie S, et al. Cumulative production curve method for the quantitative evaluation on the effect of oilfield development measures: A case study of the nitrogen injection pilot in Yanling oilfield, Bohai Bay basin. Petroleum Exploration & Development. 2016;**43**(4):672-678

[25] Yu S, Zhang S, Agbemabiese L, et al. Multi-stage goal programming models for production optimization in the middle and later periods of oilfield development. Annals of Operations Research. 2017;**255**:421-437

Chapter 4

Drilling Fluids for Deepwater Fields: An Overview

Luis Alberto Alcázar-Vara and
Ignacio Ramón Cortés-Monroy

Additional information is available at the end of the chapter

http://dx.doi.org/10.5772/intechopen.70093

Abstract

The increasing oil demand around the world along with the depletion of onshore and shallow water oil reserves have forced the oil companies moving into the development of deepwater subsea hydrocarbon reservoirs. Drilling fluids play a key role in all drilling operations, but they get a greater relevance in deepwater environments where the technological challenges of drilling at these extreme conditions generate significant operational risks as well as very high costs during the development of this kind of fields. The operational issues and concerns related to the drilling fluid design and application for deepwater fields are generally well known: narrow pore/fracture pressure gradient margins, wellbore stability, clay swelling, gas hydrates formation, formation damage, salt formations, lost circulation, stuck pipe, cuttings transport and environmental and safety aspects. Therefore, the present chapter aims to give an overview on the main challenges and research related to drilling fluid design and application for deepwater fields through the revision of the state of the art of the current and innovative technological solutions reported in literature.

Keywords: drilling fluids, deepwater, clay swelling, gas hydrates, flat rheology, lost circulation, salt formations

1. Introduction

Exploration and production operations in deepwater and ultra-deepwater fields around the world have suffered important and critical changes over the last years. Deepwater is generally considered as any water depth greater than 1500 ft., whereas waters deeper than 7000 ft. move into the ultra-deepwater category [1]. New records for water depth and measured depth in deepwater are being set regularly. In general, developments of deepwater fields are carried out under conditions of high-costs, high-risks, and long-duration projects; thus, they are usually less sensitive to short-term fluctuations in oil prices than onshore developments [1–3].

Drilling is a primordial and critical stage in the success of exploration of deepwater fields. The overriding drilling objectives are to reach the target safely in the shortest possible time and at the lowest possible cost, with required additional sampling and evaluation constraints dictated by the particular application. Drilling itself is a much larger share of total well costs in offshore development than in onshore development. Key cost drivers for offshore drilling include water depth, well depth, reservoir pressure and temperature, field size, and distance from shore. Drilling fluids can represent from 15 to 18% of the total cost of well drilling but may cause 100% of drilling problems [4–10].

In this way, new deepwater discoveries around the world become challenging tasks of technical, operational, environmental, and economic issues, where many of those tasks are focused on the selection, development, and application of drilling and completion fluids technologies [1–5].

The operational issues and concerns related to the drilling fluid design and application for deepwater fields are generally well known: narrow pore/fracture pressure gradient margins, wellbore stability, clay swelling, gas hydrates formation, formation damage, salt formations, lost circulation, stuck pipe, cuttings transport, and environmental and safety aspects [1–11]. The design, selection, and application of the right fluid system or additives require balancing each of these issues with regard to their impact on the deepwater drilling operation.

There are several works reported in literature about topics reviewed in the present chapter; however, some of them just deal about challenges on deepwater drilling operations whereas others just deal about drilling fluids systems to control some of the operational problems found during deepwater drilling. Therefore, the present chapter aims to give a general overview on the main challenges and research related to drilling fluid design and application for deepwater fields through the revision of the state of the art of the current and innovative technological solutions reported in literature, where the drilling fluid systems and additives used to treat and control these problematic and challenging tasks are also carefully reviewed.

2. Challenges and new advances for deepwater drilling fluids

2.1. Wellbore stability and clay swelling

The geological aspects to consider during deepwater drilling are very different from those found on land and in shallow water. Generally, geological formations found in deepwater fields are relatively young and very reactive [11–15]. In this way, it is generally accepted that highly reactive shale formations are intrinsic to deepwater drilling, where their interaction with drilling fluids is the main factor in wellbore instability, which is considered one of the major causes of troubles, waste of time, and over costs during drilling [16–18]. In this way, problematic shales are responsible to origin more than 90% of wellbore instability problems [18]. Moreover, it has been reported that shales (principally clays) represent 75% of all formations drilled by the oil and gas industry [16].

Wellbore instability is mainly due to the clays dispersion into ultra-fine colloidal particles, which has a direct effect on the drilling fluid properties and performance [19]. Generally, wellbore stability is not a concern for most shale formations when drilling is carried out by

http://dx.doi.org/10.5772/intechopen.70093

using oil-based and synthetic-based drilling fluids. However, the use of these drilling fluids is limited due to high costs and environmental restrictions particularly for deepwater operations [16, 19]. Thus, growing environmental concerns currently require the replacement of oil-based fluids by environmentally friendly water-based drilling fluids, which can interact with shales promoting undesirable clay-swelling phenomena [20].

Wellbore instability problems caused by clay swelling have been widely reported: sloughing shales, hole closure causing tight hole, cave-ins leading to fill on trips, and problems when running casing. In addition, other important problems have been identified such as cuttings accumulation leading to reduced hole-cleaning efficiency, buildup of thick cuttings beds, and reduced rates of penetration arising from balling of the drill bit with sticky clay [13, 15]. Moreover, wellbore instability can result in the loss of the drilling assembly, well side-tracks or in the worst case total abandonment of the well. Consequently, these problems can significantly reduce drilling rates as well as increase considerably the costs of exploration and production [15]. Therefore, minimizing and controlling shale-fluid interactions during deepwater drilling become one of the most important challenges for the design and selection of deepwater drilling fluids.

2.1.1. Clay mineralogy

Clay minerals account for about 50–60 wt.% of most shales; thus, physical properties and behavior of shale interacting with a drilling fluid rely on the type and amount of clay in the shale. Clays are naturally occurring minerals formed by the weathering and decomposition of igneous rocks. They are layered minerals, classified among the phyllosilicates, consisting of stacks of negatively charged two-dimensional aluminosilicate layers [15]. There is a great variety of different clay minerals, which differ in their composition, layer arrangement, and substitutions. However, for shale-stability purpose, the most relevant clay minerals are kaolinite, illite, smectite, and chlorite. Their main characteristics have been reported [11, 21] and are shown in **Table 1**.

Clay mineral	Chemical elements	Morphology	Surface area (m^2/gm)	Typical range of CEC (meq/100 g)	Layer thickness (Å)
Kaolinite	$Al_4[Si_4O_{10}](OH)_8$	Stacked plate or sheets	20	3–10	7
Chlorite	$(Mg, Al, Fe)_{12}[(Si, Al)_8O_{20}](OH)_{16}$	Plates, honeycomb, cabbage-head rosette, or fan	100	10–40	14
Illite	$(K_{1\text{-}1'5}Al_4[Si_{7\text{-}6'5}Al_{1\text{-}1'5}O_{20}](OH)_4)$	Irregular with elongated spines or granules	100	20–40	10
Smectite	$(1/2Ca,Na)_{0,7}(Al, Mg, Fe)_4[(Si, Al)_8O_{20}]\cdot nH_2O$	Irregular, wavy, wrinkled sheets, webby, or honeycomb	700	80–150	12–14

Table 1. Characteristics of clay minerals involved in shale stability.

Kaolinite is commonly considered as a highly stable mineral clay, non-swelling, presenting a relatively small surface area and a low adsorptive capacity (cation exchange capacity (CEC) = 3–10 meq/100 g) as shown in **Table 1**. Kaolinite clay can be easily dispersed in water-base drilling fluids [11, 18]. Chlorite mineral is generally considered as non-swelling clay. Chlorite minerals contain a layer of alumina sandwiched between two layers of silica and a layer of magnesium or iron oxide, and without interlayer water. Illite clay minerals are similar to chlorite in reactivity, presenting a low adsorptive and swelling/shrinking capacity and properties intermediate between kaolinite and smectites as shown in **Table 1**. It has been reported that some older shale rocks with a high degree of diagenesis contain only chlorite and illite as clay components. Most of these shales are relatively unreactive but some of them can hydrate and slough [11, 18]. **Table 1** shows also properties of smectite minerals which present higher adsorptive capacity (cation exchange capacity = 80–150 meq/100 g), indicating higher reactivity and swelling potential as it has been widely reported in literature [11–18, 20]. Smectite minerals include a variety of clays such as montmorillonite, hectorite, and beidellite. In addition, mixed-layer clay minerals, such as illite-smectite and chlorite-smectite, have also been found and reported [11, 21]. Several experimental techniques, such as X-ray diffraction (XRD), spectroscopy, and microscopy, are used in order to identify and characterize mineral clays. **Figure 1** shows scanning electron microscopic (SEM) photos of typical clay minerals described earlier.

Studies reported in literature about clay swelling and inhibition are most often focused upon smectite clays due to their well-known swelling potential and the frequency with which they are found during drilling operations [15]. However, it has been also reported [13] that kaolinite and illite shales can be highly unstable when drilled promoting bit-balling problems, suggesting that interlayer expansion cannot be considered as a universal causative mechanism of shale instability. Nevertheless, the tendency of sodium-saturated smectites to swell macroscopically has been generally identified as the principal source of shale instability that can potentially lead to collapse of the wellbore. Therefore, a deeper understanding about the mechanisms involved in the interactions between water-based drilling fluids and mineral clays is a key issue to get success during deepwater drilling.

2.1.2. Swelling mechanisms

Exposed to aqueous solution, clay minerals will adsorb water molecules and swell. The phenomenon is also known as clay hydration. The hydration of the clays is a function of the extent

Figure 1. SEM photos of typical clay minerals involved in shale instability. From left to right: kaolinite, chlorite-illite, and smectite (montmorillonite). *Images reproduced from the "Images of Clay Archive" of the Mineralogical Society of Great Britain & Ireland and The Clay Minerals Society (www.minersoc.org/gallery.php?id=2).*

and location of layer charge, the interlayer cation species, the water activity, the temperature, the external pressure, and the salinity of the bulk solution [22]. Crystalline and osmotic swelling are the two main mechanisms reported in literature to explain clay hydration phenomena [13–15, 20–23]. Both mechanisms are described below.

Crystalline swelling. This mechanism, also called surface hydration, can occur in all types of clay minerals when they are exposed to concentrated brine or aqueous solutions with high content of divalent or multivalent cations. The mechanism is carried out through the stepwise formation of integer-layer or mixtures of integer-layer hydrates. The process is thermodynamically comparable to phase transitions. Several water-molecules layers may line up in order to build a quasi-crystalline structure between unit layers, resulting in an increased interlayer spacing [15, 18]. The principal action force of this mechanism is the adsorption energy of water on the surface of clays. The volume and thickness of water adsorbed on the clay surface will depend on the hydration energy of exchangeable cations and the charge density on the surface of clay. In addition, type, size and charge of exchangeable cations present in the interlayer have a significant impact on swelling process. It has been reported that the presence of Ca^{2+}, Mg^{2+}, and H^{+} exchangeable cations in montmorillonite clay increases their interlayer attractive force, resulting in a thinner hydrated film and whit a directional and regular water molecules arrangement on the clay. On the other hand, for Na^{+} exchangeable cations, the interlayer attractive force decreases, the hydrated film becomes thick, and the arrangement of water molecules on the clay is not directional and regular [13, 15, 23]. Therefore, sodium montmorillonite has a higher swelling capacity than calcium montmorillonite according to the mechanism described earlier. As reported in Ref. [15], molecular simulation studies have confirmed the stepwise mechanism of crystalline swelling described earlier observing that adsorbed water molecules form distinct layers in the interlayer region. Typical interlayer spacings recorded in the crystalline swelling mechanism lie in the range of 0.9–2 nm [15]. **Figure 2a** shows the crystalline swelling mechanism described earlier.

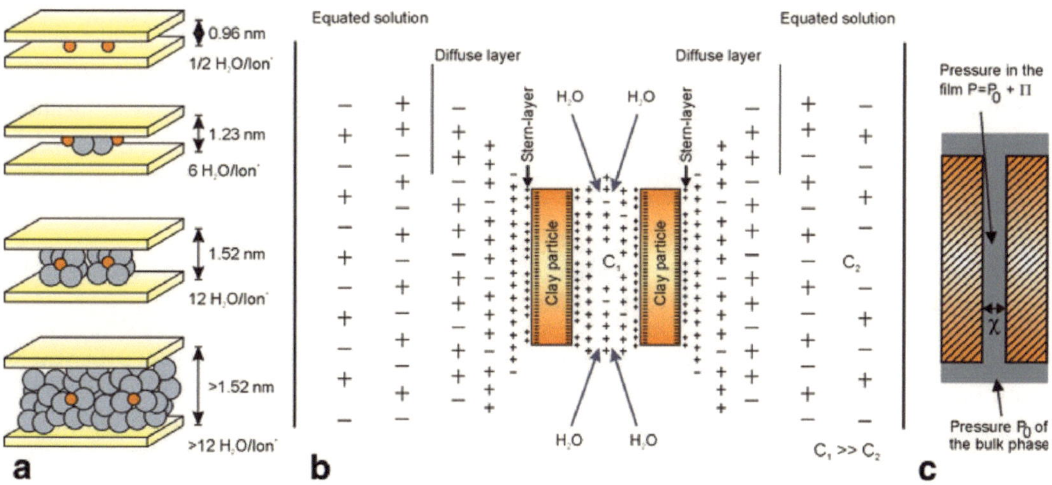

Figure 2. Swelling mechanisms [24]: (a) crystalline swelling and (b) osmotic swelling.

Osmotic swelling. This mechanism can occur only in clay minerals containing exchangeable cations in the interlayer region. A sketch of the mechanism is depicted in **Figure 2b**. As observed, when the concentration of cations in the interlayer is higher than that of the surrounding water, water molecules diffuses into the interlayer in order to dilute its ion concentration, restoring cation equilibrium [15, 23]. This phenomenon creates an osmotic repulsive pressure between the clay particles. The osmotic pressure is mainly related to the difference of the ion concentrations between the interlayer and surrounding water. In this way, the distance between clay particles increases greatly, and then the clay swelling is carried out [15, 23]. Compared with crystalline swelling, this type of swelling promotes larger volume increases with typical interlayer spacings of 2–13 nm [15]. In fact, osmotic equilibrium of the semipermeable membrane on clay particles is considered a key factor to influence the hydration film of clay swelling. Thus, it can be stated that osmotic hydration is the major factor of clay swelling. It has been identified that the tendency of sodium montmorillonite clay to swell through this osmotic mechanism is the main cause of wellbore instability that can potentially lead to collapse of the wellbore [13–15]. On the other hand, K^+-saturated smectite clay minerals do not swell through this mechanism and form crystalline hydrates even in aqueous suspension. Thus, the K^+ ion can be used to inhibit the swelling of sodium montmorillonite clay minerals [15].

2.1.3. Shales characterization

The main methods developed for shale characterization including shale-fluid interactions and clay-swelling inhibition deal with composition, reactivity (swelling), mechanical, and physicochemical properties of shales (or clay). Some of the most important experimental methods reported are described below.

X-ray diffraction (XRD). This is a basic tool in the mineralogical analysis of shales. This experimental technique has been widely used to determine clay mineral composition of shale samples, degree of crystallization and swelling profiles of clay minerals [25].

Small-angle X-ray scattering (SAXS). This technique is useful for the characterization of the microstructural and swelling properties of clay samples [15].

Small-angle neutron scattering (SANS). This is a useful and versatile technique for the study of in-homogeneities of both crystalline and amorphous structures on the sub-micron scale, with capabilities to measure samples with high degree of swelling (up to interlayer d-spacings of 2000 Å) [15].

X-ray fluorescence (XRF). It is an experimental method used to determine chemical composition (major and trace elements) of rocks, minerals, sediments, and fluids.

Computer tomography (CT). The technique is generally suitable for visualization from meter to millimeter scale. Shale applications include viewing full-diameter core sections to determine orientation relative to bedding, presence of fractures and nodules, density studies for highly interbedded interval, and quality assessment of prepared plug samples [26].

High-resolution micro-CT. This is a technique with similar principles and shale applications of the conventional CT, designed for smaller samples and employing a shorter distance between source and detector, which allows much higher resolution [26].

Scanning electron microscopy. This is an experimental method for high-resolution imaging of surfaces. It allows characterizing morphology, pore structure, and clay microstructure in shales samples. It can be also useful for studies of shale-fluid interactions [27].

Cation exchange capacity and methylene blue test (MBT). These are standard methods used to determine the reactivity (swelling capacity) of a clay or a shale sample. CEC is defined as the ability of clay minerals to absorb cations in such form that they can be easily exchanged for other cations present in an aqueous solution. Additional information about the reactivity of the shale can be obtained, if the exchangeable cations are identified and quantified [28].

Linear swelling tests (SLTs). This is a standard technique used widely to evaluate shale swelling by linear displacement [29]. Some experimental apparatus have the capability to evaluate swelling by this method at reservoir (high-pressure and high-temperature) and dynamic conditions.

Mercury injection porosimetry. This is a standard method for characterizing pore throat size distribution from micron to nanoscale. For shale samples, mercury is able to penetrate within and between the coarse rigid grains as well as the clay intergrain areas and secondary minerals [26].

Gravimetric swelling test (GST). This is an experimental method used to measure water and ion motion during shale/mud interaction in order to determine compatibility between shales and drilling fluids [29].

Capillary suction time (CST). This is an experimental test for determination of filtration properties and salt concentration optimization [16]. It is used primarily to determine filter-cake permeability, but data have been also used to study shales reactivity in filter cakes and the effect of brine composition on clays in a filter cake.

Thermal gravimetric analysis (TGA). This technique allows characterization of the different types of water (free, interlayer, bound, and crystalline) in a shale sample as well as to identify the type of clay content in shale samples [30].

Hot-rolling dispersion test. This technique is widely used in optimizing drilling fluid. This test provides an assessment of the inhibition of shale cuttings exposed to a drilling fluid evaluating in this way the risk of dispersion or swelling in the wellbore [16].

Pressure transmission test. This method can be used for confined or unconfined shale samples. The experimental apparatus allows evaluating shale-drilling fluid interaction and estimates shale permeability, coefficient of reflectivity (membrane efficiency) as well as ionic diffusion coefficient [31].

Rheological tests. This is an experimental method useful for the evaluation of rheological behavior of clays in aqueous suspensions to study clay-fluid interactions. It allows characterizing colloidal behavior of clay-fluid systems through rheological properties such as thixotropy, viscoelasticity, and yield stress [32].

Triaxial test. This technique is useful for pore pressure measurements and compressive stress/strain behavior in shale samples [15]. It also allows characterizing reactivity behavior of shales by measuring the swelling pressures and strains that result from exposure of shale cores to different test fluids. Preserved shale cores can be used [33].

Dielectric analysis. These experimental methods are useful to quantify swelling clay content and to determine specific surface areas in clays and shales samples. In addition, wettability and anisotropy properties of shales can be also determined by using dielectric techniques [26].

2.1.4. Additives in drilling fluids for shale stabilization

Over the past decades, the demands for effective shale stabilizers have never stopped. Especially, with the development of shale gas and deepwater fields all over the world, such demands have never become as urgent as today. Understanding of the behavior and responses of shale-reactive formations to drilling fluids with chemical additives has been an important challenge in the oil industry for many years because of the several and complex chemical and physical phenomena present in these types of formations. Different types of chemical additives for shale stabilization have been used in the oil industry and reported in literature [11, 14, 15, 17, 34–42]. The number of commercial shale stabilizers is impressive. Each of them has a particular mechanism by which they can inhibit the swelling, disintegration, and dispersion of clay minerals interacting with water. However, most of these mechanisms are based on the change of the ionic strength and the transport behavior of the fluids into the clays, where the cations and anions present in the additives determine their capability and efficiency for clay-swelling inhibition [18]. Some authors have reported three main mechanisms to reduce clay swelling: ion exchange, coating of the clay particles by stabilizers, and modification of surface affinity toward water [43]. In this way, clay-swelling inhibitors are classified as temporary and permanent shale inhibitors [42]. Temporary shale inhibitors prevent swelling and migration of clays but are easily removed by the formation-produced fluids following the treatment [41]. Simple inorganic salts are the most common temporary shale inhibitors. However, most recent advances in shale stabilization have been focused on the area of permanent clay stabilizer additives [41]. **Table 2** shows a summary of the main additives reported in literature for shale stabilization.

2.2. Gas hydrates

Gas hydrate formation is another severe challenge to deepwater drilling fluid technology. Gas hydrate formation has been identified as a potential shallow hazard facing deepwater drilling since the mid-1980s [44, 45]. For deepwater drilling, conditions such as low temperatures and high pressures promote hydrate formation in the drilling fluid when gas is present, causing wellbore plugging, blockage in pipelines, or/and blowout preventers, thus leading to serious safety problems and increasing operation costs [46–48]. In addition, gas hydrates decomposition also causes serious problems such as wellbore instability, lost circulation, and blowout [45, 46]. In this way, for deepwater drilling operations, gas hydrate formation is not only an economic issue but also more importantly a safety issue [46]. Thus, the problem that needs to be solved is to avoid gas hydrates formation during well-controlled situations or

http://dx.doi.org/10.5772/intechopen.70093

Group of shale stabilizers	Main characteristics	Shale stabilizer additives	References
Salts	Primary chemical of choice for clay stabilization. Capabilities to influence swelling and osmotic pressures, viscosities of filtrate and shale stability	Potassium chloride Sodium chloride Concentrated brines ($CaCl_2$, $CaBr_2$, $ZnCl_2$, $MgCl_2$, $MgBr_2$, $ZnBr_2$) Formate and acetate salts	[11, 14, 15]
High-molecular-weight polymers	Shale stabilization through clay particle encapsulation, inhibiting swelling, and dispersion	Polyacrylamides Polyvinylpyrrolidones Acrylate copolymers	[11, 14, 15, 34]
Low-molecular-weight polymers	Clay swelling and dispersion inhibited through the intercalation of inhibitor species into the interlayer of clay minerals	Polyglycerols Polyglycols Polypropylene oxides	[14, 15, 34, 35]
Charged polymers	Swelling-inhibitive effect through their adsorption onto clay particle surfaces. Classified in cationic, anionic, and amphoteric. Synthetic or natural in origin	Polymeric quaternary amines Celluloses Starches Charged polyacrylamides	[15, 34–36]
Non-polymeric amines	Clay swelling inhibition through the principle of substitution of cationic species for a sodium ion in the clay lattice	Mono-cationic amines Oligomeric cationic amines	[36]
Carbonaceous additives	Very limited effect on shale stabilization and without effect on swelling pressure	Asphaltenes Gilsonites Graphites	[14]
Silicates	Inexpensive and usually recommended for all shale-stabilization uses, including formations in situ fractured	Sodium silicate Potassium silicate	[14, 17]
Saccharides	Environmentally friendly low-molecular-weight viscosifiers and reducers of hydraulic flow of water in shales	Methylated saccharides	[14, 37]
Nanomaterials	Capabilities of shale permeability reduction during drilling which stops fluid invasion, inhibits swelling, and improves wellbore stability. Environmentally friendly	Nanoparticles	[38–41]

Table 2. Additives for shale stabilization.

to minimize the impact of hydrate formation, such as eliminate potential hydrate blockages [47]. Gas hydrates are crystalline inclusion compounds formed of hydrogen-bonded water molecules as hosts and gas molecules entrapped in the water cavities as guests [49, 50]. Gas hydrates can be only stable at high pressures and low temperatures (as observed in **Figure 3**), conditions found during deepwater drilling operations.

Most gas hydrates are formed from methane; however, gas molecules such as ethane, propane, and 1-butane, and inorganic gases such as nitrogen, hydrogen, and carbon dioxide are also able to be enclathrated into the water lattices. The three most common gas hydrate structures reported are cubic structure I (sI), cubic structure II (sII), and hexagonal structure H (sH) [49, 50]. Nucleation, growth, and agglomeration are the main stages of phase transitions associated with hydrate plug formation [49].

Gas hydrate problems found in deepwater operations become inaccessible or impractical for most conventional preventive methods. In fact and as reported some years ago, the annual costs to clean gas hydrate blockage might exceed $100 million at a rate near $1 million per mile of affected lines [45, 51]. In this way, injection of inhibitor chemicals is the main method generally used to avoid the formation of gas hydrates during deepwater drilling operations [45–50].

2.2.1. *Hydrate inhibitors*

According to the inhibition mechanism, hydrate inhibitors are classified as thermodynamic hydrate inhibitors (THIs) and low-dosage hydrate inhibitors (LDHIs) [45].

The main action mechanism of the thermodynamic hydrate inhibitors (THI) is to delay hydrogen bonding of water molecules forming hydrate structures, modifying in this way the

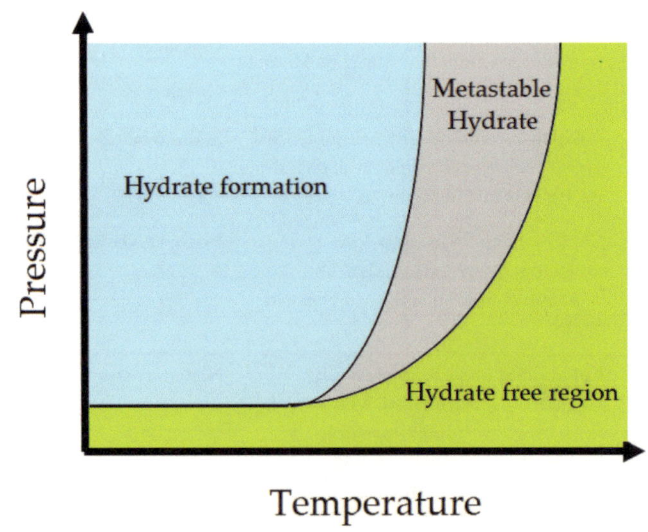

Figure 3. Typical phase diagram of hydrate stability zones.

hydrate–liquid–vapor equilibrium at a given pressure, temperature, gas composition, and water salinity [50]. However, thermodynamic hydrate inhibitors have a strong but limited capability to inhibit the formation of gas hydrates due to higher concentration of these inhibitors which may be required for deepwater drilling operations, which results in higher drilling fluid density, increasing operational costs as well as the logistical and environmental concerns [48–50].

Low-dosage hydrate inhibitors are a more recent technology for preventing hydrate plugging [52]. LDHIs can be subdivided into two basic categories: kinetic hydrate inhibitors (KHIs or KIs) and antiagglomerants (AAs).

The main action mechanism of the KHIs is to delay mainly the nucleation but also the growth of the gas hydrates, extending the hydrate induction time, which is the most critical factor for field operations, to exceed the residence time of the reservoir fluid [47]. Thus, KHIs play a role mainly as gas hydrate *anti-nucleators* [52] where thermodynamic conditions of hydrate formation are usually not significantly affected, as KHIs do not modify the hydrate–liquid–vapor equilibrium [45].

Antiagglomerant inhibitors prevent gas hydrate crystals from increasing their size, which results in the formation of smaller particles dispersed during residence time in the subsea pipeline allowing the generation of transportable slurries [49]. A polar, hydrate-philic head and a hydrophobic, fatty chain form usually the structures of antiagglomerant inhibitors. In addition, due to their surfactant nature, they will accumulate at the water/oil interface, just where hydrates first begin to form [52]. The application of low-dosage hydrate inhibitors to prevent hydrate plugs in deepwater fields has been tested and reported as successful and is now a well-established technology [52]. In **Figure 4**, the main additives reported as hydrate inhibitors are shown [53–74].

2.3. Rheological behavior

Rheological properties of drilling fluids are key parameters in offshore operations, especially in the extreme and complex conditions found during deepwater drilling. The term "flat rheology" used to describe a drilling fluid is a concept recently introduced to the oil industry and refers in general to "constant" or "continuous" rheological properties [46, 75–78] as shown in **Figure 5**.

The broad range of exposure temperatures typically found during deepwater drilling operations greatly influences rheological behavior of drilling fluids; particularly their viscosity and yield point properties can be affected. This can promote lost circulation and high equivalent circulating density (ECD) increasing difficulty of pressure control [46, 75, 76]. Thus, in order to avoid critical ECDs, drilling fluid rheology should be controlled. In general, a thinner fluid yield lowers ECDs. However, cuttings removal and barite suspension issues should also be considered for the rheological design of drilling fluids [77]. Therefore, one of the main challenges to design drilling fluids for deepwater operations is to effectively balance fluid rheology for equivalent circulating density, hole cleaning, and barite suspension simultaneously [77].

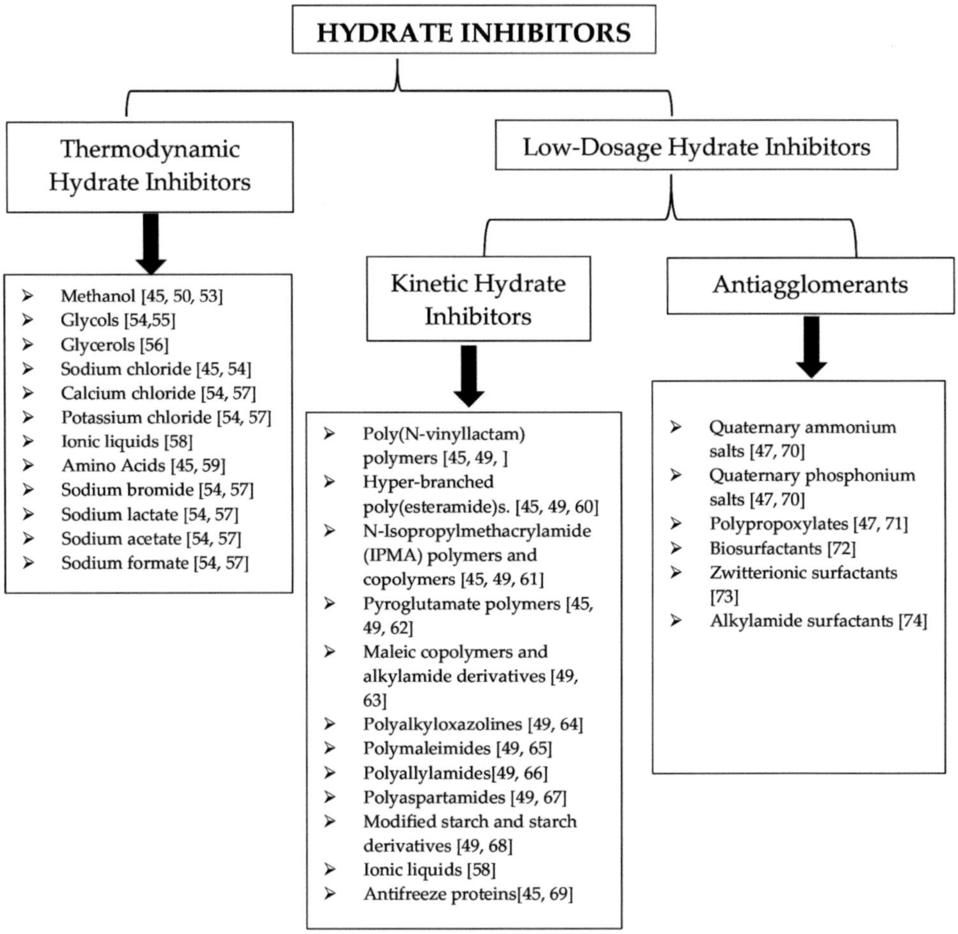

Figure 4. Additives for hydrate inhibition.

Flat-rheology properties for drilling fluids can be obtained through the usage of mixtures or packages of several specific additives such as emulsifiers, rheology modifiers, and viscosifiers. Typical components for a flat-rheology synthetic-based mud (SBM) have been reported [77, 78]. The main components are usually the following:

- Organophilic clay
- Emulsifier
- Wetting agent
- Fluid loss control
- Polymeric rheology modifier
- Viscosifier

The emulsifier additive helps to minimize the impact of drill solids on the rheological properties of the synthetic fluid. Wetting agents or chemical thinners can be used to reduce yield point

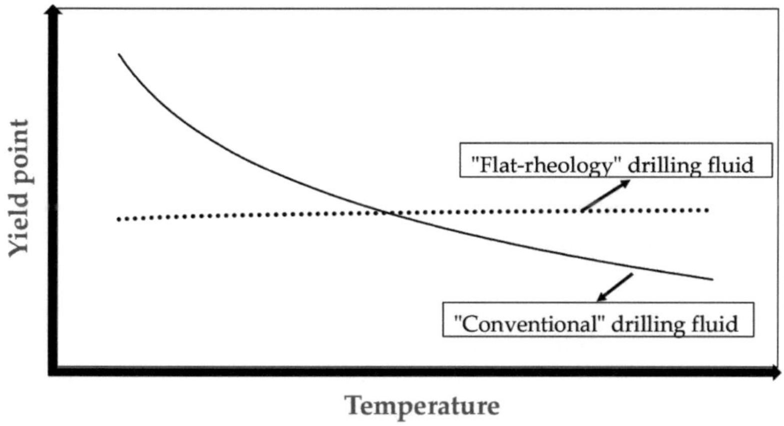

Figure 5. Representative rheological behavior of "flat-rheology" and "conventional" drilling fluids.

in invert-emulsion fluids as well as to reduce the solids tolerance of the fluid. Organophilic clays used at a minimal concentration provide an optimal rheology modification. The polymeric rheology modifier also reduces viscous properties at low temperatures while increasing them at high temperatures. Finally, the viscosifier provides the desired enhancement in overall viscosity and suspension capacity [77, 78].

High performance has been reported using "flat rheology" approach for drilling fluids, which can be achieved by using accurate combinations of emulsifier, wetting agent, rheology modifiers, and supplementary viscosifiers. However, the ability to control a flat-rheology profile can be also influenced by other parameters and phenomena, which cannot always be controlled such as temperature-pressure variations, interactions of rheological modifiers with drill solids, shear-rate variations in the annulus, salinity effects, and changes in the concentration of rheological modifiers during circulation of drilling mud [77, 78].

Another key rheological parameter for deepwater drilling fluids is the "gel strength." It is defined as the shear stress of the drilling fluid measured at a very low value of shear rate after it has been set for 10 min [79] and it is considered as a measure of the degree of thixotropy present in the drilling fluid. Gel strength determines the ability of the drilling fluid to suspend drilled cuttings and other solid additives along the length of the drill pipe/borehole annulus when the circulation of the drilling fluid is stopped during tripping or in any other operation [79]. However, for deepwater fields, drilling fluid can suffer shear degradation processes due to increased depths and high pipe shear, then its rheological properties, including the gel strength, can be severely affected; therefore, its capabilities to suspend and transport drill cuttings are reduced considerably [79, 80].

2.4. Lost circulation

Lost circulation is one of the most troublesome problems for deepwater drilling operations. Lost circulation (or lost returns) is an undesired event where a smaller amount of drilling fluid is returned from the wellbore than is pumped into it, thus drilling mud is lost into the formation [81, 82]. Lost circulation is a major cause of nonproductive time (NPT) in drilling, which can significantly raise operational costs. It has been reported that more than 12% of NPTs are due to problems of lost circulation in drilling areas of the Gulf of Mexico (GoM) [81]. Worldwide, the impact of lost circulation on well construction has been estimated to be around two to four billion dollars annually in nonproductive times, drilling fluid loss, and materials used to stem the losses [83]. **Figure 6** shows candidate formations for lost circulation events [84].

Lost circulation is carried out through one of two basic mechanisms: invasion and fracturing. Invasion refers to fluid loss to formations that are cavernous, vugular, fractured, or unconsolidated, whereas fracturing mechanism refers to the fluid loss due to hydraulic fracturing from excessive induced pressures [11]. However, most fluid losses are due to hydraulic-driven fractures covering from the wellbore to the far-field region [81]. Thus, lost circulation will be carried out mainly through fracturing mechanism, where the mud pressure in the wellbore promotes or creates new fractures or opens preexisting fractures on the wellbore wall [81]. In this way, formations with a narrow mud weight window, the safe drilling margin between pore pressure (or collapse pressure) and fracture pressure, have a greater propensity to suffer lost circulation problems.

For deepwater formations, water depth can cause a lower fracture pressure resulting in a narrow mud weight window, making it very challenging to maintain the needed wellbore pressure, and increasing considerably the propensity of lost circulation [81, 85]. Therefore, minimizing and controlling lost circulation is another important challenge to deepwater drilling fluid technology.

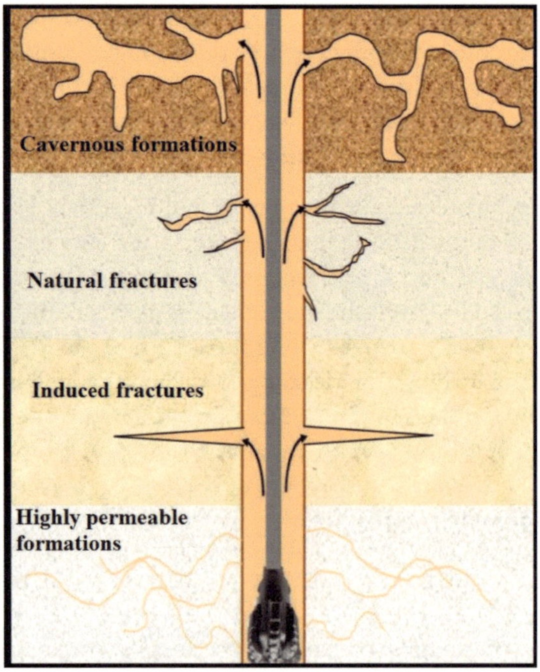

Figure 6. Schematic classification of lost circulation [84].

Some authors have identified some main concerns associated with lost circulation in deepwater environments. These issues are the following: drilling salt formations, identifying loss zones, controlling seepage losses, running casing and cementing, excessive rates of penetration, wellbore breathing, drilling fluid rheology, inadequate shoe tests, synthetic-based fluid compressibility, and well control [84].

In order to solve and control lost circulation problems, the following treatments are usually employed: lost circulation materials (LCMs), settable materials, and blends of the two [82].

Settable materials are usually pumped in a liquid state and solidify downhole, sealing the thief zone. Examples of these kinds of materials are bentonite-oil-mud systems, cement, gunk, and cross-linked systems. Thus, these materials usually need some extra preparation time and some setting time downhole before they can reduce and stop lost circulation [82].

Lost circulation materials (LCMs) have been widely used to avoid or stop losses. These materials are pumped downhole in order to bridge and seal fractures and voids, thereby stopping losses [82]. LCMs can be classified according to their physical and chemical properties as well as their action mechanism [84]. Physical properties are mainly size and appearance, whereas chemical properties include solubility in acids, swellability, and reactivity with other chemicals [84]. Their performance, however, commonly declines with the circulation time, which is related to the decrease in the average size of the solid components of these materials. This phenomenon is known as "shear degradation" [86].

Figure 7 shows a summary of the treatments employed for lost circulation. Finally, in order to adequately choose and design the optimal treatment to solve and control lost circulation,

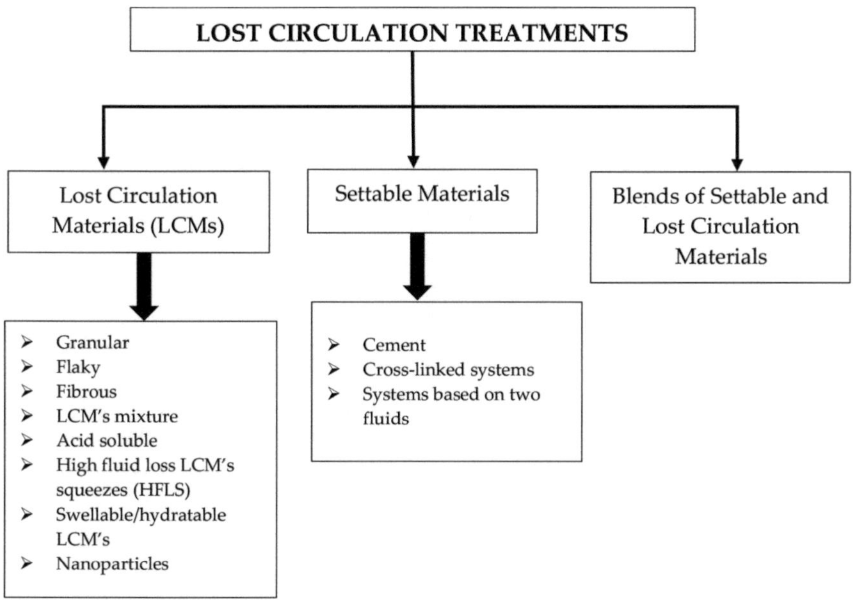

Figure 7. Classification of lost circulation treatments.

available data should be processed, analyzed, and used. Required information include the amount of loss, the kind of loss mechanism, quantitative data about fractures and pores such as fracture apertures, spacing, and pore throat size; geological setting (pay zone, shale, unconsolidated sand, gravel, etc.) and for deepwater formations, the mud weight window is a critical parameter that should be determined [81–85].

2.5. Salt formations

In many deepwater plays around the world, salt formations overlie prolific reservoirs containing a significant amount of hydrocarbons [87]. Successful drilling of the salt layers is not easy, and it is considered a challenge in deepwater drilling operations due to the complex salt behavior. The Gulf of Mexico (GoM) is the most active deepwater region in the world where salt is a dominant structural element that increases drilling risks and affects long-term well integrity [88].

Salt formations are considered efficient traps of hydrocarbons as observed in **Figure 8**. These traps were developed due to local faulting and bending processes in formations nearby the salt layers and upward migration of salt layers displacing other sediments [87, 88]. The chemistry of salts found in these formations can vary significantly. However, typical salts reported [11, 88] are the following: halite (NaCl), Sylvite (KCl), Bischofite ($MgCl_2{\cdot}6H_2O$), Carnalite ($KMgCl_3{\cdot}6H_2O$), Polyhalite ($K_2MgCa_2(SO_4)_4{\cdot}2H_2O$), and Tachydrite ($CaCl_2{\cdot}MgCl_2{\cdot}12H_2O$).

In addition, salt formations can contain other evaporate minerals such as gypsum ($CaSO_4{\cdot}2H_2O$), anhydrite ($CaSO_4$), kieserite ($MgSO_4{\cdot}H_2O$), limestone ($CaCO_3$), or dolomite ($CaMg(CO_3)_2$) associated with their structure [11, 88]. Salt presents exceptional and problematic characteristics. One of the most critical is its capability to deform or creep. Thus, it can exhibit plastic flow at certain temperature and pressure through other geological rock beds under stress, which results in the reduction of wellbore size, wellbore closure, drill string sticking as well as in casing collapse [11, 87]. The creep rate of salt will depend on

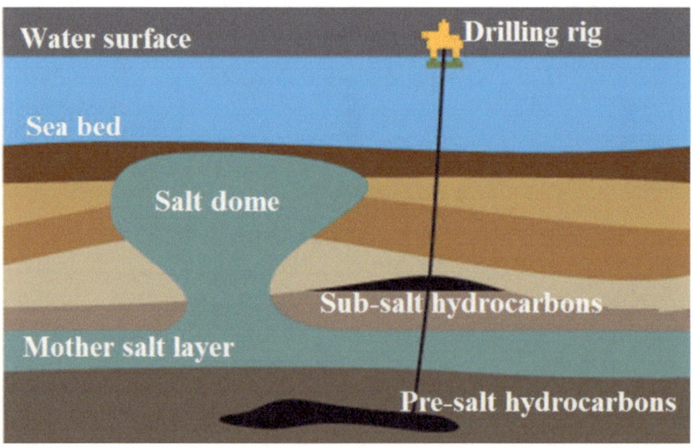

Figure 8. Typical geological salt dome formation [90].

several factors such as depth, overburden pressures, temperature, minerology and presence of impurities, moisture content, local, and regional geomechanical stresses [87]. In addition, salt dissolution during drilling can promote hole enlargement due to the increase of rates of penetration [11, 89]. Other salt-drilling hazards are sutures and inclusions, rubble zones, and tar, as reported [87].

Lost circulation problems are often found in salt formations [82, 91]. The thief zone at the base of the salt promotes drastic lost circulation and well control problems, often resulting in loss of the interval or the entire well [90]. Controlling losses in this kind of formations is extremely problematic. Thus, treatments to control lost circulation problems in salt formations during drilling can last for weeks, affecting considerably operation costs, particularly for deepwater drilling operations [82, 91].

Drilling fluids play a key role to carry out a successful drilling operation in salt formations found in deepwater fields. Key properties of muds need to be controlled when drilling salt formations are mainly density, salinity, and rheology [87]. About density property, the key role of mud weight in drilling salt formations is to minimize the creep rate. It has been reported that increasing mud weight can efficiently control salt creep and thus prevent wellbore size reduction, wellbore closure, and drill string sticking [92]. Salinity of drilling fluid is important in order to minimize salt dissolution, which is necessary to maintain the drilling mud salinity at or near saturation with respect to the drilled salt formation [87]. In addition, salt dissolution can be influenced by the flow regime [87, 89]. Rheological properties of muds can be modified during drilling of salt formations as the drill cuttings interact with the drilling fluid, affecting the drilling performance. However, rates of salt dissolution can decrease as the viscosity increases [87]. **Table 3** shows the main types of fluids commonly used to drill salt formations including deepwater plays as reported [87].

Type of drilling fluid	Advantages	Disadvantages
Riserless water-base fluids	• They can be designed and formulated for different salinity levels ranging from freshwater to supersaturation • Cost typically lower than other fluid types	• Limitations to get optimal performance. • Additional equipment can be required to maintain acceptable rheological properties.
High-performance water-base fluid	• Good hydrate-inhibition properties. • Similar performance to synthetic-base fluids when designed properly	• Environmentally less recyclable from one well to another and may require biocides depending on the formulation
Synthetic-base fluids	• High stability in terms of contamination tolerance • Superior performance in terms of rates of penetration (ROP) • Higher capacity for hydrate inhibition	• Very expensive

Table 3. Drilling fluids for salt formations.

2.6. Environmental and safety aspects

Safety and environmental concerns have long been a priority for deepwater operations. Waste-drilling mud is produced inevitably and has a negative impact on marine ecological environment during offshore oil exploration and development, resulting in a serious damage to the marine environment and harming the people's health [93–95]. Thus, preventing pollution and minimizing environmental impact in a cost-effective way are the challenging tasks confronting the industry of drilling fluids nowadays [11].

The Macondo incident in the Gulf of Mexico in 2010 has been widely studied, which led to the deaths of 11 workers on the transocean's deepwater horizon drilling rig as well as the release of an estimated 4.9 million barrels of oil [96]. Thus, drilling fluid companies must now comply with new offshore safety and environmental regulations [93]. The best available techniques and best environmental practices based on the waste management hierarchy of avoidance, reduction, reuse, recycling, recovery, and residue disposal can be also applied to the management of waste-drilling mud produced during deepwater operations [11, 93]. Usually for waste disposal during offshore operations, there are three basic options: discharge, haul to shore, or grind and inject [11].

The basic items, present in waste-drilling mud, having potential to cause environmental damage are heavy metals, salts compounds, organic wastes, acid or bases, and suspended solids. Toxicity tests are used to determine the combined effects of pollution on test organisms [11]. In addition, hazardous effects of additives such as defoamers, descalers, thinners, viscosifiers, lubricants, stabilizers, surfactants, and corrosion inhibitors on marine and human life have been reported. Such effects range from minor physiological changes to reduced fertility and higher mortality rates [97]. In this way, the oil and gas industry anticipates that the zero discharge of oil-contaminated drilling wastes will soon be the global standard.

Therefore, another of the great challenging tasks in deepwater operations is the development of novel environmentally friendly drilling fluids with better or similar performance, efficiency, and cost than oil-based drilling fluids. Several researchers and companies have reported new formulations of drilling fluid with minimal but not zero environmental impact [97]. **Table 4** shows some examples of this kind of drilling fluids.

Environmentally friendly drilling fluid system	Main characteristics	References
Silicate-based drilling fluids	• Good environmental compatibility • High performance for drilling reactive shales • Potential to damage the formation	[98]
High-density HPHT water-based fluid system	• Excellent fluid-loss control and capabilities to generate thermally stable rheology. • Chrome-free fluid	[99]

Environmentally friendly drilling fluid system	Main characteristics	References
Low-salinity glycol water-based drilling fluid	• For reactive shale formations • Performance limited to the presence of electrolytes	[100]
Zirconium citrate-based drilling fluid	• Environmentally friendly • Good rheological stability at high temperature • Performance affected by solids absorption	[101]
Water-based drilling fluids with eco-friendly polymers	• Eco-friendly polymers derived from tamarind gum and tragacanth gum • Cheaper than conventional polymers • Favorable rheological properties and low potential of formation damage	[102]
Formulations of water-soluble polymer amphoteric cellulose ether	• Low cost • Environmentally friendly • Potential to damage the formation	[103]
Starch additives for water-based muds	• Environment friendly fluid loss additives • Low manufacturing cost	[104]
Oil-based drilling fluid based on vegetable oils	• Derived from palm oil and groundnut oil • Highly biodegradable and good eco-toxicological properties	[105]
Polymeric potassium-silicate drilling fluid	• Excellent borehole-stability properties • Environmentally friendly	[106]
Formulations of hyperbranched polyglycerols	• Clay-inhibitive properties • Potential to be used as an environmental friendly inhibitor additive in WBFs	[107]
Offshore-drilling fluid system	• Good rheological properties, temperature tolerance, and collapse prevention performance • Low chromaticity, nontoxicity, and little effect on marine environment	[95]
High-performance water-based drilling fluid	• High performance on shale stabilization • High rate of penetration • Eco-friendly fluid	[108]
Nanoparticle-based drilling fluids	• For reactive shale formations • Stability in rheological properties • Environmentally friendly	[109]

Table 4. Some examples of environmentally friendly drilling fluids.

3. Conclusions

Large volumes of the world's future oil reserves are located in deep and ultra-deep water fields. Advances in exploration and production of these fields over the last years indicate that as soon as one deepwater record is broken, another surpasses it. Technological challenges of drilling at these extreme conditions generate significant operational risks as well as very high costs. Exploration of new technology frontiers of deep and ultra-deep water drilling will increase inevitably the demand and thus the devolvement of innovative technological solutions related to the design and application of drilling fluids for this kind of challenging and complex formations.

In this chapter, an overview about the main challenges facing the deepwater drilling fluids industry around the world was presented. The main concerns about deepwater drilling fluids reviewed were the following: wellbore stability and clay swelling, gas hydrates, rheological behavior, lost circulation, salt formations, environmental, and safety aspects. The design, selection, and application of the right fluid system or additives will require balancing each of these issues with regard to their impact on the deepwater drilling operation.

Therefore, the future of deepwater technologies must be focused toward the design and development of innovative materials with high performance, low cost, and with sustainability characteristics such as environmentally friendly and zero impact on the environment.

Acknowledgements

The authors express gratitude to the Instituto Mexicano del Petróleo (IMP) for both providing the research facilities and granting permission to publish the present chapter. L.A.A.-V. thanks both the IMP (Fondo Sectorial CONACyT-SENER Hidrocarburos) and the Direccion de Catedras CONACYT for the financial support provided for the writing and publication of this chapter.

Author details

Luis Alberto Alcázar-Vara[1*] and Ignacio Ramón Cortés-Monroy[2]

*Address all correspondence to: albertof18@gmail.com

1 CONACYT, Mexican Institute of Petroleum (IMP), México D.F., México

2 Mexican Institute of Petroleum (IMP), México D.F., México

References

[1] Zamora M, Broussard PN, Stephens MP. The top 10 mud-related concerns in deepwater drilling operations. In: SPE International Petroleum Conference and Exhibition in Mexico; January 2000; Society of Petroleum Engineers. 2000

[2] McLean A, Wilde A, Zamora M, Rafferty M. The top 10 mud-related concerns in deepwater drilling operations–revisited after 10 years. In: SPE International Petroleum Conference and Exhibition; Houston; 2010

[3] EIA. Trends in U.S. Oil and Natural Gas Upstream Costs. Energy International Agency, Washington, DC, USA: March 23, 2016

[4] Hossain ME, Al-Majed AA. Fundamentals of Sustainable Drilling Engineering. John Wiley & Sons, New Jersey, USA; 2015

[5] Leffler WL, Pattarozzi R, Sterling G. Deepwater Petroleum Exploration & Production: A Nontechnical Guide. PennWell Books, Tulsa, Oklahoma, USA; 2011

[6] Hodder M. Drilling fluids design for deepwater wells. Petroleum Engineer International. 1998;**71**(3):25-32

[7] Caenn R, Chillingar GV. Drilling fluids: State of the art. Journal of Petroleum Science and Engineering. 1996;**14**(3-4):221-230

[8] Garside R, Snell RO, Cook H. Deepwater technology and deepwater developments. In: The Eleventh International Offshore and Polar Engineering Conference. International Society of Offshore and Polar Engineers. January 2001

[9] Rocha LAS, Junqueira P, Roque JL. Overcoming deep and ultra deepwater drilling challenges. In: Offshore Technology Conference. January 2003

[10] Darley, HC, Gray GR. Composition and Properties of Drilling and Completion Fluids. Gulf Professional Publishing, Houston, Texas, USA; 1988

[11] M.I. Corporation. M.I. Drilling Fluids Engineering Manual, Revision No. A-1. Houston, TX; 2002

[12] Cameron C, Baroid H. Deepwater drilling fluids—What's new?. In: AADE-05-NTCE-79. 2005. pp. 1-6

[13] Wilson MJ, Wilson L. Clay mineralogy and shale instability: An alternative conceptual analysis. Clay Minerals. 2014;**49**(2):127-145

[14] Van Oort E. On the physical and chemical stability of shales. Journal of Petroleum Science and Engineering. 2003;**38**:213-235

[15] Anderson RL, Ratcliffe I, Greenwell HC, Williams PA, Cliffe S, Coveney PV. Clay swelling—A challenge in the oilfield. Earth Science Reviews. 2010;**98**:201-216

[16] Khodja M, Canselier JP, Bergaya F, Fourar K, Khodja M, Cohaut N, Benmounah A. Shale problems and water-based drilling fluid optimisation in the Hassi Messaoud Algerian oil field. Applied Clay Science. 2010;**49**(4):383-393

[17] Van Oort E, Hale AH, Mody FK. Transport in shales and the design of improved water-based shale drilling fluids. In: SPEDC, APE Annual Technical Conference and Exhibition; 25-28 September 1996; New Orleans. 1996

[18] Karpiński B, Szkodo M. Clay minerals–mineralogy and phenomenon of clay swelling in oil & gas industry. Advances in Materials Science. 2015;**15**(1):37-55

[19] Khodja M, Bergaya F, Canselier JP, Khodja-Saber M, Cohaut N. Drilling Fluid Technology: Performances and Environmental Considerations. INTECH Open Access Publisher, Rijeka, Croatia; 2010

[20] Hensen EJ, Smit B. Why clays swell. The Journal of Physical Chemistry B. 2002;**106**(49): 12664-12667

[21] Faruk C. Mechanism of Clay Swelling from Reservoir Formation Damage—Fundamentals, Modeling, Assessment, and Mitigation. Elsevier, Houston, Texas, USA; 2000

[22] de Carvalho Balaban R, Vidal ELF, Borges MR. Design of experiments to evaluate clay swelling inhibition by different combinations of organic compounds and inorganic salts for application in water base drilling fluids. Applied Clay Science. 2015;**105**:124-130

[23] Hu X, Hu S, Jin F, Huang S, editors. Physics of Petroleum Reservoirs. Springer Mineralogy. Springer, Berlin/Heidelberg, Germany; 2016. p. 506. ISBN:3662532840, 9783662532843

[24] Ruedrich J, Bartelsen T, Dohrmann R, Siegesmund S. Moisture expansion as a deterioration factor for sandstone used in buildings. Environmental Earth Sciences. 2011;**63**(7-8):1545-1564

[25] Hillier S. Quantitative analysis of clay and other minerals in sandstones by X-ray powder diffraction (XRPD). In: Worden RH, Morad S, editors. Clay Mineral Cements in Sandstones. Vol. 34. International Association of Sedimentologists Special Publication, Oxford, UK; 2003. pp. 213-251

[26] Josh M, Esteban L, Delle Piane C, Sarout J, Dewhurst DN, Clennell MB. Laboratory characterisation of shale properties. Journal of Petroleum Science and Engineering. 2012;**88**:107-124

[27] Cotecchia F, Cafaro F, Guglielmi S. Microstructural changes in clays generated by compression explored by means of SEM and image processing. Procedia Engineering. 2016;**158**:57-62

[28] Kahr G, Madsen FT. Determination of the cation exchange capacity and the surface area of bentonite, illite and kaolinite by methylene blue adsorption. Applied Clay Science. 1995;**9**(5):327-336

[29] Zhang J, Chenevert MM, Talal A, Sharma MM. A new gravimetric-swelling test for evaluating water and ion uptake of shales. In: SPE 89831. SPE Annual Technical Conference and Exhibition; September 26-29, 2004; Houston (TX). 2004

[30] Santos H, Gupta A. Application of thermal analysis to characterize clay-rich formations. In: Presented at the Petroleum Conference of the South Saskatchewan Section of the Petroleum Society of CIM; 19-22 October, 1997; Regina, SK, Canada. 1997

[31] Muniz ES, Fontoura SAB, Lomba RFT. Development of equipment and testing methodology to evaluate rock-drilling fluid interaction. In: Paper 599, GulfRock04. 6th North America Rock Mechanics Symposium (NARMS); Houston, TX. 2004

[32] Paineau E, Michot LJ, Bihannic I, Baravian C. Aqueous suspensions of natural swelling clay minerals. 2. Rheological characterization. Langmuir. 2011;**27**(12):7806-7819

[33] Santos H, Diek A, Da Fontoura S, Roegiers JC. Shale reactivity test: A novel approach to evaluate shale-fluid interaction. International Journal of Rock Mechanics and Mining Sciences. 1997;**34**(3-4):268-e1

[34] Fink JK, editor. Clay Stabilization. Chapter 3, Petroleum Engineer's Guide to Oil Field Chemicals and Fluids. Gulf Publishing Company, Houston, Texas, USA; 2012. pp. 125-137

[35] Quintero L. An overview of surfactant applications in drilling fluids for the petroleum industry. Journal of Dispersion Science and Technology. 2002;**23**(1-3):393-404

[36] Patel AD. Design and development of quaternary amine compounds: Shale inhibition with improved environmental profile. In: SPE International Symposium on Oilfield Chemistry. Society of Petroleum Engineers; January 2009

[37] Simpson JP, Walker TO, Jiang GZ. Environmentally acceptable water-base mud can prevent shale hydration and maintain borehole stability. In: Paper IADC/SPE 27496 presented at the IADC/SPE Drilling Conference; 15-18 February 1994; Dallas. 1994

[38] Akhtarmanesh S, Shahrabi MA, Atashnezhad A. Improvement of wellbore stability in shale using nanoparticles. Journal of Petroleum Science and Engineering. 2013;**112**:290-295

[39] Hoelscher KP, De Stefano G, Riley M, Young S. Application of nanotechnology in drilling fluids. In: SPE International Oilfield Nanotechnology Conference and Exhibition. Society of Petroleum Engineers; January 2012

[40] Jain R, Mahto V, Sharma VP. Evaluation of polyacrylamide-grafted-polyethylene glycol/silica nanocomposite as potential additive in water based drilling mud for reactive shale formation. Journal of Natural Gas Science and Engineering. 2015;**26**:526-537

[41] Sameni A, Pourafshary P, Ghanbarzadeh M, Ayatollahi S. Effect of nanoparticles on clay swelling and migration. Egyptian Journal of Petroleum. 2015;**24**(4):429-437

[42] Gomez SL, Patel A. Shale inhibition: What works?. In: SPE International Symposium on Oilfield Chemistry. Society of Petroleum Engineers; April 2013

[43] Zaltoun A, Berton N. Stabilization of montmorillonite clay in porous media by high-molecular-weight polymers. SPE Production Engineering. 1992;**7**(02):160-166

[44] McConnell DR, Zhang Z, Boswell R. Review of progress in evaluating gas hydrate drilling hazards. Marine and Petroleum Geology. 2012;**34**(1):209-223

[45] Rogers R. Offshore Gas Hydrates: Origins, Development, and Production. Gulf Professional Publishing, Houston, Texas, USA; 2015

[46] Hu Y, Yue Q, Liu S, Fu Z, Liang S. Research on deepwater synthetic drilling fluid and its low temperature rheological properties. Petroleum Science. 2011;**8**(4):485-489

[47] Kelland MA, Mønig K, Iversen JE, Lekvam K. Feasibility study for the use of kinetic hydrate inhibitors in deep-water drilling fluids. Energy & Fuels. 2008;**22**(4):2405-2410

[48] Zhao X, Qiu Z, Huang W. Characterization of kinetics of hydrate formation in the presence of kinetic hydrate inhibitors during deepwater drilling. Journal of Natural Gas Science and Engineering. 2015;**22**:270-278

[49] Ke W, Kelland MA. Kinetic hydrate inhibitor studies for gas hydrate systems: A review of experimental equipment and test methods. Energy & Fuels. 2016;**30**(12):10015-10028

[50] Fink JK, editor. Gas Hydrate Control—Chapter 13. Petroleum Engineer's Guide to Oil Field Chemicals and Fluids. Gulf Publishing Company, Houston, Texas, USA; 2012. pp. 391-426

[51] Sivaraman R. Flow assurance: Understanding and controlling natural gas hydrate. In: Gas TIPS, Summer 2002, A Publication of Gas Technology Institute, Vol. 8. U.S. Dept. of Energy and Hart Publications, Inc.; 2002. p. 3

[52] Kelland MA. History of the development of low dosage hydrate inhibitors. Energy & Fuels. 2006;**20**(3):825-847

[53] Anderson FE, Prausnitz JM. Inhibition of gas hydrates by methanol. AIChE Journal. 1986;**32**(8):1321-1333

[54] Ebeltoft H, Majeed Y, Sœrgärd E. Hydrate control during deepwater drilling: Overview and new drilling-fluids formulations. SPE Drilling & Completion. 2001;**16**(01):19-26

[55] Cha M, Shin K, Kim J, Chang D, Seo Y, Lee H, Kang SP. Thermodynamic and kinetic hydrate inhibition performance of aqueous ethylene glycol solutions for natural gas. Chemical Engineering Science. 2013;**99**:184-190

[56] Wu HJ, Englezos P. Inhibiting effect of triethylene glycol and glycerol on gas hydrate formation conditions. Journal of Chemical & Engineering Data. 2006;**51**(5):1811-1813

[57] Lugo R, Dalmazzone C, Audibert A. U.S. Patent No. 7,709,419. Washington, DC: U.S. Patent and Trademark Office; 2010

[58] Tariq M, Rooney D, Othman E, Aparicio S, Atilhan M, Khraisheh M. Gas hydrate inhibition: A review of the role of ionic liquids. Industrial & Engineering Chemistry Research. 2014;**53**(46):17855-17868

[59] Sa JH, Lee BR, Park DH, Han K, Chun HD, Lee KH. Amino acids as natural inhibitors for hydrate formation in CO_2 sequestration. Environmental Science & Technology. 2011;**45**(13):5885-5891

[60] Klomp UC. International Patent Application WO/2001/077270; 2001

[61] Chua PC, Kelland MA, Ishitake K, Satoh K, Kamigaito M, Okamoto Y. Kinetic hydrate inhibition of poly(nisopropylmethacrylamide)s with different tacticities. Energy Fuels. 2012;**26**:3577-3585

[62] Leinweber D, Feustel M. U.S. Patent Application 20090054268; 2009

[63] Klug P, Kelland MA. International Patent Application WO/1998/023843; 1998

[64] Villano LD, Kommedal R, Fijten MWM, Schubert US, Hoogenboom R, Kelland MA. A study of the kinetic hydrate inhibitor performance and seawater biodegradability of a series of poly(2-alkyl-2-oxazoline)s. Energy Fuels. 2009;**23**(7):3665-3675

[65] Colle KS, Costello CA, Talley LD. Canadian Patent Application 96/2178371; 1996

[66] Colle KS, Costello CA, Berluche E, Oelfke RH, Talley LD. International Patent Application WO/1996/041834; 1996

[67] Chua PC, Sæbø M, Lunde A, Kelland MA. Dual kinetic hydrate inhibition and scale inhibition by polyaspartamides. Energy Fuels. 2011;**25**(11):5165-5172

[68] Lee JD, Wu H, Englezos P. Cationic starches as gas hydrate kinetic inhibitors. Chemical Engineering Science. 2007;**62**:6548-6555

[69] Walker VK, Zeng H, Ohno H, Daraboina N, Sharifi H, Bagherzadeh SA, Alavi S, Englezos P. Antifreeze proteins as gas hydrate inhibitors. Canadian Journal of Chemistry. 2015;**93**(8):839-849

[70] Maccioni F, Passucci C. Torque moment as indicator of low dosage hydrates inhibitors: Effects on multiphase systems. Experimental study on quaternary ammonium and phosphonium compounds. In: Proceedings of the 7th International Conference on Gas Hydrates (ICGH 2011); July 17–21, 2011; Edinburgh, UK

[71] Kelland MA, Svartås TM, Andersen LD. Gas hydrate anti-agglomerant properties of polypropoxylates and some other demulsifiers. Journal of Petroleum Science and Engineering. 2009;**64**(1):1-10

[72] York JD, Firoozabadi A. Comparing effectiveness of rhamnolipid biosurfactant with a quaternary ammonium salt surfactant for hydrate anti-agglomeration. The Journal of Physical Chemistry B. 2008;**112**(3):845-851

[73] Kelland MA, Svartaas TM, Øvsthus J, Tomita T, Chosa JI. Studies on some zwitterionic surfactant gas hydrate anti-agglomerants. Chemical Engineering Science. 2006;**61**(12):4048-4059

[74] Kelland MA, Svartaas TM, Øvsthus J, Tomita T, Mizuta K. Studies on some alkylamide surfactant gas hydrate anti-agglomerants. Chemical Engineering Science. 2006;**61**(13):4290-4298

[75] Hilfiger MG, Thaemlitz CJ, Moellendick E. Investigating the chemical nature of flat rheology. In: SPE Deepwater Drilling and Completions Conference. Society of Petroleum Engineers; September 2016

[76] Schlemmer RP, Phoon G. A new generation associative polymer extends temperature stability of deepwater drilling fluid. In: International Petroleum Technology Conference. January 2011

[77] Van Oort E, Friedheim J, Toups B. New flat-rheology synthetic-based for improved deep-water drilling. In: Paper SPE 90987 Presented at the SPE Annual Technical Conference and Exhibition; 26-29 September 2004; Houston, TX, USA

[78] Mullen GA, Tanche-Larsen PB, Clark DE, Giles A. The pro's and con's of flat rheology drilling fluids. In: Paper AADE-05-NTCE-28 Presented at the AADE 2005 Drilling Fluids Conference; April 2005; Wyndam Greenspoint, Houston, TX

[79] Shah SN, Shanker NH, Ogugbue CC. Future challenges of drilling fluids and their rheological measurements. In: AADE Fluids Conference and Exhibition; April 2010; Houston, TX

[80] Bybee K. Drilling-fluid rheology under deepwater drilling conditions. Journal of Petroleum Technology. 1999;**51**(11):36-39

[81] Feng Y, Gray KE. Review of fundamental studies on lost circulation and wellbore strengthening. Journal of Petroleum Science and Engineering. 2017;**152**:511-522. http://dx.doi.org/10.1016/j.petrol.2017.01.052

[82] Lavrov A. Lost Circulation: Mechanisms and Solutions. Gulf Professional Publishing; 2016

[83] Cook J, Growcock F, Guo Q, Hodder M, Van Oort E. Stabilizing the wellbore to prevent lost circulation. Oilfield Review. 2011;**23**:26-35

[84] Alsaba M, Nygaard R, Hareland G. Review of lost circulation materials and treatments with an updated classification. In: AADE National Technical Conference and Exhibition; April 2014; Houston, TX; 2014. pp. 15-16

[85] Power D, Ivan CD, Brooks SW. The top 10 lost circulation concerns in deepwater drilling. In: SPE Latin American and Caribbean Petroleum Engineering Conference. Society of Petroleum Engineers; January 2003

[86] Valsecchi P. On the shear degradation of lost-circulation materials. SPE Drilling & Completion. 2014;**29**(03):323-328

[87] Amer A, Dearing H, Jones R, Sergiacomo M. Drilling through salt formations: A drilling fluids review. In SPE Deepwater Drilling and Completions Conference. Society of Petroleum Engineers; September 2016

[88] Fredrich JT, Fossum AF, Hickman RJ. Mineralogy of deepwater Gulf of Mexico salt formations and implications for constitutive behavior. Journal of Petroleum Science and Engineering. 2007;**57**(3):354-374

[89] Willson SM, Driscoll P, Judzis A, Black A, Martin W, Ehgartner B, Hinkenbein T. Drilling salt formations offshore with seawater can significantly reduce well costs. In: SPE 87216, IADC/SPE Drilling Conference; 1-3 March 2004; Dallas, TX

[90] The Vallourec Oil & Gas Magazine. Available from: http://www.connection-mag.com/?p=1123 [Accessed: January 18, 2017]

[91] Sanders WW, Williamson RN, Ivan CD, Powell D. Lost circulation assessment and planning program: Evolving strategy to control severe losses in deepwater projects. In: SPE/IADC Drilling Conference. Society of Petroleum Engineers; January 2003

[92] Aburto M, D'Ambrosio P. The evolution of hole opening while drilling practices to enlarge salt and subsalt sections in the Gulf of Mexico. In: AADE-2009NTCE-09-02, AADE National Technical Conference; April 2009; New Orleans

[93] Cummings R, Garcia C, Hawthorn A, Holicek R, Dribus JR, Haslin L. Beyond deep—The challenges of ultradeep water. Oilfeld Review. Winter 2014/2015;**26**(4):34-45

[94] Khodja M, Khodja-Saber M, Canselier JP, Cohaut N, Bergaya F. Drilling Fluid Technology: Performances and Environmental Considerations, Products and Services; from R&D to Final Solutions. In: Fuerstner I, editor. ISBN: 978-953-307-211-1, InTech, Rijeka, Croatia; 2010. Available from: http://www.intechopen.com/books/products-and-services--from-r-d-to-final-solutions/drilling-fluid-technology performances-and-environmental-considerations

[95] Xie SX, Jiang GC, Chen M, Li ZY, Mao H, Zhang M, Li YY. Evaluation indexes of environmental protection and a novel system for offshore drilling fluid. Petroleum Science and Technology. 2014;**32**(4):455-461

[96] Crone TJ, Tolstoy M. Magnitude of the 2010 Gulf of Mexico oil leak. Science. 2010; **330**(6004):634-634

[97] Apaleke AS, Al-Majed AA, Hossain ME. Drilling fluid: State of the art and future trend. In: North Africa Technical Conference and Exhibition. Society of Petroleum Engineers; January 2012

[98] Van Oort E, Ripley D, Ward I, Chapman JW. Silicate-based drilling fluids: Components, costeffective and benign solutions to wellbore stability problem. IDAC/SPE; 1996

[99] Tehrani A, Young S, Gerrard D, Fernandez J. Environmentally friendly water based fluid for HT/HP drilling. In: SPE International Symposium on Oilfield Chemistry. Society of Petroleum Engineers; January 2009

[100] Brady ME, Craster B, Getliff JM, Reid PI. Highly inhibitive, low-salinity glycol water-base drilling fluid for shale drilling in environmentally sensitive locations. In: SPE International Conference on Health, Safety, and Environment in Oil and Gas Exploration and Production. Society of Petroleum Engineers; January 1998

[101] Nicora LF, Burrafato G. Zirconium citrate: A new generation dispersant for environmentally friendly drilling fluids. In: IADC/SPE Asia Pacific Drilling Technology. Society of Petroleum Engineers, Jakarta, Indonesia; January 1998

[102] Sharma VP, Mahto V. Studies on less expansive environmentally safe polymers for development of water based drilling fluids. In: SPE Asia Pacific Oil & Gas Conference and Exhibition. Society of Petroleum Engineers; January 2006

[103] Warren B, van der Horst P, Stewart W. Application of amphoteric cellulose ethers in drilling fluids. In: International Symposium on Oilfield Chemistry. Society of Petroleum Engineers; January 2003

[104] Amanullah M, Yu L. Environment friendly fluid loss additives to protect the marine environment from the detrimental effect of mud additives. Journal of Petroleum Science and Engineering. 2005;**48**(3):199-208

[105] Dosunmu A. Development of environmentally friendly oil based mud using palm-oil and groundnut-oil. In: Nigeria Annual International Conference and Exhibition; Nigeria: Society of Petroleum Engineers; January 2010

[106] Hector MS, Ramirez RL, Heliodoro PP. Novel application of ecological system drilling fluid accomplishes significant advances in the hole stability problems. In: SPE International Petroleum Conference and Exhibition; Mexico: Society of Petroleum Engineers; January 2002

[107] Teixeira GT, Lomba RFT, Francisco ADDS, da Silva JFC, Nascimento RSV. Hyperbranched polyglycerols, obtained from environmentally benign monomer, as reactive clays inhibitors for water-based drilling fluids. Journal of Applied Polymer Science. 2014;**131**(12):40384/1- 40384/7

[108] Witthayapanyanon A, Leleux J, Vuillemet J, Morvan R, Pomian A, Denax A, Bland R. High performance water-based drilling fluids-an environmentally friendly fluid system achieving superior shale stabilization while meeting discharge requirement offshore Cameroon. In: SPE/IADC Drilling Conference. Society of Petroleum Engineers; March 2013

[109] Sharma MM, Chenevert ME, Guo Q, Ji L, Friedheim J, Zhang R. A new family of nanoparticle based drilling fluids. In: SPE Annual Technical Conference and Exhibition. Society of Petroleum Engineers; January 2012

Evaluation of Different Correlation Performance for the Calculation of the Critical Properties and Acentric Factor of Petroleum Heavy Fractions

Dacid B. Lacerda, Rafael B. Scardini,
André P. C. M. Vinhal, Adolfo P. Pires and
Viatcheslav I. Priimenko

Additional information is available at the end of the chapter

http://dx.doi.org/10.5772/intechopen.71166

Abstract

The characterization of petroleum fluids is fundamental for the calculation of their thermodynamic properties. Laboratory experiments are able to identify a limited number of pure components present in a sample. All remaining species, the so called "cut", are characterized by its molecular weight and density. The thermodynamic calculations performed using cubic equations of state require the critical properties and the acentric factor, which are unknown for the petroleum "cut." In this chapter, different correlations are used to calculate the critical properties and the acentric factor of the "cut" fraction. The performance of the correlations is evaluated through the comparison of a simulated pressure-volume-temperature (PVT) experiment using an equation of state and experimental data of two reservoir fluids.

Keywords: correlations of critical properties, petroleum heavy fractions, PVT experiments, thermodynamics, phase equilibrium

1. Introduction

Petroleum is a complex mixture of several chemical components, mainly hydrocarbons. In addition to the hydrocarbons, it may also contain some inorganic contaminants such as carbon dioxide (CO_2), nitrogen (N_2) and hydrogen sulfide (H_2S).

The physical properties of reservoir fluids are related to the concentration of their components. Some properties such as bubble point pressure, oil formation volume factor, solubility ratio, oil density, gas formation volume factor and gas specific gravity are of particular interest in black oil reservoir engineering studies. These properties are generally obtained in laboratory using reservoir fluids samples. These experiments seek to replicate the isothermal recovery path of an oil field.

Pressure-volume-temperature (PVT) experiments are carried out in liquid mixtures of hydrocarbons and during the pressure reduction steps, dissolved gas is released. There are two types of experiments that simulate the constant temperature depletion of a reservoir fluid, "flash" and "differential". In a "flash" experiment, the overall composition of the system is kept constant, whereas in a "differential" experiment the gas phase is removed from the system at each pressure step.

These experiments can be simulated using equations of state in order to evaluate different recovery schemes without carrying out one experiment for each possible scenario, especially in the cases of enhanced oil recovery techniques. The correct identification of the species and their concentrations is fundamental for the success of the simulation. However, in any laboratory test, the heavy fraction of the oil, the so-called "cut," is characterized by its molecular weight and density. In order to simulate the experiment, correlations are necessary to calculate the critical properties and the acentric factor of the "cut" from its molecular weight and density. In this chapter, different correlations are used to determine the critical properties and acentric factor of the heavy fraction of two reservoir fluids, the PVT experiments of these fluids are simulated using an equation of state and the results are compared with laboratory data.

2. Methodology

Cubic equations of state are widely used to describe the volumetric properties of pure substances and mixtures in the petroleum industry. Furthermore, these models can be used for equilibrium calculations and to estimate PVT properties, such as bubble point pressure, oil formation volume factor, solubility ratio, oil specific weight, gas formation volume factor and gas density.

The input parameters of the most used equations of state are the acentric factor, ω, critical temperature, T_c, and critical pressure, P_c, of the mixture components. These parameters are tabulated for a large number of chemical compounds, but for the heavy fractions of petroleum fluids, they are determined from correlations. Most correlations are functions of density, γ, molecular mass, M, and/or normal boiling temperature, T_b, of the fractions [1].

Edmister [2] proposed a correlation to estimate the acentric factor of pure liquids and petroleum fractions. This correlation is given by:

$$\omega = \frac{3\left[\log\left(\frac{P_c}{14.7}\right)\right]}{7\left[\frac{T_c}{T_b} - 1\right]} - 1 \tag{1}$$

Cavett [3] presented equations to estimate the critical temperature and pressure of hydrocarbon fractions. These correlations are a function of the normal boiling point and the API gravity:

$$T_c = a_0 + a_1(T_b) + a_2(T_b)^2 + a_3(API)(T_b) + a_4(T_b)^3 + a_5(API)(T_b)^2 + a_6(API)^2(T_b)^2 \quad (2)$$

$$\begin{aligned}\log(P_c) = {} & b_0 + b_1(T_b) + b_2(T_b)^2 + b_3(API)(T_b) + b_4(T_b)^3 + b_5(API)(T_b)^2 \\ & + b_6(API)^2(T_b) + b_7(API)^2(T_b)^2\end{aligned} \quad (3)$$

T_b: normal boiling temperature [°F].

The coefficients of Eqs. (2) and (3) are shown in **Table 1**.

Kesler and Lee [4] developed correlations for critical temperature and pressure, molecular weight and acentric factor of oil fractions. These expressions are given by:

$$\begin{aligned}\ln(P_c) = {} & 8.3634 - \frac{0.0566}{\gamma} - \left(0.24244 + \frac{2.2898}{\gamma} + \frac{0.11857}{\gamma^2}\right)10^{-3}T_b \\ & + \left(1.4685 + \frac{3.648}{\gamma} + \frac{0.47227}{\gamma^2}\right)10^{-7}T_b^2 - \left(0.42019 + \frac{1.6977}{\gamma^2}\right)10^{-10}T_b^{-3}\end{aligned} \quad (4)$$

$$T_c = 341.7 + 811.1\gamma + (0.4244 + 0.1174\gamma)T_b + (0.4669 - 3.26238\gamma)\frac{10^5}{T_b} \quad (5)$$

$$\omega = -7.904 + 0.1352K_w - 0.007465K_w^2 + 8.359T_{br} + (1.408 - 0.01063K_w)\frac{1}{T_{br}}, \quad (6)$$

for $T_{br} > 0.8$, and

$$\omega = \frac{-ln\left(\frac{P_c}{14.7}\right) - 5.92714 + \frac{6.09648}{T_{br}} + 1.28862\,ln(T_{br}) - 0.169347T_{br}^6}{15.2518 - \frac{15.6875}{T_{br}} - 13.4721\,ln(T_{br}) + 0.43577T_{br}^6}, \quad (7)$$

for $T_{br} \leq 0.8$.

i	a_i	b_i
0	768.0712100000	2.82904060
1	1.7133693000	$0.94120109 \times 10^{-3}$
2	−0.0010834003	$-0.30474749 \times 10^{-5}$
3	−0.0089212579	$-0.20876110 \times 10^{-4}$
4	$0.3889058400 \times 10^{-6}$	$0.15184103 \times 10^{-8}$
5	$0.5309492000 \times 10^{-5}$	$0.11047899 \times 10^{-7}$
6	$0.3271160000 \times 10^{-7}$	$-0.48271599 \times 10^{-7}$
7		$0.13949619 \times 10^{-9}$

Table 1. Coefficients of Cavett [3] correlation.

θ	a	b	c	Average deviation (%)	Max deviation (%)
M	-4.56730×10^{-5}	2.19620	−1.0164	2.6	11.8
T_c	24.27870	0.58848	0.3596	1.3	10.6
P_c	-3.12281×10^{9}	−2.31250	2.3201	3.1	−9.3
V_c	-7.52140×10^{-3}	0.28960	−0.7666	2.3	−9.1

Table 2. Coefficients of Riazi and Daubert [6] correlation.

In Eq. (6), K_w is the Watson characterization factor, given by:

$$K_w = \frac{(T_b)^{\frac{1}{3}}}{\gamma} \tag{8}$$

Standing [5] developed the following correlations based on experimental data:

$$T_c = 608 + 364 \log(\mathrm{M} - 71.2) + [2450 \log(\mathrm{M}) - 3800] \log(\gamma) \tag{9}$$

$$P_c = 1188 - 431 \log(\mathrm{M} - 61.1) + [2319 - 852 \log(\mathrm{M} - 53.7)](\gamma - 0.8) \tag{10}$$

Riazi and Daubert [6] proposed a simple two-parameter equation to calculate the physical properties of pure components and mixtures of hydrocarbons:

$$\theta = a T_b^b \gamma^c \tag{11}$$

where θ is the property (T_c, P_c, v_c or M), v_c is the critical volume (ft^3/lb) and *a* to *c* are constants for the correlation of each property.

Table 2 shows the coefficients of Eq. (11) and the errors in the estimative of each property.

Sim and Daubert [13] determined expressions to calculate the critical pressure and critical temperature of petroleum "cuts," given by:

$$P_c = 3.48242 \times 10^9 T_b^{-2.3177} \gamma^{2.4853} \tag{12}$$

$$T_c = exp\left(3.99347 T_b^{0.08615} \gamma^{0.04614}\right) \tag{13}$$

Twu [7] proposed a set of correlations, based on the perturbation-expansion theory with normal paraffins as the reference state, to determine the critical properties of heavy fractions of hydrocarbons. The method is based on the selection of a normal paraffin with normal boiling temperature, T_{bPi}, identical to the normal boiling temperature of the C_{n+} fraction. Once the normal paraffin is chosen, the heavy fraction critical properties are calculated through the following steps:

1. Calculate the critical properties of the normal paraffin

$$T_{cPi} = T_{bC+} + \left[A_1 + A_2 T_{bC+} + A_3 T_{bC+}^2 + A_4 T_{bC+}^3 + \frac{A_5}{(A_6 T_{bC+})^{13}}\right] \tag{14}$$

$$P_{cPi} = \left[A_1 + A_2\left(1 - \frac{T_{bC+}}{T_{cPi}}\right)^{0.5} + A_3\left(1 - \frac{T_{bC+}}{T_{cPi}}\right) + A_4\left(1 - \frac{T_{bC+}}{T_{cPi}}\right)^2 + A_5\left(1 - \frac{T_{bC+}}{T_{cPi}}\right)^4\right] \tag{15}$$

$$\gamma_{Pi} = \left[A_1 + A_2\left(1 - \frac{T_{bC+}}{T_{cPi}}\right) + A_3\left(1 - \frac{T_{bC+}}{T_{cPi}}\right)^3 + A_4\left(1 - \frac{T_{bC+}}{T_{cPi}}\right)^{12}\right] \tag{16}$$

$$v_{cPi} = \left[1 - A_1 + A_2\left(1 - \frac{T_{bC+}}{T_{cPi}}\right) + A_3\left(1 - \frac{T_{bC+}}{T_{cPi}}\right)^3 + A_4\left(1 - \frac{T_{bC+}}{T_{cPi}}\right)^{14}\right]^{-8} \tag{17}$$

The constants of the Eqs. (14)–(17) are shown in **Table 3**.

Property	A_1	A_2	A_3	A_4	A_5	A_6
T_{cPi}	0.5332272	0.191017×10^{-3}	0.779681×10^{-7}	$-0.284376 \times 10^{-10}$	0.959468×10^{-2}	0.01
P_{cPi}	3.83854	1.19629	34.8888	36.1952	104.193	—
γ_{Pi}	0.843593	−0.128624	−3.36159	−13749.5	—	—
v_{cPi}	−0.419869	0.505839	1.56436	9481.7	—	—

Table 3. Constants used in the calculation of the critical properties of the normal paraffin [7].

2. Calculate the heavy petroleum fraction properties from the equations

$$T_{cCn+} = T_{cPi}\left[\frac{1 + 2f_{Ti}}{1 - 2f_{Ti}}\right] \tag{18}$$

$$f_{Ti} = \left\{\exp\left[5\left(\gamma_{pi} - \gamma_{Cn+}\right)\right] - 1\right\}\left[\frac{A_1}{T_{bCn+}^{0.5}} + \left(A_2 + \frac{A_3}{T_{bCn+}^{0.5}}\right)\left\{\exp\left[5\left(\gamma_{pi} - \gamma_{Cn+}\right)\right] - 1\right\}\right] \tag{19}$$

$$v_{cCn+} = v_{cPi} = \left[\frac{1 + 2f_{vi}}{1 - 2f_{vi}}\right]^2 \tag{20}$$

$$f_{vi} = \left\{\exp\left[4\left(\gamma_{pi}^2 - \gamma_{Cn+}^2\right)\right] - 1\right\}\left[\frac{A_1}{T_{bCn+}^{0.5}} + \left(A_2 + \frac{A_3}{T_{bCn+}^{0.5}}\right)\left\{\exp\left[4\left(\gamma_{pi}^2 - \gamma_{Cn+}^2\right)\right] - 1\right\}\right] \tag{21}$$

$$P_{cCn+} = P_{cPi}\left(\frac{v_{cPi}}{v_{cCn+}}\right)\left[\frac{1 + 2f_{Pi}}{1 - 2f_{Pi}}\right]^2 \tag{22}$$

$$f_{Pi} = \left\{ \exp\left[0.5\left(\gamma_{pi} - \gamma_{Cn+}\right)\right] - 1 \right\} \left[\left(A_1 + \frac{A_2}{T_{bCn+}^{0.5}} + A_3 T_{bCn+} \right) + \left(A_4 + \frac{A_5}{T_{bCn+}^{0.5}} + A_6 T_{bCn+} \right) \left\{ \exp\left[0.5\left(\gamma_{pi} - \gamma_{Cn+}\right)\right] - 1 \right\} \right] \quad (23)$$

where T_{cPi} is the critical temperature of the normal paraffin (°R), T_{cC+} is the critical temperature of the heavy petroleum fraction (°R), P_{cPi} is the critical pressure of the normal paraffin (psia), P_{cC+} is the critical pressure of the heavy oil fraction (psia), γ_{Pi} is the density of the normal paraffin, γ_{C+} is the density of the heavy petroleum fraction, v_{cPi} is the critical volume of the normal paraffin (ft^3/lbmol) and v_{cC+} is the critical volume of the heavy oil fraction (ft^3/lbmol).

Constants utilized in Eqs. (18)–(23) are presented in **Table 4**.

Riazi and Daubert [8] developed a general correlation given by the following expression:

$$\theta = a\theta_1^b \theta_2^c \exp\left(d\theta_1 + e\theta_2 + f\theta_1\theta_2\right) \quad (24)$$

In Eq. (24), θ_1 and θ_2 are two parameters accounting for the molecular forces and molecular size of a component. It was found that (T_b, γ) and (M, γ) are appropriate sets of input parameters for the correlation. Based on these results, two expressions were proposed:

$$\theta = aT_b^b \gamma^c \exp\left(dT_b + e\gamma + fT_b\gamma\right) \quad (25)$$

$$\theta = aM^b \gamma^c \exp\left(dM + e\gamma + fM\gamma\right) \quad (26)$$

Tables 5 and **6** present the coefficients for Eqs. (25) and (26).

Property	A_1	A_2	A_3	A_4	A_5	A_6
f_{Ti}	−0.362456	0.0398285	−0.948125	—	—	—
f_{vi}	0.466590	−0.182421	3.01721	—	—	—
f_{Pi}	2.53262	−46.19553	−0.00127885	−11.4277	252.14	0.00230535

Table 4. Constants used in the calculation of the critical properties of the heavy oil fraction C_{n+} [7].

θ	a	b	c	d	e	f
M	581.96	−0.97476	6.51274	0.000543076	9.53384	0.00111056
T_c	10.6443	0.81067	0.53691	−0.000517470	−0.54444	0.00035995
P_c	6162000	−0.4844	4.0846	−0.00472500	−4.8014	0.0031939
V_c	0.0006233	0.7506	−1.2028	−0.00146790	−0.26404	0.001095

Table 5. Coefficients of Riazi and Daubert [8] correlation as a function of T_b and γ.

http://dx.doi.org/10.5772/intechopen.71166

θ	a	b	c	d	e	f
T_c	544.4	0.2998	1.0555	−0.000134780	−0.61641	
P_c	45203	−0.8063	1.6015	−0.000180780	−0.3084	
V_c	0.01206	0.20378	−1.3036	−0.00265700	0.5287	0.0026012
T_b	6.77857	0.401673	−1.58262	0.00377409	2.984036	−0.00425288

Table 6. Coefficients of Riazi and Daubert [8] correlation as a function of M and γ.

The coefficients were adjusted to experimental data of 38 pure hydrocarbons with carbon number in the range of 1–20, including paraffins, olefins, naphthenes and aromatics with molecular weight between 70 and 300, and boiling point in the range of 80–650°F.

Magoulas and Tassios [9] correlated the critical properties of heavy fractions through the following expressions:

$$T_c = -1247.4 + 0.792M + 1971\gamma - \frac{27000}{M} + 707.4 \tag{27}$$

$$ln(P_c) = 0.01901 - 0.0048442M + 0.13239\gamma + \frac{227}{M} - \frac{1.1663}{\gamma} + 1.2702ln(M) \tag{28}$$

$$\omega = -0.64235 + 0.00014667M + 0.021876\gamma - \frac{4.559}{M} + 0.21669ln(M) \tag{29}$$

Riazi and Alsahhaf [10] developed a correlation for the acentric factor:

$$\omega = -\left[0.3 - exp\left(-6.252 + 3.64457M^{0.1}\right)\right] \tag{30}$$

Sancet [11] proposed correlations for P_c and T_c as a function of the molecular weight of the heavy oil fractions:

$$P_c = 82.82 + 653\exp\left(-0.007427\text{M}\right) \tag{31}$$

$$T_c = -778.5 + 383.5\ln\left(\text{M} - 4.075\right) \tag{32}$$

From the aforementioned correlations, 22 sets were created and used to calculate the critical properties and acentric factor of the heavy fraction of two reservoir fluids (**Table 7**). These sets were evaluated by comparing the experimental data and simulated PVT analysis of these fluids. The PVT experiment was simulated using the Peng-Robinson equation of state [12]:

$$P = \frac{RT}{v-b} - \frac{a}{v(v+b) + b(v-b)} \tag{33}$$

where R is the universal gas constant, T is the temperature and v is the molar volume.

Set	P_c and T_c	ω	Set	P_c and T_c	ω
1	Cavett [3]	Riazi and Alsahhaf [10]	12	Twu [7]	Riazi and Alsahhaf [10]
2	Cavett [3]	Edmister [2]	13	Twu [7]	Edmister [2]
3	Standing [5]	Riazi and Alsahhaf [10]	14	Riazi and Daubert [8]	Riazi and Alsahhaf [10]
4	Standing [5]	Edmister [2]	15	Riazi and Daubert [8]	Edmister [2]
5	Kesler and Lee [4]	Kesler and Lee [4]	16	Riazi and Daubert [8]	Riazi and Alsahhaf [10]
6	Kesler and Lee [4]	Riazi and Alsahhaf [10]	17	Riazi and Daubert [8]	Edmister [2]
7	Kesler and Lee [4]	Edmister [2]	18	Magoulas and Tassious [9]	Magoulas and Tassious [9]
8	Riazi and Daubert [6]	Riazi and Alsahhaf [10]	19	Magoulas and Tassious [9]	Riazi and Alsahhaf [10]
9	Riazi and Daubert [6]	Edmister [2]	20	Magoulas and Tassious [9]	Edmister [2]
10	Sim and Daubert [13]	Riazi and Alsahhaf [10]	21	Sancet [12]	Riazi and Alsahhaf [10]
11	Sim and Daubert [13]	Edmister [2]	22	Sancet [12]	Edmister [2]

Table 7. Sets used to characterize the properties of heavy oil fractions.

3. Results

In order to choose the most appropriate correlation for the calculation of the critical properties of heavy fractions of a reservoir fluid, 22 sets of correlations were used in the estimative of properties of the "cut" of fluids A and B [14]. The compositions of the fluids A and B are presented in **Tables 8** and **9**, respectively.

Fluid A contains approximately 40% of methane and 6% of hydrocarbons with 20 or more carbons, while fluid B contains 13% of heavy fractions and 27% of methane.

The PVT experiments were simulated using the Peng-Robinson equation of state. The results were compared with the experimental data, and the total relative deviations are shown in **Tables 10** and **11**.

The best correlation for fluids A and B is chosen based on the sum of the average deviation for each property (**Tables 10** and **11**). For fluid A, the correlation of Twu [7] led to the best results in terms of the critical properties while for the fluid B it was the correlation of Kesler and Lee [4]. For both fluids, the correlation of Riazi and Alsahhaf [10] was chosen to calculate the acentric factor (**Table 12**).

Figures 1–6 compare the experimental and calculated PVT data of fluid A using the Peng-Robinson equation of state with the critical pressure and temperature calculated from the correlation of Twu [7] and the acentric factor calculated from the Riazi and Alsahhaf [10] correlation. It is possible to note the excellent agreement between the experimental and simulated data. **Figures 7–10** show the same results for fluid B, with similar performance.

Component	Mole fraction (%)	Component	Mole fraction (%)	Component	Mole fraction (%)
N_2	0.390	nC_5	2.150	C_{13}	1.590
CO_2	0.300	C_6	2.790	C_{14}	1220
C_1	40.200	C_7	4.280	C_{15}	1.250
C_2	7.610	C_8	4.310	C_{16}	1.000
C_3	7.950	C_9	3.080	C_{17}	0.990
iC_4	1.190	C_{10}	2.470	C_{18}	0.920
nC_4	4.080	C_{11}	1.910	C_{19}	0.600
iC_5	1.390	C_{12}	1.690	$C_{20}+$	6.640

Table 8. Fluid A composition [14].

Component	Mole fraction (%)	Component	Mole fraction (%)	Component	Mole fraction (%)
H_2S	0.383	iC_5	1.937	C_{12}	0.02285
N_2	0.450	nC_5	2.505	C_{13}	2.364
CO_2	2.070	C_6	3.351	C_{15}	1.752
C_1	26.576	C_7	4.311	C_{16}	1.589
C_2	7.894	C_8	4.133	C_{17}	1.492
C_3	6.730	C_9	0.03051	C_{18}	1.263
iC_4	1.485	C_{10}	0.02033	C_{19}	0.812
nC_4	3.899	C_{11}	0.02635	$C_{20}+$	12.962

Table 9. Fluid B composition [14].

Set	Average B_o (%)	Average R_s (%)	Average ρ_o (%)	Average B_g (%)	Average Z_g (%)	Average γ_g (%)	Average P_b (%)	Total deviation (%)
1	1.98	5.64	6.00	2.33	2.55	1.74	11.07	31.31
2	2.75	9.69	7.87	2.04	2.08	1.04	22.43	47.91
3	5.21	15.60	23.08	1.72	1.14	1.40	38.58	86.74
4	7.36	15.89	18.63	2.53	2.79	1.71	8.83	57.73
5	2.06	6.88	6.48	2.25	2.42	1.54	14.90	36.51
6	2.01	6.08	6.13	2.29	2.48	1.53	12.61	33.13
7	2.42	8.42	7.20	2.13	2.22	1.23	19.27	42.88
8	5.97	11.28	15.58	2.08	2.09	0.40	16.98	54.38
9	5.12	9.58	15.81	2.18	2.22	0.74	18.37	54.04

Set	Average B_o (%)	Average R_s (%)	Average ρ_o (%)	Average B_g (%)	Average Z_g (%)	Average γ_g (%)	Average P_b (%)	Total deviation (%)
10	1.15	7.33	2.48	2.67	2.98	2.36	6.90	25.88
11	1.73	9.08	3.53	2.31	2.48	1.69	13.84	34.67
12	1.96	3.95	3.64	2.52	2.79	1.45	7.32	23.61
13	2.25	9.06	5.48	2.19	2.31	1.31	18.88	41.48
14	2.67	4.24	4.69	2.42	2.63	0.83	8.54	26.01
15	2.84	9.83	7.23	1.89	1.77	0.59	23.75	47.91
16	5.75	14.71	23.21	1.84	1.65	0.86	29.23	77.24
17	6.96	21.15	27.12	1.18	1.32	1.36	25.00	84.08
18	3.20	11.82	8.44	1.88	1.78	1.10	27.97	56.20
19	1.75	6.04	5.57	2.40	2.64	2.24	11.90	32.54
20	2.55	8.99	6.99	2.04	2.10	1.09	20.77	44.52
21	4.53	13.00	20.59	1.91	1.73	0.79	30.85	73.39
22	4.63	13.49	20.85	1.83	1.64	0.73	32.29	75.46

Table 10. Deviations between measured and calculated properties for fluid A.

Set	Average B_o (%)	Average R_s (%)	Average ρ_o (%)	Average Z_g (%)	Average P_b (%)	Total deviation (%)
1	0.56	10.87	5.17	2.65	0.56	19.80
2	1.73	8.50	7.18	3.14	17.86	38.41
3	3.52	26.85	28.74	3.60	38.09	100.80
4	3.41	29.22	27.19	3.29	28.32	91.42
5	1.30	8.00	3.60	2.74	5.44	21.07
6	0.59	9.16	2.82	2.60	3.54	18.70
7	1.80	7.39	4.28	2.92	11.61	27.99
8	4.42	28.11	19.98	3.23	11.99	67.73
9	3.64	26.31	20.31	3.18	15.25	68.69
10	0.53	10.68	4.32	2.58	6.35	24.47
11	2.21	5.79	1.54	2.83	7.63	20.00
12	0.97	7.90	2.04	2.50	21.16	34.57
13	2.95	4.92	0.81	2.76	7.61	19.04
14	0.81	8.78	1.61	2.71	16.27	30.18
15	3.51	6.76	2.18	3.27	17.56	33.28

http://dx.doi.org/10.5772/intechopen.71166

Set	Average B_o (%)	Average R_s (%)	Average ρ_o (%)	Average Z_g (%)	Average P_b (%)	Total deviation (%)
16	4.24	31.31	28.49	3.31	27.69	95.04
17	4.69	25.44	32.52	3.40	49.36	115.40
18	2.17	9.25	10.62	3.09	24.93	50.05
19	0.82	12.16	7.85	2.60	2.57	25.99
20	1.95	9.84	10.62	3.19	25.10	50.71
21	3.02	26.67	26.45	3.43	32.53	92.10
22	3.16	26.63	26.78	3.48	34.50	94.55

Table 11. Deviations between measured and calculated properties for fluid B.

Fluid	Set	P_c and T_c	ω
Fluid A	12	Twu [7]	Riazi and Alsahhaf [10]
Fluid B	6	Kesler and Lee [4]	Riazi and Alsahhaf [10]

Table 12. Best correlations for the critical properties and the acentric factor.

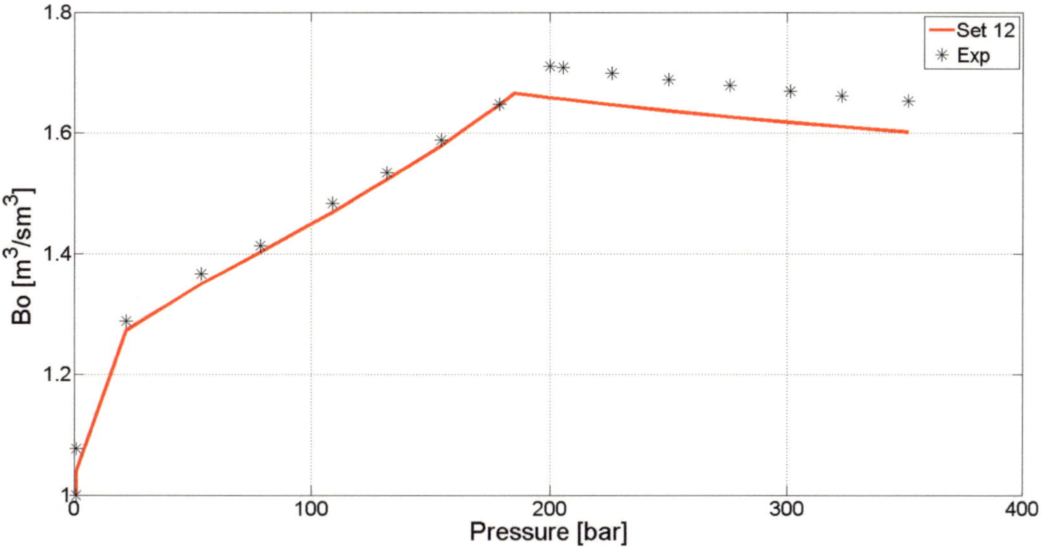

Figure 1. Fluid A—oil formation-volume-factor.

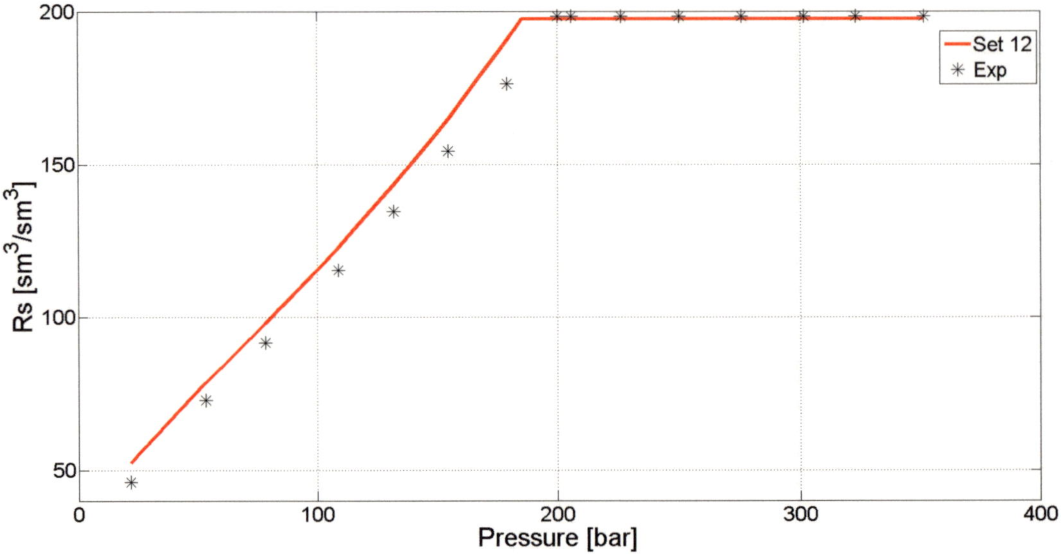

Figure 2. Fluid A—solubility ratio.

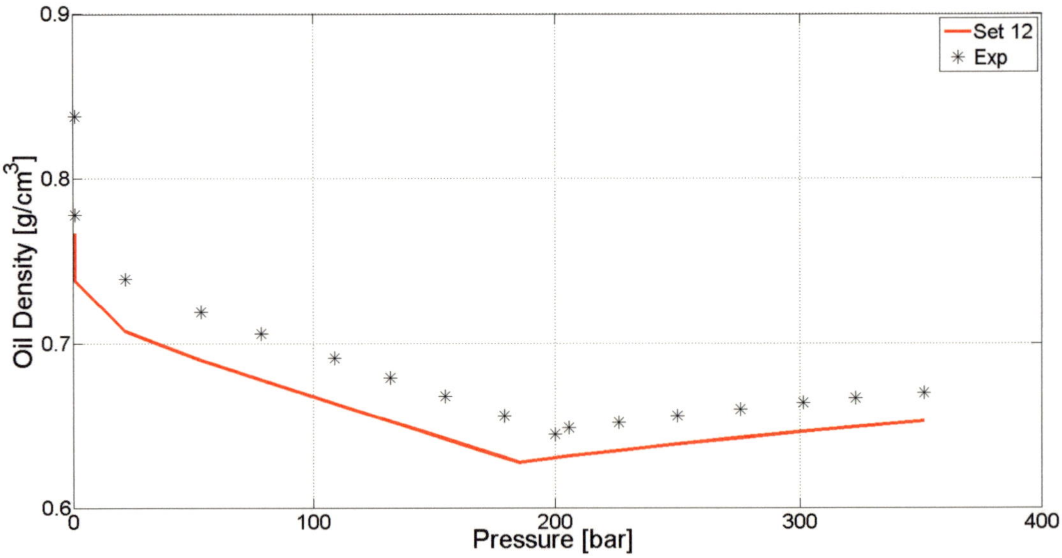

Figure 3. Fluid A—oil density.

http://dx.doi.org/10.5772/intechopen.71166

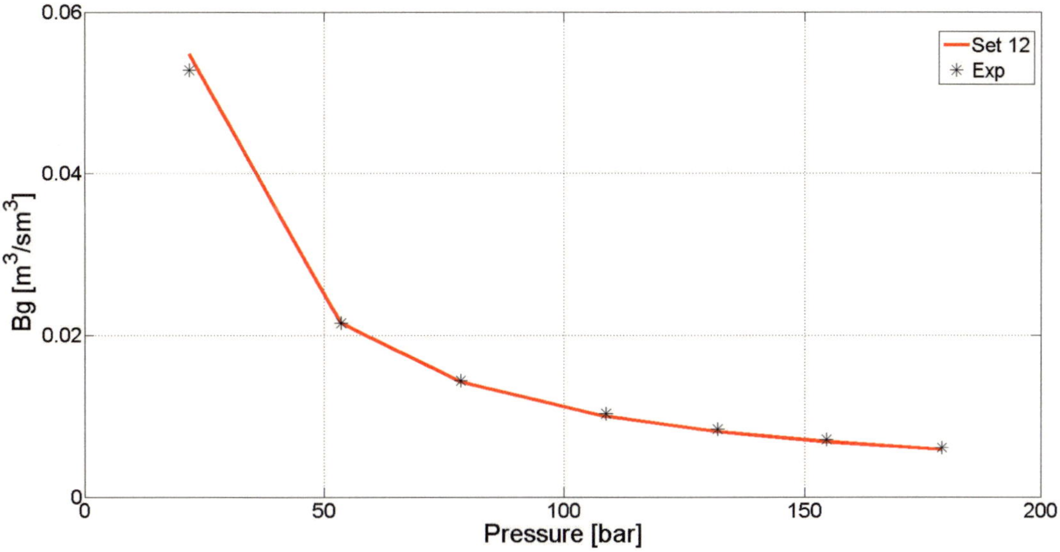

Figure 4. Fluid A—gas formation-volume-factor.

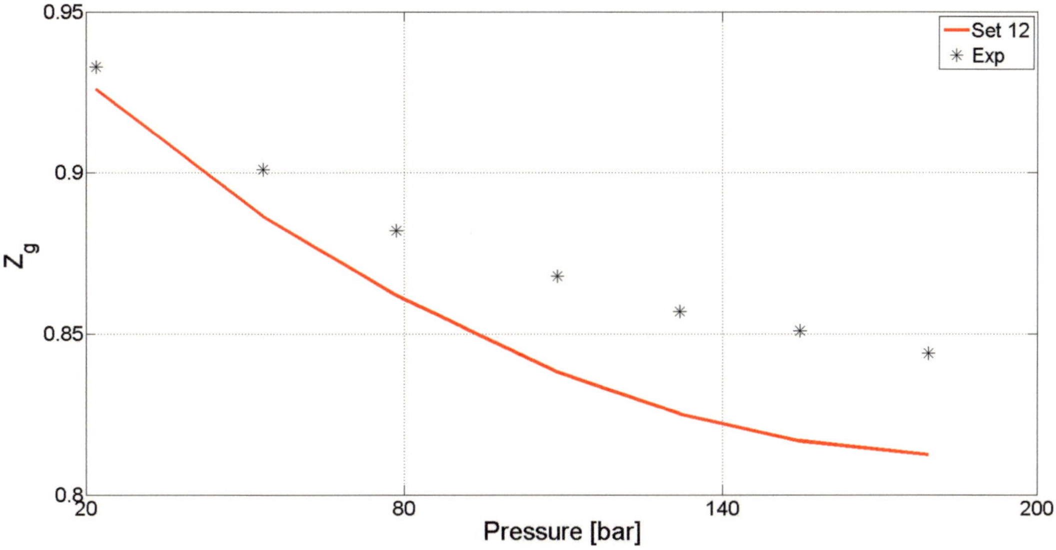

Figure 5. Fluid A—gas compressibility factor.

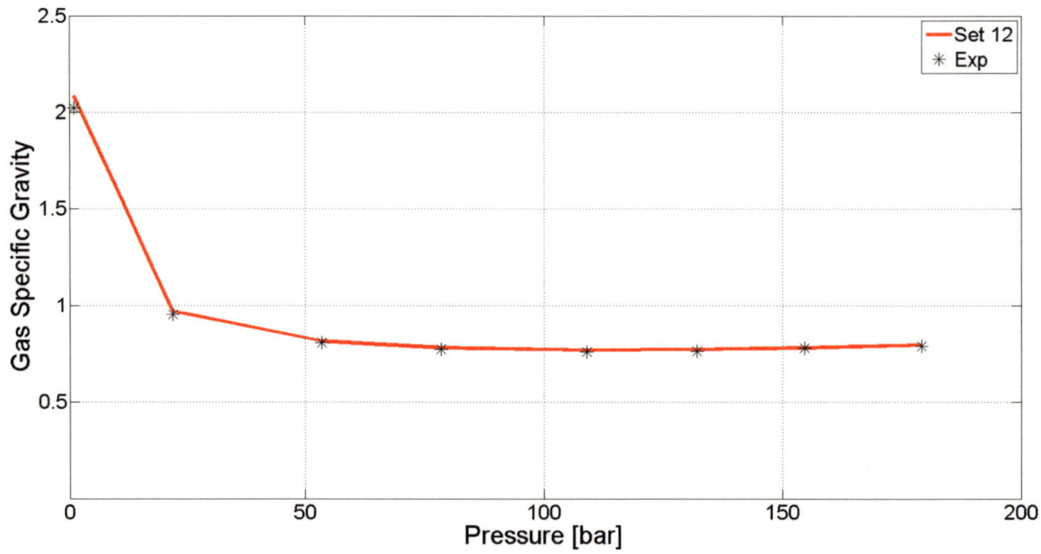

Figure 6. Fluid A—gas specific gravity.

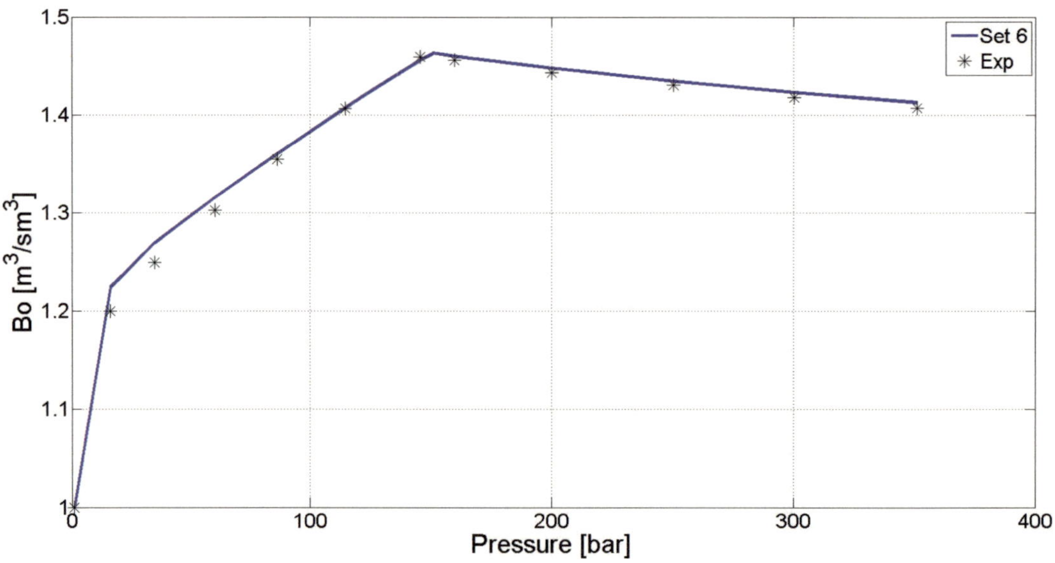

Figure 7. Fluid B—oil formation-volume-factor.

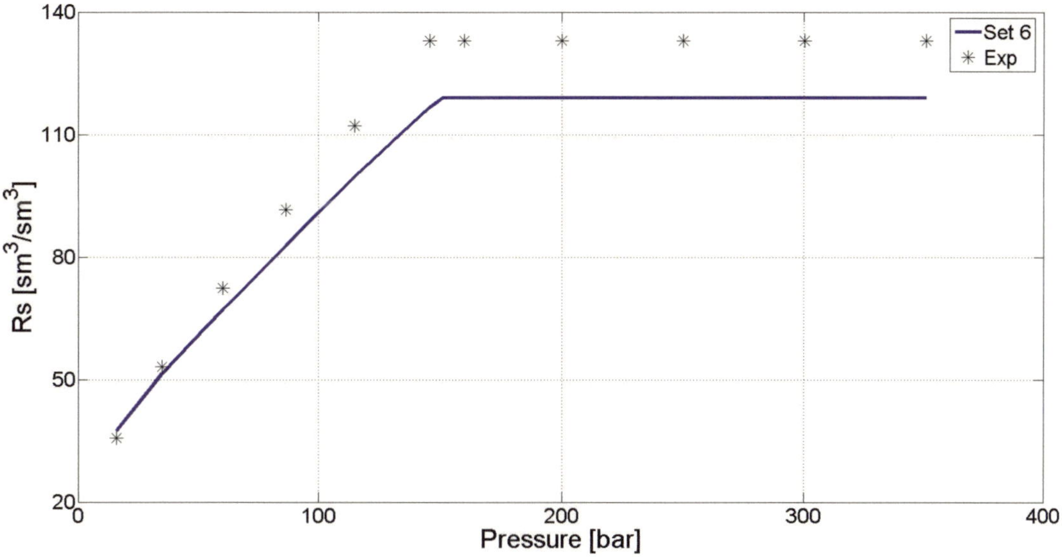

Figure 8. Fluid B—solubility ratio.

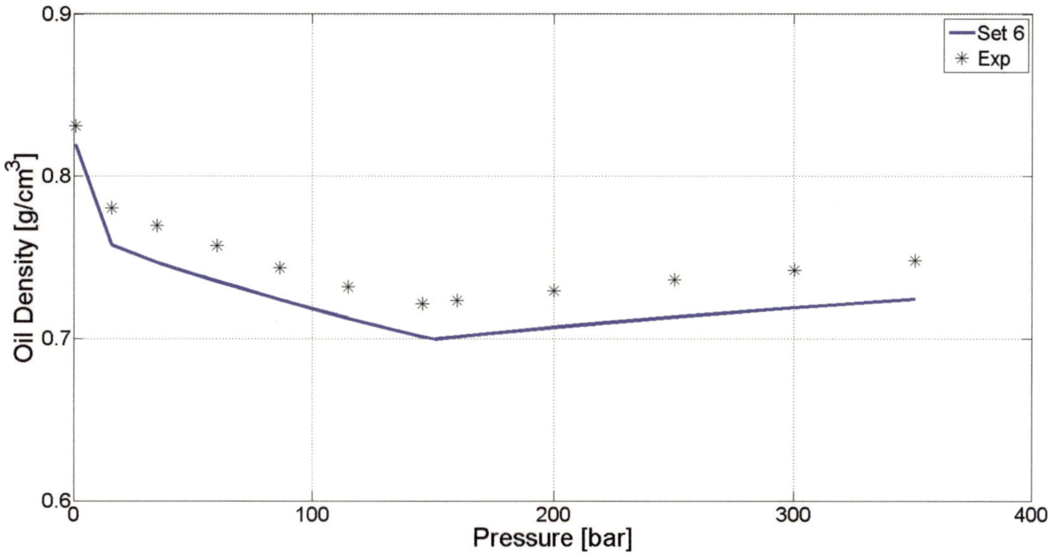

Figure 9. Fluid B—oil density.

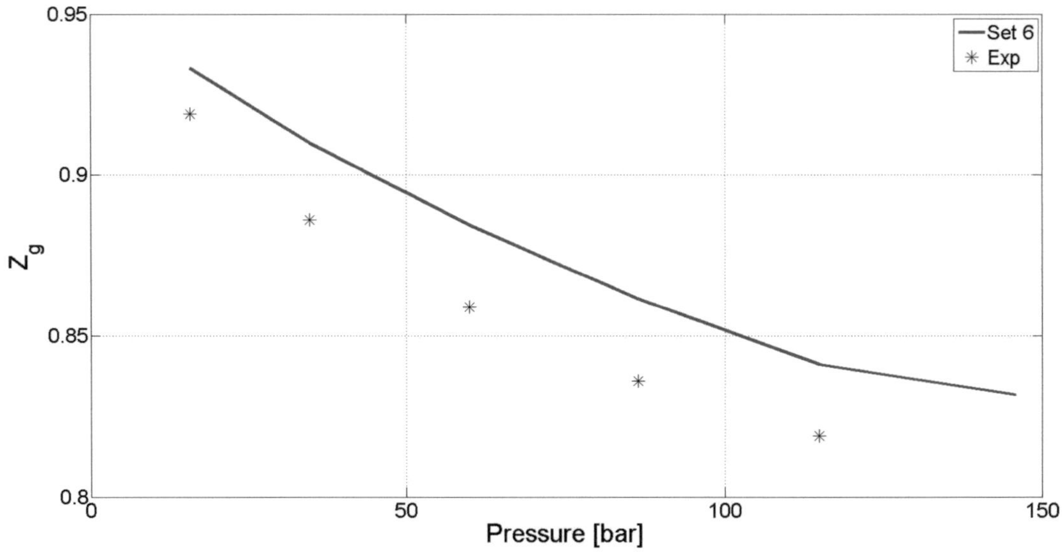

Figure 10. Fluid B—gas compressibility factor.

4. Conclusions

Overall, 22 sets of correlations to calculate the critical properties and acentric factors of heavy oil fractions were tested in order to simulate a PVT experiment of two reservoir fluids. The results obtained from the simulation were compared with experimental laboratory PVT data. The Twu [7] correlation showed the best performance for one case, while the Kesler and Lee [4] correlation led to better results for the other fluid. In both cases, the Riazi and Alsahhaf [10] correlation was used to calculate the acentric factor. These results show the importance of evaluating the correlations for each reservoir fluid since no model for the critical properties and acentric factor calculations can be applied in any case.

Nomenclature

B_g: gas formation volume factor [m^3/sm^3]

B_o: oil formation volume factor [m^3/sm^3]

K_w: Watson's characterization factor

M: molecular weight

P_b: bubble point pressure [psia]

P_c: critical pressure [psia]

T_b: normal boiling temperature [°R]

T_c: critical temperature [°R]

R_s: solubility ratio [m^3/sm^3]

Z_g: gas compressibility factor

v_c: critical volume [ft^3/lbmol]

ω: acentric factor

ρ_o: oil density [g/cm^3]

γ_g: gas specific gravity

Author details

Dacid B. Lacerda, Rafael B. Scardini, André P. C. M. Vinhal, Adolfo P. Pires* and Viatcheslav I. Priimenko

*Address all correspondence to: adolfo.puime@gmail.com

Laboratório de Engenharia e Exploração de Petróleo, Universidade Estadual do Norte Fluminense Darcy Ribeiro, Macaé, RJ, Brazil

References

[1] Danesh A. PVT and Phase Behavior of Petroleum Reservoir Fluids. Amsterdam: Elsevier Science; 1998

[2] Edmister WC. Applied hydrocarbon thermodynamics, part 4: Compressibility factors and equations of state. Petroleum Refinery. 1958;**37**:173-175

[3] Cavett RH. Physical data for distillation calculations—Vapor-liquid equilibria. API Proc., Sec. III. 1962;**42**

[4] Kesler MG, Lee BI. Improve prediction of enthalpy of fractions. Hydrocarbon Processing. 1976;**55**:153-158

[5] Standing MB. Volumetric and Phase Behavior of Oil Field Hydrocarbon Systems. Dallas: Society of Petroleum Engineers; 1977

[6] Riazi MR, Daubert TE. Simplify property predictions. Hydrocarbon Processing. 1980;**23**: 115-116

[7] Twu CH. An internally consistent correlation for predicting the critical properties and molecular weights of petroleum and coal-tar liquids. Fluid Phase Equilibria. 1984;**16**(2): 137-150

[8] Riazi MR, Daubert TE. Characterization parameters for petroleum fractions. Industrial Engineering & Chemistry Research. 1987;**26**:755-759

[9] Magoulas K, Tassios D. Thermophysical properties of *n*-alkanes from C_1 to C_{20} and their prediction for higher ones. Fluid Phase Equilibria. 1990;**56**:119-140

[10] Riazi MR, Alsahhaf TA. Physical properties of heavy petroleum fractions and crude oils. Fluid Phase Equilibria. 1996;**117**(1):217-224

[11] Sancet GF. Heavy fraction C_{7+} characterization for PR-EOS. In: SPE Annual Technical Conference and Exhibition. Anaheim, 2007

[12] Peng DY, Robinson DB. A new two-constant equation of state. Industrial Engineering Chemistry Fundamentals. 1976;**15**:59-64

[13] Sim WJ, Daubert TE. Prediction of vapor-liquid equilibria of undefined mixtures. Industrial & Engineering Chemistry Process Design and Development. 1980;**19**:380-393

[14] Pedersen KS, Christensen PL. Phase Behavior of Petroleum Reservoir Fluids. Boca Raton: Taylor & Francis; 2006

Chapter 6

Characterization of Crude Oils and the Precipitated Asphaltenes Fraction using UV Spectroscopy, Dynamic Light Scattering and Microscopy

Ernestina Elizabeth Banda Cruz,
Nohra Violeta Gallardo Rivas, Ulises Páramo García,
Ana Maria Mendoza Martinez and
José Aarón Melo Banda

Additional information is available at the end of the chapter

http://dx.doi.org/10.5772/intechopen.70108

Abstract

Analysis of crude oil composition provides important information that impacts on the recovery, handling, and transportation of hydrocarbons. Crude characterization also provides data in the analysis of geochemistry of the source of origin. Crude oil characterization by optical methods is usually difficult because of its dark color; however, those characterizations are crucial because they give information that can affect some analysis procedures. Ultraviolet-visible (UV-vis) spectroscopy is a simple and practical technique that allows the characterization of crude oil through dilution in solvents. A comparative study of crude oil solutions contrasted with their asphaltene fractions was performed. Each solution was analyzed in triplicate, on a UV-vis spectrophotometer. Calibration curves for both raw solutions showed no significant variations, indicating stability. Additionally, the results of dispersion and migration phenomena indicated stability only for crude oil solutions. The aggregate size dispersion was different for each type of crude and varied with respect to time. Scanning electron microscopy (SEM) and transmission electron microscopy (TEM) showed the type of morphology present for each type of asphaltene.

Keywords: heavy oil, asphaltenes, UV-vis spectroscopy, aggregation, dynamic light scattering, SEM, TEM

1. Introduction

1.1. General aspects of crude oil

The name crude oil is used for naturally occurring and unprocessed petroleum. Crude oils are mixtures of many componentes, hydrocarbons (mostly alkanes, cycloalkanes, and aromatics) and other organic compounds containing nitrogen, oxygen, sulfur, and traces of metals [1].

Traditionally, these compounds are classified in several complex groups. The most popular is the method of separation into four groups: saturates, aromatics, resins, and asphaltenes (SARA) [2].

In refineries, one of the main criteria for control process is the content of asphaltene. **Table 1** shows the composition differences normally found in conventional oils, heavy oils, and residues.

1.2. Importance of the characterization of crude oil and asphaltenes

Asphaltenes are petroleum hydrocarbons with extremely complex molecular structure containing sulfur (0.3–10.3%), oxygen (0.3–4.8%), nitrogen (0.6–3.3%), and metal elements such as Fe, Ni, and V in small quantities [4, 5].

The association of metal porphyrins with asphaltenes is probably the reason behind the great difficulty in extracting or separating metal porphyrins from oil or waste oil. Metal porphyrins entrapment by asphaltenes was predicted long ago when the presence of free radicals in asphaltenes was discussed [6]. The porphyrins carry nickel and vanadium in the same proportions as established in the depositional environment. During generation and migration, these porphyrin structures are incorporated into the crude oil and carried along to the reservoir, preserving this information about the proportions of nickel and vanadium in the source rock. The nickel and vanadium content are also important in establishing a value for the oil. Nickel and vanadium will poison catalysts during the refining process. As such, high concentrations of nickel and vanadium in crude oil will reduce the market value of the oil [7].

On the other hand, the aromatic moiety, through π-π^* and dipole interactions, was thought to be one of the dominant contributors to asphaltenes self-association. Hence, proper characterization

Sample	Typical composition range (wt%)		
	Asphaltene	Resin	Oil fraction[1]
Conventional oil	<0.1–12	3–22	67–97
Heavy oil	11–45	14–39	24–64
Residue	11–29	29–39	<39

[1]Correspondent to the fraction composed of saturate and aromatic.

Table 1. Composition for conventional oil, heavy oil and residue (Adapted from Speight Ref. [3]).

of the asphaltene molecular structure and its aromaticity is fundamental in understanding the self-association phenomenon [8].

The compositional analysis of crude oils provides important information that impacts petroleum exploitation at every step along the value chain, from exploration through production, transportation, and refining. In upstream operations, geochemical inversion of crude oil to the provenance of petroleum is invaluable in petroleum system analysis, especially where source-bed information may be difficult or impossible to obtain during exploration. In reservoir management, crude oil composition is essential for fluid correlation and the construction of functional, effective static reservoir models [9].

Heavy oils usually display a greater content of asphaltenes and resins than conventional oils, which directly impacts recovery, transport and refining processes. Colloidal properties of asphaltenes and resins have been the subject of intense debate in the literature [10].

In refineries, some issues include the reduction of the overall rates of hydrotreating reactions, deactivation of catalysts form surface deposition, and increased coke formation (asphaltenes are a coke precursor). In general, asphaltene precipitation limits the ultimate level of conversion. From this operational definition, asphaltenes are anticipated as a group of complex compounds, which are highly polydispersed and cannot be absolutely prescribed by some simple physicochemical parameters. There are fewer large aggregates and narrower distributions once asphaltenes are in infinitely diluted systems of higher temperatures and better solvents; however, they still exhibit a real polydispersity. The average molecular weight (MW) is not necessarily a good parameter to characterize asphaltenes, simply because asphaltenes are defined through their solubility in aliphatic hydrocarbons [11]. Asphaltenes, usually derived either from coal or from petroleum vacuum distillation residues, are wide spread, and a class of compounds that are by their solubility, insoluble in n-heptane, and soluble in toluene [12, 13].

1.3. Types of characterizations for crude oil and asphaltenes

Depending on the origin of the crude oil from which they are precipitated, asphaltenes can exhibit wide differences in composition and structure. The amount and length of the alkyl side chains as well as the number of aromatic rings may change in such a way that the variations of molecular weight and aromaticity factor of asphaltenes are notorious [14].

Crude oil is a complex mixture of compounds difficult to characterize using optical methods because of its dark color. Ultraviolet-visible (UV-vis) spectroscopy is simple, practical, and inexpensive. It allows the characterization of crude oil through dilution in solvents like cyclohexane. Recently, a study was conducted on the development of a spectroscopic method for the quantification of functional groups characteristic of asphaltenes, making direct quantification from oil a simple and practical method [15]. The strongest absorption of metal porphyrin occurs in the vicinity of 425 nm and is called Soret band [16]. In addition to the well-defined amounts of energy to increase its vibrational and rotational moiety, a molecule can also absorb some energy to increase the excitation of its electrons. The energy changes involved are considerably greater than those involved in vibrational and rotational energy changes and correspond to radiation in the ultraviolet (200–400 nm) and visible (400–750 nm) regions [17].

Other research about combustion characterization and kinetics of four different origin crude oil samples was determined using thermogravimetry-differential thermal analysis (TGA-DTA) and thermogravimetry-Fourier transform infrared (TGA-FTIR) and thermogravimetry-mass spectrophotometry (TGA-MS) techniques [18].

Four crude oil samples from the oil fields Intisar A, Intisar D, and Intisar E (Sirte Basin, Libya) were investigated in order to define depositional environment, lithology, thermal maturity, and geologic age of the corresponding source rocks. Saturated biomarkers (n-alkanes, isoprenoids, steranes, and triterpanes) were determined using gas chromatography-mass spectrometry (GC-MS) and gas chromatography-mass spectrometry-mass spectrometry (GC-MS-MS). Aromatic hydrocarbons (phenanthrene, methylphenanthrenes, methyl dibenzothiophenes, and trimethyl naphtalenes) were analyzed by GC-MS [19].

***A large amount of research is currently oriented to the study of asphaltene aggregation in different solvents [20]. The aggregation of asphaltene solutions in toluene was investigated. The particle diameter grows with increasing concentrations of asphaltene, and the increase rate varies with asphaltene concentration. This trend proves the presence of interactions that promote the formation of aggregates. With increasing concentration of asphaltene the formation of small particles is no longer dominant since, at higher concentrations, strong attractive interactions from small particles may cause the formation of larger particles [21].

Most research has been focused on native asphaltenes. Some researchers determined the molecular weight, density, and solubility parameter distributions of asphaltenes from thermo and hydrocracked oils [22]. Other researchers precipitated asphaltene fractions from heavy oil using a series of solvents that are similar from the thermodynamic viewpoint but are discriminated by their solubility parameters [2].

Asphaltene aggregation is usually discussed by means of the appearance of flocculation, which is regulated by thermodynamics of phase transition and can be observed by optic and spectroscopic techniques. However, growth and precipitation of aggregates not only depend on thermodynamic parameters (e.g., concentration and temperature) but also kinetic factors [23].

Recently, a study was conducted on the development of a spectroscopic method for the quantification of functional groups characteristic of asphaltenes, making direct quantification from oil a direct and practical method [24]. Nondiluted light and heavy crude oils and their blends were studied with additives of pyrolysis oil and Kemelix, both acting as asphaltene dispersants [25].

The classic form used to study the composition of an extremely complex mixture such as oil is to separate it into discrete simple fractions that can be analyzed, but the main objective of this work was to develop a spectroscopic method for the quantification of functional groups characteristic of asphaltenes, making a direct and practical method for quantification from oil. This investigation was conducted as a comparative study of two heavy crude oil solutions (HC1 and HC2) and asphaltene precipitated (AHC1 and AHC2) using UV-vis spectroscopy and dynamic light scattering (DLS). Particle size of the aggregates was performed by scanning electron microscopy (SEM) and high-resolution transmission electron microscopy (TEM).

2. Methodology

2.1. Materials

Cyclohexane with 99.9% of purify supplied by Tedia, analytical balance, OHAUS Adventurer Pro, AV264C model and UV-vis Spectrophotometer, model GBC Cintral 303 were used for the quantification of characteristic functional asphaltene groups in crude oil. A dynamic light scattering (DLS) Zetasizer Nano Malvern for determining the particle size and Heavy Fuel Formulation Classic Turbiscan and software Turbisoft Classic for analyzing crude oil migration phenomena were used.

Table 2 shows SARA fraction of HC1 and HC2 heavy crudes determined by ASTM-D2007-98 with 18°API density, which was used in this work.

2.2. Extraction of asphaltene from heavy crude oils

Asphaltenes were precipitated by ASTM D2007-80 standard method (American Society for Testing and Materials [ASTM], 1983).

2.3. UV-visible spectroscopy analysis of the crude-solvent solutions

Beginning with a standard 100 mg L^{-1} solution of crude in cyclohexane, dilutions of 10–100 mg L^{-1} were prepared. Successively and in triplicate, each dilution was analyzed on a UV-vis spectrophotometer to obtain the different absorbance (A) signals vs wavelength (λ), in the range of 200–450 nm. The information was graphed to obtain an absorbance vs concentration curve, the equation of a straight line and the value of the coefficient of determination (R^2). For each concentration, the highest absorbance peak was sought, which was important for the calibration curves.

2.4. UV-visible spectroscopy analysis of the asphaltenes-solvent solutions

Based on a standard solution of 100 mg L^{-1} asphaltenes in cyclohexane, solutions 10–100 mg L^{-1} were prepared. Successively and in triplicate, each solution was analyzed on a UV-vis spectrophotometer for the different absorbance signals (A) versus wavelength (λ), in the range of 200–600 nm. The obtained data were plotted to graph a curve to show variations of absorbance

SARA fraction	HC1	HC2
Saturated	21.66	23.69
Aromatic	30.57	38.74
Resins	24.35	14.06
Asphaltenes	23.41	23.51

Table 2. SARA fraction in weight % of the analyzed crude oils.

versus concentration. The equation of a line was determined, and the coefficient of determination (R^2) was established. For each concentration, the highest absorbance peak was sought, which was important for the calibration curves.

2.5. Dynamic light scattering analysis (DLS) of crude oils and precipitated asphaltenes solutions

DLS analysis was conducted to determine the state of aggregation and stability of crude oil and asphaltenes in cyclohexane from a standard solution of 100 mg L^{-1}, performing triplicate measurements. For both solutions, ultrasonic vibration was applied for 20 min.

2.6. Crude oils migration phenomena analysis

Migration phenomena in crude oil were determined by measuring the variation of the transmittance versus the height of the signal for each sample. These measurements were carried out as a function of time, as the mechanism of this instrument is to take readings of transmitted and backscattered light at a maximum sample height of 80 mm. This scan can be repeated with a programmable frequency to obtain a macroscopic fingerprint of the sample. It was a 2-h analysis (12 scans at intervals of 5 min) and then, 1 min each 24 h for a period of 30 days. Experiments were performed at room temperature (~25°C).

2.7. Scanning electron microscopy and energy dispersive X-ray spectroscopy

Microstructure and surface elemental characterization of asphaltenes were performed through scanning electron microscopy (SEM) by using a JEOL JSM 6390-LV. Samples were dispersed and adhered to a double-sided graphite tape placed over a brass sample holder.

2.8. High-resolution transmission electron microscopy

The shape and size of asphaltene nanoparticles were examined through high-resolution transmission electron microscopy (HTEM) by using a JEOL JEM-2200FS equipment. Asphaltenes suspensions in ethanol were prepared by applying 5 min of ultrasonication, and one drop of each solution was placed on the grid and left to dry for 10 min with an incandescent light lamp.

3. Results and discussion

3.1. UV-vis spectroscopy analysis of crude-solvent and asphaltene-solvent solutions

By using UV-vis spectroscopy, the analyses of crude-solvent and asphaltene-solvent solutions were possible. Calibration curves were constructed for both solutions. In approximately 120 min of UV-vis spectroscopy analysis, no significant changes were detected.

Figures 1 and **2** show the wavelength versus absorbance plot for crude-solvent solutions in cyclohexane, varying their concentration from 10 to 70 mg L^{-1} for HC1 and HC2 oils. In the operating UV-vis region of the test, three signals were observed and identified: at approximately

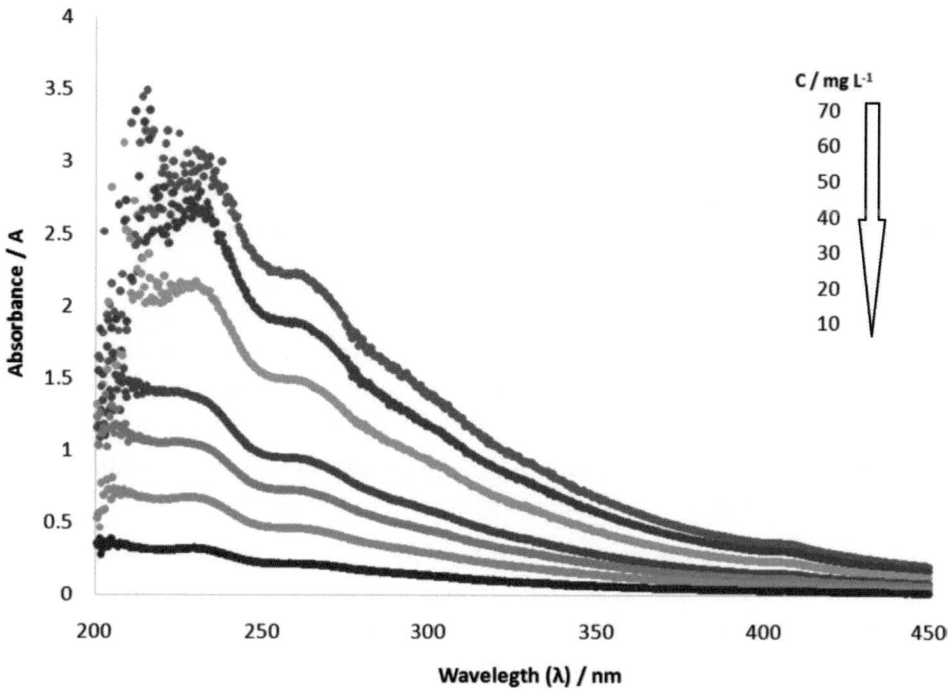

Figure 1. UV spectrum for HCl oil solutions, at different concentrations (decreasing order from 70 to 10 mg L^{-1}).

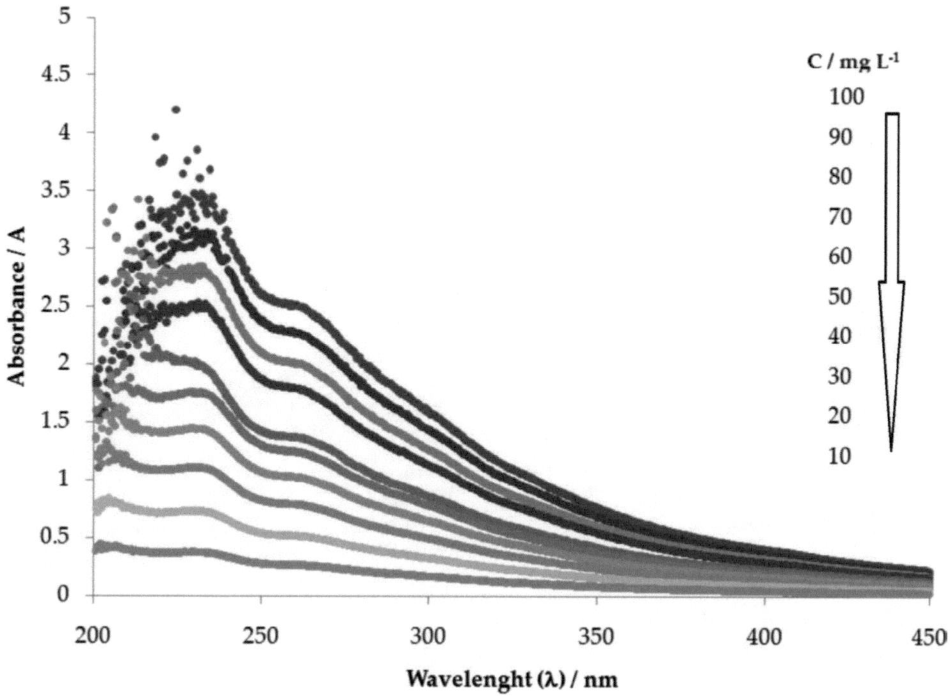

Figure 2. UV spectrum for HC2 oil solutions, at different concentrations (decreasing order from 100 to 10 mg L^{-1}).

230 nm, the signal corresponding to benzenic compounds; at 260 nm, the one of naphthenic compounds; and at 410 nm, the Soret band, showing information of metal porphyrin compounds [26–30] for HC1 oil and only the first two for HC2 oil. It was noticed that, as the oil concentration in the solutions decreased, the intensity of the three signals did the same, as effect of concentration for both crudes. Nevertheless, maximum absorbance signals could not be detected neither above 70 mg L^{-1} concentrations for HC1 oil due to noise nor below 10 mg L^{-1} because of the low concentration of HC1 oil, which precludes the detection of species by this technique [26].

Due to a smaller amount of resins in the HC2 oil composition, it was possible to display signals from 100 mg L^{-1}. For HC2 oil solutions above the concentration of 100 mg L^{-1}, it was not possible to detect signals due to the presence of noise. Below the concentration of 10 mg L^{-1}, no signs were present.

Figures 3 and **4** show the absorbance versus concentration graph for asphaltene solutions (AHC1 and AHC2), varying the concentration from 10 to 100 mg L^{-1}.

For AHC1 (**Figure 3**), three signals were detected: the one attributed to benzenic compounds at 235 nm, naphtenic compounds at 262 nm, and the Soret band at 410 nm, which corresponds to metal porphyrins present in asphaltene solutions [26, 30, 31]. For AHC2 solutions (**Figure 4**), only the two signals were detected at 235 and 262 nm. With this information, calibration curves for each compound and their respective linear equations were calculated.

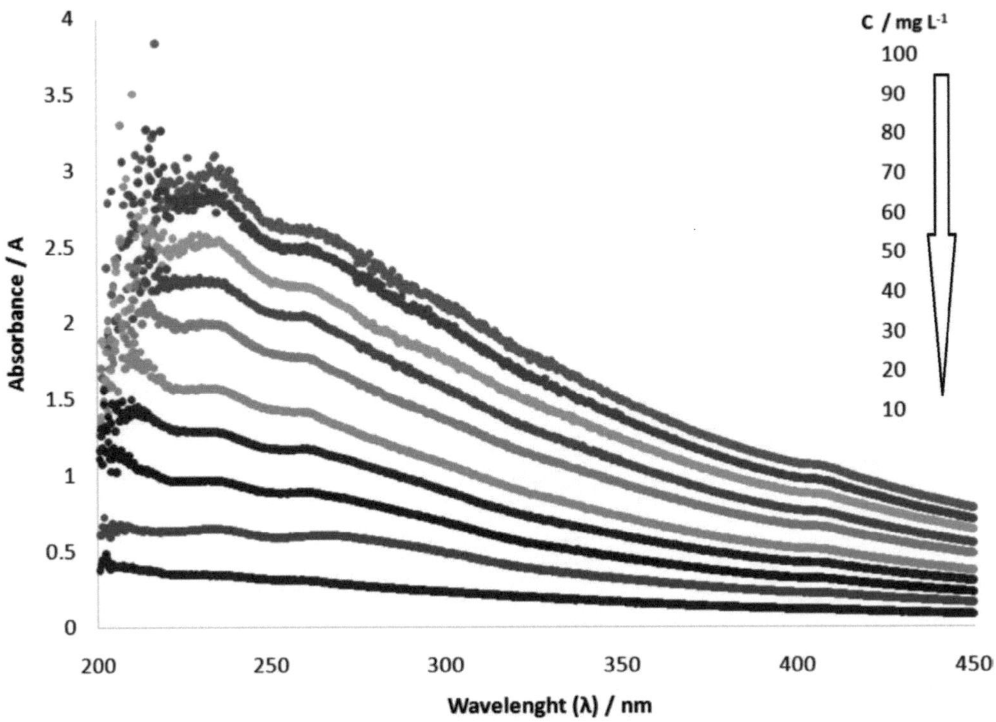

Figure 3. UV spectrum for AHC1 solutions, at different concentrations (decreasing order from 100 to 10 mg L^{-1}).

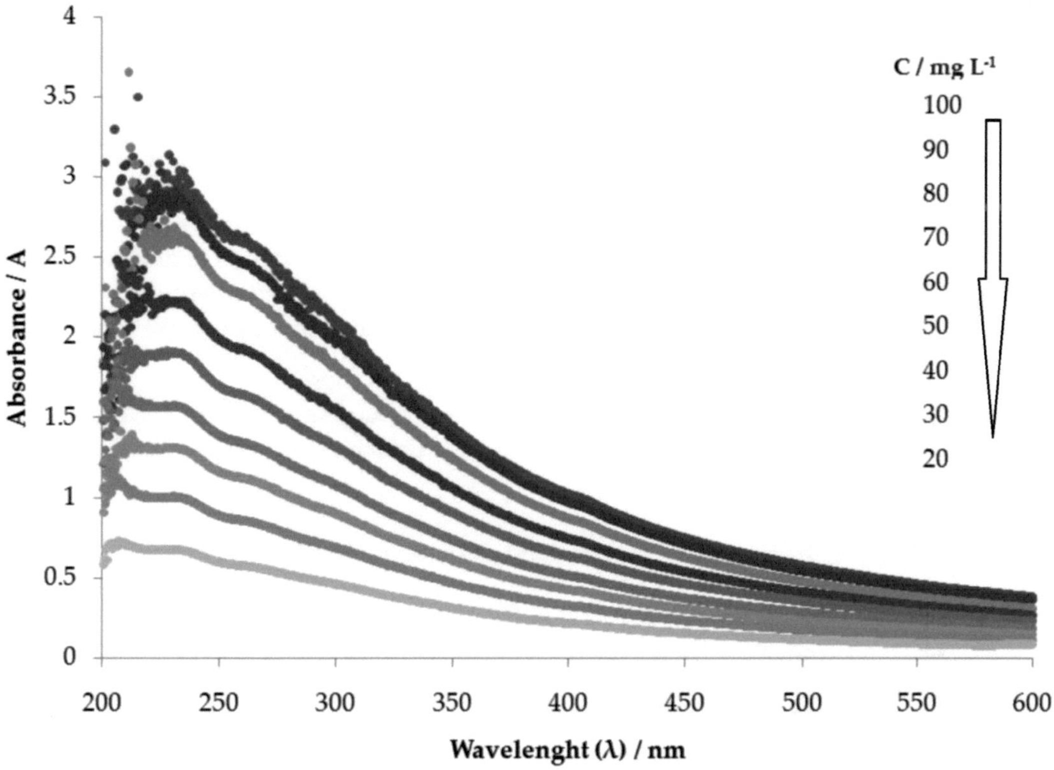

Figure 4. UV spectrum for AHC2 solutions, at different concentrations (decreasing order from 100 to 20 mg L^{-1}).

For concentrations under 50 mg L^{-1} of AHC1 solutions, no Soret band was found, and similarly, no signals were detected for samples prepared at concentrations below 10 mg L^{-1}, due to the low concentration of asphaltenes; therefore, it was not possible to detect species by this method. As mentioned earlier, signal detection was impossible for HC1 oil solutions at concentrations exceeding 70 mg L^{-1}; however, for asphaltene solutions, the three signals were found, even above those concentrations. This phenomenon is attributed to the fact that crude is a complex mixture of various compounds including saturates, asphaltenes, resins, and aromatics, and as the saturation of a crude solution increases, noise becomes more evident in the spectrum, hindering the signal recognition. It was also observed that the amplitude of the maximum absorbance signals for crude solutions was more intense compared to absorbance maximum for asphaltene solutions. That is to say, higher absorbance maximum was observed when working with whole crude, because of the overlapping of individual absorbances, corresponding to many of the different components of crude. That complex composition includes the saturate fraction, which is nonpolar and is composed of linear alkanes, branched alkanes, and cycloalkanes [30]. The saturates tend to absorb strongly in the deep UV region, with high-frequency electronic transitions, as their electrons are tightly bound and require more incident energy to be excited. Meanwhile, aromatic hydrocarbons form structures of one or more rings, where multiring structures are often referred to as polycyclic aromatic hydrocarbons. Similar to saturates, a redshift is noted as complexity increases, and more rings are added to the molecule [32, 33]. Bathochromic shift exists in signals of benzenic and naphtenic compounds for both asphaltene solutions at longer wavelengths because of more interactions between asphaltenes caused by the absence of resins.

It was possible to obtain signals at the concentration of 100 mg L^{-1} for AHC2 solutions. For concentrations below 20 mg L^{-1}, it was not possible to detect signals.

Figure 5(a) and **(b)** show the calibration curves of HC1 and HC2 oil solutions, respectively. Coefficient values for each compound are presented, since little dispersion of data was found. The HC1 oil solutions at concentration of 40 mg L^{-1} and HC2 oil solutions at concentration of 60 mg L^{-1} appear to be slightly deviated, which can be attributed to the presence of aromatics that favor aggregation in crude.

Figure 6(a) and **(b)** show the calibration curves of asphaltene solutions for HC1 and HC2, respectively. Coefficient values for each compound are presented. No dispersion of data was found.

3.2. Study of aggregate dispersion in crude and asphaltene solutions

DLS analysis was carried out in order to evaluate the average size of crude and asphaltene solutions in cyclohexane over time. **Figure 7(a)** shows the behavior of average size particles (Z_{AV}) over time (t) for HC1 oil solutions in cyclohexane at 100 mg L^{-1}. Z_{AV} was plotted from t = 0 min, when the particle size was 51.16 nm, to t = 100 min, when the average aggregate size showed an increase to 72.27 nm. In general, this behavior corroborates the instability of the unfractionated HC1 oil, although that its components coexist in equilibrium, which permits to perform analyses with good and reproducible results. Recently, a study was conducted to evaluate the aggregation state of Mexican crude oil solutions from two different sources and SARA compositions, by using dynamic light scattering (DLS) [34].

Figure 7(b) shows Z_{AV} (nm) behavior of HC2 oil solutions versus time. During the first 40 min, there was a Z_{AV} increase from 173.16 to 542 nm. After 40 min, Z_{AV} stabilized in the range

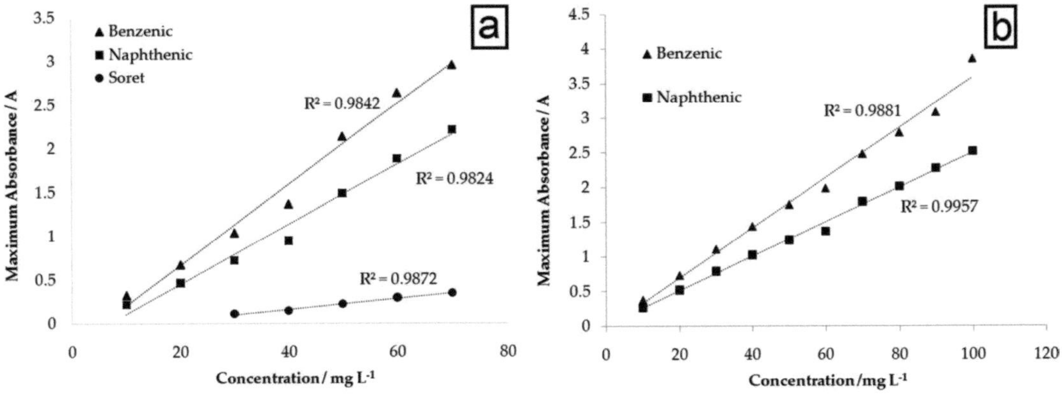

Figure 5. Calibration curves for benzenic, naphthenic and petroporphyrinic compounds in heavy crude oil solutions. (a) HC1 and (b) HC2.

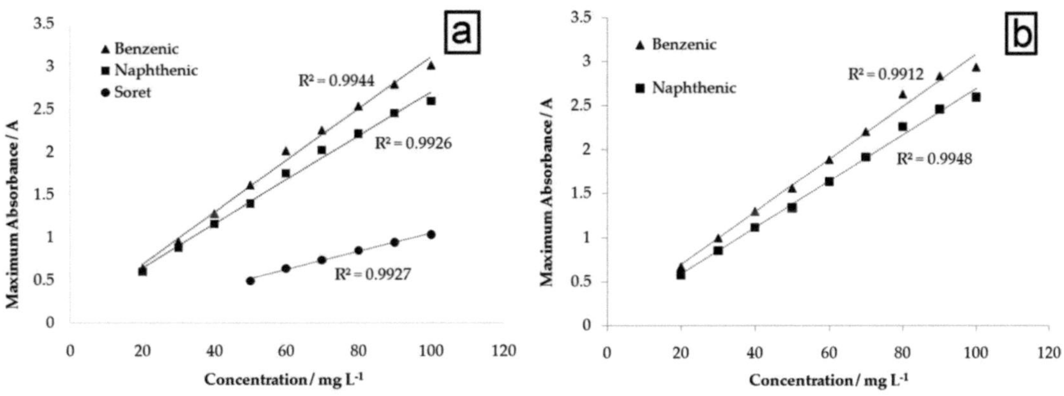

Figure 6. Calibration curves for benzenic, naphthenic and petroporphyrinic compounds in asphaltene solutions. (a) AHC1 and (b) AHC2.

of 563–624 nm. Molecules of the asphaltene fraction are dispersed in crude oil as stabilized colloids, and they can form aggregates by steric effect according to their sizes, polar characteristics, and solvent polarity [35].

The solutions of HC2 oil showed bigger size of aggregates in comparison with the solutions of HC1 oil, due to a smaller amount of resins present in the HC2 oil as confirmed by SEM. Banda et al. [34] studied different crude oil concentrations by using DLS. Results indicate that the aggregation state of the studied solutions is influenced by crude oil composition, specifically by the amount of resins. The resins are less soluble in n-alkanes than in cyclohexane and therefore are more likely to interact with asphaltenes to keep them in the form of small aggregates [5].

In the case of asphaltene solutions for both oils, the average aggregate size changed so rapidly that it was impossible to determine it. The aggregate size instability of asphaltenic solutions is a result of the absence of resins, which are known to favor stability [1, 37]. Therefore, the average aggregate size was no longer uniform, and unlike the behavior observed in unfractionated crude oil, the values corresponding to asphaltene solutions did not fit any tendency [1, 25–38].

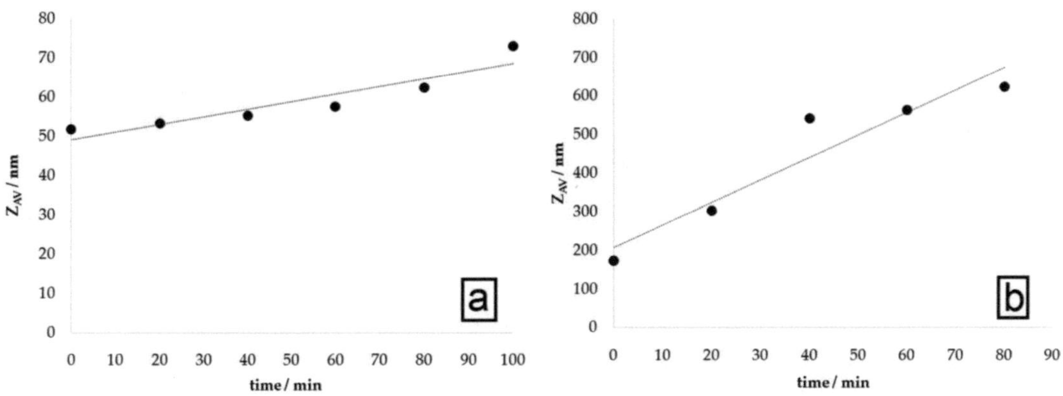

Figure 7. The average aggregate size (Z_{AV}) over time for (a) HC1 and (b) HC2 oil solutions.

3.3. Scanning electron microscopy (SEM)

Figure 8(a) and **(b)** shows a SEM micrograph of AHC1 asphaltenes. In the figures, a slightly porous surface can be observed. It is possible to appreciate agglomerates of various shapes and sizes. The larger agglomerates correspond to asphaltenes. At the top of the **Figure 8(a)**, large agglomerate of asphaltene is presented with lamellar morphology. On the surface, smaller agglomerates with smooth surface that correspond to resins can be noted.

Figure 8(c) and **(d)** shows a SEM micrograph of AHC2 asphaltenes, where a surface with lower precipitation degree is presented. SEM micrographs show stacked sheets of asphaltenes over the surface.

Two types of asphaltenes extracted from crude oil of different sources and SARA compositions were studied, by using SEM and HTEM. Significant differences in morphology and aggregate size for each type of asphaltene were noted. Results from these techniques reveal particulated low-porosity smooth surfaces due to the presence of resin, as well as the determination of the characteristic elements found in asphaltenes [34].

3.4. High-resolution transmission electron microscopy (HTEM)

Figure 9(a) and **(b)** shows AHC1 crude oil asphaltenes HTEM micrographs. The presence of asphaltene aggregates is constituted by nanometric particles. The presence of agglomerates of various sizes and shapes is found. Differences in image contrast are probably because of mate-

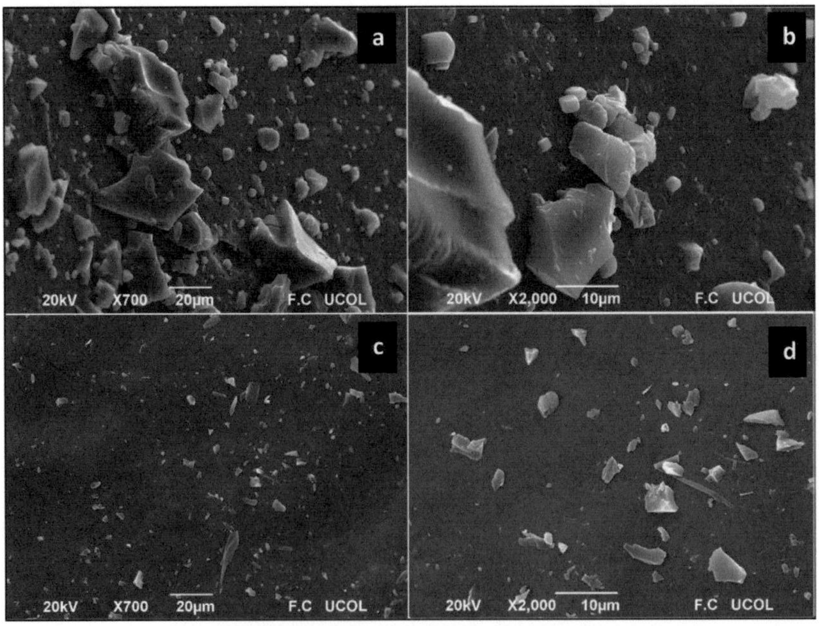

Figure 8. Morphology of (a and b) AHC1 and (c and d) AHC2 crude oil asphaltenes.

http://dx.doi.org/10.5772/intechopen.70108

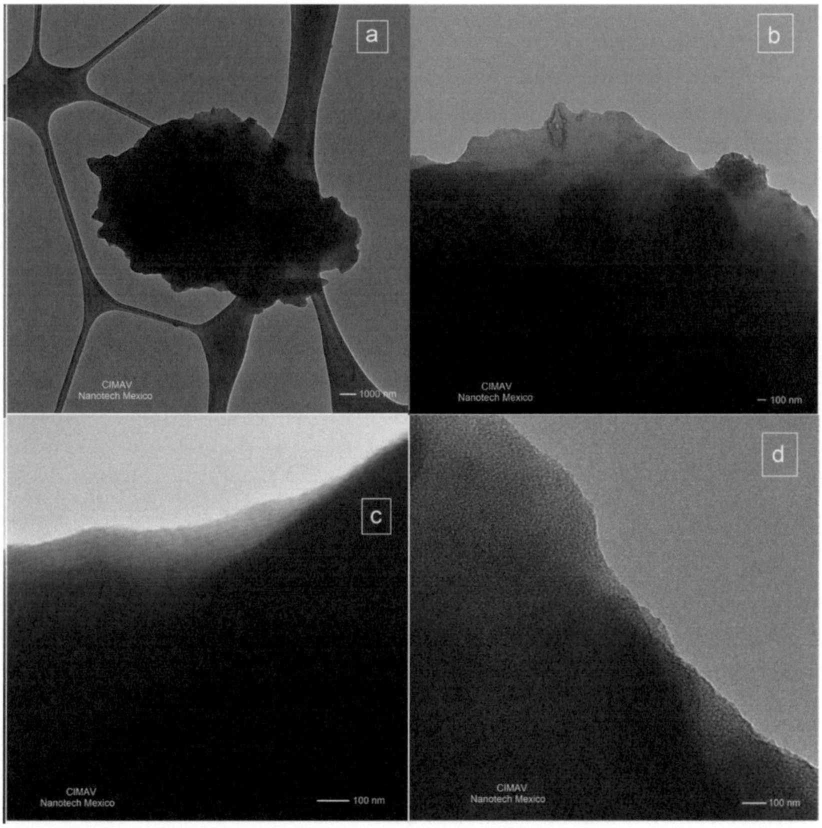

Figure 9. Morphology of AHC1 (a and b) and AHC2 (c and d) crude oil asphaltenes.

rial superposition for both images [39]. **Figure 9(a)** shows the grid of Cu/C with the sample. Agglomerates of approximately 1000 nm in size are presented. At the top of **Figure 9(b)**, layered structures are presented. In the center of micrograph, agglomerates with regular edges and approximately 500 nm sizes are found. At the bottom of the figure, there are agglomerates greater than 100 nm size.

Figure 9(c) and **(d)** shows AHC2 crude oil asphaltenes HTEM micrographs. **Figure 9(c)** proves the existence of crystalline zones in AHC2 asphaltene particles. Reticular stripes appear at the bottom of the image, indicating the nanocrystalline limits of the stacked layers [13]. **Figure 9(d)** shows the aggregates are constituted by nanometric particles. The presence of lamellar morphology is observed. There are stacked sheets at the top of the image, and there are agglomerates of approximately 200 nm. In the center of the micrograph, agglomerates are observed with a lamellar morphology and approximately 300 nm size. HTEM indicated the presence of asphaltene aggregates constituted by nanometric particles and asphaltene stacking as well as ultrafine nanocrystalline-oriented structures. Dark zone at the bottom of micrograph shows asphaltene stacking. The jagged surfaces along the edge of the structure correspond to nanocrystal-oriented structures.

3.3. Analysis of migration phenomenon in heavy crude oil

Furthermore, in **Figure 10**, the graph of percent transmittance versus cell length for stability assessment and crude oils separate fronts of this study is presented. Generally and conclusively, both crude oils were very stable. There was no change of percent transmittance (>8%) within 30 days of analysis. No phenomena of measurable migration (separation, flocculation, coalescence, or sedimentation) were observed [36, 40]. To confirm the results yielded by the Turbiscan instrument, the migration speed of the agglomerates of HC1 was determined to be 3.69×10^{-6} mm min^{-1} and for HC2 of 8.683×10^{-7} mm min^{-1} (**Table 3**).

Low values of migration velocity and high sample stability were observed in **Figure 10**, which allow us to establish that the oils are stable over a wide range of time (there is no separation of components, mainly resins and asphaltenes). Studies conducted by the working group have shown that the rates of sedimentation of asphaltenes of various Mexican crudes in toluene and n-heptane are of the order of 5.94×10^{-2} mm min^{-1}. These studies showed monodispersion and stability, because the asphaltenes tend to form large aggregates quickly, which facilitates precipitation [27].

Recently, a study of migration rate in a period of 30 days of two crudes from different sources and SARA composition. For both crude oils are very stable and did not show changes in the percentage of transmittance during the 30 days of analysis [15].

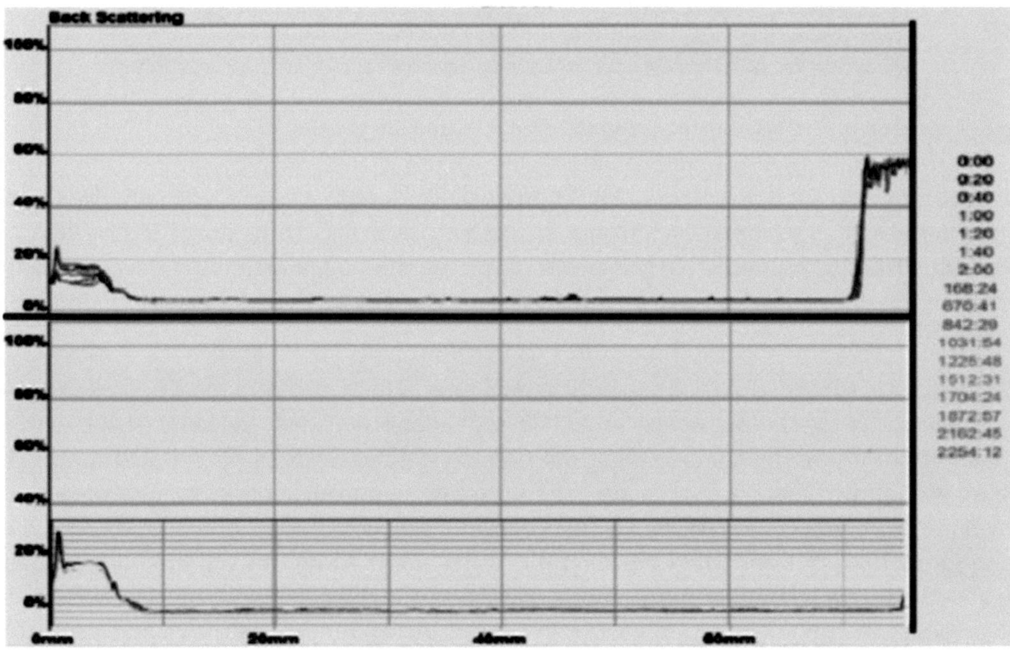

Figure 10. Curve of percent transmittance versus cell length (mm). Stability analysis of HC1 and HC2 oils obtained from the Turbisoft and Migration softwares of Classic Turbiscan software.

Sample	Oil	Migration velocity, mm min^{-1}
1	HC1	3.69×10^{-6}
2	HC2	8.68×10^{-7}

Table 3. Migration velocity values of HC1 and HC2 oils.

4. Conclusions

UV-vis spectroscopy is an important tool that can be used to characterize crude oil solutions and their asphaltenes fractions. Characteristic signals of benzenic, naphthenic, and Soret compounds, which correspond to petroporphyrins, were identified in solutions of crude oil and extracted asphaltenes, using UV-vis spectroscopy. Bathochromic shift exists in signals of benzenic and naphtenic compounds for both asphaltene solutions at longer wavelengths because of more interactions between asphaltenes caused by the absence of resins. It was also observed that the amplitude of the maximum absorbance signals for crude solutions was more intense compared to absorbance maximum for asphaltene solutions. Higher absorbance maximum were observed for whole crude, because of the overlapping of individual absorbances, corresponding to many of the different components of crude. Stability in solutions was also established under test conditions using UV-vis spectroscopy.

DLS analysis indicated a slight increase in aggregate particle size during the first 100 min for solutions of HC1 oil. For HC2 solutions, Z_{AV} values were unstable during the first 80 min. The amount of resins present in each crude affects the aggregate size and the stability of the solutions with respect to time. Z_{AV} of asphaltene solutions changes rapidly due to the absence of resins for both crudes.

SEM showed a lamellar morphology in both asphaltenes. A greater quantity of resins was observed on the surface of AHC1. It was possible to observe differences in the morphology of the resins for AHC1 and AHC2. HTEM indicated the presence of asphaltene aggregates constituted by nanometric particles and asphaltene stacking as well as ultrafine nanocrystalline-oriented structures in AHC1 and AHC2.

A migration study was carried out for 30 days, indicating stability of asphaltenes in both crudes, backing up the results obtained in all the undertaken characterizations.

Acknowledgements

E. E. Banda Cruz postdoctoral scholarship holder thanks to SENER-CONACYT-Hidrocarburos Project (Mexico) No. 177007. Furthermore, we appreciate the funding provided by the Tecnológico Nacional de México with the project No. 5713.16-P and the Fis. Darío Pozas of the Universidad de Colima (México) for the study of SEM microscopy and Dr. Ivan Estrada of Laboratorio de Nanotecnologia, CIMAV (México) for HTEM analysis.

Author details

Ernestina Elizabeth Banda Cruz, Nohra Violeta Gallardo Rivas*, Ulises Páramo García, Ana Maria Mendoza Martinez and José Aarón Melo Banda

*Address all correspondence to: nohvigari@itcm.edu.mx

Centro de Investigación en Petroquímica, División de Estudios de Posgrado e Investigación, Instituto Tecnológico de Cd. Madero, Prol. Bahía de Aldhair y Av. De las Bahías, Parque de la Pequeña y Mediana Industria, Altamira, Tamaulipas, México

References

[1] Langevin D, Argillier JF. Interfacial behavior of asphaltnes. Advances in Colloid and Interface Science. 2016;**233**:83-93. DOI: 10.1016/j.cis.2015.10.005

[2] Ilyin S, Arinina M, Polyakova M, Bondarenko G, Konstantinov I, Kulichikhin V, Malkin A. Asphaltenes in heavy crude oil: Designation, precipitation, solutions, and effects on viscosity. Journal of Petroleum Science and Engineering. 2016;**147**:211-217. DOI: 10.1016/j.petrol.2016.06.020

[3] Santos R, Loh W, Bannwart A and Trevisan O. An Overwiew of heavy oil properties and its recovery and tranportation methods. Brazilian Journal of Chemical Engineering. 2014;**31**:571-590. dx.doi.org/10.1590/0104-6632.20140313s00001853

[4] Nezhad ER, Heidarizadeh F, Sajjadifar S, and Abbasi Z. Dispersing of petroleum asphaltenes by acidic ionic liquid and determination by UV-visible spectroscopy. Journal of Petroleum Science and Engineering. 2013;**2013**:203036

[5] Acevedo S, Castillo J, Fernández A, Goncalves S, and Ranaudo M. A study of multilayer adsorption of asphaltenes on glass surfaces by photothermal surface deformation. relation of this adsorption to aggregate formation in solution. Energy & Fuels. 1998; **12**:386-390

[6] Acevedo S, Guzman K, Labrador H, Carrier H, Bouyssiere B, and Lobinski R. Trapping of metallic porphyrins by asphaltene aggregates: A size exclusion microchromatography with high-resolution inductively coupled plasma mass spectrometric detection study. Energy Fuels. 2012;**26**:4968-4977

[7] Dembicki H. Interpreting Crude Oil and Natural Gas Data. Practical Petroleum Geochemistry of Explaration and Production. 2017:135-188. DOI: 10.1016/B978-0-12-803350-0.00004-0

[8] Rezaee NE, Heidarizadeh F, Sajjadifar S, Abbasi Z. Dispersing of petroleum asphaltenes by acidic ionic liquid and determination by UV-visible spectroscopy. Journal of Petroleum Engineering. 2013;**2013**:1-5. DOI: 10.1155/2013/203036

[9] Bissada K, Tan J, Szymczyk E, Darnell M, Mei M. Group-type characterization of crude oil and bitumen. Part I: Enhanced separation and quantification of saturates, aromatics, resins and asphaltenes (SARA). Organic Geochemistry. 2016;**95**:22-28. DOI: 10.1016/j.orggeochem.2016.02.007

[10] Santos RG, Loh W, Bannwart AC and Trevisan OV. An overview of heavy oil properties and its recovery and transportation. Brazilian Journal of Chemical Engineering. 2014;**31**:03:571-590. DOI: 10.1590/0104-6632.20140313s00001853

[11] Zewen L, Jing Z, Patrice C, Chupeng Y. Discussion on the structural features of asphaltene molecules. Energy & Fuels. 2009;**23**:6272-6274. DOI: 10.1021/ef901126m

[12] Goual L. Petroleum asphaltenes, crude oil emulsions-composition stability and characterization. In: Abdul-Raouf M, El-Sayed, editors. InTech; 2012. DOI: 10.5772/35875. ISBN: 978-953-51-0220-5

[13] Li K, Vasiliu M, McAlpin C, Yang Y, Dixon D, Batzle M, Liberatore M, Herring A. Further insights into the structure and chemistry of the Gilsonite asphaltene from a combined theoretical and experimental approach. Fuel. 2015;**157**:16-20. DOI: 10.1016/j.fuel.2015.04.029

[14] Leyva C, Ancheyta J, Berrueco, Millán M. Chemical characterization of asphaltenes from various crude oils. Fuel Processing Technology. 2013;**106**:734-738. DOI: 10.1016/j.fuproc.2012.10.009

[15] Banda EE, Padrón SI, Gallardo NV, Rivera JL, Páramo U, Díaz NP, Mendoza AM. Crude oil UV spectroscopy and light scattering characterization. Petroleum Science and Technology. 2016;**34**:732-738. DOI: 10.1080/10916466.2016.1161646

[16] Ze F, Yujia X, Fang H, Pingle L, He'an L. Catalytic oxidation of cyclohexane to KA oil by zinc oxide supported manganese 5, 10, 15, 20-tetrakis(4-nitrophenyl)porphyrin. Molecular Catalysis A: Chemical 2015;**410**:221-225. DOI: 10.1016/j.molcata.2015.09.027

[17] El-Bassoussi AA, El Sayed SM, El Anmed MH, Basta JS, and Attia ESK. Characterization of some local petroleum residues by spectroscopic techniques. Petroleum Science and Technology. 2010;**28**:430-444

[18] VersanKok MK, Varfolomeev MA, Nurgaliev DK. Crude oil characterization using TGA-DTA, TGA, FTIR and TGA-MS techniques. Journal of Petroleum Science and Engineering. 2017;**154**:534-542. DOI: 10.1016/j.petrol.2016.12.018

[19] Faraj MAM, Knudsen TŠ, Nytoft HP, Jovan B. Organic geochemistry of crude oils from the Intisar oil field (East Sirte Basin, Libya). Journal of Petroleum Science and Engineering. 2016;**147**:605-616. DOI: 10.1016/j.petrol.2016.09.030

[20] Caballero V, Castillo J, Ranaudo M, and Gutiérrez H. Estudio de la cinética de agregación de asfaltenos de crudos furrial y cerro negro mediante dispersión de luz. Revista de la Facultad de Ingeniería U.C.V (online). 2013;**28**(3):47-53

[21] Xu J, and Liu H. The growth and development of asphaltene aggregates in toluene solution. Petroleum Science and Technology. 2015;**155**:33:23-24

[22] Powers DP, Sadeghi H, Yarranton HW, Berg FGA. Regular solution based approach to modeling asphaltene precipitation from native and reacted oils: Part 1, molecular weight, density, and solubility parameter distributions of asphaltenes. Fuel. 2016;**178**:218-233. DOI: 10.1016/j.fuel.2016.03.027

[23] Marczak W, Dafri D, Modaressi A, Zhou H and Rogalski H. Physical state and aging of flucculated asphaltenes. Energy Fuels. 2007;**21**:1256-1262

[24] Wargadalam V, Norinaga K, Lino M. Size and Shape of a coal asphaltene studied by viscosity an diffusion coefficient measurements. Fuel 2002;**81**:1403-1407. DOI: 10.1016/S0016-2361(02)00055-8

[25] Goual L, Sedghi M, Wang X, Zhu Z. Asphaltene aggregation and impact of alkylphenols. Langmuir. 2014;**30**:5394-5403. DOI: 10.1021/la500615k

[26] Merola MC, Carotenuto C, Gargiulo V, Stanzione F, Ciajolo A, Minale M. Chemical – Physical analysis of rheologically different samples of a heavy crude oil. Fuel Processing Technology. 2016;**148**:236-247. DOI: 10.1016/j.fuproc.2016.03.001

[27] Banda EE, Padrón SI, Gallardo NV, Páramo U, Díaz NP, Melo JA. Physicochemical characterization of heavy oil and the precipitated asphaltenes fraction using UV spectroscopy and dynamic light scattering. Journal of Engineering Technology. 2017;**6**:49-58

[28] Groenzin H, Mullins OC. Molecular size and structure of asphaltenes from various sources. Energy & Fuel. 2000;**14**:677-684. DOI: 10.1021/ef990225z

[29] Mogollón R, Rodríguez W, Larrota C, Ortiz C, Torres R. Fraccionamiento y Desmetalización Biocatalítica de asfaltenos de crudo castilla. Ciencia, Tecnología y Futuro. 2002;**1**: 109-221. ISSN 0122-5383

[30] Adebiyi FM, Thoss V. Organic and elemental elucidation of asphaltene fraction of Nigerian crude oils. Fuel. 2014;**118**:426-431. DOI: 10.1016/j.fuel.2013.10.044

[31] Spiecker PM, Gawrys KL, Trail CB, Kilpatrick PK. Effects of petroleum resins on asphaltene aggregation and water-in-oil emulsion formation. Colloids and Surfaces A: Physicochemical and Engineering Aspects. 2003;**220**:9-27. DOI: 10.1016/S0927-7757(03)00079-7

[32] Sakthivel S, Gardas RL, Sangwai JS. Spectroscopic investigations to understand the enhanced dissolution of heavy crude oil in the presence of lactam, alkyl ammonium and hydroxyl ammonium based ionic liquids. Journal of Molecular Liquids. 2016;**221**:323-332. DOI: 10.1016/j.molliq.2016.05.062

[33] Ovalles C, Rechsteiner Jr CE. Analytical Methods in Petroleum Upstream Applications. Florida: CRC Taylor & Francis; 2015. pp. 207-208. ISBN: 9781482230864

[34] Banda EE, Rivas NV, Páramo U, Estrada IA, Pozas D, Reyes J. Crude oil aggregation by microscopy and dynamic light scattering. Petroleum Science and Technology. 2016;**34**:1812-1917. DOI: 10.1080/10916466.2016.1230754

[35] Acevedo S, Castro A, Negrin J, Fernandez A, Escobar G, Piscitelli V. Relations between asphaltene structures and their physical and chemical propesties: The Rosary-type structure. Energy & Fuels. 2007;**21**:2165-2175

[36] Pereira JC, Delgado J, Briones A, Guevara M, Scorzz C, Salage L. The effect of solvent nature and dispersant performance on asphaltene precipitation from diluted solutions of instable crude oil. Petroleum Science and Technology. 2011;**29**:2432-2440. DOI: 10. 1080/10916461003735061

[37] Murillo HJ, Garcia CI, Dominguez J, Lopez RC, Duran V, Aburto J. Aggregation behavior of heavy crude oil-ionic liquids solutions by fluorescence spectroscopy. Energy Fuel. 2009;**23**:4584-4592. DOI: 10.1021/ef 9004175

[38] Akbarzadeh K, Hammami A, Kharrat A, Zhang D, Allenson S, Creek J, Mullins, OC. Asphaltenes—problematic but rich in potential. Oilfield Review. 2007;**19**:22-43

[39] Salmón SV, Herrera RU, Valdez MA, Lira CG. Efffect of the concentration of ionic surfactants on the electrokinetic behavior of asphaltene precipitated from a maya mexican crude oil. Revista Mexicana de Ingeniería Química. 2010;**9**(3):343-357

[40] Andrews AB, Guerra RE, Mullins OC, Sen PN. Diffusivity of asphaltene molecules by fluorescence correlation spectroscopy. Journal of Physical Chemistry A 2006;**110**:8093-8097. DOI: 10.1021/jp062099n

Chapter 7

Density Anomalies in Crude Oil Blends Reflect Multiple Equilibrium States of Asphaltene Colloidal Aggregates

Igor N. Evdokimov, Aleksey A. Fesan and
Aleksandr P. Losev

Additional information is available at the end of the chapter

http://dx.doi.org/10.5772/intechopen.71813

Abstract

Density measurements revealed anomalies of nonideality (maxima of excess density) at some compositions in binary blends of light and heavy crude oils from diverse origins. By IR absorption measurements, density anomalies were attributed to increased contents of suspended asphaltene colloidal-sized particles/aggregates in the blends. By comparison with a database of world's native crude oils, it was concluded that density anomalies may correspond to different equilibrium structural states of asphaltene colloids that occur at several specific asphaltene contents, apparently common for petroleum media of any origin.

Keywords: crude oil blends, nonideality, excess density, volumetric shrinkage, asphaltene aggregates, structural transformations

1. Introduction

A common approach frequently used in the petroleum industry for either transportation or refining purposes is blending of heavy crude oils with lighter ones [1–5]. Therefore, studying the effect of blending on the physical and chemical properties of produced petroleum fluids has become increasingly important. It was soon realized that blends of multicomponent crude oils are nonideal systems, for which ideal "mixing rules" developed for binary mixtures of pure chemical substances [6–8] are not applicable. For example, in ideal mixtures, both mass and volume are additive parameters and the total volume of the binary blend V_{ideal} is equal to the sum of the volumes of the components:

$$V_{ideal} = V_1 + V_2 \tag{1}$$

Consequently, the density of the ideal binary blend:

$$\rho_{ideal} = \rho_1 \phi_1 + \rho_2 \phi_2, \tag{2}$$

where ϕ_1 and ϕ_2 are volume fractions of the components.

In binary blends of crude oils, deviations of the measured volume V_{blend} and density ρ_{blend} from the above ideal mixing rules are usually characterized [9–14] by such interrelated dimensionless quantities as "volumetric shrinkage":

$$S = \frac{V_{ideal} - V_{blend}}{V_{ideal}} \tag{3}$$

and "excess density":

$$\rho_{ex} = \frac{\rho_{blend} - \rho_{ideal}}{\rho_{ideal}} \tag{4}$$

where, by definition:

$$\rho_{ex} = S/(1 - S) \tag{5}$$

Empirical testing of crude oil blends by a number of major oil companies has produced vast amounts of volumetric shrinkage and excess density data [10]. An API Measurement Committee correlated the data collected over several decades and in 1996 released equations, published in the API Manual of Petroleum Measurement Standards [15], which served as the accepted industry standard for over 20 years. In particular, the standard equation for volumetric shrinkage (in % units):

$$S = 2.69 \times 10^4\, C\,(100 - C)^{0.819} \left(\frac{1}{dL} - \frac{1}{dH}\right)^{2.28} \tag{6}$$

where C is the concentration, in liquid volume % of the lighter oil in the blend (i.e., $C = 100\,\phi$); dL and dH are the densities (in kg/m^3) of lighter and heavier oil, respectively.

The authors of Eq. (6) made no attempts to account for the presence of any specific molecular fractions in the blended oils. Consequently, for all blends, the $S(C)$ dependencies are similar dome-shaped functions peaked at $C \approx 55\%$, while the densities of particular oils affect only the magnitude of this peak. In spite of its over-simplified nature, Eq. (6) is still frequently employed by various research groups for approximating experimental data [9, 14, 16, 17].

More recent tendencies in discussing the measured nonideal properties of crude oil blends are attempts to reveal specific molecular substances responsible for nonideality. The most discussed nonideal petroleum constituents in crude oil blends are asphaltenes [18–27]. In particular, it was suggested that volumetric shrinkage/excess density may result from an increase in the equilibrium content of asphaltene colloidal aggregates dispersed in the blend, which occur at some specific asphaltene contents [23–27].

In this work, we describe a study of the nonideality phenomena in three blends of native crude oils of diverse origin by experimental density measurements. This study is supplemented by IR optical absorption experiments, which provide the direct proof of the crucial role of asphaltene aggregation.

2. Diversity of studied native crude oils: preparation of oil blends

To ensure that the observed nonideal effects are not specific for particular crude oils, in the described series of experiments, we employed blends of light and heavy native crude oils collected from the well heads at several Russian oilfields geographically separated by thousands of kilometers (cf. **Table 1**).

	Heavy oil	**Light oil**
Blend #1		
Region	Samara	Volgograd
Oilfield; well #	Kalmayursk.; 238	Korobkovsk.; n/a
Location	53°12′10″N; 50°08′27″E	50°19′00″N; 44°48′00″E
Density, 20°C, kg/m^3	963.9	820.3
Viscosity, 20°C, cSt	3732	4.6
Asphaltenes, wt%	15.48	0.064
Resins, wt%	35.91	16.00
Paraffins, wt%	1.53	9.00
Sulfur, wt%	2.5	0.3
Metals (V + Ni), wt%	0.175	n/a
Solids, wt%	0.587	0.06
Blend #2		
Region	Tatarstan	Yugra
Oilfield; well #	Aznakayevsk.; 24,534	Pogranichn.; 43P
Location	54°32′16″N; 52°47′54″E	63°11′57″N; 75°27′02″E
Density, 20°C, kg/m^3	893.2	818.3
Viscosity, 20°C, cSt	39.3	2.94
Asphaltenes, wt%	4.19	0.14
Resins, wt%	20.57	3.82
Paraffins, wt%	0.89	2.59
Sulfur, wt%	2.28	0.28
Metals (V + Ni), wt%	0.076	0.018
Solids, wt%	0.07	0.006

	Heavy oil	Light oil
Blend #3		
Region	Komi	Yugra
Oilfield; well #	Usinsk.; n/a	Potochn.; 401
Location	66°10′38″N; 57°21′14″E	61°15′15″N; 75°12′44″E
Density, 20°C, kg/m³	955.3	840.8
Viscosity, 20°C, cSt	962.7	7.37
Asphaltenes, wt%	10.87	0.46
Resins, wt%	9.45	2.34
Paraffins, wt%	5.45	5.21
Sulfur, wt%	1.09	0.56
Metals (V + Ni), wt%	0.0111	0.0137
Solids, wt%	0.46	0.04

Table 1. The diverse properties of native crude oils employed in the studied blends.

Samples of blends with various oil volume fractions ϕ (determined with an accuracy $\Delta\phi \leq 0.01$) were prepared in 100 ml dark glass vessels. Each prepared sample was intermixed manually with a glass rod and was aged at room temperature over a standard period of 3 days for equilibration. Before density measurements, samples were again homogenized by stirring with a glass rod for 2 minutes.

3. Density measurements reveal multiple "excess density" peaking in crude oil blends of varying compositions

Densities of all samples of Blend #1 were determined using a standard 10 ml pycnometer. In the studied density range of 818–964 kg/m^3 (cf. **Table 1**), an accuracy of density measurement was ≤1 kg/m^3. For samples of Blend #2 and Blend #3 measurements were performed in an oscillating U-tube densitometer (model VIP-2M, produced by TERMEX, Tomsk, Russia) with an accuracy of ±0.3 kg/m^3. All measurements were performed at 24 ± 0.5°C.

Experimental values of "excess density" ρ_{ex} were calculated according to Eq. (4) on the basis of the measured densities of the parent crude oils and of samples of oil blends with various compositions. These experimental values are shown in the graphs of **Figure 1** by filled circles connected by straight-line segments. For comparison, continuous curves in **Figure 1** illustrate the predictions of the standard API model in Eq. (6) converted to ρ_{ex} units by Eq. (5).

Immediately evident is a strong quantitative disagreement of model predictions with experimental data (note that in **Figure 1**, all model data are multiplied by the factor of 10). A fundamental

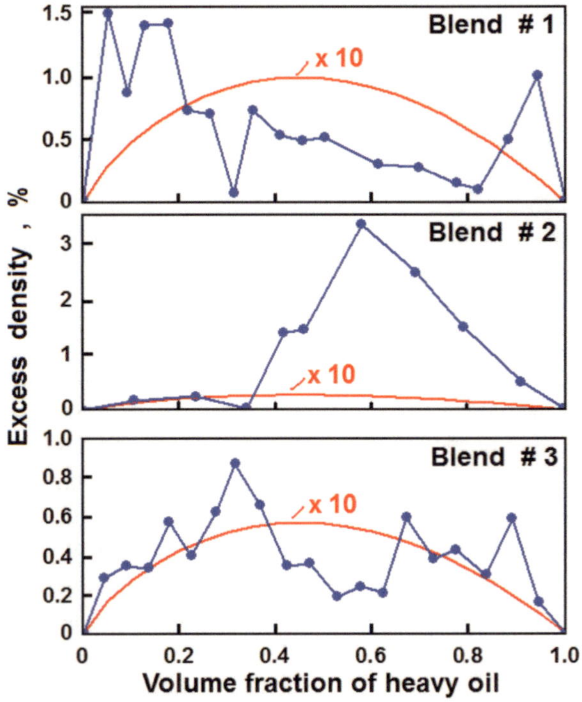

Figure 1. Filled circles connected by straight-line segments—experimental values of excess density in the studied crude oil blends. Continuous curves—excess densities (multiplied by 10) calculated according to the standard API model.

qualitative difference of the model with experimental results appears even more significant. Namely, as noted in Introduction, the API model predicts a single maximum of ρ_{ex}, roughly at equal contents of light and heavy oil in any blend. In contrast, experimental data in **Figure 1** reveal multiple peaks of the measured excess density for all studied blends with diverse properties of the parent light and heavy oils.

In accordance with previous publications [23–27], we suggest that the main mechanism behind the observed multiple extrema of ρ_{ex}, not accounted for by the standard API model, is an increase in the equilibrium content of asphaltene colloidal aggregates dispersed in the blend, which occur at some specific asphaltene contents. A direct experimental verification of this assumption by IR absorption measurements is described in the following section.

4. IR absorption experiments confirm close relationship of excess density with the content of asphaltene colloidal aggregates in crude oil blends

Infrared optical absorption spectra were measured in the 650–4000 cm^{-1} range with 2 cm^{-1} increment using a model FT-801 FTIR Spectrometer (Simex, Novosibirsk, Russia). For convenience of presentation/discussion, the values of "transmittance" T output of the instrument (in %) were converted to the values of "absorption" (in %) calculated as (100 − T).

Figure 2 show experimental IR absorption spectra for the parent light and heavy oil components of Blend #3 (cf. **Table 1**). By literature analysis, we could not reliably distinguish in these complex spectra any characteristic peaks representative of asphaltene colloidal aggregates.

To solve this problem, we utilized a well-known method of changing the aggregation state of asphaltenes by dilution of crude oils with *n*-heptane [28–30]. At strong dilutions (e.g., at the ratio of 1/40 v/v, as in the standard ASTM D 6560 method [28]), asphaltenes precipitate, flocculate, and fall out of a solution as solid deposits. At small dilutions, the precipitated asphaltenes form a conglomeration of colloidal species suspended in solution, but no solid deposits may be formed even in course of lengthy experiments [29, 30]. In our studies, we diluted the parent heavy oil, employed as a component in Blend #3, with *n*-heptane at the ratio of 1/4 (v/v). The diluted oil was stored for 2 days in quiescent conditions to ensure gravitational spatial segregation of asphaltene colloidal particles (no solid deposits were registered even at the end of this period). Samples from the lower and the upper layers of diluted crude oil (enriched and deficient in asphaltene colloidal particles) were then extracted with syringe, and the respective IR absorption spectra were measured. Specific contribution of asphaltene colloids to IR absorption was revealed by calculating the difference between both spectra and smoothing the difference spectrum by 11-point sliding data window. In the difference spectrum, as illustrated at the upper part of **Figure 3**, the most prominent group of absorption peaks above a continuous background (dashed line) was registered in the range of 2800–3200 cm^{-1}, with the main absorption peak of intensity I_A at 3041 cm^{-1}. By literature analysis [31, 32], absorption at 3040–3050 cm^{-1} may be ascribed to excitation of the aromatic sp^2=C–H stretching vibrations. Note that in various previous experiments, this =C–H absorption band was regarded as one of the principal structural parameters of asphaltenes [33–36]. Note also that in the IR spectra of the original crude oils and of their blends, the discussed peak at 3041 cm^{-1} has a very small intensity compared to the main absorption bands—cf. the lower part of **Figure 3**.

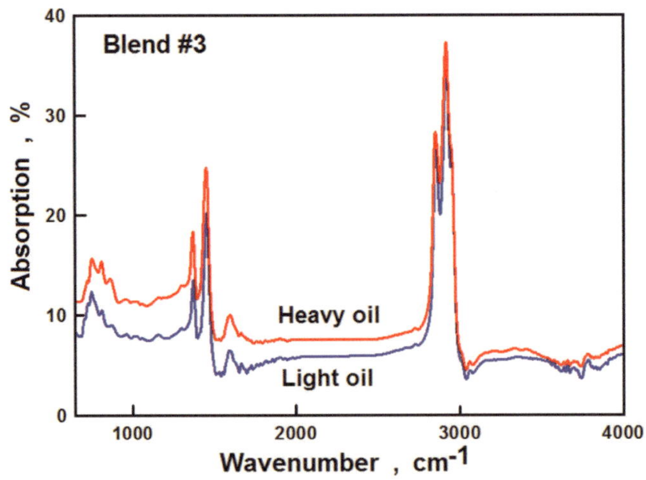

Figure 2. IR absorption spectra of the parent light and heavy oils, employed as components of Blend #3.

http://dx.doi.org/10.5772/intechopen.71813

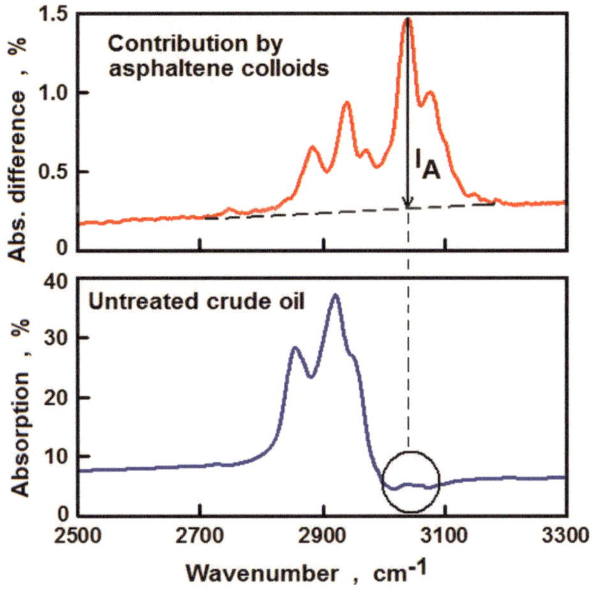

Figure 3. Above: specific contribution to IR absorption by asphaltene colloids (cf. text). Below: the same part of absorption spectrum for the parent heavy crude oil.

On the basis of the above arguments, we concluded that changes in the content of colloidal-sized asphaltenes in crude oil blends with varying compositions may be characterized by intensity increments of the 3041 cm^{-1} absorption peak ΔI_A (measured with respect to the I_A value in the lighter crude oil). **Figure 4** shows a good correspondence of ΔI_A variations in Blend #3 with the measured excess density variations from **Figure 1**. Hence, IR experiments confirm the above suggestion that peaking of ρ_{ex} is caused by the growth of the equilibrium contents of asphaltene colloidal aggregates suspended in crude oil blends. Moreover, as shown in the following section, maxima of excess density apparently occur at some specific asphaltene contents in all blends, independent of the origin of the blended crude oils.

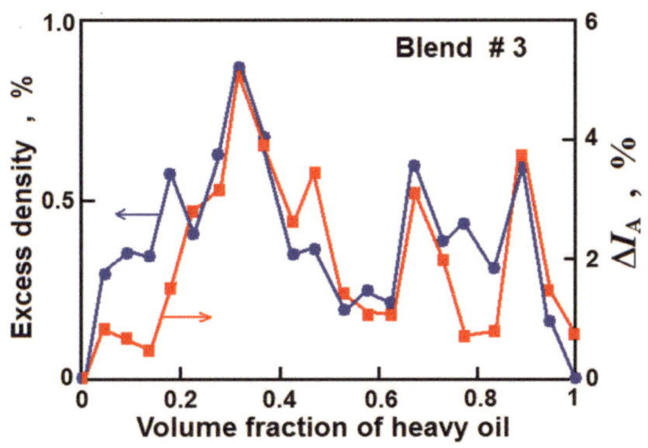

Figure 4. Close relationship of excess density variations in crude oil Blend #3 with content of colloidal-sized asphaltenes, characterized by intensity increments ΔI_A of the IR absorption peak at 3041 cm^{-1}.

5. Density maxima in the studied crude oil blends and in the database of world's native crude oils are observed at the same special asphaltene contents in petroleum media

Previously, we have analyzed distributions of density/specific gravity in a database of several hundreds world's native crude oils from a diversity of geographical/geological locations [23]. **Figure 5** shows the scatter plot of specific gravity (*SG*) versus *Log* of asphaltene content (C_A) in this database (filled triangles). The analysis of these data revealed that (1) minimum values of *SG* are randomly distributed along some smooth curve and (2) distribution of maximum *SG* values is not random, and these tend to concentrate near certain asphaltene contents. To highlight these points, maximum and minimum values of *SG* in the scatter plot were evaluated in consecutive equal intervals of $\Delta Log(C_A) = 0.05$ and the results were connected by continuous lines, as illustrated in **Figure 5**. It is clearly seen that anomalous peaking of specific gravity/density is observed in native crude oils with some particular asphaltene contents C_A* listed in **Table 2**. In Ref. [23], these anomalies were attributed to different structural states (structural phases) of asphaltene colloids suspended in crude oils and the reliability of the values of respective asphaltene contents C_A* was demonstrated by the analysis of databases from other publications [37–39].

Additional analysis of excess density measurements from **Figure 1** in terms of asphaltene content has shown that in all blends peaking of ρ_{ex} apparently occurs at the same characteristic asphaltene contents C_A* revealed in native crude oils. The results of this analysis are illustrated in **Figures 6–8**. The experimental excess density values from **Figure 1** are denoted by filled circles. Filled squares denote maximum values of specific gravity from a database of world's native crude oils (**Figure 5**). A fairly good qualitative coincidence of both data sets is observed in all cases.

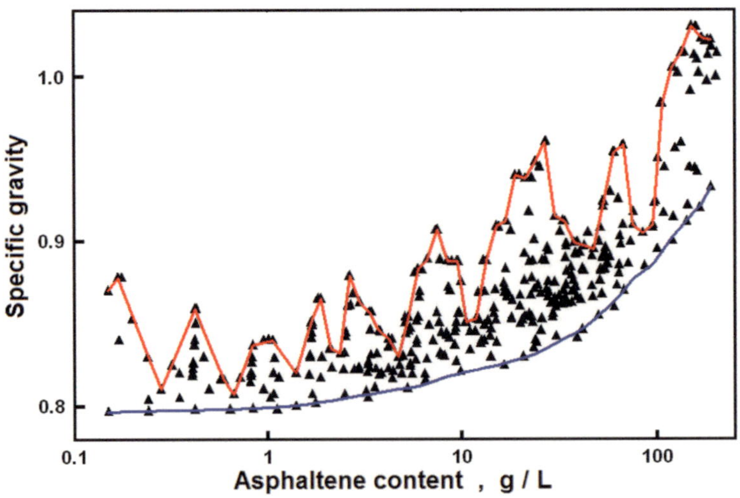

Figure 5. Data points: the scatter plot of specific gravity (density) in a database of world's native crude oils from a diversity of geographical/geological locations. Continuous lines: analysis of maximum and minimum values of specific gravity in this database. Note specific gravity peaking at some characteristic asphaltene contents.

No.	3a	3a	3b	4	5	6
C_A^*, g/L	1.88	2.66	7.50	26.6	66.8	149.6

Table 2. Characteristic asphaltene contents C_A^* (numbered as in Ref. [23]) at which maximum density anomalies are observed in native crude oils (cf. **Figure 5**).

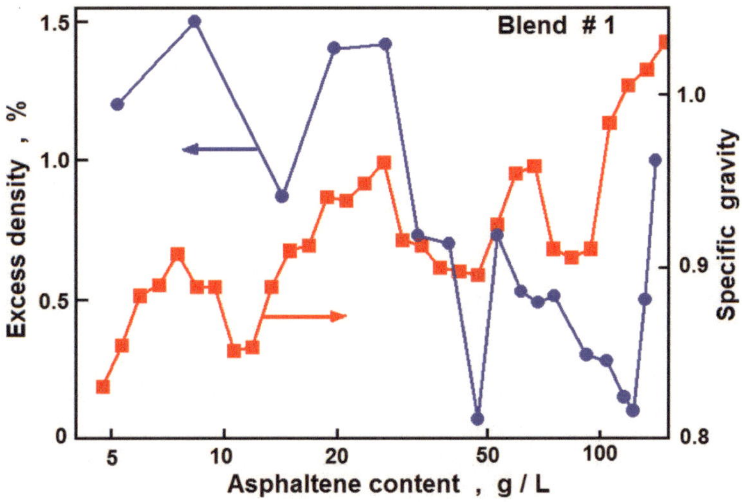

Figure 6. Comparison of the experimental excess density data for Blend #1 (filled circles) with the maximum values of specific gravity from a database of world's native crude oils (filled squares).

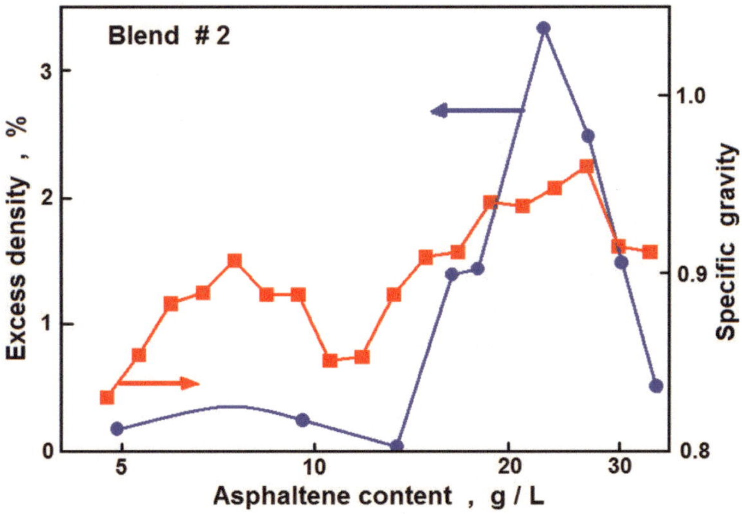

Figure 7. Comparison of the experimental excess density data for Blend #2 (filled circles) with the maximum values of specific gravity from a database of world's native crude oils (filled squares).

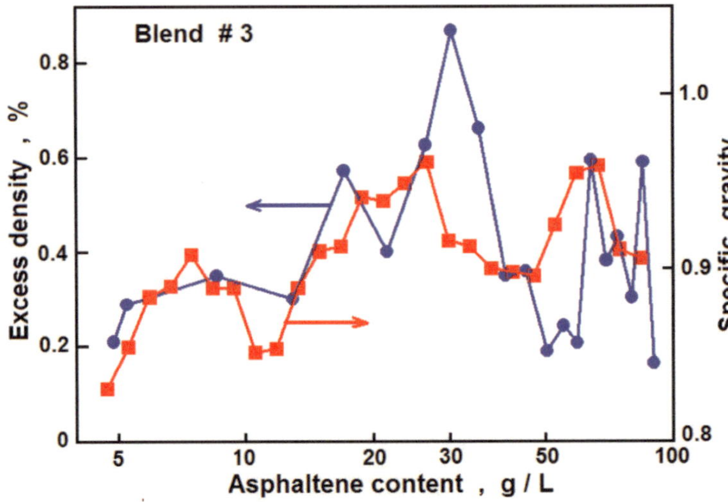

Figure 8. Comparison of the experimental excess density data for Blend #3 (filled circles) with the maximum values of specific gravity from a database of world's native crude oils (filled squares).

Summarizing, **Figures 6–8** indicate that the observed multiple nonideal anomalies in the properties of equilibrated crude oil blends (multiple peaks of excess density/volumetric shrinkage values) apparently result from step-like changes between several equilibrium structural phases of suspended asphaltene colloids, which are observed between several particular asphaltene contents, apparently common for petroleum media of any origin.

6. Plausible constitution of multiple structural phases of asphaltenes. Apparent resemblance of asphaltene transformations to those of block copolymers

In spite of intensive experimental and theoretical studies, association and aggregation mechanisms of asphaltene colloids are still not well characterized and are subjects of ongoing debate [22–25, 29, 30]. In particular, for decades, the prevailing aggregation paradigm (now dismissed as erroneous) has been that at a "critical micelle concentration" ("CMC" ≈ 2–10 g/L) of asphaltenes in native petroleum, or in "good" solvents, such as toluene, there is a *one-step* transition from a structural phase of single asphaltene molecules to a structural phase colloidal asphaltene "micelles" of 4–6 molecules [40]. The development of more accurate experimental techniques provided substantial evidence of asphaltene aggregation at concentrations much lower than previously quoted "CMC." Hence, the aggregation paradigm has been changed quantitatively, though not qualitatively. The most popular current model is that at a "critical nanoaggregate concentration" ("CNAC" ≈ 100–200 mg/L) of asphaltenes in native crude oils, or in "good" solvents, there is a *one-step* transition from a phase of individual asphaltene monomers to a phase of colloidal "nanoaggregates" of 4–6 monomers (further clustering of primary aggregates is usually regarded as a dynamic random process and is not discussed in terms of a phase transformations) [41, 42].

- The dismissed paradigm [40]:

$$\text{MOLECULE} \xrightarrow{CMC} \text{COLLOIDAL MICELLE}$$

- The current paradigm [41, 42]:

$$\text{MONOMER} \xrightarrow{CNAC} \text{COLLOIDAL NANOAGGREGATE}$$

Apparently, due to their simplicity, the above *one-step* (two-phase) paradigms have been and still are the ones, almost exclusively employed for description of asphaltene aggregation phenomena and for interpretation of experimental data. Regardless of the difference in suggested molecular mechanisms, the above paradigm closely resembles the idealized model of one-step aggregation ("micellization") with just two indigenous solute phases developed for solutions of primitive amphiphilic surfactants like sodium dodecyl sulfate (SDS) [43]. It is now well known that much more common in nature are complex self-organizing amphiphilic species, e.g., block copolymers, which possess a wide array of indigenous colloidal phases in solutions [44].

However, all evidence for the close resemblance of petroleum asphaltenes to block copolymers, both in their molecular structure and in their propensity for self-assembly into wide array of nano-sized and micro-sized structures, still remains virtually unnoticed. In fact, the existence of several aggregation stages below and above CNAC has been repeatedly demonstrated by various experiments (cf. a review in Ref. [45]). It appears that the corresponding characteristic concentrations C_A^* only marginally depend on the geographical/geological origin of asphaltenes or on the type of solvents/crude oils.

Historically, asphaltenes are defined not as members of some particular family of chemical species, but as a fraction of petroleum, which is soluble in toluene ("good solvent"), and resemblance of asphaltene molecular architecture to that of block copolymers for the first time was suggested in 1994 by I.A. Wiehe [46], whereas in 2002, W. Loh noticed a parallel between some aggregation processes for block copolymers and asphaltenes [47]. Various molecular models of asphaltenes have been developed [22, 29, 30, 41]; all these models agree that asphaltene molecules contain the following principal "building blocks" [46]: (1) rigid flat condensed systems of aromatic rings with an affinity for "good solvents" and (2) flexible linear alkyl side chains, which may contain some heteroatoms, and have an affinity for "nonsolvents." In the "continental-type" models, aromatic regions are large, whereas in "archipelago-type" models, aromatic blocks are much smaller. Aggregation of such "multiblock" molecules may proceed via π-π stacking and H-π bonding, involving aromatic blocks, as well as via polar and hydrogen-bonding interactions between side-chain blocks. By virtue of these noncovalent interactions, asphaltenes in solutions as well as in native petroleum apparently exhibit multiple structural phases—a well-known property of block copolymers [44].

Recently, it was experimentally proven [48–52] that basic asphaltene molecules (unimers, monomers) typically include very small, 1–3 ring, aromatic systems, in contrast to the popular notions about the predominance of large multiring fused systems [41, 42]. These basic molecules become predominant equilibrium species only after dissolution of solid asphaltenes in "good" solvents (benzene, toluene, etc.) to asphaltene concentrations $C \leq 0.5–0.6$ mg/L. Aggregation of basic asphaltene molecules commences at $C = 0.6–0.7$ mg/L, much smaller than the above-discussed CNAC, and the predominant equilibrium species become polydisperse "primary" asphaltene aggregates, which may contain up to 10–12 basic molecules associated mainly in a head-to-tail manner by noncovalent interactions. On the basis of these experimental results, it concluded that the adequate structural description of primary asphaltene aggregates may be that suggested in Ref. [53]. The authors of this publication described that such aggregates as supramolecular assemblies of molecules, combining cooperative binding by Brønsted acid-base interactions, hydrogen bonding, metal coordination complexes, and interactions between cycloalkyl and alkyl groups to form hydrophobic pockets, in addition to aromatic π-π stacking. They suggested a range of aggregate architectures, which almost certainly occur simultaneously, including porous networks and host-guest complexes. The latter may include organic clathrates, in which occluded guest molecules stabilize the assembly of a cage, as methane does in gas hydrates [53]. With increasing asphaltene concentration (still below CNAC), several structural phases of soft, voluminous, primary aggregates are observed in solutions, which may be accompanied by a release of some occluded "guests" [51]. Finally, at $C \approx 100–130$ mg/L, primary aggregates shrink and may be described as "solid-like" asphaltene colloidal nanoparticles [52]. Interpretations of some structural asphaltene phases above CNAC by various authors were reviewed in Refs [23–25, 45, 54].

Plausible types of some of the multiple structural transformations of asphaltenes are schematically illustrated in **Figure 9**.

In **Figure 9**, characteristic asphaltene contents C_A^* are numbered as in Ref. [23] (cf. also **Table 2**):

1. At asphaltene contents of 0.6–0.7 mg/L, apparently there is a transition from a solution of small asphaltene basic molecules to a solution of voluminous primary aggregates.
2. At concentrations of 100–130 mg/L, soft primary aggregates are transformed to solid-like "nanocolloids" with individual nanoparticles 2–4 nm in diameter.

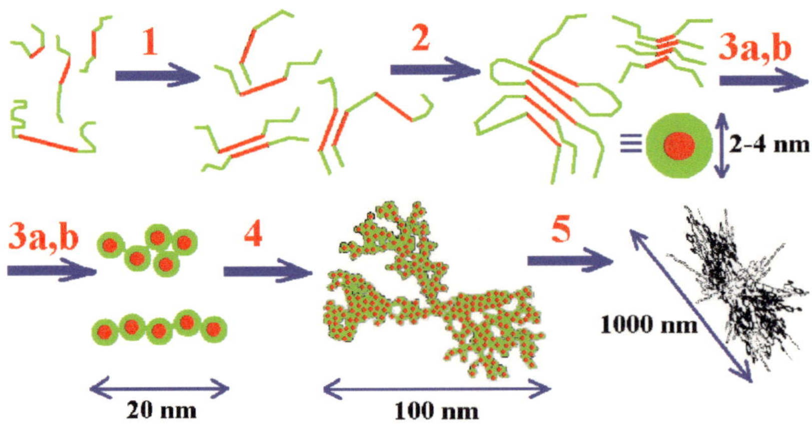

Figure 9. Plausible structural phase transformations of asphaltenes at some characteristic asphaltene contents, numbered as in Ref. [23].

3. This group of two close characteristic concentrations (3a,b) in the range of 1.7–8 g/L (cf. **Table 2**) may reflect the appearance of elongated colloidal clusters of solid-like nanoparticles, 15–20 nm in length.

4. At concentrations of 25–30 g/L, elongated clusters presumably form fractal aggregates ≥100 nm in size.

5. At 55–70 g/L, there start to appear micron-sized asphaltene flocs still suspended in the liquid media.

To conclude this section, it should be emphasized that revealing multiple asphaltene structural phases in solutions and in native petroleum was largely facilitated by a seemingly trivial but a crucial improvement in analysis of experimental data. Namely, previously overlooked details of analyzed correlations become perceptually well separated only at data plots with log concentration scales, common in surfactant chemistry, while in conventional petroleum studies, linear-scale plots are still employed [29, 30, 41].

7. Conclusion

In contrast to the predictions of the standard API model, the described density measurements revealed multiple anomalies of excess density at several compositions in all studied blends of light and heavy crude oils from diverse origins. It was suggested that density anomalies are caused by increased contents of suspended asphaltene colloidal-sized particles/aggregates in the blends. In IR absorption experiments, the most intense absorption peak representative of asphaltene colloids appeared to be that at 3041 cm^{-1}, by literature analysis, ascribed to excitation of the aromatic $sp^2{=}C{-}H$ stretching vibrations. Intensity variations of this peak correlated with variations of excess density, supporting the conclusion of the governing role of asphaltene colloids in the observed density anomalies. By comparison with an extended database of world's native crude oils, it was further concluded that different density anomalies may correspond to different equilibrium structural states of asphaltene colloids, which occur at several specific asphaltene contents, apparently common for petroleum media of any origin.

It is hoped that the insight gained into the nonideal behavior of the densities of crude oils, upon mixing and the explanations proposed, will assist in a better scientific understanding of the mechanism of the changes that may occur in mixtures. This should help oil producers and refiners in a better utilization of the existing oil resources, i.e., crude oils and petroleum products.

Acknowledgements

We acknowledge the support from the Nedra-Test Research and Testing Establishment, LLC, Moscow, Russia, in purchasing the experimental equipment.

Author details

Igor N. Evdokimov*, Aleksey A. Fesan and Aleksandr P. Losev

*Address all correspondence to: physexp@gubkin.ru

Department of Physics, Gubkin Russian State University of Oil and Gas, Moscow, Russia

References

[1] Gateau P, Hénaut I, Barré L, Argillier JF. Heavy oil dilution. Oil & gas Science and technology - rev. IFP. 2004;**59**(5):503-509. DOI: 10.2516/ogst:2004035

[2] Todd CM. Downstream planning and innovation for heavy oil development - a producer's perspective. Journal of Canadian Petroleum Technology. 1988;**27**(1):79-86

[3] Martínez-Palou R, Mosqueira ML, Zapata-Rendón B, Mar-Juárez E, Bernal-Huicochea C, Clavel-López JC, Aburto J. Transportation of heavy and extra-heavy crude oil by pipeline: A review. Journal of Petroleum Science and Engineering. 2011;**75**(3-4):274-282. DOI: 10.1016/j.petrol.2010.11.020

[4] Gary JH, Handwerk GE, Kaiser MJ. Petroleum Refining Technology and Economics. 5th ed. Boca Raton, FL: CRC Press; 2007

[5] Al-Besharah JM. The Effect of Blending on Selected Physical Properties of Crude Oils and their Products [PhD Thesis]. Birmingham, UK: The University of Aston in Birmingham; 1989

[6] Reid RC, Prausnitz JM, Sherwood TK. The Properties of Gases and Liquids. 3rd ed. New York: McGraw-Hill, Inc.; 1977

[7] Benson SW, Cruickshank FR, Golden DM, Haugen GR, O'Neal HE, Rodgers AS, Shaw R, Walsh R. Additivity rules for the estimation of thermochemical properties. Chemical Reviews. 1969;**69**(3):279-324. DOI: 10.1021/cr60259a002

[8] Nielsen LE. Predicting the Properties of Mixtures: Mixture Rules in Science and Engineering. New York: Marcel Dekker, Inc.; 1978

[9] Arenas FG, Garcia CA, Orozco LM. Models for Estimating the Shrinkage Factor in the Mixtures of Diluents with Colombian Heavy Crude Oils. Paper SPE-153554-MS. Presented at the SPE Latin America and Caribbean Petroleum Engineering Conference. Mexico: Mexico City; 2012. DOI: 10.2118/153554-MS

[10] James JH. Shrinkage losses resulting from liquid hydrocarbon blending. In: Proceedings of the 87th Annual International School of Hydrocarbon Measurement. Vol. 1. Oklahoma City, Oklahoma, USA: Curran Associates, Inc.; 2012. pp. 452-458

[11] Ashcroft SJ, Booker DR, Turner JCR. Volumetric behavior of mixtures of crude oils and light hydrocarbons. Journal of the Institute of Energy. 1992;**65**(464):131-136

[12] Shanshool J, Talib E. Fitted shrinkage formula of spiked crude oil. Petroleum Science and Technology. 2002;**20**(3-4):435-439. DOI: 10.1081/LFT-120002111

[13] Saryazdi F. Density Prediction for Mixtures of Heavy Oil and Solvents [MSci Thesis]. Calgary, Alberta, Canada: University of Calgary; 2012

[14] Shanshool J, Habobi N, Kareem S. Volumetric behavior of mixtures of different oil stocks. Petroleum & Coal. 2011;**53**(3):223-228

[15] API Manual of petroleum measurement standards (MPMS). Chapter 12. Calculation of petroleum quantities. In: Section. Vol. 3. Volumetric shrinkage resulting from blending light hydrocarbons with crude oil. Washington, D.C.: American Petroleum Institute; 1996

[16] Nengkoda A. The Role of Crude Oil Shrinkage in Heavy Mix Light Crude in Main Oil Pipeline: Case Study Oman. Paper SPE 148925. Presented at the SPE Heavy Oil Con_ ference and Exhibition. Kuwait: Kuwait City; 2011. DOI: 10.2118/148925-MS

[17] Slaiman EN, Al-Qamaje HM, Atta SZ. Composition and temperature dependence of excess volume of heavy oil-stocks mixtures + (gas oil or toluene or reformate). Al-Khwarizmi. Engineering Journal. 2011;**7**(3):59-67

[18] Rodríguez S, Ancheyta J, Guzmán R, Trejo F. Experimental setups for studying the compatibility of crude oil blends under dynamic conditions. Energy & Fuels. 2016;**30**(10):8216-8225. DOI: 10.1021/acs.energyfuels.6b01698

[19] Bassane JFP, Sad CMS, Neto DMC, Santos FD, Silva M, Tozzi FC, Filgueiras PR, de Castro EVR, Romão W, Santos MFP, da Silva JOR, Lacerda Jr V. Study of the effect of temperature and gas condensate addition on the viscosity of heavy oils. Journal of Petroleum Science and Engineering. 2016;**142**:163-169. DOI: 10.1016/j.petrol.2016.02.006

[20] Gabrienko AA, Martyanov ON, Kazarian SG. Effect of temperature and composition on the stability of crude oil blends studied with chemical imaging in situ. Energy & Fuels. 2015;**29**(11):7114-7123. DOI: 10.1021/acs.energyfuels.5b01880

[21] Weihe IA, Kennedy RJ. The oil compatibility model and crude oil incompatibility. Energy & Fuels. 2000;**14**(1):56-59. DOI: 10.1021/ef990133+

[22] Wiehe IA. Asphaltene solubility and fluid compatibility. Energy & Fuels. 2012;**26**(7):4004-4016. DOI: 10.1021/ef300276x

[23] Evdokimov IN. The importance of asphaltene content in petroleum - the revision of some persistent stereotypes. Petroleum Science and Technology. 2010;**28**(7):756-763. DOI: 10.1080/10916460902804739

[24] Evdokimov IN. The importance of asphaltene content in petroleum II - multi - peak viscosity correlations. Petroleum Science and Technology. 2010;**28**(9):920-924. DOI: 10.1080/10916460902937018

[25] Evdokimov IN. The importance of asphaltene content in petroleum III - new criteria for prediction of incompatibility in crude oil blends. Petroleum Science and Technology. 2010;**28**(13):1351-1357. DOI: 10.1080/10916460903096731

[26] Wang Z, Liu H, Sun X, Ji S, Guo A, Chen K. Compatibility of heavy blends evaluated by fouling and its relationship with colloidal stability. Petroleum Science and Technology. 2015;**33**(6):686-693. DOI: 10.1080/10916466.2015.1007380

[27] de la Cruz JLM, Cedillo-Ramírez JC, Aguirre-Gutiérrez AJ, García-Sánchez F, Aquino-Olivos MA. Incompatibility determination of crude oil blends from experimental viscosity and density data. Energy & Fuels. 2015;**29**(2):480-487. DOI: 10.1021/ef501512b

[28] ASTM. D6560-00; IP 143/01: Standard Test Method for Determination of Asphaltenes (Heptane Insolubles) in Crude Petroleum and Petroleum Products. West Conshohocken, PA: American Society for Testing and Materials; 2000

[29] Andersen SI, Speight JG. Asphaltene precipitation and incipient flocculation in mixed solvents. Preprints – Division of fuel chemistry. American Chemical Society. 1992;**37**(3): 1335-1341

[30] Speight JG. Petroleum asphaltenes - Part 1: Asphaltenes, resins and the structure of petroleum. Oil & gas Science and technology - Rev. IFP. 2004;**59**(5):467-477. DOI: 10.2516/ogst:2004032

[31] Bellamy LJ. The Infra-Red Spectra of Complex Molecules. New York: John Wiley & Sons; 1957

[32] Silverstein RM, Webster FX. Spectrometric Identification of Organic Compounds. 6th ed. New York: John Wiley & Sons; 1998

[33] Almusallam AS, Shaaban M, Nettem K, Fahim MA. Delayed aggregation of asphaltenes in the presence of alcohols by dynamic light scattering. Journal of Dispersion Science and Technology. 2013;**34**(6):809-817. DOI: 10.1080/01932691.2012.704737

[34] Guillen MD, Iglesias MJ, Domingues A, Blanco CG. Semi quantitative FTIR analysis of a coal tar pitch and its extracts and residues in several organic solvents. Energy & Fuels. 1992;**6**(4):518-525. DOI: 10.1021/ef00034a025

[35] Apicella B, Alfè M, Barbella R, Tregrossi A, Ciajolo A. Aromatic structures of carbonaceous materials and soot inferred by spectroscopic analysis. Carbon. 2004;**42**(8-9):1583-1589. DOI: 10.1016/j.carbon.2004.02.010

[36] Santamaria A, Yang N, Eddings E, Mondragon F. Chemical and morphological characterization of soot and soot precursors generated in an inverse diffusion flame with aromatic and aliphatic fuels. Combustion and Flame. 2010;**157**(1):33-42. DOI: 10.1016/j.combustflame.2009.09.016

[37] Katz BJ, Robison VD. Oil quality in deep-water settings: Concerns, perceptions, observations, and reality. AAPG Bulletin. 2006;**90**(6):909-920. DOI: 10.1306/01250605128

[38] Mansoori GA. Asphaltene Deposition and its Control [Internet Publication]. 1995. Available from: http://tigger.uic.edu/~mansoori/Asphaltene.Deposition.and.Its.Control_html [Accessed 2017-08-15]

[39] Alboudwarej H, Felix J, Taylor S, Badry R, Bremner C, Brough B, Skeates C, Baker A, Palmer D, Pattison K, Beshry M, Krawchuk P, Brown G, Calvo R, Triana JAC, Hathcock R, Koerner K, Hughes T, Kundu D, de Cardenas JL, West, C. Highlighting heavy oil. Oilfield Review. 2006;**18**(2):34-53

[40] Sheu EY. Physics of asphaltene micelles and microemulsions - Theory and experiment. Journal of Physics: Condensed Matter. 1996;**8**(25A):A125-A141. DOI: 10.1088/0953-8984/8/25A/009

[41] Lisitza NV, Freed DE, Sen PN, Song Y-Q. Study of asphaltene nanoaggregation by nuclear magnetic resonance (NMR). Energy & Fuels. 2009;**23**(3):1189-1193. DOI: 10.1021/ef800631a

[42] Rane JP, Harbottle D, Pauchard V, Couzis A, Banerjee S. Adsorption kinetics of asphaltenes at the oil-water interface and nanoaggregation in the bulk. *Langmuir*. 2012; **28**(26):9986-9995. DOI: 10.1021/la301423c

[43] Hartley GS. Aqueous Solutions of Paraffin Chain Salts. A Study in Micelle Formation. Hermann et Cie: Paris, France; 1936

[44] Hadjichristidis N, Pispas S, Floudas G. Block Copolymers: Synthetic Strategies, Physical Properties, and Applications. Hoboken, NJ, USA: John Wiley & Sons, Inc.; 2003

[45] Evdokimov IN. Characterization of asphaltenes and crude oils by near-UV/visible absorption spectroscopy. In: Duncan JA, editor. Asphaltenes: Characterization, Properties and Applications. New York: Nova Science Publishers, Inc.; 2010. pp. 1-46

[46] Wiehe IA. The pendant-core building block model of petroleum residua. Energy & Fuels. 1994;**8**(3):536-544. DOI: 10.1021/ef00045a003

[47] Loh W. Block copolymer micelles. In: Hubbard A, editor. Encyclopedia of Colloid and Surface Science. New York, USA: Marcel Decker Inc.; 2002. pp. 802-813

[48] Evdokimov IN, Fesan AA. Multi-step formation of asphaltene colloids in dilute solutions. Colloids and Surfaces A: Physicochemical and Engineering Aspects. 2016;**492**:170-180. DOI: 10.1016/j.colsurfa.2015.11.072

[49] Evdokimov IN, Fesan AA, Losev AP. New answers to the optical interrogation of asphaltenes. Monomers and primary aggregates from steady state fluorescence studies. Energy & Fuels. 2016;**30**(6):4494-4503. DOI: 10.1021/acs.energyfuels.6b00027

[50] Evdokimov IN, Fesan AA, Losev AP. New answers to the optical interrogation of asphaltenes. Complex states of primary aggregates from steady state fluorescence studies. Energy & Fuels. 2016;**30**(10):8226-8235. DOI: 10.1021/acs.energyfuels.6b01943

[51] Evdokimov IN, Fesan AA, Losev AP. Occlusion of foreign molecules in primary asphaltene aggregates from near-UV–visible absorption studies. Energy & Fuels. 2017; **31**(1):1370–1375. DOI: 10.1021/acs.energyfuels.6b02826

[52] Evdokimov IN, Fesan AA, Losev AP. Asphaltenes: Absorbers and scatterers at near-ultraviolet–visible–near-infrared wavelengths. Energy & Fuels. 2017;**31**(4):3878–3884. DOI: 10.1021/acs.energyfuels.7b00114

[53] Gray MR, Tykwinski RR, Stryker JM, Tan X. Supramolecular assembly model for aggregation of petroleum asphaltenes. Energy & Fuels. 2011;**25**(7):3125-3134. DOI: 10.1021/ef200654p

[54] Evdokimov IN. "Fine phase transformations" in petroleum - the basis for emerging nanotechnologies. In: Montclaire KL, editor. Petroleum Science Research Progress. New York, USA: Nova Science Publ; 2008. pp. 235-260

Chapter 8

A Realistic Look at Nanostructured Material as an Innovative Approach for Enhanced Oil Recovery Process Upgrading

Lezorgia Nekabari Nwidee, Ahmed Barifcani, Maxim Lebedev, Mohammad Sarmadivaleh and Stefan Iglauer

Additional information is available at the end of the chapter

http://dx.doi.org/10.5772/intechopen.72672

Abstract

With the continuous rise in energy demand and decline in reserves, the Petroleum Industries are constantly in search of inventive and novel approaches to optimize hydrocarbon recovery despite several decades of deployment of conventional and enhanced strategies. This chapter presents an in-depth analysis of nanomaterial (nanoparticles), their unique characteristics and potentials in relation to smart field development, enhanced oil recovery (EOR) and CO_2 geosequestration. The particles surface functionalities, unique size dependent property, adsorption, and transport behavior were scrutinized. The materials precise role in enhancing reservoir parameters that influences rock–fluid interactions, and reservoir fluid distribution and displacement such as permeability, wettability, interfacial tension, and asphaltene aggregate growth inhibition were evaluated. The study argues that the application of nanoparticle based fluids as novel EOR approach offers more holistic measures, potentials, and opportunities than micro and macro particles and can stimulate the continuous evolution of EOR processes even under harsh reservoir conditions, thus, offering better benefits over conventional surface-active agents. We believe this study will significantly impact the understanding of EOR with respect to nanoparticles, which is crucial for augmenting reservoir processes and to accelerate the realization of nanoparticles for EOR and CO_2 sequestration processes at industrial scale.

Keywords: EOR, nanoparticles, adsorption, Asphaltene, wettability, IFT, permeability, viscosity, CO_2

1. Overview

It is a fact that the period of cheap hydrocarbon recovery is waning as hydrocarbons are currently been explored in remote regions under harsh reservoir conditions. The accessible hydrocarbon resources are constantly declining with a corresponding increase in energy demand which invariably contributes to the price irregularities in the oil and gas industry. Globally, the growth in energy demand appears to be predominant amidst feasible unconventional energy options (renewable energy). Unarguably, newer energy sources, such as nuclear, wind, geo-thermal and solar are effective measures for addressing energy shortage. However, they are yet to be ample alternatives for substituting the role of oil in meeting the ever-rising energy demand. To date, these demands are currently being substantiated via hydrocarbon sources especially petroleum – oil is still the most valuable product with great global economic impact. Currently, diverse conventional strategies (waterflooding), and enhanced methods (use of chemicals, gases, and microbes) are been deployed in oil fields, regardless, these challenges persist. More efficient, yet cost-effective, environmentally friendly, and novel alternatives are constantly being searched for process feasibility, and to meet the current global high-energy demand. Interestingly, these challenges can be averted through innovative scientific and engineering approaches such as nanotechnology, specifically, via the use of nanostructured materials. Nanostructured materials in the form of nanoparticles or carbon nanotubes have attracted attention for consideration as effective solutions for resolving challenges in the oil and gas industry. Nanoparticles are considered the simplest forms of nanostructured materials categorize as metallic, metal oxides or magnetic nanoparticles; whereas, the carbon nanotubes are carbon allotropes with a cylindrical nanostructure in the form of a tube which can be single, double, or multiple walled nanotubes [1, 2]. These materials have vast application potentials industrially, however, the choice of a specific material type is highly dependent on the application process.

Research has shown that diverse potentials exist for the application of nanostructured materials in the oil and gas industry for smart field development (**Table 1**). The carbon nanotube materials display great unique properties such as stiffness, strength, and tenacity in comparison to other fiber materials, which lacks one or more of these characteristics. Also, the thermal and electrical conductivity of carbon nanotubes is higher in comparison to other conductive materials [1]. Nanoparticles (NP) are characterized by unique thermal, mechanical, and chemical properties thus particularly interesting for the oil and gas industry. For instance, oil and gas tools are manufactured with nanomaterial functionalities; the inner surfaces of pipelines and production facilities can be coated with nano-based paints; nanoparticle based catalysts are been used in the oil and gas downstream for oil refining and petrochemical activities. Currently, complex reservoir rock heterogeneity and microstructures, and the corresponding effect on multiphase flow as well as the reservoir depth information can now be effectively characterized using sensors fabricated with nanoparticle materials. Superparamagnetic nanoparticles can be effectively used for conductive heat generation [3]. Nanoparticles dispersed in solvents are also been used for drilling, exploration, production, and enhanced oil recovery processes. Such fluids can be used for resolving shallow water flow challenges, drilling and production issues such as borehole instability linked to poorly consolidated and

http://dx.doi.org/10.5772/intechopen.72672

Processes	Material type	Applications	References
Drilling and completion			
Drilling Bits	Carbon nanomaterial - nano-diamond polycrystalline diamond compact (PDC) Technology	• Suitable for functionalization of PDC cutters for drill bits • Effective for the manufacture of advanced drill bits that has good resistance to abrasion and corrosion upon nano-coating	[1, 46, 47]
Logging while drilling (LWD)	Nanostructured glass-ceramics	• Effective neutron detectors - Li-6 scintillation nanostructured glass ceramics shows better efficiency than ordinary Li-6 scintillation materials	[1]
Cementing	Carbon nanotube (CNT)	• Suitable as reinforcement materials for cement instead of conventional fibers	[1]
Drilling fluids	NP	• Control of fluid loss and wellbore stability • Design of drilling fluids suitable for harsh reservoir conditions • Nanoparticle based drilling fluids can be effective for elimination or reduction of drilling problems such as pipe sticking, equipment wear and tear, rig instability and poor hole cleaning issues	[1, 4]
Exploration			
Geothermal production	NP	• Nanoscale metals suitable for delineation of ore deposits during geochemical exploration	[1, 19, 48]
Exploration			
Imaging tool	NP	• Hyperpolarized silicon Np deliver innovative tools for measurement and imaging oil exploration	[1, 29, 49]
Production			
Hydrate Recovery	NP	• Nickel-iron Np can be used for dissociation of hydrates	[1, 29]
Viscoelastic surfactant stimulation fluid (VES)	NP	• Np associates with VES micelles for stabilization of fluid viscosity	[1, 29]
Reservoir characterization and management			
EOR	NP	• Advanced EOR approach - suitable for surface/interfacial modifications and oil recovery	[23, 24, 31, 50, 78, 79]
Inhibition of inorganic scales	NP	• Inhibition can be achieved over long distances from the injection point using synthesized nanoparticles	[42]
Hydrocarbon detection	NP	• Suitable for detection of hydrocarbon in formation rocks	[1, 47, 51]
Oil microbe detection tool	Nano optical fiber	• Suitable for reservoir penetration and for transmission of laser light which enables the detection of stranded or bypassed oil	[1, 47, 52]
Carbon capture and storage	NP/ Nanomembranes	• Enhances improved gas streams separation and for contamination removal from oil	[1, 19, 47]

Table 1. Nanostructured material applications for smart field development in the oil and gas industry.

or reactive shale zones, reduction of the frictional resistance between pipes and borehole well, and for prevention of loss circulation of drilling mud [4]. Despite the relevance of nanoparticles in diverse oil and gas engineering processes, reservoir engineering specifically enhanced oil recovery (EOR) have attracted the most attention in recent times. EOR is mainly driven by the possibilities of increasing the overall oil displacement efficiency which is greatly dependent on microscopic and macroscopic displacement efficiencies [5, 6]. Unarguably, EOR mechanisms augment additional oil recovery potentials than the primary and secondary methods. However, the successful deployment of EOR requires an understanding of the mechanisms that control efficient fluid displacement and distribution, as well as, overall optimal recovery efficiencies. A number of processes and technologies are used in this regard to upsurge or maintain recovery from existing fields which includes the injection of steams (Thermal EOR), fluid(s) (Chemical EOR) and gases into a reservoir (Gas EOR), and recently microbes (Microbial EOR). These processes facilitate effective displacement of oil towards the producing well, thus, production from matured fields are enhanced as the injected fluid, gases, or microbes interact with the formation rock and oil systems thereby creating a favorable condition for better recovery. Comprehensive breakdown of these processes is provided in our previous work [6]. In the face of diverse EOR technological advancements, there exist several challenges that impede process efficiency. The key process challenges among others are:

- Thermal EOR (TEOR): high operational cost, and environmental unfriendliness owing to emissions from surface steam generation.
- Chemical EOR (CEOR): high chemical cost, chemical mismatch, porous media pore-plugging and chemical trapping issues.
- Gas EOR (GEOR): separation issues, inadequate sources of gas on basis of geographical availability, and mobility control issues.
- Microbial EOR (MEOR): complex process mechanisms, and poor oil recovery potential.

Among these challenges is the need for more economically feasible approach that will reduce operational cost and improve the quality and amount of recoverable oil to meet the global market demands with less environmental footprints.

2. Nanoparticles

Nanoparticles are nano-sized structures with at least a phase of one or more dimensions in the nanometer size range (typically 1–100 nm). The uniqueness of the properties of the material dates to several centuries (ninth century AD). The first optical property of nanoparticle was discovered from a luster technology on a glazed ceramic in Mesopotamia at the time, which later became popular in Egypt, Spain, and Persia. An archeomaterial of this material indicted the optical properties were as a result of the metallic nanoparticles that were dispersed on the outermost layers of the glaze [7]. With the advent of time, practicalities of nanoparticle have become even more prevalent owing to their physical and chemical properties. Nanoparticles cut across a wide variety of materials such as; (a) materials with surface protrusions spatially separated by distances in the order of nanometers; (b) porous materials with particle sizes in the nanometer range or nanometer sized metallic clusters dispersed within a porous matrix; (c) polycrystalline materials with nanometer sized crystallites. Quantization of the electronic states of nanoparticles, and the possible state manipulation through shape and size control are the key drivers for research in nanoscience or nanotechnology. At nano-scale, the property changes that occur are based on the nature of the material and the applied mechanisms. For instance, when metallic nanoparticles transition from bulk metal to a much smaller metallic nanoparticle, the bulk metal energy continuum changes to produce even more discrete energy levels – an indication of a decrease in the density of the electronic states [8].

2.1. Characteristics of nanoparticles

Nanoparticle application though in its early stage have emerged a novel EOR pathway owing to its unique properties and transport behavior that can address the underlying mechanisms that influence reservoir behavior and the corresponding effect on efficient recovery. Nanoparticles are considered suitable for EOR due to their strong particle surface interaction with solvents, which helps to overcome density difference that make materials float or sink in a system This material has an extensive range of unique characteristics for varied functionalities such as surface plasmon resonance, superior catalytic activity, intrinsic reactivity, quantum confinement, great adsorption affinity and dispersibility [9–11]. Interestingly, the particles surface functionalities can be engineered with potentials for subsurface engineering applications.

Comparison of the exceptional properties of nanoparticles to their bulk material counterpart shows that nanoparticles display quite different properties and superior behavior. Its applications enable the creation of new composites with unique properties, which allows for innovative technological advancements. Typically, nanoparticles are characterized by thermal, mechanical, physical, and chemical properties. These properties span through material's grain boundary, particle size, surface area per unit volume, size of the particles, purity, thermal conductivity etc. On basis of their thermal properties, nanoparticle based fluids exhibit high thermal conductivity and high surface area than conventional base-fluids. The thermal conductivity of conventional base liquid can be improved by addition of nanoparticles as heat transfer occurs at the surface of the particles [12]. The high surface area increases the thermophysical properties and the small particle size enhances their potentials as absorber fluids due to high potential of large surface coverage in heat transfer processes. Lee et al. [12], Choi [13], and Eastman et al. [14] reported that nanoparticle based fluids exhibit higher thermal conductivity with great dependency on factors such as the material type, size, shape, surface area, particle volume fraction, base fluid material and temperature. For example; the thermal conductivity of ethylene glycol – a conventional base fluid with thermal conductivity of approximately 0.258 (W/m K), can be significantly enhanced (20% increase) upon addition of about 4% volume CuO of 35 nm particles. On the mechanical properties, nanoparticles exhibit high ductility, high average hardness, and maximum strain in comparison to the other composite materials, thus, considered suitable for improving fracture toughness. However, the particles geometry can influence the mechanical properties of nanoparticles such as tensile modulus, tensile strength, and fracture toughness [15]. Zhang and Singh [16] showed that the addition of 4.5% volume fraction of well-bonded Al_2O_3 particles to unsaturated polyester led to a 100% increase in fracture toughness. It has also been reported [17, 18] that nanoparticles are suitable filler materials for improving the mechanical properties of composites. The addition of small amount of rattan nanoparticles (0–5%) to polymer matrix (polypropylene) as fillers enhanced the mechanical performance of composites.

With respect to the physical and chemical properties, these span through shape, size, structure, quantum confinement, lattice parameter and symmetry, charge, solubility, and surface coating [19–21]. The enhanced physiochemical properties and superior performance of nanoparticles is mainly based on the overriding effect of surface forces from the particles atomic scale size rather than the mass forces. The physical and chemical properties of the bulk materials are typically constant irrespective of the material size; and the number of surface atoms is substantially smaller than atoms in the bulk phase. This is not the case with nanoparticles owing to the small particle size, high surface area, and unique transport behavior that enhances surface performance. Nanoparticles yield applications that are even more extensive with the particle size approaching nanoscale. The particles coalesce into the structure of materials and display strength. The atoms of conventional materials are located in the interior of the particle, whereas, the atoms of nanoparticles are located on the particles surface [22–25]. A dramatic change in properties such as thermal resistance, chemical and catalytic activities, melting point, internal pressure, magnetism, and optical behavior occurs as the atoms at the surface of the material become even more significant owing to the large surface area to volume ratio. The surface area to volume ratio appears exponentially larger, and the number of surface atoms becomes even more substantial with respect to that in the bulk phase [2, 10, 11, 26–29]. Amanullah and Al-Tahini [4] reported a comparison between the surface area to volume ratio of spherical particles of the same nanomaterial in micrometer, millimeter, and nanometer. Nanoparticle surface area to volume ratio is over a million times higher than the conventional scales and the particle surface area is also higher than the bulk particles – micro and macro materials of the same mother source, which enhances their increased potentials even with a significantly low concentration of the materials in the system. Such high surface area increases the particles surface energy which leads to structural transitioning [30]. This behavior allows for favorable particle adsorption at the surface boundaries and permits high particles contact tendencies with nearby materials or solvent molecules in suspensions [23], and the particles boundaries facilitate their enhanced quantum effects. A key benefit of nanoparticles in EOR is the ability to alter reservoir rock surfaces – where strongly hydrophobic surfaces can be rendered hydrophilic upon contact with nano – based systems [23, 24, 31–38] thus interesting for engineering and subsurface processes, EOR, and carbon geo-sequestration process efficiency.

2.2. Nano-based smart fluids

Nanoparticle based fluids otherwise called smart fluids are fluid systems containing nano-additives, and the systems exist in the form of suspensions or emulsions known as nanofluids, nano-catalyst suspensions or nanoemulsions. The nano-emulsions are emulsions that exhibit unique characteristics such as great stability in comparison to micro-emulsions. Anton et al. [99] established that nano-emulsions are kinetically stable irrespective of temperature changes

http://dx.doi.org/10.5772/intechopen.72672

in comparison to other emulsions that are thermodynamically unstable systems with free energy of emulsion formation greater than zero. Since nanoparticles are active emulsifiers of oil and water [39, 40], preparation of ultra-stable emulsion is attainable by the control of the degree of surface charges. The emulsions are not subject to gravity driven separation as the dispersed droplets are relatively small size which eliminates separation processes owing to phase density difference. The particle dispersals are typically mixtures of water, oil, surfactant, or brine. It has been reported that nano-emulsions upsurge additional oil recovery [41]. Zhang et al. [42] reported that nanoparticles form stable emulsions, however, the level of stability depends on the oil polarity as the formation of oil-water stable emulsions are more favorable to non-polar oils and water oil emulsions are preferable with polar oil. Another relevant system suitable for EOR processes in porous media is the nano-based catalysts suspensions, which demonstrates high efficiency for heavy oil recovery. The presence of nanocatalyst in porous media has been reported to enhance bitumen conversion to lighter products [10, 43, 45]. Nanofluids are colloidal suspensions of solid nanoparticles or nanofibers consisting of a carrier medium and solid phase. A key advantage of nanofluids over conventional fluids in EOR is that nanoparticles improve the property of its dispersals even at very low particle concentrations in the suspension. The particles in suspension enhance the performance of the carrier fluid during production owing to high dependency on the unique electrical, rheological, and magnetic properties of the nanoparticles. Since the carrier fluids are normally conventional heat transfer liquids, the particles also enhance the thermal conductivity and convective heat transfer performance of this base liquids as its thermal conductivities are typically an order-of-magnitude higher than that of the base fluids [9, 63]. The nano-based smart fluids display significantly high functional abilities in different systems while reducing the overall fluid cost irrespective of high cost of individual additives [4].

3. Definite role of nanoparticles in porous media applications for process enhancement

3.1. Adsorption/transport potentials

Nanoparticles injected into porous media can exhibit different phenomenal behaviors; adsorption, desorption, transport, and blocking inclinations. The interactions between nanoparticles and the walls of the porous media can lead to adsorption if the force of attraction is higher than the repulsive with the reverse leading to desorption, however, adsorption and desorption is a dynamic balance process that is controlled by the total force existing between the porous media wall and nanoparticle. Blocking can occur if the nanoparticle diameter appears bigger than the size of the pore throat or due to particle to particle aggregation at the pore throat which can inhibit efficient transport of nanoparticle based fluids through the porous media [44, 54–59]. Generally, nanoparticles regardless of their small particle sizes are easily susceptible to aggregation in liquid suspensions, which promotes surface adsorption of neighboring particles or molecules. This behavior is due to the particles large surface-to-volume ratio and poor wettability in the metal matrix [2, 60, 61, 63]. However, a recent study on nanoparticle adsorption and transport behavior shows that nanoparticle based fluids can efficiently flow through the porous media without excessive adsorption and retention inclinations with an equilibrium adsorption estimation of 1.27 mg/g for 5000 ppm nanofluid [62]. More so, when the particle size of nanomaterial is reduced to near or less than the wavelength of electrons conductivity, the property of the material changes. Since oil is typically recovered from formation rocks with micron-sized pores, nanoparticles are much smaller fine particles with unique size dependent physical and chemical properties that enhance easy penetration through the reservoir pores since the particle size is much smaller than the conventional rock pores [22–24]. Thus, fluid flow through the reservoir pore spaces can be significantly improved with little or no negative impact on the formation permeability, or porous media pore-plugging or chemical trapping related issues that are commonly associated with conventional EOR processes. Shen et al. [63] and Zhang et al. [64] also reported the excellent transport, diffusion, retention, and releasing properties of nanoparticles in porous media.

3.2. Asphaltene growth inhibition

Nanoparticles exhibit strong chemical stability, irrespective of harsh reservoir conditions and great selective adsorption behavior in the presence of heavy components in crude oil such as asphaltenes. Asphaltenes are common petroleum macromolecules which poises challenges such as well bore plugging, adsorption, pipeline deposition, and sedimentation issues. Asphaltenes [65, 66] exist in forms of solid deposits and usually self-associate with resins in crude oil and are often problematic during production as they can plug wellbore tubing and valves, and contribute to undesirable formation rock surface coating. Such asphaltenes – polydispersed [67] can exist in the form of colloidal suspensions or micelles which form deposits when the wettability conditions of the reservoirs are altered. A study by Franco et al. [68] on the adsorption behavior of nanoparticles on heavy components in crude oil such as resins and asphaltenes shows that nanoparticles in contact with crude oil is capable of adsorbing resins and (or) asphaltenes, however, nanoparticles exhibit great selectivity and preferential affinity for adsorption of asphaltenes (n-C_7) compared to resins. This behavior is also confirmed by an earlier report by Nassar et al. [56, 57], which shows that nanoparticles have great affinity for asphaltenes—where alumina nanoparticles adsorbed asphaltenes more readily than micro porous alumina particles with similar acid content. Taborda et al. [69] investigated nanoparticles effect on adsorption isotherms of n-C_7 asphaltenes at 298 K temperature in complex crude oil systems. The authors compared the performance of different ranges of SiO_2 and Al_2O_3 nanoparticles. The SiO_2A nanoparticles showed better adsorption capacity than others as the adsorption potential to capture n-C_7 asphaltenes is in the order of $SiO_2A > SiO_2 > Al_2O_3$. The acidic silica exhibited better adsorptive capacity especially at low concentrations owing to increased surface acidity of the nanoparticles as n-C_7 asphaltenes adsorption increases concurrently with increase in surface acidity [70].

Heavy oils are typically known for their high asphaltene content (> 5 wt.%), such asphaltenes [71, 72] forms viscoelastic network of large size nanoaggregate clusters and promote formation damage as a result of self-association, and if sub-saturated oil reservoirs exist at pressures above the bubble point. The asphaltene becomes excluded from the oil matrix and then forms deposits on the reservoir rock. This behavior causes blockages of the porous media and wettability challenges. Reduction of large asphaltene aggregates using nanoparticles by breaking the attachment points within the viscoelastic network is thus necessary to improve their configuration and distribution in the oil matrix. Taborda et al. [69] examined the behavior of n-C_7 asphaltenes in the presence and absence of nanoparticles as a function of time at 298 K temperature. An initial increase in the mean size of asphaltene aggregates (d_{asp}) occurred with time, followed by a decrease in size in the absence of nanoparticles with a notable stabilization of approximately 300 min. This was attributed to the aggregation-fragmentation forces that exist under shear conditions which has a direct growth / reduction effect on d_{asp}. Similar behavior was also observed in the presence of the diverse nanoparticles (Al_2O_3, SiO_2 and SiO_2A) tested, however, a lower d_{asp} and faster stabilization (180 min) was attained in a decreasing order of $SiO_2A < SiO_2 < Al_2O_3$. Franco et al. [73] conducted an asphaltene displacement and adsorption test to inhibit formation damage using nanoparticles at reservoir conditions. The authors reported a significant increase in oil production level owing to the presence of nanoparticles. 80% increase in oil recovery occurred after the injection of nanoparticle based fluids in damage systems owing to efficient inhibition of the growth of asphaltene aggregates and the swift capture of asphaltene through adsorption.

3.3. Permeability improvement

Permeability plays a major role in oil recovery processes as adequate permeability can improve reservoir fluid displacement. High permeable formations permit high fluid displacement alongside other essential factors, such as capillary pressure and formation wettability. Relative permeability is considered the ratio of the phase permeability to the absolute permeability k. The phase permeability primarily exists as a water, gas, or oil phase as shown in Eqs. (1)–(3):

$$\text{Water phase relative permeability} : k_{rw} = k_w/k \tag{1}$$

$$\text{Gas phase relative permeability} : k_{rg} = k_g\ /k \tag{2}$$

$$\text{Oil Phase relative permeability} : k_{ro} = k_o/k \tag{3}$$

Nanoparticles have the potentials of enhancing the reservoirs permeability for achieving higher recovery factor. Nanoparticles can efficiently expand the pore radius, enable the reduction of flow resistivity of the water injected into the pores, promote water permeability enrichment, promote reduction in injection pressure and enhance injection volume as well as recovery potentials. Wang et al. [74] reported that the injection of 0.5 wt.% of SiO_2 nanoparticle concentration led to over 10% increase in oil recovery. The small particle size of this material enabled penetration into the low-permeability layers while averting hydration occurrences from water-rock contact, and promoted injection and subsequent augmentation of oil recovery even in low permeable regions. Guzmán et al. [75] performed a core flooding test in bauxite porous media, relative permeability was measured in the presence and absence of fracturing fluid containing nanoparticles. Injection of fracturing fluids into reservoirs can induce formation damage, where the original K_{ro} values appears lower than the base system due to some degree of formation damage. However, with optimized fracturing fluid containing nanoparticles, K_{ro} becomes higher than the base system with mobility range of 0.42–0.80. Formation damage reduction from 55 to 16% occurred which was about 71% reduction upon addition of nanoparticles to the optimized fracturing fluid, thus, an indication that the presence of nanoparticles improved the fluids performance and inhibits the associated formation damage which may be due to the porous media wettability alteration as nanoparticle inclusion changes the rock wettability from oil-wet to water-wet.

3.4. Wettability alteration

Oil recovery exhibits great dependency on the formation wettability as it controls the fluid displacements of the wetting and non-wetting phase at the pore scale. Wettability is however influenced by the degree to which the crude oil polar components deposits or adsorbs on the mineral rock surface, hence, an understanding of fluid–rock interaction is vital for efficient EOR processes. Quantification of reservoir wettability in relation to nanoparticles are analyzed by contact angle, imbibition etc.

3.4.1. Contact angle

Contact angle is an effective wettability measurement approach for evaluating rock surface wetting as it enables the valuation of the balance of forces at the contact line. Theoretically, this has been demonstrated by the Young-Laplace equation (Eq. (4); **Figure 1**). Wetting phenomenon is closely related to the reservoir rock surfaces and the imbibing fronts, as such if a liquid wet a solid, the change in wettability can be dramatic even for a slight surface modification of the surface property of the solid. Wetting via nano-modification have been reported in several literatures [23, 24, 31, 36, 37, 76, 77, 80–85], despite the disparity in test methods, a clear trend in nanoparticles wetting efficiency in changing hydrophobic surfaces towards favorable hydrophilic conditions were reported. For example: Ju and Fan [76] tested the wetting inclination of lipophobic and hydrophilic polysilicon nanoparticles in hydrophobic sandstone formation. The authors reported that the presence of nanoparticles caused a change in wettability to hydrophilic state owing to efficient particle adsorption on the surface of the rock. Nwidee et al. [24] assessed nanoparticles surface modification effect on solid-wetting in hydrophobic limestone formation over a wide range of temperatures (22–70°C) using Zirconium oxide—ZrO_2 and Nickel oxide—NiO; concentrations - 0.05 wt.%). To optimize the wetting characteristics of the nanoparticles, appropriate amount of surfactant (cationic - $C_{16}TAB$) was used, which facilitated the reduction in the solid-nanofluid-oil three-phase contact angle and maximized the structural force from the nanoparticles confinement in the wedge film. The nanoparticles formed a wetting wedge in form of an ordered structure near the three-phase contact line of the droplet on the rock surfaces, which stimulated significant wetting and spreading effect with ZrO_2 system exhibiting better surface wetting over NiO ($ZrO_2C_{16}TAB$ > $NiOC_{16}TAB$) while maintaining great stability at elevated temperature (**Figures 2** and **3**). Cao et al. [77] performed

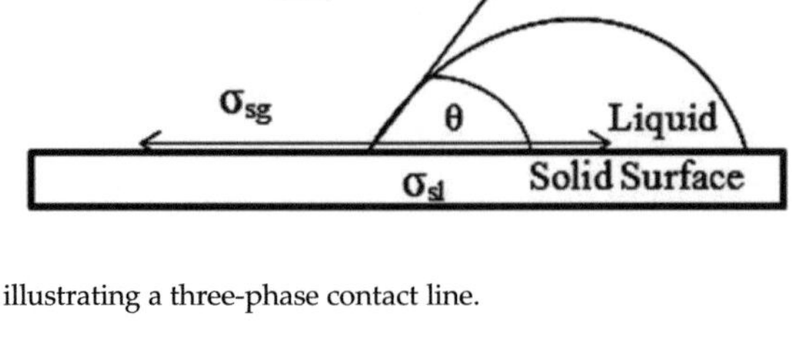

Figure 1. Young's equation illustrating a three-phase contact line.

a wetting investigation in the presence of zirconium oxide, and silicon oxide dispersed in surfactant, alkali and brine solutions. The authors reported that the nanofluids displayed great stability under harsh reservoir conditions of high pressure and high temperature.

$$\sigma_{sg} = \sigma_{sl} + \sigma_{lg}\, cos\theta \tag{4}$$

where σ_{sg} is the solid surface free energy; σ_{sl} is the IFT between the solid and liquid; σ_{lg} is the liquid surface tension; θ is the contact angle.

Notably, although nanoparticles promote formation rock wetting, the extent to which wetting occurs is dependent on the level of contact established between the rock surface and the wetting phase, which can impact the rather immobile fluid on the rock surface while mobilizing such fluids towards the flow lines.

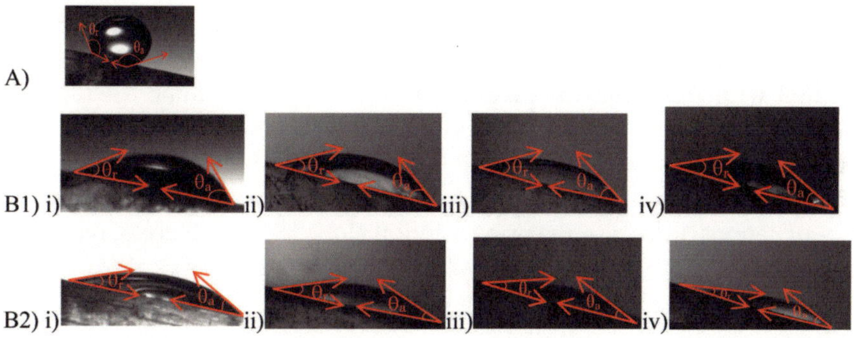

Figure 2. The receding (θ_r) and advancing (θ_a) contact angle images of water droplets (probe liquid) before and after exposure to nanofluid (A) oil modified rock (before nano-modification - high θ - indicates the surface's hydrophobicity - 152° θ_a); (B) nano modified at different temperatures (22–70°C) (B1- $NiOC_{16}TAB$; B2 - $ZrO_2C_{16}TAB$ - low θ - indicates hydrophilic conditions) which represents strong interaction with the rock surface thus exhibiting low surface energies and inclination to wet; I–IV); see graphical representation (**Figure 3**) for θ - values [24]©.

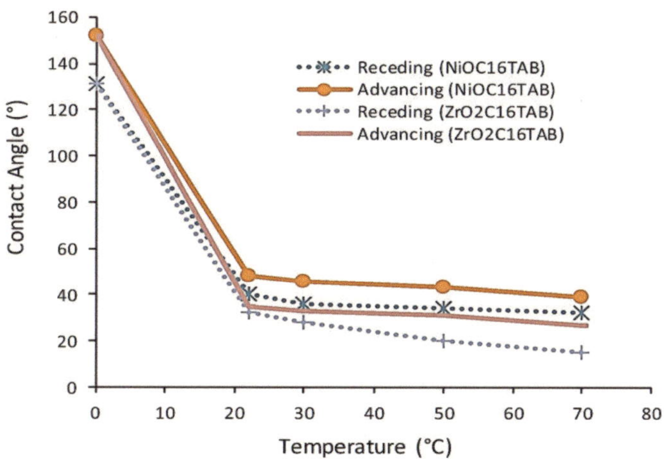

Figure 3. θ_R and θ_A depicting wettability change of NiO, ZrO_2 (0.05 wt.%), $C_{16}TAB$ (0.5 wt.%) blends in toluene (1-hour Exposure time) at different temperatures [24]©.

3.4.2. *Imbibition*

Imbibition enables an assessment of the displacement capacity of a wetting phase over a non-wetting phase. When immiscible fluids in the rock pore spaces are in contact, there exists possibilities for pressure discontinuity across the interfaces which shows high dependency on interfacial tension, the porous media pore spaces, and the wetting angle. Capillary pressure (the pressure variation between two fluid phases in contact - Eq. (5)) and wetting phase saturation plays a key role in imbibition processes and exhibits great dependency on wettability. For example; if a formation exhibits hydrophobic wetting characteristics, water imbibition in such systems is typically inhibited owing to negative capillarity. Whereas, if a formation exhibits hydrophilic wetting characteristics, such promotes better oil displacement from the rock as water readily imbibes into the rock pores and the presence of positive capillary pressure promotes this behavior. In a typical complex crude oil system, Zhang et al. [86] performed an imbibition test under high temperature, pressure, and high salinity condition using IIT nanofluid to simulate reservoir condition. The nanoparticle dispersion was reported as insensitive to electrolytes or temperature and stable in harsh reservoir environment. Core plugs saturated with crude oil was placed into a glass jar containing the test solution (IIT nanofluid and brine) to test their potentials to displace oil from the cores at 55°C. As the fluids imbibe into the core, oil is expelled which floats on the top of the glass jar. The recovered oil was recorded as a function of percentage of the original oil in place versus time in the presence and absence of nanoparticles. For the brine solution alone in the absence of nanoparticles only about 17% of the crude oil was displaced, whereas, the oil-wet cores exposed to the nanofluid formulations displaced approximately 50% of the crude oil. Youssif et al. [87] conducted a tertiary recovery core flooding test using silica nanofluid after imbibition by water flooding. Black oil (32.5 API and 4.6 cp) from the North Sea was used in the flooding experiments. Oil saturated cores were subjected to imbibition by initial saturation with brine to displace oil at injection rate of 0.5 ml/min and a pore volume of 1.77 PV. Eq. (6) was used for the calculation of the displacement efficiency. The injection of silica nanofluid in the core enhanced oil production with increase in particle concentration (0.01–0.1 wt.%), then decrease in the amount of oil at much higher concentrations (0.5 wt.%) occurred after the optimal concentration (0.1 wt.%), which was attributed to possible permeability impairment. At 0.1 wt.% silica concentration, an increase in the oil recovery factor from 53.1% (waterflooding alone - **Figure 4**) to 66.40% following the nanofluid injection occurred. Better recovery occurred in comparison to water flooding alone, with an initial oil in place increase of 13.28% in the presence of the nanoparticle and low injection rate of the nanofluid decreased permeability impairment.

$$P_c = P_n - P_w \tag{5}$$

where P_c is the capillary pressure (Pa), P_n is the non-wetting pressure (Pa), P_w is the wetting pressure (Pa).

$$\mathrm{E_D} = [1 - \mathrm{S_{or2}}/\mathrm{S_{Or1}}] \times 100 \tag{6}$$

where $\mathrm{E_D}$ is the displacement efficiency, $\mathrm{S_{or}}$ is the residual oil saturation, $\mathrm{S_{or1}}$ is the Sor after brine flooding, $\mathrm{S_{or2}}$ is the Sor after nanofluid flooding.

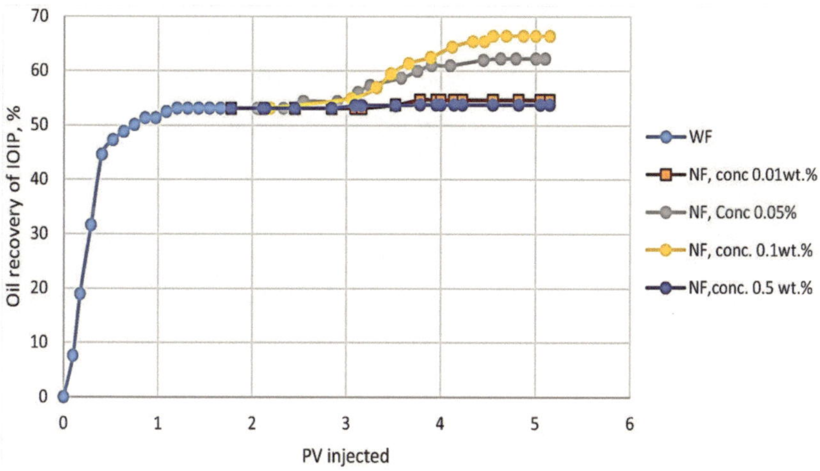

Figure 4. Oil recovery vs. pore volume (PV) injected for oil-wet Berea sandstone before and after nanofluid exposure [87]©.

3.4.3. SEM and AFM imaging

Mechanistic surface imaging tools such as scanning electron microscopy (SEM) and atomic-force microscopy (AFM) are used to characterize porous media wetting, adsorption characteristics of reservoir rocks and their surface roughness. Morphology and roughness are key parameters that influence rock surface wetting. Assessing wettability by contact angle usually does not account for surface roughness and the pore rock structures that impact liquid–solid contact lines. For improved knowledge of wetting and rock surface behavior, it is imperative that these factors are put into considerations. SEM is an approach for micro-analysis of solid materials with high accuracy. SEM produces high-resolution images for even the smallest structures at nano-scale as high magnifications are used during image processing. SEM investigations on rock substrates with respect to wetting have been reported by several researchers [88–90]. AFM also enables imaging of the surface of the samples at extremely high resolutions using a sharp tip for probing the surface features of the samples. These tools enable micro and nano-scale evaluation of solid surfaces, the topographic changes of such surfaces, and their surface roughness. The surface roughness is generally estimated using Eq. (7) and (8). **Figure 5** shows example cases of SEM and AFM images in wetting evaluations [23]. Treating rock surfaces with nanoparticles based fluids have been shown to improve surface wetness confirmed via contact angle examinations owing to efficient adsorption of nanoparticles on rock surfaces and surface roughness increase. Several wetting studies have been performed via this approach and confirms the effectiveness thereof on diverse rock surfaces [23, 24, 46–48, 91–94]. For example, the adsorption and surface roughness behavior of blends of nanoparticles (ZrO_2 and NiO) and surfactants (cetyltrimethylammonium bromide (C_{16}TAB-cationic) and triton X-100 (non-ionic)) were mechanistically assessed by exposing rock substrates to different nanoparticle-surfactant fluids (NiO/C_{16}TAB, NiO/TX-100, ZrO_2/C_{16}TAB, ZrO_2/TX-100) [24]. The treatment of the substrates with surfactant - nanoparticle based fluids enabled favorable adsorption of the particles on the surface of the rock owing to electrostatic interactions, with a corresponding increase in surface roughness (mean

Figure 5. SEM (a) and AFM(b) images of ZrO_2 nanoparticle as uniform nano-spherical particle with mean particle size of 28 nm depicting surface adsorption on calcite substrates [23]©.

surface roughness, S_a and root mean square, RMS). The $C_{16}TAB$ - nanoparticle systems exhibited better surface adsorption, higher surface roughness, and wetting inclinations than the TX-100 - nanoparticle systems in the order of $ZrO_2/C_{16}TAB$ (Sa −120 nm; RMS -200 nm) > $NiO/C_{16}TAB$ (Sa −100 nm; RMS -190 nm) > NiO/TX-100 (Sa −78 nm; RMS - 180 nm) > ZrO_2/TX-100 (Sa - 60 nm; RMS - 83 nm) [24]. Similarly, the irreversibly adsorption behavior of silica nanoparticles through SEM and AFM have also been reported [37]. Nano-modified substrates were exposed to different cleaning fluids at temperatures of 296 and 323 K. The authors reported that SiO_2 nanoparticle homogeneously adsorbed on the rock surface with increase in temperature and rock surface roughness (RMS from 18–580 nm at 296 k; 2700 nm at 323 K) which also stimulated better wetting.

$$S_a = \frac{1}{N} \sum_{j=1}^{N} /Zj/ \tag{7}$$

where N is the total data points in the measurement; Z_j is the vertical deviations measured from the average height of the surface.

$$RMS = \sqrt{\frac{\Sigma(Zj)^2}{N}} \tag{8}$$

where N is the data points; Z_j is the vertical deviation of j- point with respect to the mean line.

3.4.4. Zeta potential

Zeta potential is a simple, rapid, direct-on-particle approach that permits the characterization of particle modification. This electro kinetic potential allows for estimation of the potential difference in colloidal systems that exist between the dispersion medium and stationary layer of the fluid that attaches to particles. It is closely related to the morphology of the particle surface and the suspension stability as the colloidal dispersions stability is dependent on the zeta potential value. The zeta potential of the rock-water interface and oil–water interface charges can be estimated based on the properties of the particles and suspension conditions (ionic type and strength, pH, and temperature). For example, in carbonate formation; pH and potential ions that forms a part of the carbonate structure such as Ca_2^+ and CO_3^{2-} enables the determination of carbonate-water interface via zeta potential [95]. With respect to nanoparticles, zeta potential enables estimation of the electric potential of particles on the inner boundary of the diffuse layers and surface charge of nanoparticles, since surface charges impact their reactivity, toxicology, and agglomeration behaviors [96, 97]. However, the agglomeration inclinations of nanoparticles can promote colloidal instability. Manan et al. [98] reported that different types of nanoparticles behaves differently owing to their zeta potential. In their study, the zeta potential of four different nanoparticles were compared and reported with varied zeta potential ranges (10–24 mv); where the zeta potential of Aluminum oxide(Al_2O_3) was 24 mv, Silicon dioxide (SiO_2) - 19 mv, Titanium dioxide (TiO_2) -13 mv, and Copper oxide 10 mv. The higher the zeta potential value whether positive or negative, the higher the stability potentials of the NP suspensions. Low zeta potential of colloidal fluids indicates that the repulsive force is lower than the forces of attraction which promotes precipitation and flocculation. However, such flocculation of nanoparticle based fluids like nano-emulsions can be prevented via steric stabilization owing to their droplet size which is sub-micrometric. Overlapping of interfacial droplet layers causes steric repulsion due to the unfavorable mixing of the stabilizing chain of the adsorbed layer which is dependent on the interfacial layer thickness, density and interactions between the interfacial layer and solvent [99–102]. Theoretically, Derjaguin and Landau [103] and Overbeek and Verwey [104] through the DLVO (Derjaguin, Landau, Verwey, and Overbeek) have shown that the stability of colloidal particles in solution is governed by their total energy of interactions (attractive and repulsive) when the particles are near each other. Nanoparticle as colloidal particle interacts with its carrier fluid or dispersal due to van der Waals attraction or electrostatic repulsion forces, of which the effects of the attraction forces can be higher than the repulsive. A comparative zeta potential tests was conducted by Ahmadi et al. [105] to examine the interactions between fines and pore surfaces and the change in total energy of the interactions due to colloidal forces of alterations in the presence of nanoparticles. The pore surfaces were coated with MgO (magnesium oxide), aluminum oxide (Al_2O_3) and silicon dioxide (SiO_2). The pore surfaces changes in potential was tested and the total interaction energy was used to estimate the effects of the nanoparticles on surface properties and fines migration reduction. Upon comparison of the effects of the different nanoparticles, the authors reported that MgO displayed the highest propensity to attach fines to the surface and significantly reduced fines migration than Al_2O_3 and SiO_2, thus considered more efficient.

3.5. Interfacial tension reduction

For two liquid phases (e.g. gas-water, gas-oil or oil–water) to be in existence, there is usually an interplay of interfacial tension (IFT), which is the force that binds the surfaces of two distinct phases. During oil recovery, reduction in such IFT is essential for better fluid mobilization and efficient recovery process. More so in the presence of negative capillary pressure, a favorable reduction in IFT can be achieved owing to a reduction in the resistance forces that impedes efficient oil production. Nanoparticles exhibit great potentials as IFT reduction agents when in combination with efficient dispersals as the particles display effective adsorption tendencies on the surface of liquids, creating separation layers that permits IFT reduction. This behavior is mainly governed by the nanoparticles surface adsorption, which is reliant on the phase concentration and efficient IFT reduction rate, however, the production of a stable process is dependent on the injected particle concentration. Several studies have been conducted in this regard [23, 68, 69, 83, 106–110]; Sharma and Sangwai [110] investigated the effect of temperature (30–90°C) on IFT of four paraffin oil (n-decane, n-hexane, n-pentane, n-heptane) using nanoparticle-polymer (NP) and nanoparticle-surfactant-polymer (NSP). Nanofluids IFT decreased with temperature increase for n-decane and n-heptane; n-hexane and n-pentane decreased with subsequent increase after reaching a minimum value owing to their low molecular weight that causes volatile losses at high temperature (>70°C). Vatanparast et al. [106] investigated the dynamic IFT of *n*-heptane/water in the presence nanoparticle (SiO_2 concentration - 0–2 wt.%) and surfactants. A sharp reduction in IFT occurred in the absence of nanoparticle and equilibrium was attained in 30 s and from a notable point of about 39 mN/m. For nanoparticle based systems, a rather gradual and steady decrease in IFT occurred over longer time interval (>1500 s). The IFT commenced at about 51 mN/m (close to pure water/heptane IFT) and decreases gradually with increase in the nanoparticle concentration - an indication of the existence of a negligible amount of free surfactant molecules in the system at this point, the nanoparticles did not promote rapid IFT reduction as with the case of the pure surfactant. The exchange of surfactant between the nanoparticle and the interface promoted decrease in IFT as each nanoparticle carries certain amount of surfactant molecules. With the increase in nanoparticle concentration, even more surfactant-nanoparticle complex occurs in the bulk phase which adsorbs to the interface, thus promoting further decrease in IFT at equilibrium condition. A much lower IFT reduction was attained compared to that of surfactant alone. Zargartalebi et al. [83] and Vashisth et al. [111] also reported better IFT reduction occurs in the presence of nanoparticles due to the adsorption of nanoparticles at the aqueous suspension/oil interface.

3.6. Heavy oil recovery enhancement

The upgrading of bitumen and heavy crude oil has remained a challenge. The associated high viscosity and density makes handling and transportation processes even more complex. The unique properties of nanoparticle make it effective as adsorbent and catalysts for enhancing heavy oil upgrading. The presence of nanoparticles in reservoirs improve oil recovery through the production of lighter components achieved by catalytic hydrocracking of heavy oil and by altering the rock surface wettability when the porous medium surface is coated with nanoparticles. Hashemi et al. [45] tested nano-catalysts suspended in heavy oil matrices containing vacuum gas oil (VGO) in oil sands at high temperature and pressure. This was done to ascertain the transport behavior of nanoparticles in oil-reservoirs during heavy oil recovery and upgrading. It was found that nanoparticles specifically, UD multimetallic nanoparticles, can be transported through oil sands porous media into heavy oil reservoir as catalysts for heavy oil upgrading. Similar effects have also been reported in literature [10, 43, 44, 112, 113].

3.7. Thermal conductivity

The thermal conductivity of a material is an essential characteristic for evaluation of the heat conduction and transmission capacity of the specific material. Nusselt number usually denotes the resistance of a flowing fluid to the heat transfer since it directly considers the thermal conductivity of the fluid and indirectly by the Prandtl number. Interestingly, nanoparticle based fluids exhibits high thermal conductivity and large surface area than conventional base-fluids, although, the physical mechanism that accounts for its potential to enhance thermal conductivity is still not well understood [53, 113–115]. The thermal conductivity of conventional base liquid can be improved by addition of nanoparticles. The high surface area increases the thermophysical properties of nanoparticle based fluids and the small particle size enhances their potentials as absorber fluids as it has high potential of large surface coverage in heat transfer processes. Lee et al. [12], Choi [13] and Eastman et al. [14] reported that nanoparticle based fluids exhibits higher thermal conductivity with great dependency on factors such as the material type, size and shape, surface charge, particle volume fraction, the base fluid containing the particles and temperature.

3.8. Stabilization of viscoelastic surfactant (VES) fluids

Viscoelastic surfactant fluids are fluids composed of low molecular weight surfactants that create elongated micelle structures, which display viscoelastic behavior for fluid viscosity increase [116–120]. Huang and Clark [121] reported that the addition of small amount of nanoparticles to VES solutions leads to a pseudo crosslink between the nanoparticle and the VES micelles owing to charge attraction and surface adsorption, as nanoparticles have high surface area. The nanoparticle stabilizes the VES micelles even at high temperature conditions, while minimizing the VES fluid leakage potential in the porous media.

3.9. Control of formation fines migration

Formation fines migration can be controlled in the presence of nanoparticles to enhance reservoir productivity and oil recovery. Nanoparticles show great potentials as possible formation damage remediation measures, as they are capable of sticking fines together while holding them in place to prevent their movement in porous media. Huang et al. [120] showed that in fracture packing operations, nanoparticles coated on proppants exhibit great potentials in holding formation fines in the proppant fractures. The high surface forces of nanoparticles enable the attachment of the particles on the proppant surface during proppant fracturing pumping treatments. Movement of the formation fines to the near-wellbore region is hindered owing to the surface forces of the nanoparticles, which captures the formation fines as it migrates through the nanoparticle treated regions.

3.10. Viscosity and reservoir sweep efficiency improvement

Improving formation viscosity and sweep efficiency is of great necessity for efficient hydrocarbon flow so as to avoid poor oil, gas or water movement or distribution through the pore spaces. The addition of nanoparticles to injection water can increase contact with more swept zones and upsurge the formations sweep efficiency. Typically, the reservoirs pores are filled with hydrocarbons, the flow of water containing nanoparticles through such pores can control the movement and swelling tendencies of the local clays and other formation fines therein. This will in-turn reduce the accumulation of formation-particles and pore throats plugging in the flow channels, which can enhance better water sweep efficiency and subsequent oil recovery [121]. Maghzi et al. [122] investigated the effect of nanoparticles on the rheological behavior of polymer in promoting oil recovery. The authors reported that the presence of low concentration of nanoparticles in the polymer suspension improved the viscosity of the system and the system exhibited better viscosity with much higher oil recovery potential than the polymer system without nanoparticle.

3.11. CO_2 storage and leakage inhibition

For several decades, the excessive use of fossil fuels has contributed immensely to the increase in the atmospheric concentration of greenhouse gases. Carbon capture and storage (CCS) [123–128] is considered the most direct carbon management strategy for long-term reduction of such greenhouse gases. However, the capturing and storage process of CO_2 in geological formations such as deep saline aquifers or depleted reservoirs pose potential leakage threats, which has raised several concerns about the feasibility and long-term storage fate of such CO_2. Interestingly, when reservoirs are primed with nanoparticle based fluids, the systems demonstrate potentials of improving storage efficiency and also inhibit CO_2 leakage tendencies thus providing containment security and storage capacity improvement. Al-Anssari et al. [129] investigated the influence of pressure on the wettability of calcites treated with and without nanoparticles in the presence of CO_2 to account for pressure variation with injection depth. An increase in pressure with increase in contact angle occurred for all surfaces tested which is attributed to an increase in intermolecular interactions between CO_2 and calcite, which increased significantly with increasing CO_2 density thus an implication of a reduction in structural and residual trapping capacities at reservoir conditions [126–128, 130–133]. At ambient condition, oil-wet calcite was weakly CO_2-wet (115° θa - 0.1 MPa and 323 K) and strongly CO_2-wet at storage conditions (148° θa - 20 MPa and 323 K) - the high contact angle is an indication of possibilities of CO_2 leakage. However, the CO_2-wet surfaces were significantly rendered water -wet (41° θa - 15 MPa and 323 K) upon exposure to nanofluids. Aminzadeh et al. [134] investigated CO_2 leakage prevention mechanism through nanoparticle application. The study proffers a remediation strategy where injection of nanoparticle dispersion into a leakage pathway after CO_2 escape has been revealed can create a passive barrier against CO_2 leakage whether driven by viscous forces such as pressure, or by buoyancy. Core floods test was performed to measure the flow pattern and pressure drop in which CO_2 or a CO_2-analogue fluid displaced brine with and without dispersed nanoparticles. The in-situ saturation distribution of the phases was captured in real time for cores positioned horizontally in a modified medical scanner, and the pressure gradient during displacement was measured using pressure transducers. CT scans was used to evaluate water saturation distribution along the core after injection of 0.1 PV n-octane with and without nanoparticles. Lateral CT scans with brine (2 wt.% salinity) as the initial fluid demonstrates non-uniform displacement front. In comparison, the system containing nanoparticle suspension depicts a more distinct and uniform front with slower displacement front and more uniform CO_2 saturation near the inlet and less gravity override. The authors emphasized that the dispersion of nanoparticles on or above potential CO_2 leakage paths such as faults, abandoned wells or fractures can allow the formation of CO_2/brine foam when CO_2 enters the path, which can avert the possibilities of leakage.

3.12. CO_2 Foamability, stabilization and mobility control

Surfactants have been used as a common approach for stabilizing CO_2 foams. CO_2-in-water foams permit the reduction of the mobility of CO_2 which upsurges better sweep efficiency. Despite the process efficiency, the key limitation with this process is that, constant regeneration of surfactant based foams is required for efficient process implementation. Nanoparticle is currently being used to stabilize CO_2 instead of surfactant owing to its unique chemical properties as it offers several advantages over bulk material especially with respect to stability. CO_2 injected with nanoparticle dispersion stabilizes nanoparticle foam, although, at a threshold shear rate, since high shear rates are usually connected with preferential flow through high permeability zones in typical field cases. This promotes the tendencies for the creation of self-guiding fluids that selectively reduces the mobility of CO_2 via foam generation in fractured and gravity override regions with rapid CO_2 flow, which normally contains less oil. In the presence of the oil, the foam breaks to facilitate high recovery via contact with the mobile CO_2. AttarHamed et al. [135] assessed nanoparticle size effect on foamability, foam stability, and synergistic effect. To evaluate the foam stability, a glassware test was conducted at ambient temperature and atmospheric pressure using amorphous hydrophilic silica nanoparticle with particle size ranges of 15, 70, and 250 nm (concentration: 0.1–1.0 wt.%) and alpha olefin sulfonate—AOS (anionic surfactant). The CO_2 foam behavior was assessed by injecting CO_2 into 400 ml prepared solution in a visual cell at a constant rate of 7.9 mL/min, thereafter, the foam volume and the break time were measured and the normalize foam height estimated using Eq. (9) and the synergistic effect were also evaluated. The authors reported that the particle size, concentration, and hydrophobicity have significant effect on foam stability. The silica nanoparticles displayed an antagonism effect on foamability with an initial foam height reduction irrespective of their particle sizes - 15, 70, and 200 nm. However, for the synergistic effect on basis of foam stability, only systems containing small size nanoparticles (15 nm) and low particle concentration (0.3 and 0.5 wt.%) displayed better effect in stabilizing CO_2 in water foams over 70 and 250 nm particle size and 1 wt.% concentration and improved CO_2 in water foams stability by 25% at an initial foam volume of 20%. On basis of the particles hydrophobicity, for high stability to be achieved, nanoparticles should be designed with efficient surface coatings for adequate hydrophobicity and formation of CO_2 in water foams.

$$\text{Normalized foam height} = \text{foam height}(t)/\text{foam height } (t = 0) \tag{9}$$

where, t is the time (minutes).

Similarly, Singh and Mohanty [136] reported that nanoparticle-surfactant solutions containing only about 0.3 wt.% nanoparticles significantly modified foam mobility ratio. The increase in nanoparticle concentration (0–5 wt.%) increased the mobility reduction factor from 4000 to 8700 and increased foam stability. It has also been experimentally proven that the presence of nanoparticle creates better foam stability than surfactant or VES fluids and even polymer, and improves mobility control; and efficient for enhancing oil recovery owing to high sweep efficiency and effective viscosity [137–142] (**Table 2**).

Nanoparticle (NP)	Nanoparticle concentrations	Dispersals/reagents	Tests	Oil/formation type	Results and remarks	References
SiO_2	100–200 mg L^{-1}	Fracturing fluid (FF)	Core flooding	Crude oil (43° API); Bauxite porous media	• Enhanced porous media wetting towards water-wet condition • Reduction of formation damage by 71% • Optimized fracturing fluid for operations and modified water relative permeability by 82%	[89]
Naked TiO_2, Fe_3O_4, Fe $(OH)_3$, Al_2O_3-coated TiO_2	0.5–100 mg L^{-1}	$FeSO_4{\cdot}7H_2O$; $NaHCO_3$; $FeCl_3{\cdot}6H_2O$; NH_4OH; $CaCO_3$; NaOH; HCl; NaCl; D1 water	Zeta potential		• Zeta potential exhibits great dependency on the nanoparticle concentrations in solution • Excessively high particle concentration can create unreliable results when in the diluted dispersion owing to multiple scattering and turbidity effects	[97]
SiO_2	150–600 ppm	Synthetic polyacrylamide (Flopaam 3630S); Colloidal dispersion gel (CDG) NaCl brine; Aluminum citrate (AlCit)	Core flooding	Sandstone cores	• No oil mobilization occurred in Berea sandstone cores in the presence of the SiO_2, and polymer alone as the capillary number based on differential pressure and viscosities associated with polymer and silica floods were below critical capillary number or oil mobilization threshold • Systems containing blends of hydrolyzed polyacrylamide (HPAM) and silica (300 ppm SiO_2 dispersed in polymer 600 ppm) and NaCl (0.5%) (600/300Si/0.5) yielded oil mobilization with a reduction in Sor (21–24%) on microscopic level and generated an oil bank on the macroscopic level	[149]
SiO_2	0.01–0.10 wt.%	$C_{20}H_{44}BrN$ surfactant; CO_2; D1 water	CO_2 foam stability and mobility control		• Blends of nanoparticle and surfactant promoted higher foam stability and better viscosity than the surfactant-stabilized foam alone • NPS solution exhibited better efficiency in mobility control than the surfactant stabilized foam	[143]
SiO_2	1–25 g/L	NH_4OH; HCl; n-heptane; Toluene; n-C7 asphaltenes; TEOS (tetraethyl orthosilicate)	Formation damage inhibition	crude oil (6.8° API, 6.9×10^5 cP viscosity)	• Increased effective permeability to oil • Promoted increase in the recovery factor by 11% • Enabled the inhibition of n-C_7 asphaltene aggregate growth especially nanoparticles with the high adsorptive capacity	[144]
TiO_2, ZrO_2, SiO_2	0.07–0.5 wt.%	Paraffin, HNO_3 or $C_2H_8N_2$; n-heptane; Toluene; asphaltenes	Asphaltene stability/	crude oil (19° API)	• Nanoparticles do not act as asphaltene precipitant in strongly acidic conditions rather as dispersant for enhancing asphaltene stability	[145]

Nanoparticle (NP)	Nanoparticle concentrations	Dispersals/reagents	Tests	Oil/formation type	Results and remarks	References
			precipitation inhibition		• TiO_2 exhibited better performance over ZrO_2, SiO_2 in stabilizing asphaltenes under acidic conditions, and contributed to a higher precipitation onset point	
SiO_2; Al_2O_3; SiO_2 with acid surface (SiO_2A)	1–4 wt.%	Tween 80 surfactant; distilled water; TEOS (tetraethyl orthosilicate); NH4OH; H_2SO_4	Viscosity reduction of heavy crude oil	crude oil (13° API; of 1.2×10^5 cP viscosity)	• Addition of nanoparticles reduced the asphaltene mean aggregate size and promoted adsorption of n-C_7 asphaltenes onto the nanoparticles dispersed in the nanofluid • Viscosity decreases with increase in the temperature of crude oil; and increase in nanofluid concentration in the mixture enhanced higher degree of viscosity reduction • Significant reduction in viscosity (>90%) of heavy oil by addition of active component - SiO_2A nanoparticles	[69]
ZrO_2; NiO	0.004–0.05 wt. %	NaCl brine; D1 water; C_{16}TAB and TX-100 Surfactant	Contact Angle/ spontaneous imbibition	Toluene; limestone	• Nanoparticle based systems act as novel surface-modifiers with better impact in promoting formation wetting. Water contact angle decreased with increase in time, concentration, salinity, surface roughness and temperature in the presence of the nanofluids • Surfactant - nanoparticle systems (ZrO_2/C_{16}TAB; NiO/ C_{16}TAB; ZrO_2/TX-100; NiO/TX-100) exhibited faster water imbibition capabilities and wetting potential than the surfactant system alone (C_{16}TAB and TX-100)	[23, 24, 31]
SiO_2	0.005–0.1 wt.%	NaCl brine; D1 water	Core flooding	Limestone cores	• Particle size influences porous media retention and transport behavior; SiO_2 with small particle size promote efficient penetration through the porous media • Nanoparticle transport can be negatively impacted by salinity owing to screening effect of salts on the repulsion forces between particles in the injected fluid which can lead to particle clustering or large agglomeration effect • Retention of the nanoparticles can be significantly reduced with temperature increase due to the surface charge changes between the SiO_2 and carbonate surface	[148]
SiO_2	0.1 wt.%		Spontaneous imbibition	Dehydrated crude oil and kerosene	• Nanofluid systems demonstrated higher recovery potential than alkaline water and NaCl solutions	[146]

Nanoparticle (NP)	Nanoparticle concentrations	Dispersals/reagents	Tests	Oil/formation type	Results and remarks	References
		alkaline water (pH 10), and NaCl brine			• Higher displacement efficiency and ultimate oil recovery was attained in the presence of nanofluid • Nanoparticle based systems recovered approximately 38% of oil within 12 days, whereas, alkaline water and NaCl solution recovered only about 12 and 6% of oil respectively	
SiO_2	0–2 wt.%.	CTAB, SDS, double distilled water	Surface tension, foamability, and foam stability		• Pure hydrophilic SiO_2 in the absence of surfactants have no effect on the surface tension/surface elasticity at air-water interface • Surface activity of the surfactant is increased significantly with increase in the nanoparticle concentration • The foam stability and foamability is greatly improved with more stable foams achieved in the presence of nanoparticles than pure surfactant solution alone	[147]

Table 2. A selection of published studies on the effect of nanoparticle based systems on process efficiency.

http://dx.doi.org/10.5772/intechopen.72672

4. Conclusion

In hydrocarbon reservoirs, fluid–rock interactions are predominant occurrences owing to the existence of surface forces between molecules. The presence of oil, water and even gas is an indication that concurrent flow of fluids at the production stage of a reservoir often occurs. Fluid displacement in reservoirs, especially when more than one fluid is involved can contribute to poor oil recovery as the oil gets trapped owing to factors such as IFT between fluids, fluids immiscibility and rock surface wetting challenges. The role of nanomaterial in addressing these issues have been evaluated; the material displays great potentials as excellent particles for the design and formulations of smart fluids for applications in high temperatures-high pressures regions. The surface functionalities and excellent properties of nanomaterial such as their enhanced thermal conductivity and stability, unique particle size, high surface area, great adsorption affinity, and unique transport behavior in porous media influences their superior efficiency over the micros and macros bulk counterpart even under harsh reservoir conditions. Nanoparticles show great potentials as effective novel tools for permeability enhancement; IFT reduction, rock wettability alteration from oil-wet to water-wet, oil mobility and recovery enhancement, and CO_2 stabilization. The study identifies a prime characteristic of nanoparticles in EOR, which is its ability to improve the property of the dispersal even at low particle concentrations in the suspensions owing to high dependency on the materials unique properties. The addition of nanoparticle changes the rheological property of the fluid and the particles in suspension enhanced the performance of the carrier fluid during production and inhibits any associated formation damage. Thus nano-smart fluids are highly recommended for EOR project design as potential stabilizer of dispersed systems for hydrocarbon production augmentation, CO_2 geo-storage, and soil de-contamination processes.

Author details

Lezorgia Nekabari Nwidee[1]*, Ahmed Barifcani[1,2], Maxim Lebedev[3],
Mohammad Sarmadivaleh[1] and Stefan Iglauer[1]

*Address all correspondence to: l.nwidee@postgrad.curtin.edu.au

1 Department of Petroleum Engineering, Curtin University, Perth, Western Australia, Australia

2 Department of Chemical Engineering, Curtin University, Perth, Western Australia, Australia

3 Department of Exploration Geophysics, Curtin University, Perth, Western Australia, Australia

References

[1] El-Diasty AI, Ragab AMS. Applications of nanotechnology in the oil and gas industry: latest trends worldwide and future challenges in Egypt. North Africa, Technical Conference and Exhibition; Cairo, Egypt; 15-17April; 2013

[2] Khalil M, Jan BM, Tong CW, Berawi MA. Advanced nanomaterials in oil and gas industry: Design, application and challenges. Applied Energy. 2017;**191**:287-310

[3] Davidson A, Huh C, Bryant SL. Focused magnetic heating utilizing superparamagnetic nanoparticles for improved oil production applications. In SPE International Oilfield Nanotechnology Conference and Exhibition; Noordwijk, The Netherlands; 12–14 June; 2012

[4] Amanullah MD, Al-Tahini AM. Nanotechnology - its significance in smart fluid development for oil and gas field application. SPE Saudi Arabia Section Technical Symposium; Al-Khobar, Saudi Arabia; 9–11 May; 2009

[5] Green DW, Willhite GP. Enhanced Oil Recovery. SPE Textbook Series. Vol. 6. Richardson: Henry L. Doherty Memorial Fund of AIME, Society of Petroleum Engineers; 1998

[6] Nwidee LN, Theophilus S, Barifcani A, Sarmadivaleh M, Iglauer S. EOR processes, opportunities and technological advancements. In: Chemical Enhanced Oil Recovery (cEOR)-a Practical Overview. InTech; 2016

[7] Sciau P, Mirguet C, Roucau C, Chabanne D, Schvoerer M. Double nanoparticle layer in a 12th century lustreware decoration: Accident or technological mastery? Journal of Nano Research. 2009;**8**:133-139. Trans Tech Publications

[8] White RJ, Luque R, Budarin VL, Clark JH, Macquarrie DJ. Supported metal nanoparticles on porous materials, methods and applications. Chemical Society Reviews. 2009;**38**(2):481-494

[9] Sabet M, Hosseini SN, Zamani A, Hosseini Z, Soleimani H. Application of nanotechnology for enhanced oil recovery: A review. Defect and Diffusion Forum. 2016;**367**:149-156

[10] Hashemi R, Nassar NN, Pereira-Almao P. Nanoparticle technology for heavy oil in-situ upgrading and recovery enhancement: Opportunities and challenges. Applied Energy. 2014;**133**:374-387

[11] Hashemi-Kiasari H, Hemmati-Sarapardeh A, Mighani S, Mohammadi AH, Sedaee-Sola B. Effect of operational parameters on SAGD performance in a dip heterogeneous fractured reservoir. Fuel. 2014;**122**:82-93

[12] Lee S, Choi SS, Li SA, Eastman JA. Measuring thermal conductivity of fluids containing oxide nanoparticles. Journal of Heat Transfer. 1999;**121**(2):280-289

[13] Choi SUS. Enhancing thermal conductivity of fluids with nanoparticles, developments and applications of non-Newtonian flows. ASME FED. 1995;**105**(99):231

[14] Eastman JA, Choi US, Li S, Thompson LJ, Lee S. Enhanced thermal conductivity through the development of nanofluids. MRS Online Proceedings Library Archive. 1996;**457**:3

[15] Lim SH, Zeng KY, He CB. Morphology, tensile and fracture characteristics of epoxy-alumina nanocomposites. Materials Science and Engineering A. 2010;**527**:5670-5676

[16] Zhang M, Singh RP. Mechanical reinforcement of unsaturated polyester by Al_2O_3 nanoparticles. Materials Letters. 2004;**58**:408

[17] Nikmatin S. Bionanocomposite produced by rattan filler as a substitute glass fiber on vehicle components PhD thesis. Bogor Agricultural University; 2012

[18] Nikmatin S, Syafiuddin A, Kueh ABH, Purwanto YA. Effects of nanoparticle filler on thermo-physical properties of rattan powder-filled polypropylene composites. Jurnal Teknologi. 2015;**77**(16):181-187

[19] Strambeanu N, Demetrovici L, Dragos D, Lungu M. Nanoparticles: Definition, classification and general physical properties. In: Nanoparticles' Promises and Risks. Springer International Publishing; 2015. pp. 3-8

[20] Cadden A. Comparative effects of particle size reduction on physical structure and water binding properties of several plant fibers. Journal of Food Science. 1987;**52**:1595-1599

[21] Rao C, Biswas K. Characterization of nanomaterials by physical methods. Annual Review of Analytical Chemistry. 2009;**2**:435-462

[22] Kim BH, Hackett MJ, Park J, Hyeon T. Synthesis, characterization, and application of ultrasmall nanoparticles. Chemistry of Materials. 2013;**26**(1):59-71

[23] Nwidee LN, Al-Anssari S, Barifcani A, Sarmadivaleh M, Lebedev M, Iglauer S. Nanoparticles influence on wetting behaviour of fractured limestone formation. Journal of Petroleum Science and Engineering. 2017;**149**:782-788

[24] Nwidee LN, Al-Anssari S, Barifcani A, Sarmadivaleh M, Lebedev M, Iglauer S. Wettability alteration of oil-wet limestone using surfactant-nanoparticle formulation. Journal of Colloid and Interface Science. 2017;**504**:334-345

[25] Fakoya MF, Subhash SN. Emergence of nanotechnology in the oil and gas industry: Emphasis on the application of silica nanoparticles. Petroleum. 2017 in press

[26] Guo D, Xie G, Luo J. Mechanical properties of nanoparticles: Basics and applications. Journal of Physics D: Applied Physics. 2014;**47**(1):013001

[27] Lin Y, Skaff H, Emrick T, Dinsmore AD, Russell TP. Nanoparticle assembly and transport at liquid-liquid interfaces. Science. 2003;**299**(5604):226-229

[28] Bell AT. The impact of nanoscience on heterogeneous catalysis. Science. 2003;**299**(5613): 1688-1691

[29] Kong X, Ohadi MM. Application of micro and nanotechnologies in the oil and gas industry-An overview of the recent progress. In SPE 138241 Presented at the Abu Dhabi International Petroleum Exhibition & Conference; Abu Dhabi, UAE; 1–4 November; 2010

[30] Tsuzuki T. Nanotechnology Commercialization. Florida USA: Pan Stanford Publishing; 2013

[31] Nwidee LN, Al-Anssari S, Barifcani A, Sarmadivaleh M, Iglauer S. Nanofluids for enhanced oil recovery processes: wettability alteration using zirconium oxide. Offshore Technology Conference (OTC-26573-MS); Kuala Lumpur, Malaysia; 22–25 March; 2016

[32] Wasan DT, Nikolov AD. Spreading of nanofluids on solids. Nature. 2003;**423**(6936):156

[33] Wasan DT, Nikolov A, Kondiparty K. The wetting and spreading of nanofluids on solids: Role of the structural disjoining pressure. Current Opinion in Colloid and Interface Science. 2011;**16**(4):344-349

[34] Chengara AN, Nikolov A, Wasan DT, Trokhymchuck A, Henderson D. Spreading of nanofluids driven by the structural disjoining pressure gradient. Journal of Colloid and Interface Science. 2004;**280**(1):192-201

[35] Mc.Elfresh P, Holcomb D, Ector D. Application of nanofluid technology to improve recovery in oil and gas wells. Paper SPE 154827-MS Presented at SPE International Oilfield Technology Conference; Noordwijk; 12–14 June; 2012

[36] Al-Anssari S, Barifcani A, Wang S, Maxim L, Iglauer S. Wettability alteration of oil-wet carbonate by silica nanofluid. Journal of Colloid and Interface Science. 2016;**461**:435-442

[37] Al-Anssari S, Wang S, Barifcani A, Lebedev M, Iglauer S. Effect of temperature and SiO_2 nanoparticle size on wettability alteration of oil-wet calcite. Fuel. 2017;**206**:34-42

[38] Hamouda AA, Karoussi O. Effect of temperature, wettability, and relative permeability on oil recovery from oil-wet chalk. Energies. 2008;**1**(1):19-34

[39] Binks BP, Lumsdon SO. Influence of particle wettability on the type and stability of surfactant-free emulsions. Langmuir. 2000;**16**(23):8622-8631

[40] Binks BP, Rodrigues JA. Inversion of emulsions stabilized solely by ionizable nanoparticles. Angewandte Chemie. 2005;**117**(3):445-448

[41] Mandal A, Bera A, Ojha K, Kumar T. Characterization of surfactant stabilized nanoemulsion and its use in enhanced oil recovery. In SPE International Oilfield Nanotechnology Conference and Exhibition; Noordwijk, The Netherlands; 12–14 June; 2012

[42] Zhang T, Davidson D, Bryant SL, Huh C. Nanoparticle-stabilized emulsions for applications in enhanced oil recovery. In SPE improved oil recovery symposium; Tulsa, Oklahoma, USA; 24–28 April; 2010

[43] Shokrlu YH, Babadagli T. Transportation and interaction of nano and micro size metal particles injected to improve thermal recovery of heavy-oil. In SPE Annual Technical Conference and Exhibition; Denver, Colorado, USA; 30 October-2 November; 2011

[44] Pereira-Almao P. In situ upgrading of bitumen and heavy oils via nanocatalysis. The Canadian Journal of Chemical Engineering. 2012;**90**(2):320-329

[45] Hashemi R, Nassar NN, Pereira-Almao P. Transport behavior of multimetallic ultradispersed nanoparticles in an oil-sands-packed bed column at a high temperature and pressure. Energy & Fuels. 2012;**26**(3):1645-1655

[46] Chakrabarty S, Chatterjee K. Synthesis and characterization of nano-dimensional nickelous oxide (NiO) semiconductor. Journal of Physical Science. 2009;**13**:245-250

[47] Ponmani S, Nagarajan R, Sangwai J. Applications of nanotechnology for upstream oil and gas industry. Journal of Nano Research. 2013;**24**:7-15

[48] Wang X, Cheng Z, Lu Y, Xu L, Xie X. Nanoscale metals in earth gas and mobile forms of metals in overburden in wide-spaced regional exploration for giant deposits in overburden terrains. Journal of Geochemical Exploration. 1997;**58**(1):63-72

[49] Song YQ, Marcus C. Hyperpolarized silicon nanoparticles: reinventing oil exploration?. International Presentation; 2007

[50] Ogolo NA, Olafuyi OA, Onyekonwu MO. Enhanced oil recovery using nanoparticle. Society of Petroleum Engineers, Oil and Gas Saudi Arabian Section Technical and Exhibition, SPE 160847; Al-Khobar, Saudi Arabia; 8–11 April; 2012

[51] Kanj MY, Rashid M, Giannelis E. Industry first field trial of reservoir nanoagents. In SPE Middle East Oil and Gas Show and Conference. Society of Petroleum Engineers; Manama, Bahrain; 25–28 September; 2011

[52] Jahagirdar SR. Oil-microbe detection tool using nano optical fibers. SPE Western Regional and Pacific Section AAPG Joint Meeting held in Bakersfield; California, USA; 31 March-2 April; 2008

[53] Yu W, France DM, Routbort JL, Choi SU. Review and comparison of nanofluid thermal conductivity and heat transfer enhancements. Heat Transfer Engineering. 2008;**29**(5): 432-460

[54] Ju B, Fan T. Experimental study and mathematical model of nanoparticle transport in porous media. Powder Technology. 2009;**192**(2):195-202

[55] Hendraningrat L, Li S, Torsæter O. A glass micromodel experimental study of hydrophilic nanoparticles retention for EOR project. SPE-159161-MS; 2012

[56] Nassar NN, Hassan A, Pereira-Almao P. Effect of surface acidity and basicity of aluminas on asphaltene adsorption and oxidation. Journal of Colloid and Interface Science. 2011;**360**(1):233

[57] Nassar NN, Hassan A, Pereira-Almao P. Effect of the particle size on asphaltene adsorption and catalytic oxidation onto alumina particles. Energy & Fuels. 2011;**25**(9):3961

[58] Li S, Hendraningrat L, Torsæter O. Improved oil recovery by hydrophilic silica nanoparticles suspension: 2-phase low experimental studies. IPTC 16707-MS presented at Beijing International Petroleum Technology Conference; China; 26–28 March; 2013

[59] Li S, Kaasa AT, Hendraningrat L, Torsæter O. Effect of silica nanoparticles adsorption on the wettability index of Berea sandstone. In Paper SCA2013–059 presented at the international symposium of the society of core analysts held in Napa Valley; California, USA; 2013. pp. 16–19

[60] Wang XQ, Mujumdar AS. A review on nanofluids-part I: Theoretical and numerical investigations. Brazilian Journal of Chemical Engineering. 2008;**25**(4):613-630

[61] Yang Y, Lan J, Li X. Study on bulk aluminum matrix nano-composite fabricated by ultrasonic dispersion of nano-sized SiC particles in molten aluminum alloy. Materials Science and Engineering A. 2004;**380**(1):378-383

[62] Yu J, An C, Mo D, Liu N, Lee R. Study of Adsorption and transportation behavior of Nanoparticles in Three Different Porous Media. SPE-153337; 2012

[63] Shen D, Zhang P, Kan AT, Fu G, Alsaiari HA, and Tomson MB. Control placement of scale inhibitors in the formation with stable Ca-DTPMP nanoparticle suspension and its transport in porous medium. In SPE International Oilfield Scale Conference. Society of Petroleum Engineers; Aberdeen, UK; 28–29 May; 2008

[64] Zhang P, Shen D, Fan C, Kan A, Tomson M. Surfactant-assisted synthesis of metal-phosphonate inhibitor nanoparticles and transport in porous media. SPE Journal. 2010;**15**(03):610-617

[65] Mullins OC, Sheu EY. Structure and Dynamics of Asphaltenes. New York: Springer; 1998

[66] Sjoblom J, Aske N, Auflem IH, Brandal O, Havre TE, Saether O, Westvik A, Johnsen EE, Kallevik H. Our current understanding of water-in-crude oil emulsions. Recent characterization techniques and high -pressure performance. Advances in Colloid and Interface Science. 2003;**100**:399-473

[67] Syunyaev RZ, Balabin RM, Akhatov IS, Safieva JO. Adsorption of petroleum asphaltenes onto reservoir rock sands studied by near-infrared (NIR) spectroscopy. Energy and Fuels. 2009;**23**(3):1230-1236

[68] Franco CA, Lozano MM, Acevedo S, Nassar NN, Cortés FB. Effects of resin I on asphaltene adsorption onto nanoparticles: A novel method for obtaining asphaltenes/resin isotherms. Energy & Fuels. 2015;**30**:264-272

[69] Taborda EA, Franco CA, Lopera SH, Alvarado V, Cortés FB. Effect of nanoparticles/nanofluids on the rheology of heavy crude oil and its mobility on porous media at reservoir conditions. Fuel. 2016;**184**:222-232

[70] Montoya T, Coral D, Franco CA, Nassar NN, Cortés FB. A novel solid–liquid equilibrium model for describing the adsorption of associating asphaltene molecules onto solid surfaces based on the "chemical theory". Energy & Fuels. 2014;**28**:4963-4975

[71] Yudin IK, Anisimov MA. Dynamic light scattering monitoring of asphaltene aggregation in crude oils and hydrocarbon solutions. Asphaltenes, Heavy Oils, and Petroleomics. 2007:439-468

[72] Mullins OC, Betancourt SS, Cribbs ME, Dubost FX, Creek JL, Andrews AB, Venkataramanan L. The colloidal structure of crude oil and the structure of oil reservoirs. Energy & Fuels. 2007;**21**(5):2785-2794

[73] Franco CA, Nassar NN, Ruiz MA, Pereira-Almao P, Cortes FB. Nanoparticles for inhibition of asphaltenes damage: Adsorption study and displacement test on porous media. Energy & Fuels. 2013;**27**(6):2899

[74] Wang K, Wang C, Sun L, Yi F. Experimental research on decompression and augmented injection effect by using active SiO2 nano-powder in middle and low permeability cores. In Power and Energy Engineering Conference (APPEEC); Asia-Pacific; 2010. pp. 1–4

[75] Guzmán JD, Pineda D, Franco CA, Botero ÓF, Lopera SH, Cortés FB. Effect of nanoparticle inclusion in fracturing fluids applied to tight gas-condensate reservoirs: Reduction of methanol loading and the associated formation damage. Journal of Natural Gas Science and Engineering. 2017;**40**:347-355

[76] Ju B, Fan T. Experimental study and mathematical model of nanoparticle transport in porous media. Powder Technology. 2009;**192**(2):195-202

[77] Cao N, Mohammed MA, Babadagli T. Wettability Alteration of Heavy-Oil/Bitumen Containing Carbonates Using Solvents, high pH Solutions and Nano/Ionic Liquids. In Offshore Technology Conference; Rio de Janeiro, Brazil; 27–29 October; 2015

[78] Alomair OA, Matar KM, Alsaeed YH. Experimental study of enhanced-heavy-oil recovery in Berea sandstone cores by use of nanofluids applications. SPE Reservoir Evaluation & Engineering. 2015;**18**(03):387-399

[79] Hendraningrat L, Torsaeter O. Unlocking the potential of metal oxides nanoparticles to enhance the oil recovery. In Offshore Technology Conference-Asia. Offshore Technology Conference; Kuala Lumpur, Malaysia; 25–28 March; 2014

[80] Suleimanov BA, Ismailov FS, Veliyev EF. Nanofluid for enhanced oil recovery. Journal of Petroleum Science and Engineering. 2011;**78**(2):431-437

[81] Muherei MA, Junin R. Equilibrium adsorption isotherms of anionic, nonionic surfactants and their mixtures to shale and sandstone. Modern Applied Science. 2009;**3**(2):158

[82] Ahmadi MA, Shadizadeh SR. Induced effect of adding nano silica on adsorption of a natural surfactant onto sandstone rock: Experimental and theoretical study. Journal of Petroleum Science and Engineering. 2013;**112**:239-247

[83] Zargartalebi M, Kharrat R, Barati N. Enhancement of surfactant flooding performance by the use of silica nanoparticles. Fuel. 2015;**143**:21-27

[84] Binshan J, Shugao D, Zhian L, Tiangao Z, Xiantao S, Xiaofeng Q. A study of wettability and permeability change caused by adsorption of nanometer structured polysilicon on the surface of porous media. SPE Asia Pacific Oil and Gas Conference and Exhibition; Melbourne, Australia, October 8–10; 2002

[85] Shafie A, Yahya N, Kashif M, Zaid HM, Soleimani H, Alnarabiji MS. The band structures of single-walled carbon nanotubes and ZnO nanoparticles used for oil recovery in water flooding system. Journal of Nano Research. 2014;**26**:101-110

[86] Zhang H, Nikolov A, Wasan D. Enhanced oil recovery (EOR) using nanoparticle dispersions: Underlying mechanism and imbibition experiments. Energy and Fuels. 2014;**28**(5): 3002-3009

[87] Youssif MI, El-Maghraby RM, Saleh SM, Elgibaly A. Silica nanofluid flooding for enhanced oil recovery in sandstone rocks. Egyptian Journal of Petroleum. 2017 in press

[88] Winkler K, Paszewski M, Kalwarczyk T, Kalwarczyk E, Wojciechowski T, Gorecka E, Pociecha D, Holyst R, Fialkowski M. Ionic strength-controlled deposition of charged nanoparticles on a solid substrate. The Journal of Physical Chemistry C. 2011;**115**(39):19096-19103

[89] Nikolov A, Kondiparty K, Wasan D. Nanoparticle self-structuring in a nanofluid film spreading on a solid surface. Langmuir. 2010;**26**(11):7665-7670

[90] Ershadi M, Alaei M, Rashidi A, Ramazani A, Khosravani S. Carbonate and sandstone reservoirs wettability improvement without using surfactants for chemical enhanced oil recovery (C-EOR). Fuel. 2015;**153**:408-415

[91] Zargari S, Ostvar S, Niazi A, Ayatollahi S. Atomic force microscopy and wettability study of the alteration of mica and sandstone by a biosurfactant-producing bacterium bacillus thermodenitrificans. Journal of Advanced Microscopy Research. 2010;**5**(2):143-148

[92] Marrese M, Guarino V, Ambrosio L. Atomic force microscopy: A powerful tool to address scaffold design in tissue engineering. Journal of Functional Biomaterials. 2017;**8**(1):7

[93] Buckley JS, Lord DL. Wettability and morphology of mica surfaces after exposure to crude oil. Journal of Petroleum Science and Engineering. 2003;**39**(3):261-273

[94] Jarrahian K, Seiedi O, Sheykhan M, Sefti MV, Ayatollahi S. Wettability alteration of carbonate rocks by surfactants: A mechanistic study. Colloids and Surfaces A: Physicochemical and Engineering Aspects. 2012;**410**:1-10

[95] Zhang P, Austad T. Wettability and oil recovery from carbonates: Effects of temperature and potential determining ions. Colloids and Surfaces A: Physicochemical and Engineering Aspects. 2006;**279**(1):179-187

[96] Thielbeer F, Donaldson K, Bradley M. Zeta potential mediated reaction monitoring on nano and microparticles. Bioconjugate Chemistry. 2011;**22**(2):144-150

[97] Wang N, Hsu C, Zhu L, Tseng S, Hsu JP. Influence of metal oxide nanoparticles concentration on their zeta potential. Journal of Colloid and Interface Science. 2013;**407**:22-28

[98] Manan MA, Farad S, Piroozian A, Esmail MJA. Effects of nanoparticle types on carbon dioxide foam flooding in enhanced oil recovery. Petroleum Science and Technology. 2015;**33**(12):1286-1294

[99] Anton N, Benoit JP, Saulnier P. Design and production of nanoparticles formulated from nano-emulsion templates- a review. Journal of Controlled Release. 2008;**128**(3):185-199

[100] Tadros T, Izquierdo P, Esquena J, Solans C. Formation and stability of nano-emulsions. Advances in Colloid and Interface Science. 2004;**108**:303-318

[101] Napper DH. Polymeric Stabilization of Colloidal Dispersions. Vol. 3. London: Academic Press; 1983

[102] Tadros TF. The Effect of Polymers on Dispersion Properties. Academic Press; 1982

[103] Derjaguin BV, Landau L. Theory of the stability of strongly charged lyophobic sols and of the adhesion of strongly charged particles in solutions of electrolytes. Acta Physicochimica URSS. 1941;**14**(6):633-662

[104] Overbeek JTG, Verwey EJW. Theory of the Stability of Lyophobic Colloids: The Interaction of sol Particles Having an Electric Double Layer. Amsterdam: Elsevier; 1948

[105] Ahmadi M, Habibi A, Pourafshary P, Ayatollahi S. Zeta-potential investigation and experimental study of nanoparticles deposited on rock surface to reduce fines migration. SPE Journal. 2013;**18**(03):534-544

[106] Vatanparast H, Javadi A, Bahramian A. Silica nanoparticles cationic surfactants interaction in water-oil system. Colloids and Surfaces A: Physicochemical and Engineering Aspects. 2017;**521**:221-230

[107] Al-Anssari S, Wang S, Barifcani A, Iglauer S. Oil-water interfacial tensions of silica nanoparticle-surfactant formulations. Tenside, Surfactants, Detergents. 2017;**54**(4):334-341

[108] Hosseini SN, Shuker MT, Hosseini Z, Tomocene TJ, Shabib-Asl A, Sabet M. The role of salinity and brine ions in interfacial tension reduction while using surfactant for enhanced oil recovery. Research Journal of Applied Sciences, Engineering and Technology. 2015;**9**(9):722-726

[109] Lan Q, Yang F, Zhang S, Liu S, Xu J, Sun D. Synergistic effect of silica nanoparticle and cetyltrimethyl ammonium bromide on the stabilization of o/w emulsions. Colloids and Surfaces A: Physicochemical and Engineering Aspects. 2007;**302**(1):126-135

[110] Sharma T, Sangwai JS. Silica nanofluids in polyacrylamide with and without surfactant: Viscosity, surface tension, and interfacial tension with liquid paraffin. Journal of Petroleum Science and Engineering. 2017;**152**:575-585

[111] Vashisth C, Whitby CP, Fornasiero D, Ralston J. Interfacial displacement of nanoparticles by surfactant molecules in emulsions. Journal of Colloid and Interface Science. 2010; **349**(2010):537-543

[112] Molina HJL. Transport of Catalytic Particles Immersed in Fluid Media through Cylindrical Geometries under Heavy Oil Upgrading Conditions. Calgary: University of Calgary; 2009

[113] Saidur R, Leong KY, Mohammad HA. A review on applications and challenges of nanofluids. Renewable and Sustainable Energy Reviews. 2011;**15**(3):1646-1668

[114] Eastman JA, Choi SUS, Li S, Yu W, Thompson LJ. Anomalously increased effective thermal conductivities of ethylene glycol-based nanofluids containing copper nanoparticles. Applied Physics Letters. 2001;**78**(6):718-720

[115] Liu MS, Lin MCC, Tsai CY, Wang CC. Enhancement of thermal conductivity with cu for nanofluids using chemical reduction method. International Journal of Heat and Mass Transfer. 2006;**49**(17):3028-3033

[116] Nehmer WL. Viscoelastic gravel-pack carrier fluid. In SPE Formation Damage Control Symposium. Society of Petroleum Engineers; 1988

[117] Brown JE, King LR, Nelson EB, Ali SA. Use of a viscoelastic carrier fluid in frac-pack applications. In SPE Formation Damage Control Symposium. Society of Petroleum Engineers; 1996, January

[118] Samuel M, Card RJ, Nelson EB, Brown JE, Vinod PS, Temple HL, Qu Q, Fu DK. Polymer-free fluid for hydraulic fracturing. In SPE Annual Technical Conference and Exhibition. Society of Petroleum Engineers; 1997

[119] Huang T, Crews JB. Nanotechnology applications in viscoelastic surfactant stimulation fluids. SPE Production & Operations. 2008;**23**(04):512-517

[120] Huang T, Crews JB, Willingham JR. Using nanoparticle technology to control fine migration. In SPE Annual Technical Conference and Exhibition. Society of Petroleum Engineers; 2008

[121] Huang TT, Clark DE. Enhancing oil recovery with specialized nanoparticles by controlling formation-fines migration at their sources in waterflooding reservoirs. Society of Petroleum Engineering Journal. 2015;**20**(4):743-746

[122] Maghzi A, Mohebbi A, Kharrat R, Ghazanfari MH. An experimental investigation of silica nanoparticles effect on the rheological behavior of polyacrylamide solution to enhance heavy oil recovery. Petroleum Science and Technology. 2013;**31**(5):500-508

[123] Iglauer S, Paluszny A, Pentland CH, Blunt MJ. Residual CO_2 imaged with X-ray microtomography. Geophysical Research Letters. 2011;**38**(21)

[124] Blunt MJ, Bijeljic B, Dong H, Gharbi O, Iglauer S, Mostaghimi P, Pentland C. Pore-scale imaging and modelling. Advances in Water Resources. 2013;**51**:197-216

[125] Iglauer S, Salamah A, Sarmadivaleh M, Liu K, Phan C. Contamination of silica surfaces: Impact on water-CO_2-quartz and glass contact angle measurements. International Journal of Greenhouse Gas Control. 2014;**22**:325-328

[126] Iglauer S, Al-Yaseri AZ, Rezaee R, Lebedev M. CO_2 wettability of caprocks: Implications for structural storage capacity and containment security. Geophysical Research Letters. 2015;**42**(21):9279-9284

[127] Iglauer S, Pentland CH, Busch A. CO_2 wettability of seal and reservoir rocks and the implications for carbon geo-sequestration. Water Resources Research. 2015;**51**(1): 729-774

[128] Rahman T, Lebedev M, Barifcani A, Iglauer S. Residual trapping of supercritical CO_2 in oil-wet sandstone. Journal of Colloid and Interface Science. 2016;**469**:63-68

[129] Al-Anssari S, Arif M, Wang S, Barifcani A, Lebedev M, Iglauer S. CO_2 geo-storage capacity enhancement via nanofluid priming. International Journal of Greenhouse Gas Control. 2017;**63**:20-25

[130] Iglauer S, Mathew MS, Bresme F. Molecular dynamics computations of brineCO_2 interfacial tensions and brine-CO_2-quartz contact angles and their effects on structural and residual trapping mechanisms in carbon geo-sequestration. Journal of Colloid and Interface Science. 2012;**386**(1):405-414

[131] Al-Yaseri AZ, Lebedev M, Barifcani A, Iglauer S. Receding and advancing (CO_2 + brine + quartz) contact angles as a function of pressure, temperature, surface roughness, salt type and salinity. Journal of Chemical Thermodynamics. 2016;**93**:416-423

[132] Arif M, Barifcani A, Iglauer S. Solid/CO_2 and solid/water interfacial tensions as a function of pressure temperature, salinity and mineral type: Implications for CO_2- wettability and CO_2 geo-storage. International Journal of Greenhouse Gas Control. 2016;**53**:263-273

[133] Iglauer S. CO_2-water-rock wettability: Variability, influencing factors, and implications for co_2 geostorage. Accounts of Chemical Research. 2017;**50**(5):1134-1142

[134] Aminzadeh B, Chung DH, Bryant SL, Huh C, DiCarlo DA. CO_2 leakage prevention by introducing engineered nanoparticles to the in-situ brine. Energy Procedia. 2013;**37**: 5290-5297

[135] AttarHamed F, Zoveidavianpoor M, Jalilavi M. The incorporation of silica nanoparticle and alpha olefin sulphonate in aqueous CO_2 foam: Investigation of foaming behavior and synergistic effect. Petroleum Science and Technology. 2014;**32**(21):2549-2558

[136] Singh R, Mohanty KK. Synergistic stabilization of foams by a mixture of nanoparticles and surfactants. Society of Petroleum Engineers; 2014

[137] Nguyen P, Fadaei H, Sinton D. Nanoparticle stabilized CO_2 in water foam for mobility control in enhanced oil recovery via microfluidic method. SPE Heavy Oil Conference-Canada; Calgary, Alberta, Canada, 10–12 June; 2014

[138] Jikich SJ. CO_2EOR: Nanotechnology for mobility control studied. Journal of Petroleum Technology. 2012;**64**(7):28-31

[139] Worthen AJ, Bagaris HG, Chen Y, Bryant SL, Huh C, Johnston KP. Nanoparticle-stabilized carbon dioxide in water foams for enhanced oil recovery. SPE Improved Oil Recovery Symposium; Tulsa, Oklahoma, USA, 16–18 April; 2012

[140] Worthen AJ, Bagaria HG, Chen Y, Bryant SL, Huh C, Johnston KP. Nanoparticle-stabilized carbon dioxide-in-water foams with fine texture. Journal of Colloid and Interface Science. 2013;**391**:142-151

[141] Emrani AS, Nasr-El-Din HA. Stabilizing CO_2 foam by use of nanoparticles. Society of Petroleum Engineers Journal of Engineering Chemistry. 2017;**28**(2):988-994

[142] Salehi M, Johnson SJ, Liang JT. Mechanistic study of wettability alteration using surfactants with applications in naturally fractured reservoirs. Langmuir. 2008;**24**(24):14099-14107

[143] Farhadi H, Riahi S, Ayatollahi S, Ahmadi H. Experimental study of nanoparticle-surfactant-stabilized CO_2 foam: Stability and mobility control. Chemical Engineering Research and Design. 2016;**111**:449-460

[144] Betancur S, Carmona JC, Nassar NN, Franco CA, Cortés FB. Role of particle size and surface acidity of silica gel nanoparticles in inhibition of formation damage by asphaltene in oil reservoirs. Industrial & Engineering Chemistry Research. 2016;**55**(21):6122-6132

[145] Mohammadi M, Akbari M, Fakhroueian Z, Bahramian A, Azin R, Arya S. Inhibition of asphaltene precipitation by TiO_2, SiO_2, and ZrO_2 nanofluids. Energy & Fuels. 2011;**25**(7):3150-3156

[146] Dai C, Wang X, Li Y, Lv W, Zou C, Gao M, Zhao M. Spontaneous imbibition investigation of self-dispersing silica nanofluids for enhanced oil recovery in low-permeability cores. Energy & Fuels. 2017;**31**(3):2663-2668

[147] Vatanparast H, Samiee A, Bahramian A, Javadi A. Surface behavior of hydrophilic silica nanoparticle-SDS surfactant solutions: I. Effect of nanoparticle concentration on foamability and foam stability. Colloids and Surfaces A: Physicochemical and Engineering Aspects. 2017;**513**:430-441

[148] Al-Anssari S, Nwidee LN, Ali M, Sangwai JS, Wang S, Barifcani A, Iglauer S. Retention of silica nanoparticles in limestone porous media. Asia Pacific Oil and Gas Conference and Exhibition held in Bali, Indonesia; 17–19 October; 2017

[149] Skauge T, Spildo K, Skauge A. Nano-sized particles for EOR. In SPE Improved Oil Recovery Symposium. Society of Petroleum Engineers; Tulsa, Oklahoma, USA; 24–28 April; 2010

Chapter 9

Lessons Learned from Our Recent Research in Chemical Enhanced Oil Recovery (C-EOR) Methods

Bing Wei, Peng Wei, Shuai Zhao and Wanfen Pu

Additional information is available at the end of the chapter

http://dx.doi.org/10.5772/intechopen.71816

Abstract

As a result of the ever-increasing global energy demand coupled with the rapid decline of the oil production, the games of enhanced oil recovery (EOR) are played in many oilfields worldwide especially in China. It was reported that EOR jobs produced 45.1 × 10^4 m^3/d of oil production rate in 2014 all over the world, proving the significance of these jobs. Due to the complex geology, chemical enhanced oil recovery (C-EOR) methods are considered the predominant technology in China and takes nearly 86% of the total EOR projects currently. This fact motivates us to develop novel and more advanced C-EOR methods for different geological types of Chinese reservoirs such as high temperature and pressure, ultralow permeability, heavy oil reservoirs, etc. Through 20 years' efforts, many advantageous C-EOR methods have been successfully developed in our group and tested in oilfields such as stabilized foam injection, nanofluid flooding, functional polymer flooding, etc. Herein, this chapter summarized the latest experimental results of three representative C-EOR methods. More attentions were given to the relationship between bulk properties and flow behaviors in porous media. The lessons learned from our research in C-EOR were also discussed in this chapter.

Keywords: enhanced oil recovery, chemical flooding, innovative EOR, nanofluid, polymer microsphere, enhanced foam

1. Nanofluid flooding

To date, hydrolyzed polyacrylamide (HPAM) is still the most widely used polymer due to its availability in large quantities with customizable properties (molecular weight, hydrolysis degree, etc.) and low manufacturing cost. However, acrylamide-based polymers are susceptible to chemical, mechanical, thermal, and microbial degradations [1]. In addition, the acrylamide monomers place detriment to the environment due to its toxicity [2, 3]. Nanofluids, which are obtained through dispersing nanoparticles in the base fluids, are attracting research attentions. The nanofluids usually exhibit high thermal conductivities, which are significant to the development of energy, and have been widely used in drug delivery solar cells, lipase immobilization, soil remediation, lubrication, and hydraulic fracturing of gas and oil. In addition, nanofluids have been introduced to the enhanced oil recovery (EOR) area due to their unique thermal properties and large surface area [4–8]. Many works have been done to study the EOR performance of nanofluids composed of different nanoparticles. Hendraningrat et al. evaluated the displacement efficiency of the nanofluid. They claimed that the nanofluid increased the oil recovery by 7–14.3% and the optimum concentration was 0.05 wt% for water-wet core [9, 10]. The stability of nanofluids was considered as an important factor, which is closely associated with the success of nanofluid flooding [11]. Roustaei et al. demonstrated that the modified silica nanoparticles were able to enhance light oil recovery by reducing interfacial tension and altering wettability [12]. The potential of hydrophobic and lipophilic nanoparticles in EOR was also verified [13–15].

Cellulose is one of the most abundant biopolymers on earth and widely contained in wood, cotton, hemp, and other renewable origins [16–19]. When cellulose is in nanosize, many distinctive properties are therefore created, for example, high strength, chemical accessibility, large surface area, etc., which accordingly cause the increasing attentions in the research community [20, 21].

1.1. Surface grafting

In the interest of further improving the properties of the nanocellulose (NC) toward EOR use, two functional NC samples were successfully prepared in our group *via* surface grafting. The chemical structures and micromorphologies of the three samples are shown in **Figure 1**. **Figure 2** shows the TG/DTG curves of the nanocellulose. It was noted that after surface grafting, the thermal stability of resultant NC (NC-KY and NC-KYSS) was slightly higher than the origin caused by the grafted AMPS and alkyl chains [16, 17, 22, 23].

1.2. Rheological properties of the nanofluids

The shear rate and mass-dependent behavior of nanofluid viscosity are shown in **Figure 3**. All the three nanofluids are pseudoplastic fluids, i.e., shear-thinning fluids. The viscosity of the nanofluid gradually increased with the concentration, which was consistent with the tendency of other relevant polymers as HPAM. After grafting, the viscosity of the nanocellulose was detracted especially NC-KYSS resulted from enlarged molecular space induced by the loaded groups. Nevertheless, a significant thickening power was observed for NC-KYSS at an industrial level as indicated in **Figure 3**. It has been well understood that viscosity of the system is the predominant component in any chemical-based EOR process. Compared with other particle-based nanofluids, the nanocellulose nanofluids are advantageous with noticeable thickening property [17].

1.3. Dispersity of nanocellulose in electrolyte

The colloidal stability of nanofluid is one of the major considerations for EOR. The presence of cations (Na^+, Ca^{2+}, and Mg^{2+}) usually leads the nanocellulose to aggregate and

http://dx.doi.org/10.5772/intechopen.71816

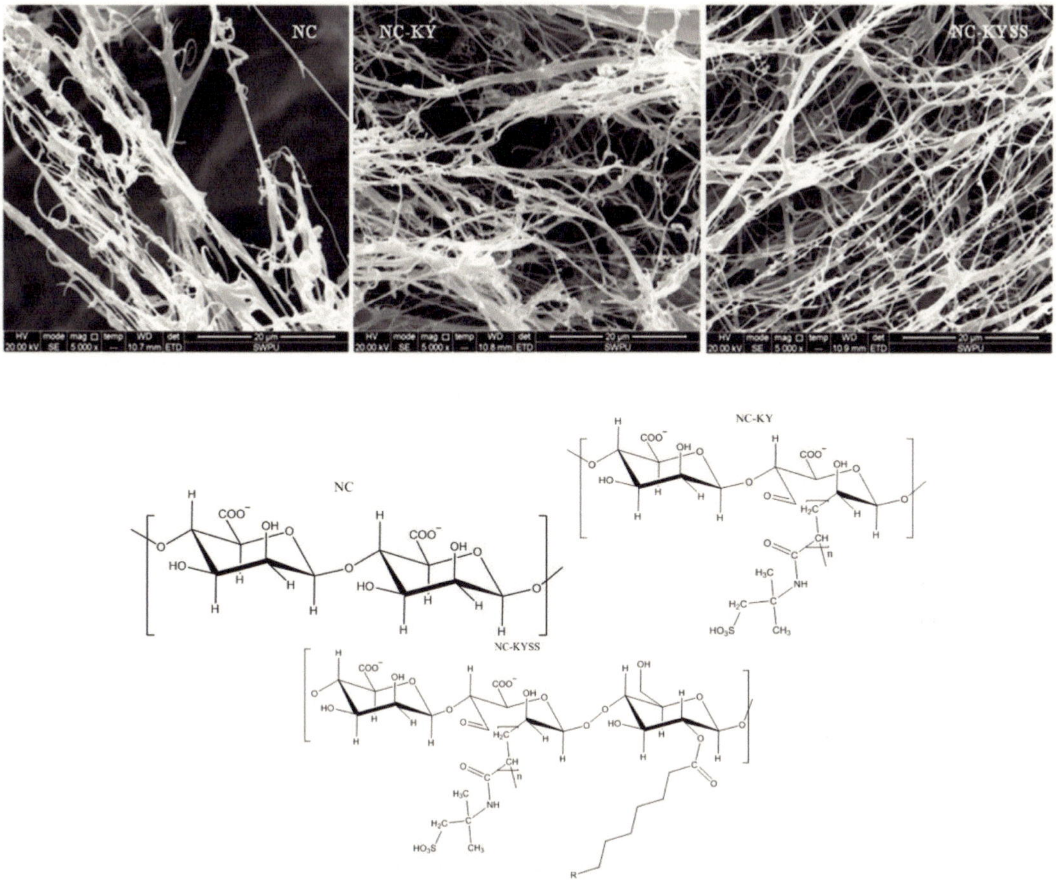

Figure 1. Chemical structure and SEM images of the nanocellulose.

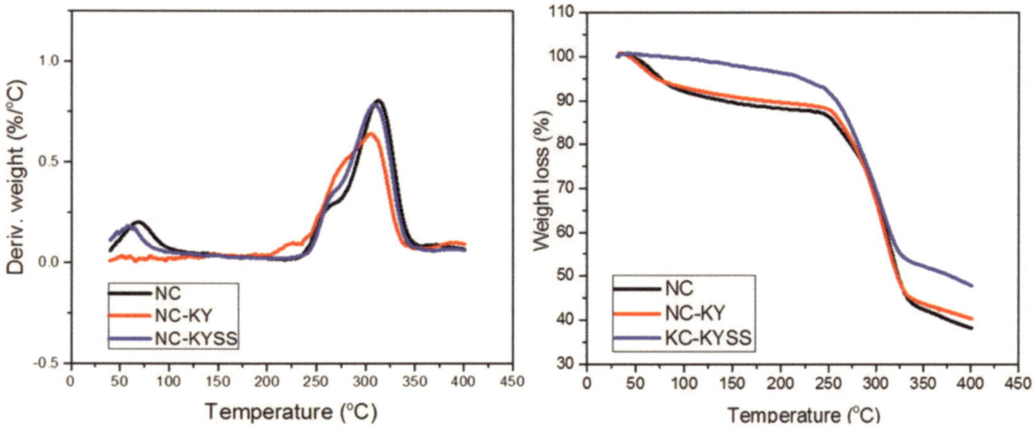

Figure 2. TG/DTG curves of the nanocellulose.

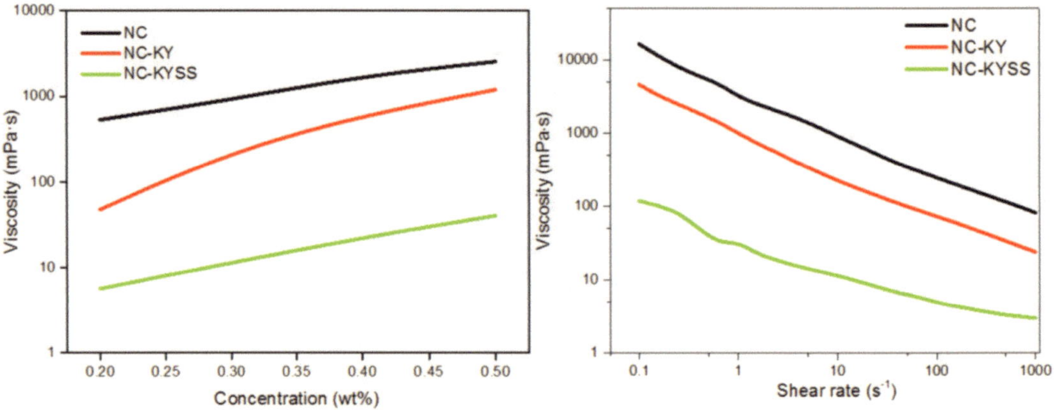

Figure 3. Rheological properties of the nanocellulose fluids.

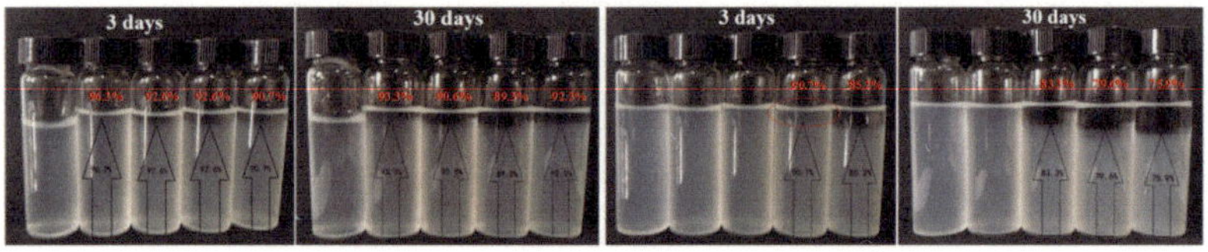

NC NC-KY

(NaCl: 0, 0.1, 0.3 and 0.7% from left to right)

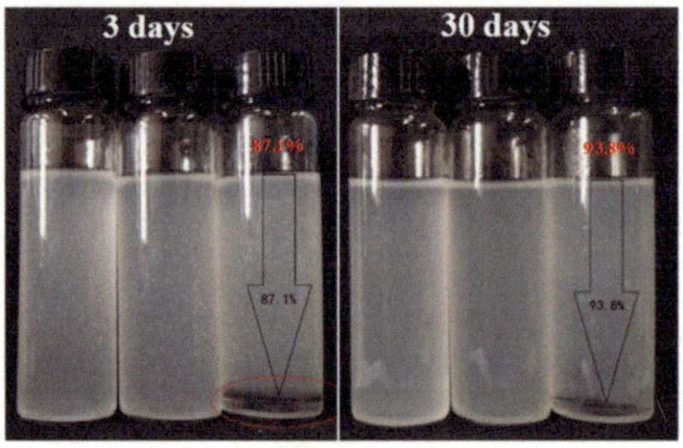

NC-KYSS

(NaCl: 1, 1.5 and 2% from left to right)

Figure 4. Images of the nanocellulose nanofluids in electrolyte.

flocculate, subsequently impacting the fluidity in porous media. **Figure 4** presents the images of the nanocellulose fluids in NaCl electrolyte with storage time. It was observed that the ungrafted nanocellulose (NC) fluid remained homogeneous for 30 days in deionized water; however, when the cation was present, nanocellulose aggregation and flocculation occurred as a result of the compressed electric double layer. The grafting of AMPS did not effectively hinder the aggregation. This fact indicated that the sole electrostatic repulsion was not sufficient to combat high salinity. In contrast, simultaneous grafting of AMPS and alkyl chains noticeably enhanced the dispersity of the nanocellulose in electrolyte caused by extra steric hindrance [17].

1.4. Microscale displacement behaviors

To study the flow behaviors in porous media and EOR efficiency of the tested nanofluids, a five-spot visual micromodel was designed and assembled in our lab. In addition to visual observation, the propagations of the displacing phase and displaced phase were captured and further analyzed using a coded MatLab program. Due to the dispersity issue, NC-KYSS fluid was first tested in the oil displacement experiment. In the homogenous sand pack model, the waterflooding oil recovery was nearly 46% of original oil in place (OOIP) at 4 PV of brine. Then, a 0.4 PV of NC-KYSS fluid slug was injected into the model to simulate EOR followed by a post waterflooding. It was shown that this nanofluid was able to further improve the oil recovery after waterflooding. In our work, approximately 6% OOIP was produced, and the water cut was reduced simultaneously, verifying the potential of this nanofluid in EOR. **Figure 5** shows the micromodel images captured during the displacement tests. From these images, the reduction of the oil saturation of the sand pack model can be visually observed. Through pixel reading, the oil recovery efficiencies of the waterflooding, nanofluid flooding, and post waterflooding were quantitatively determined as shown in **Figure 6**.

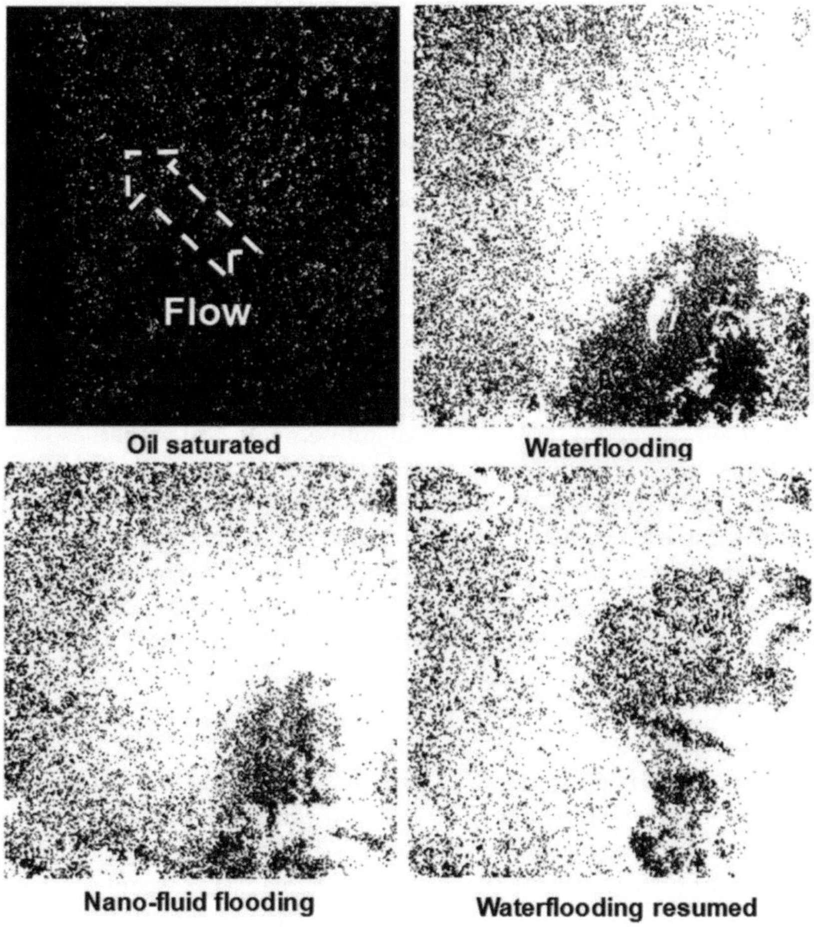

Figure 5. MatLab-released images of the microdisplacement test.

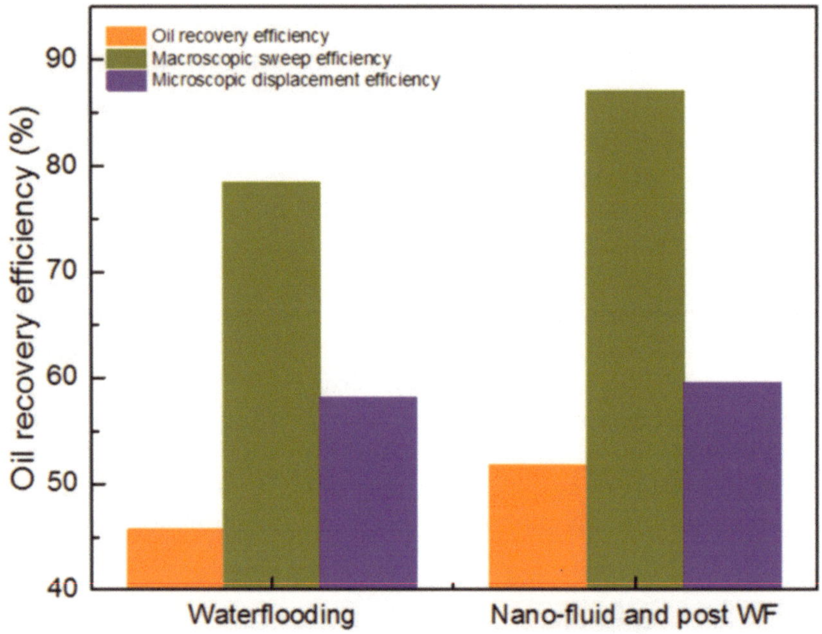

Figure 6. Oil recoveries in the stages of waterflooding and nanofluid flooding.

The results proved that the macroscopic sweep and microscopic displacement efficiencies of the nanofluid flooding were increased by 9.2 and 1.3%, respectively, on the basis of waterflooding [22].

2. Stabilized foam flooding

The EOR efficiency of a foam injection is largely governed by foam stability. However, due to the complex reservoir conditions (pressure, temperature, geological properties, crude, etc.), the generated bubbles in porous media easily rupture particularly when they are in direct contact with hydrocarbons, which are considered as antifoaming agents and detrimental to foam stability [24, 25]. Substantial efforts have been made toward foam stability enhancement from both theoretical and experimental levels [26]. The methods that utilize chemical agents to prevent film drainage have been successfully proposed. The widely used agents can be categorized into two groups, i.e., polymer and nanoparticle. Sydansk observed that adding polymer to the aqueous phase resulted in the improvement of the foam viscosity and stability. Shen et al. compared the stability of HPAM- and xanthan gum-enhanced foam system during migrating in porous media [26]. An associative polymer was also employed to stabilize foam lamella through the molecular interactions with surfactants [27]. Using starch particle to improve the physical properties of foam was suggested by Zhang et al. The results indicated that the film drainage was considerably mitigated owing to starch participation [28]. As for nanoparticle, silica is the most widely used nanoparticle according to reports [29, 30]. Farhadi et al. examined the stability and mobility control of a silica-stabilized CO_2 foam system. It was proved that this system produced uniform and small foams with high apparent viscosity compared to the surfactant-only system

http://dx.doi.org/10.5772/intechopen.71816

[31]. Hydrophobic silica was also adopted to stabilize a conventional foam system, and the resultant foam system was more rough than that of PEG-coated nanosilica-stabilized foam [32]. In addition to silica, the possibility of aluminum, fly ash, and $CaCO_3$ as foam stabilizers was also assessed [33–35]. Based on the previous works, three possible mechanisms for foam film stabilization were elucidated by Horozov, including monolayer of bridging particles, bilayer of close-packed particles, and network of particle aggregates [36].

2.1. Xanthan gum-stabilized foam

2.1.1. Alteration of foam morphology

As shown in **Figure 7**, CHSB foam displays small and uniform bubble sizes, whereas polymer-enhanced foam presents the large and coarse sizes in the initial state. Afterward, a significant difference is also observed after 10 min, i.e., the bubbles of CHSB foam change into a form of large and polyhedral shape. By contrast, the change behavior of polymer-enhanced foam is very small. Due to the large viscous force, the polymer-surfactant mixtures can create thick and stable liquid films delaying coalescence, which thus promotes the stability of polymer-enhanced foam [37, 38].

2.1.2. Foam microstructure in oil-bearing

In this section, the effect of crude oil on foam properties was shown in **Figure 8**. As observed from **Figure 8(a)**, two systems maintained the favorable foam properties in the presence of 15% oil; nevertheless, the foam stability gradually decreased as the oil content continued to increase. In contrast, the foam volume of the xanthan-enhanced

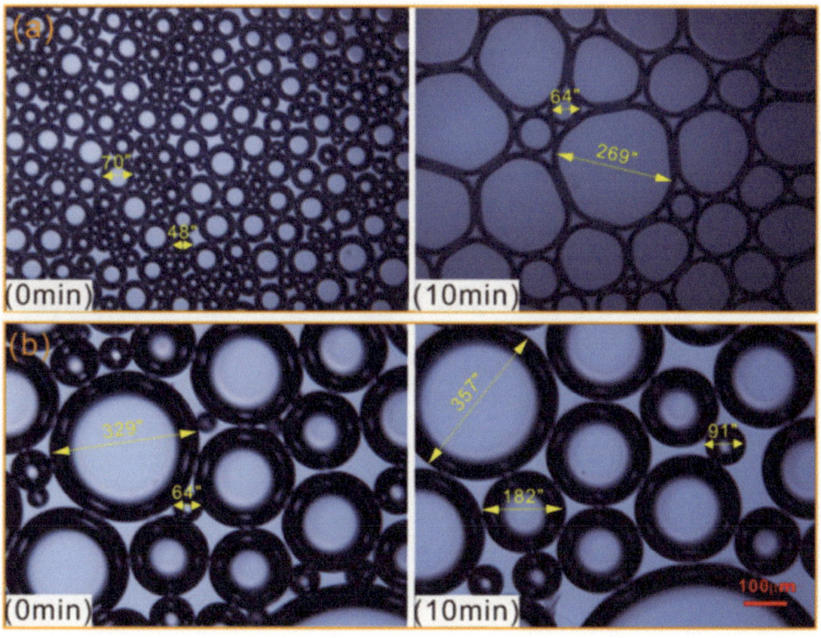

Figure 7. Microstructures of CHSB foam (a) and xanthan-enhanced foam (b) changes over time.

foam increased even when 20% of oil was present, indicating the stabilizing effect of polymer in oil-bearing environment. As shown in **Figure 8(b)**, the half-life ratio was close to 8 in the absence of oil. When the oil was added to the foam system, this ratio nearly remained unchanged within 10% of oil content and then gradually increased to about 10 with the increase of oil content to 20%.

To further evaluate the effect of crude oil on foam, the microstructures of oil foam were visually observed using LM. As shown in **Figure 9(a)**, the oil droplets tended to self-assemble at the plateau borders of CHSB foam with liquid drainage, and the deposited oil droplets merged into the larger drops at a fast rate before finally forming an oil band at the plateau border. From **Figure 9(b)**, the oil droplets of the xanthan-enhanced foam had a smaller size

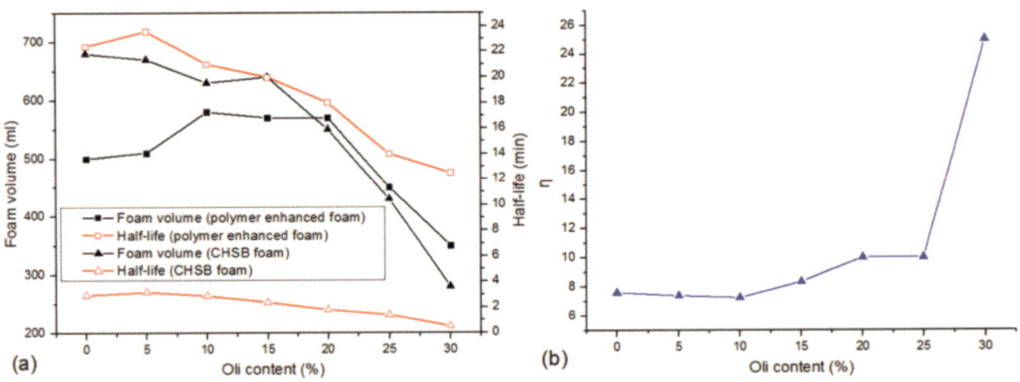

Figure 8. Foam properties as a function of oil content, (a) variations of foam volume and half-time, (b) the half-life ratio (η) of xanthan-enhanced foam to CHSB foam.

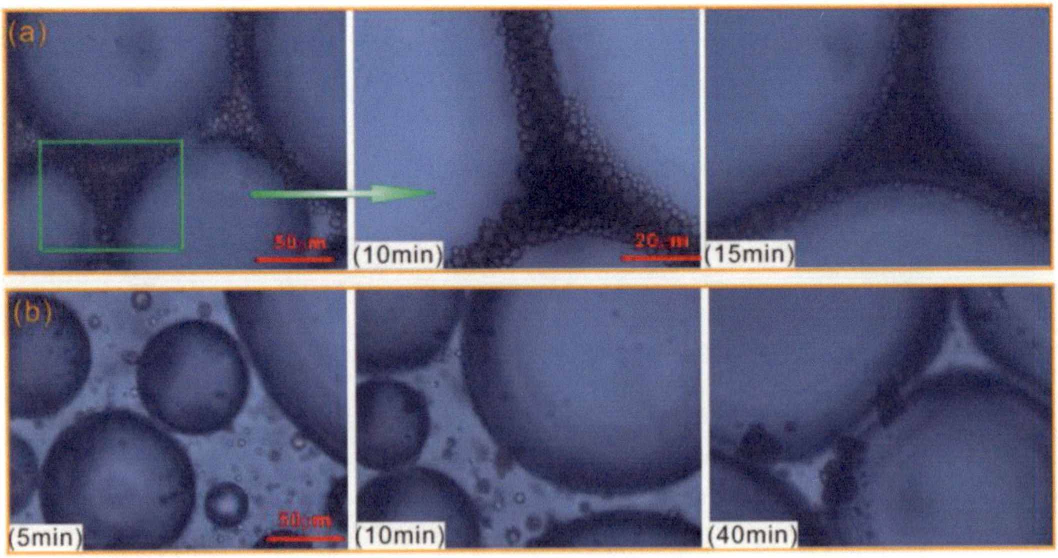

Figure 9. Microphotographs of oil drops as a function of time and CHSB foam (a) and xanthan-enhanced foam (b).

and were uniformly dispersed in the matrix of foam films (i.e., oil in water emulsions). Furthermore, the accumulated oil droplets distributed in foam films instead of plateau borders. As a result, the crude oil can be emulsified into small and stable particles in the foam films.

2.1.3. *Foam injection in the heterogeneous model*

As shown in **Figure 10**, the oil recovery of high-permeable core is 30–42% after waterflooding, but with the increase of permeability ratio, the oil recovery of low-permeable core decreased, which indicated that the swept volume of the low permeability core was quite low. The remarkable variations of division ratio and oil recovery were observed in heterogeneous cores as a result of the injection of the xanthan-enhanced foam. Since foam gradually established the high flow resistance in the high-permeable core, the subsequent flow was diverted to the low-permeable core displacing the residual oil. Moreover, foam has been proven to perform excellent in the area of low oil saturation and large spatial scale. Therefore, the large amount of crude oil was produced from both the high-permeable core and the low-permeable core, which thus led to the facts that the enhanced oil recovery increased as the heterogeneity gets stronger especially in the low-permeable core.

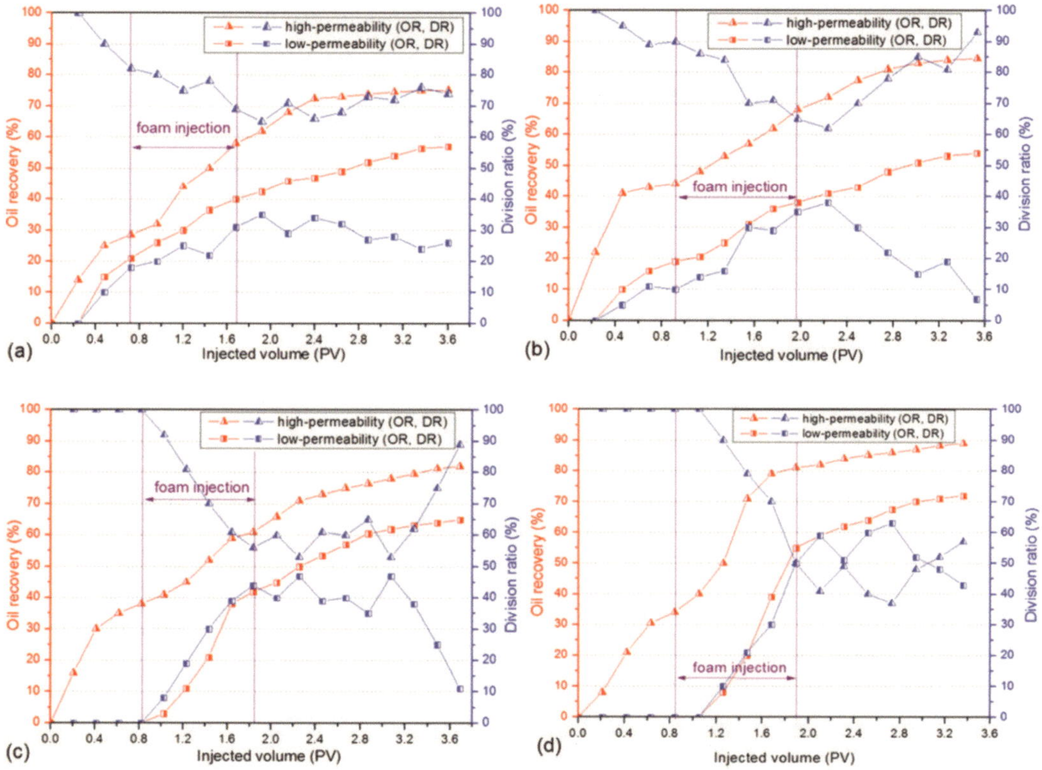

Figure 10. Changes of oil recovery (OR) and division ratio (DR) in the high and low permeability of heterogeneous cores. The different permeability ratios are 1.5 (a), 3 (b), 6 (c), and 12 (d), and xanthan enhanced foam is injected.

Consequently, the significant flow diversion and oil increase were achieved by conducting polymer-enhanced foam flooding in the heterogeneous formation. The larger permeability ratio can promote the advantage of foam to carry out mobility reduction and oil recovery enhancement.

2.2. Nanocellulose-stabilized foam

2.2.1. Foam decay in the absence and presence of crude oil

Figure 11 shows the foam decay profiles of the four systems in the absence and presence of oil. In the bulk tests, the criterion of half decay time, which was defined as the time that the first half foam volume lasts, was frequently used to quantify the stability of the prepared foam. Normally, long half decay time represents high foam stability and vice versa. The incorporation of the nanocellulose fluids notably extended the foam life from generation to completely vanish. The enhancement was more significant when oil was present, as revealed by the profiles shown in the right. These profiles also demonstrated that oil was detrimental to foam stability. After nanofluid stabilization, the half decay time or stability of the foam was considerably promoted. The hypothesized mechanisms of foam lamella stabilization are as follows. Due to the synergism of the nanocellulose molecules, the close-packed foam lamella is further strengthened, which consequently renders the generated bubbles less probability to rupture. Moreover, the substantial OH- groups on the nanocellulose surface are also capable of impeding liquid drainage owing to the hydrogen bonding interactions between water, nanocellulose, and surfactant molecules in the film lamella. The close-packed foam lamella leads the foam volume to shrink due to the increased liquid viscosity and density, whereas the molecular interactions at the foam lamella can stabilize the foam film and accordingly extend the foam life [22].

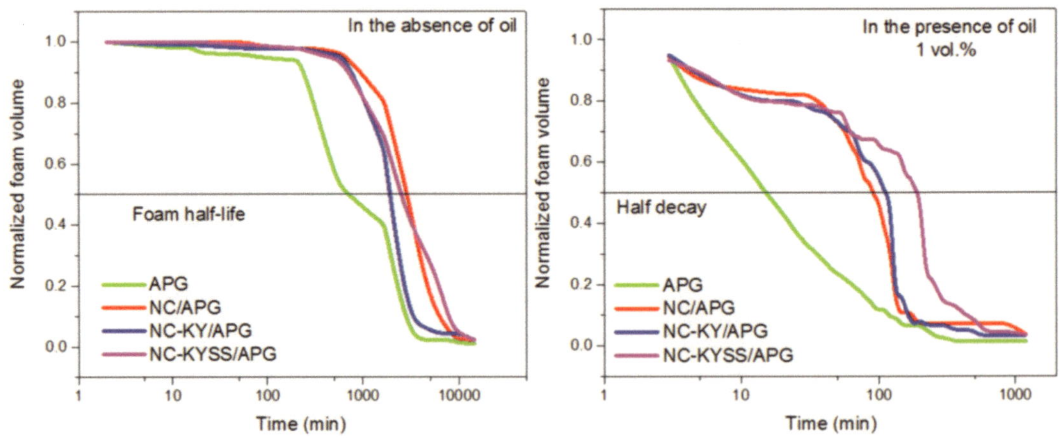

Figure 11. Foam decay profiles of four systems (APG: foaming agent).

2.2.2. *Morphology of the stabilized foam*

The participation of NC-KYSS made the resultant bubbles smaller than the surfactant-only foam, which subsequently led to thicker foam lamella due to the stabilizing effect as shown in **Figure 12**. This effect enables the prepared foam to retard inter-bubble gas diffusion. When hydrocarbons were present, the improvement of the foam lamella thickness was more significant. The microscopic observations were highly in agreement with the results obtained in the bulk cylinder tests. The coarsening rate of the foam bubbles was accelerated when hydrocarbons were involved and consequently resulted in large bubbles. Although the hydrocarbons have exerted a detriment to foam morphology, considerably thick foam lamella was constructed due to the collaboration of NC-KYSS. The reasons behind to interpret this fact have been discussed above [22].

2.2.3. *Microdisplacement behaviors*

Using the same digital method as established in Section 1.4, the flow patterns of oil and foam in porous media were captured as shown in **Figure 13**. The residual oil saturation (S_{or}) represented by the black pixel indicated the limited oil recovery of waterflooding stage. As foam injection proceeded, an "oil bank" was constructed in the front of foam as a result of the mobilized residual oil. However, due to the defoaming/antifoaming effect, the conventional foam quickly channeled/fingered through the established oil bank leaving oil-bearing area untouched. In contrast, the injection of the KYSS-stabilized foam created a tough and "piston-like" displacing front owing to the prominent oil tolerance and accordingly promoted the macroscopic sweep efficiency and microscopic displacement efficiency as shown in **Figure 14** [22].

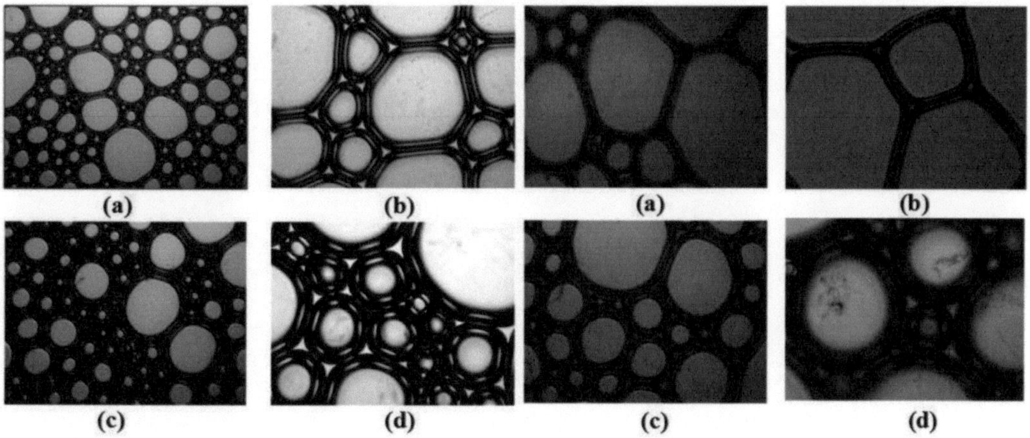

Figure 12. Microscopy images of the foam (a and b: APG foam at 40 and 100×; c and d NC-KYSS-stabilized foam at 40 and 100×).

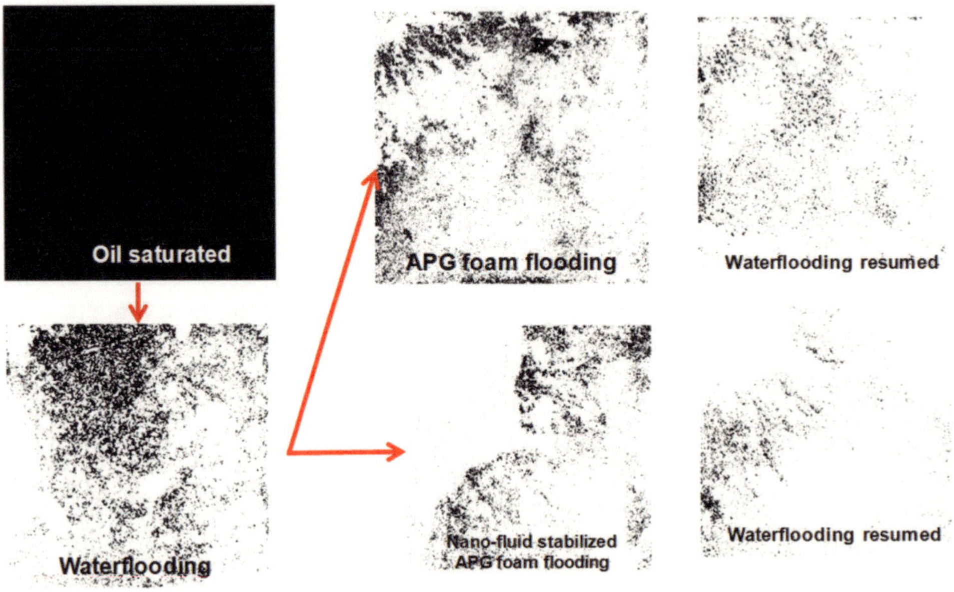

Figure 13. MatLab-released images of the foam flooding.

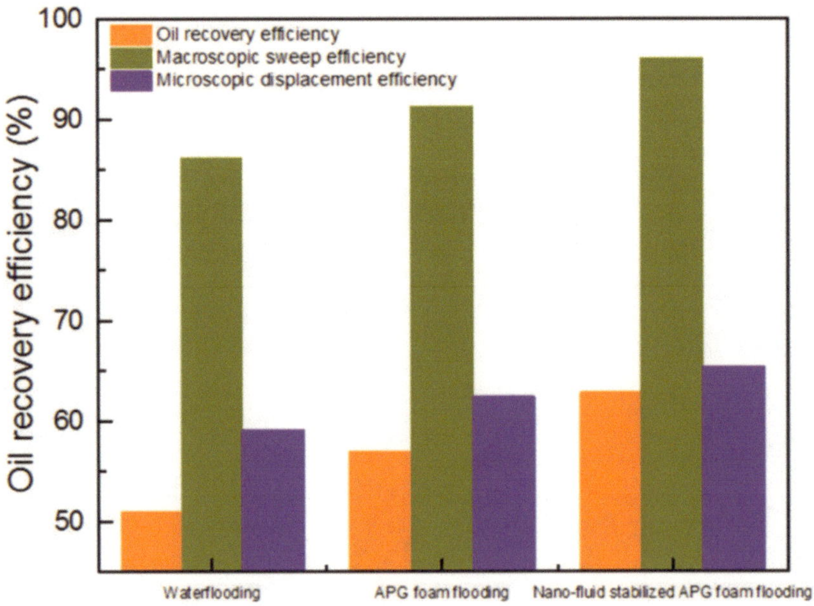

Figure 14. Oil recovery efficiencies of the foam flooding.

3. Microsphere injection

After extensive waterflooding, most of the oilfields worldwide have entered the middle and high water cut stage; crude oil production decreases significantly. Meanwhile, the exploration of new oil and gas resources is becoming more and more difficult. In order to maintain the oil production of the aged oilfield, to find an economical and effective enhanced oil recovery (EOR) technique has become the focus of researchers.

In the last decade, a new dispersion gel called polymer microspheres (PMs) was employed for high temperature and salinity reservoirs to achieve conformance control. The PMs that match well with pore-throat size can be synthesized via the polymerization of polymer monomer, crosslinking agent, initiator, and active agent. The associated physical properties and dynamic displacement behaviors of PMs synthesized by inverse emulsion polymerization in our laboratory would be discussed briefly in next sections aiming at rendering quite a few evidences guiding its application in actual reservoirs.

3.1. Physical properties of PMs

As shown in **Figure 15**, the PMs were micron-grade particles with milky white and high sphericity. The initial particle size distribution of PMs synthesized by inverse emulsion polymerization in our laboratory was in the range of 2.364 (d_{10}) to 26.212 μm (d_{90}), and its median diameter (d_{50}) was 9.996 μm (see **Figure 16**) [39].

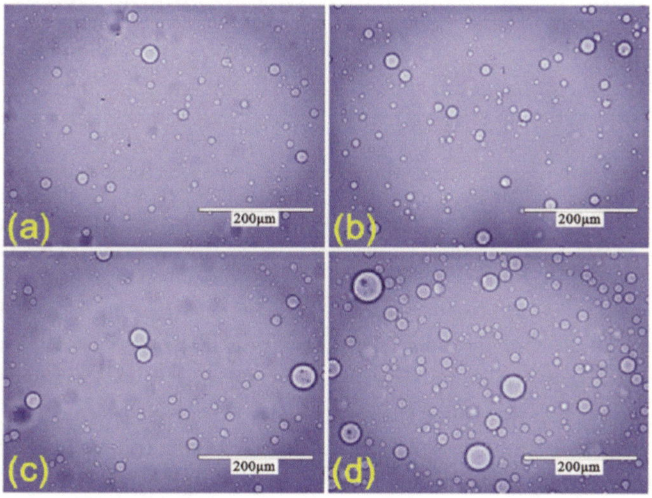

Figure 15. Dispersion state of PM system at different times: (a) initial state, (b) 10 minutes, (c) 20 minutes, and (d) 30 minutes.

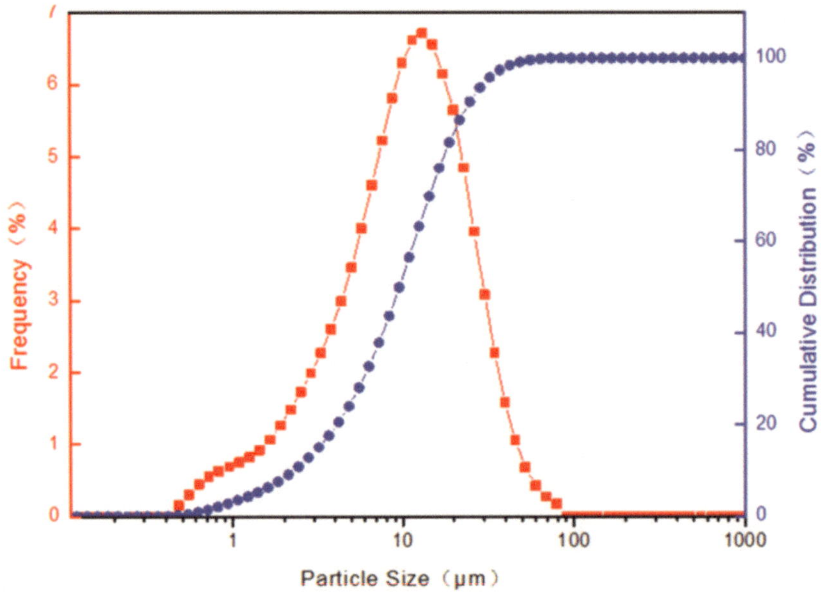

Figure 16. Initial particle size distribution of PMs in the injection water.

Figure 15(a)–(d) shows the morphology of PMs swelling for 15 days in injection water at 25, 50, 75, and 104°C, respectively. The PMs kept high sphericity and could well disperse in water with no precipitation and agglomeration within the temperature ranging from 25 to 104°C. The results show that the PMs could withstand temperature of 104°C and salinity of 25,000 mg/L. This renders PMs' potential in high temperature and salinity reservoirs.

The hydration swelling property of PMs is represented by the expansion ratio, which can be calculated by Eq. (1):

$$ER = (D_2 - D_1)/D_1 \tag{1}$$

where ER represents expansion ratio and D_1 and D_2 are median diameter of PMs before and after expansion, respectively, μm.

The changing curves of the expansion ratio of PMs under different salinity conditions at 25°C are shown in **Figure 17**. At the same salinity, the expansion ratio of PMs increased with time; the expansion rate in the initial stage was fast, then gradually slowed down, and leveled up after swelling for 7 days. At the same hydration time, the expansion ratio of PMs decreased with the salinity. In particular, the swelling ratio of PMs in the deionized water was much larger than that in the brine.

Figure 18 displays the apparent viscosity of 0.5 wt% PM system in the deionized water, NaCl solution with salinity of 10,000 and 20,000 mg/L, reservoir injection water, respectively, after swelling 10 days at 25°C. In the range of low shear rate, the apparent viscosity of PM system

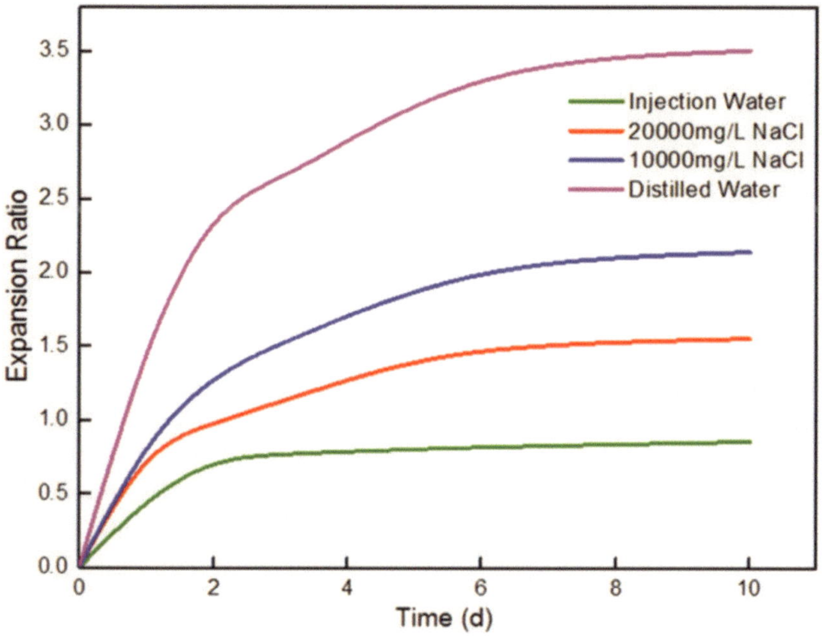

Figure 17. Influence of salinity on the expansion ratio of PMs.

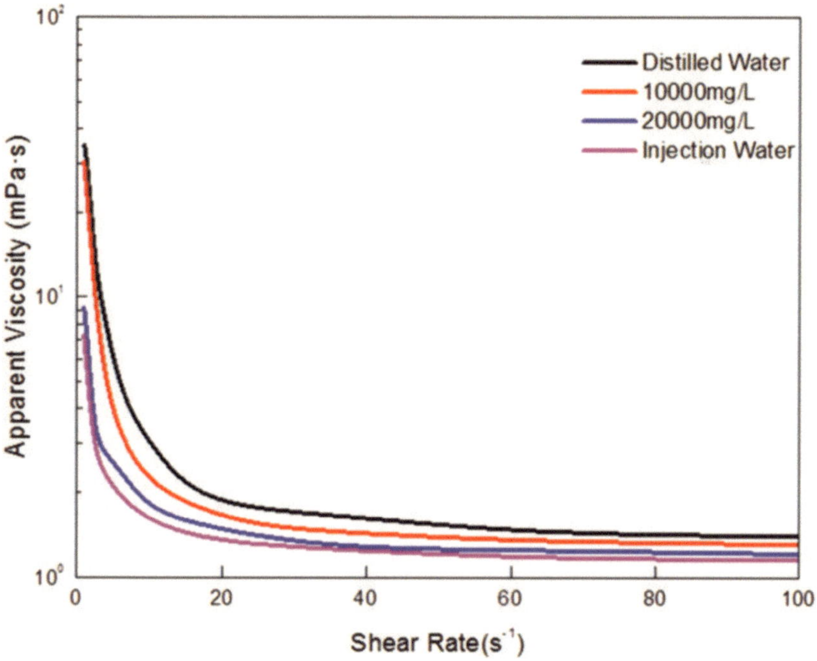

Figure 18. Apparent viscosity of PM system versus the shear rate (25°C).

decreased significantly with the shear rate. At the high shear rate, the apparent viscosity of PM system increased or decreased slightly with the shear rate, which is close to the viscosity of water. So the original water injection pipeline could be directly used to inject the PM system. The dynamic elasticity of four types of PMs (without solvent) mentioned in this section versus frequency is shown in **Figure 19**. For the same PMs, when the oscillation frequency was increased from 0.1 to 100 Hz, the elastic modulus of PMs gradually increased. At the same frequency, the elastic modulus of PMs increased with the salinity, which is related with the charge of PMs [39].

3.2. Dynamic displacement behaviors of PMs

The fluid distribution of the visual model, in which diagonal upper and lower area represents low and high-permeable zone, respectively, is shown in **Figure 20**. During PMF, the PM system mainly flowed into the high-permeable zone. In the SWF stage (**Figure 20(d)**), the low-permeable zone was significantly activated evidencing these PMs can achieve effective plugging for high-permeable region as well as allow the subsequent fluid to flow into the low-permeable zone rich in remaining oil [40].

3.3. Matching relationship between microsphere diameter and pore-throat size

As shown in **Figure 21**, the migration type with β >1.99 (the ratio of median diameter of PMs and average pore-throat diameter of core) is defined as "complete plugging," in which the pressure drop increased noticeably with the injection of PM system. "High-efficiency plugging" arises

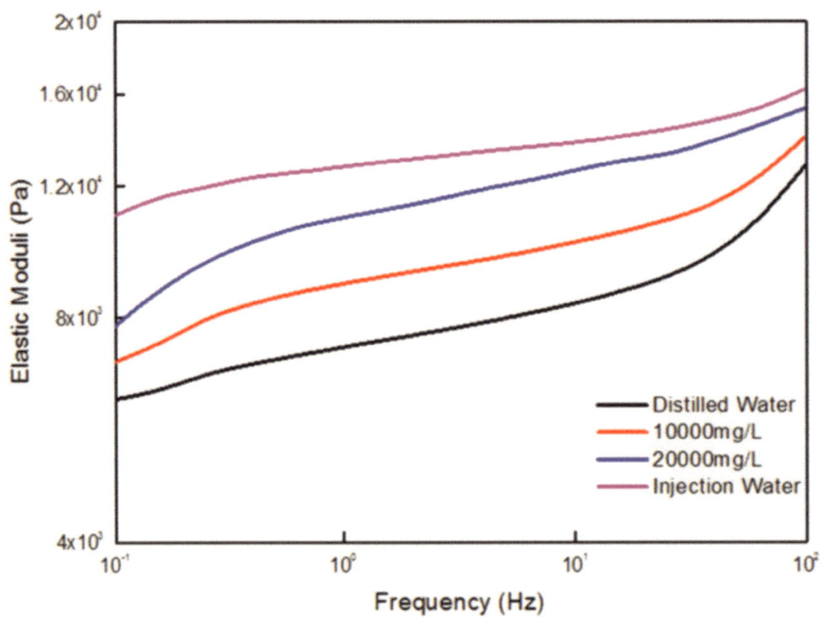

Figure 19. Elastic modulus of PMs versus the frequency (25°C).

http://dx.doi.org/10.5772/intechopen.71816

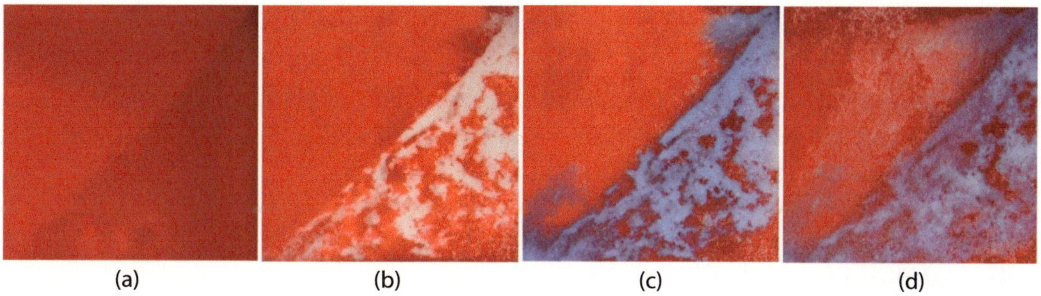

Figure 20. Fluid distribution of the visual model (a) initial oil saturation, (b) WF, (c) PMF, and (d) SWF.

when β is in the range of 1.02 and 1.99. In this case, the differential pressure first increased with the injection of PMs until a peak pressure was achieved. This demonstrated the plugging of the pore throats. Afterward, the pressure curve tended to fluctuate with the injection continued, indicating that the PMs might break up and thus pass through the pore throats and then deeply migrate forward into the core until the next block-breakthrough cycle occurred. The differential pressure gradually decreased before finally leveling off during SWF. The resistance coefficient and blocking rate were more than 21.9 and 80.9%, respectively. In the case of 0.59 < β < 1.02, the migration was regarded as "low-efficiency plugging," in which the differential pressure varied similarly with "high-efficiency plugging" with exception of the causing lower pressure. The relevant resistance coefficient and blocking rate were 8.3–21.9 and 20–80.9%, respectively. "Smooth pass" occurred when β is smaller than 0.59. The blocking rate was less than 20% including the reduction of cross-sectional area caused by the adsorption of PMs on the pore surfaces. Hence, the majority of the PMs passed the pore throats successfully without any plugging effect [39].

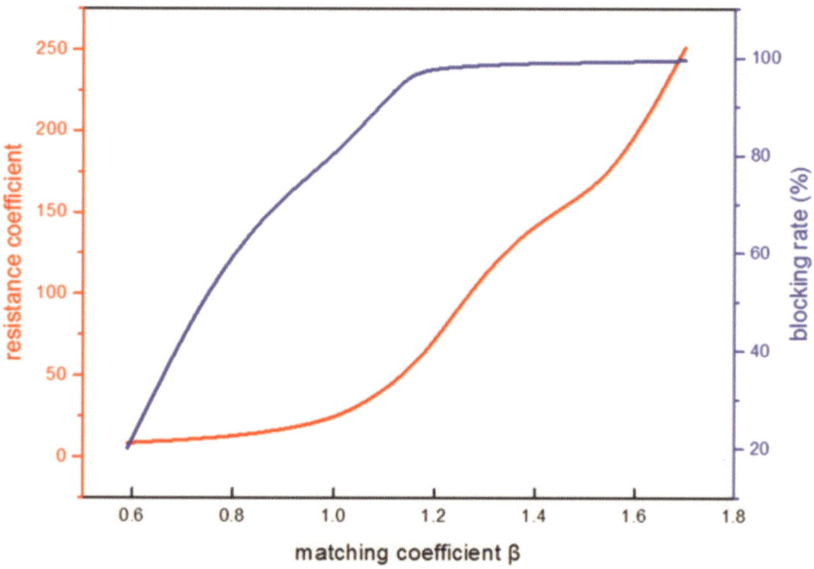

Figure 21. Resistance coefficient and blocking rate versus matching coefficient.

Acknowledgements

The authors gratefully acknowledge the financial support provided by the Canadian Foundation for Innovation (CFI), National Key Basic Research Program of China (2015CB250904), and PetroChina Innovation Foundation (2015D-5006-0212).

Author details

Bing Wei*, Peng Wei, Shuai Zhao and Wanfen Pu

*Address all correspondence to: bwei@swpu.edu.cn

State Key Laboratory of Oil and Gas Reservoir Geology and Exploitation, Southwest Petroleum University, Chengdu, China

References

[1] Wei B, Romero-Zerón L, Rodrigue D. Evaluation of two new self-assembly polymeric systems for enhanced heavy oil recovery. Industrial and Engineering Chemistry Research. 2014;**53**(43):16600-16611

[2] Kusanagi K, Murata S, Goi Y, Sabi M, Zinno K, Kato Y, Togashi N, Matsuoka T, Liang Y. Application of cellulose nanofiber as environment-friendly polymer for oil development. In: Paper SPE 176456 presented at SPE/IATMI Asia Pacific Oil & Gas Conference and Exhibition

[3] El Ela MA, Sayyouh H. An integrated approach for the application of the enhanced oil recovery projects. Journal of Petroleum Science and Engineering. 2014;**3**:176-188

[4] Cheraghian G, Hendraningrat L. A review on applications of nanotechnology in the enhanced oil recovery part A: effects of nanoparticles on interfacial tension. International Nano Letters. 2016;**1**:1-10

[5] Wilson A. Molecular-dynamics study examines effect of nanoparticles on oil/water flow. Journal of Petroleum Technology. 2013;**65**(02):148-151

[6] Yang X, Liu ZH. A kind of nanofluid consisting of surface-functionalized nanoparticles. Nanoscale Research Letters. 2010;**5**(8):1324-1328

[7] Kwak K, Kim C. Viscosity and thermal conductivity of copper oxide nanofluid dispersed in ethylene glycol. Korea-Australia Rheology Journal. 2005;**17**(2):35-40

[8] Khairul M, Shah K, Doroodchi E, Azizian R, Moghtaderi B. Effects of surfactant on stability and thermo-physical properties of metal oxide nanofluids. International Journal of Heat and Mass Transfer. 2016;**98**:778-787

[9] Hendraningrat L, Li S, Torsæter O. A coreflood investigation of nanofluid enhanced oil recovery. Journal of Petroleum Science and Engineering. 2013;**111**:128-138

[10] Buckley JS, Fan T. Crude oil/brine interfacial tensions. Petrophysics. 2007;**48**(03):175-185

[11] Hendraningrat L, Torsæter O. A stabilizer that enhances the oil recovery process using silica-based nanofluids. Transport Porous Media. 2015;**108**(3):679-696

[12] Roustaei A, Saffarzadeh S, Mohammadi M. An evaluation of modified silica nanoparticles' efficiency in enhancing oil recovery of light and intermediate oil reservoirs. Egyptian Journal of Petroleum. 2013;**22**(3):427-433

[13] Onyekonwu MO, Ogolo NA. Investigating the use of nanoparticles in enhancing oil recovery. SPE Paper 140744 Present at Nigeria Annual International Conference and Exhibition. Society of Petroleum Engineers. 2010

[14] Ehtesabi H, Ahadian MM, Taghikhani V. Enhanced heavy oil recovery using TiO_2 nanoparticles: Investigation of deposition during transport in Core plug. Energy & Fuels. 2014;**29**(1):1-8

[15] Lv YZ, Wang J, Yi K, Wang W, Li C. Effect of oleic acid surface modification on dispersion stability and breakdown strength of vegetable oil-based Fe_3O_4 nanofluids. Integrated Ferroelectrics. 2015;**163**(1):21-28

[16] Wei B, Li Q, Jin F, Li H, Wang C. The potential of a novel nano-fluid in enhancing oil recovery. Energy & Fuels. 2016;**30**(4):2882-2891

[17] Li Q, Wei B, Xue Y, Wen Y, Li J. Improving the physical properties of nano-cellulose through chemical grafting for potential use in enhancing oil recovery. Journal of Bioresources and Bioproducts. 2016;**4**(1):186-191

[18] Eichhorn SJ, Dufresne A, Aranguren M, Marcovich NE, Capadona JR, Rowan SJ, Weder C, Thielemans W, Roman M, Renneckar S. Review: Current international research into cellulose nanofibres and nanocomposites. Journal of Materials Science. 2010;**45**(1):1-33

[19] Sathvika T, Manasi M, Rajesh V, Rajesh N. Microwave assisted immobilization of yeast in cellulose biopolymer as a green adsorbent for the sequestration of chromium. Chemical Engineering Journal 2015;**279**:38-46

[20] Janardhnan S, Sain MM. Isolation of cellulose microfibrils-an enzymatic approach. BioResources. 2006;**1**(2):176-188

[21] Klemm D, Schumann D, Kramer F, He08ler N, Hornung M, Schmauder HP, Marsch S. Nanocelluloses as innovative polymers in research and application. Advances in Polymer Science. 2006;**205**:49-96

[22] Wei B, Li H, Li Q, Wen Y, Sun L, Wei P, Pu WF, Li YB. Stabilization of foam lamella using novel surface-grafted nanocellulose-based nanofluids. Langmuir. 2017;**33**(21):5127-5139

[23] Li QZ, Wei B, Lu LM, Li YB, Wen YB, Pu WF, Li H, Wang CY. Investigation of physical properties and displacement mechanisms of surface-grafted nano-cellulose fluids for enhanced oil recovery. Fuel. 2017;**207**(13):352-364

[24] Nikolov AD, Wasan DT, Huang DW, Edwards DA. The effect of oil on foam stability: Mechanisms and implications for oil displacement by foam in porous media. Paper SPE 15443 Presented at SPE Annual Technical Conference and Exhibition. Society of Petroleum Engineers. 1986

[25] Xu X, Saeedi A, Liu K. Laboratory studies on CO_2 foam flooding enhanced by a novel amphiphilic ter-polymer. Journal of Petroleum Science and Engineering. 2015;**138**:153-159

[26] Shen C, Nguyen QP, Huh C, Rossen WR. Does polymer stabilize foam in porous media? Paper SPE 99796 Presented at SPE/DOE Symposium on Improved Oil Recovery. Society of Petroleum Engineers. 2006

[27] Hernando L, Bertin H, Omari A, Dupuis G, Zaitoun A. Polymer-enhanced foams for water profile control. Paper SPE 179581 Presented at SPE Improved Oil Recovery Conference. Society of Petroleum Engineers. 2016

[28] Zhang YQ, Chang ZD, Luo WL, Gu SN, Li WJ, An JB. Effect of starch particles on foam stability and dilational viscoelasticity of aqueous-foam. Chinese Journal of Chemical Engineering 2015;**23**(1):276-280

[29] Khajehpour M, Etminan SR, Wassmuth F, Bryant S. Nanoparticles as foam stabilizer for steam-foam process. Paper SPE 179826 Presented at SPE EOR Conference at Oil and Gas. 2016

[30] Emrani AS, Nasr-El-Din HA. Stabilizing CO_2-foam using nanoparticles. Paper SPE 174254 Presented at SPE European Formation Damage Conference and Exhibition. 2013

[31] Farhadi H, Riahi S, Ayatollahi S, Ahmadi H. Experimental study of nanoparticle-surfactant-stabilized CO_2 foam: Stability and mobility control. Chemical Engineering Research & Design. 2016;**111**:449-460

[32] Worthen A, Chen Y, Bryant S, Bagaria H, Huh C, Johnston K. Nanoparticle stabilized carbon dioxide in water foams for enhanced oil recovery. Paper SPE 154285 Presented at SPE Improved Oil Recovery Symposium. Society of Petroleum Engineers. 2012

[33] Zhou Y, Li Y, Yuan J. The stability of aluminum foams at accumulation and condensation stages in gas injection foaming process. Colloids and Surfaces, A: Physicochemical and Engineering Aspects. 2015;**482**:468-476

[34] Eftekhari AA, Krastev R, Farajzadeh R. Foam stabilized by fly-ash nanoparticles for enhancing oil recovery. Industrial and Engineering Chemistry Research. 2015;**54**(50):12482-12491

[35] Cui ZG, Cui YZ, Cui CF, Chen Z, Binks BP. Aqueous foams stabilized by in situ surface activation of $CaCO_3$ nanoparticles via adsorption of anionic surfactant. Langmuir. 2010;**26**(15):12567-12574

[36] Horozov TS. Foams and foam films stabilised by solid particles. Current Opinion in Colloid & Interface Science. 2008;**13**(3):134-140

[37] Pu WF, Wei P, Sun L. Stability, CO 2 sensitivity, oil tolerance and displacement efficiency of polymer enhanced foam. RSC Advances. 2017;**7**(11):6251-6258

[38] Pu WF, Wei P, Sun L. Experimental investigation of viscoelastic polymers for stabilizing foam. Journal of Industrial and Engineering Chemistry. 2017;**47**:360-367

[39] Zhao S, Pu WF, Wei B, Xu XG, A comprehensive investigation of polymer microspheres (PMs) migration in porous media: EOR implication. Submitted to Fuel

[40] Pu WF, Zhao S, Wei B. Research on the profile control and displacement mechanisms of polymer microspheres. Submitted to Energy Fuels

Chapter 10

Numerical Study of Low Salinity Water Flooding in Naturally Fractured Oil Reservoirs

Reda Abdel Azim, Sara Faiz, Shaik Rahman,
Ahmed Elbagir and Nour Al Obaidi

Additional information is available at the end of the chapter

http://dx.doi.org/10.5772/intechopen.71299

Abstract

Due to the increase of the activities in the oil industries, higher interest has been given to enhance the recover the trapped oil and produce more oil from the matured reservoirs. Worldwide, enhanced oil recovery (EOR) is implemented in most reservoirs to recover additional amounts of oil that are not recovered during secondary recovery by water flood or gas injection. Recently, a numerous techniques such as thermal, miscible, immiscible and chemical has proposed to enhanced oil recovery and to increase the producible oil from oil reservoirs. The suitability and the success of a specific EOR process are highly sensitive to reservoir and fluid characteristics, recovery efficiency, availability of injected fluids, and costs. One of the common techniques which have been proposed recently is low salinity water flooding where the sea water with a controlled salinity and salt content is used to alter the rock wettability or enhance the fine migration and resulted in higher oil production. This study aims to investigate the possibility of using low salinity water flooding in naturally fractured reservoirs. The wettability changes are taking into account in terms of oil/water relative, saturation and capillary pressure as these parameters play a key role during the simulation of brine injection. The results show that the oil recovery significantly increases specially for water wet reservoirs as the reason behind is the decreasing water production after the breakthrough of the low saline brines.

Keywords: enhance oil recovery, low salinity water injection, naturally fractured reservoir

1. Introduction

Enhanced oil recovery (EOR) is a technique that used to recover trapped oil left in reservoirs after primary and secondary recovery methods. Although, EOR is a challenging process where a variety of parameters play an important role such as harsh environments; EOR processes are applied worldwide in most reservoirs to retrieve additional amounts of oil which it cannot be recovered during secondary recovery by waterflood or gas injection.

Due to the expected increase in energy consumption worldwide which is assumed to exceed 50% by the end of 2030, different path of renewable resources has been proposed to meet this growth and energy demand. Primary and secondary oil recovery which is considered as predominate energy resource produce only 15–30% of the oil original in place (OOIP) [1] and the percentage of the recovered oil is mainly depend on the compressibility of the fluids and reservoir initial pressure. More than 50% of the reserved oil subsequent to the OOIP will be trapped which in some cases is amenable to tertiary or enhance-oil-recovery process [2]. The difficulty of oil recovery was due to the chemical equilibrium between the crude oil, formation water and the rock structure and characteristic. The distribution of the oil and water within the porous rock is depend on the contact between the rock surface and the two fluids or in the other hand is depend on the rock wetting properties.

EOR stages encompass a variety of processes, including miscible flooding processes, chemical flooding process, thermal flooding processes, and microbial flooding processes. Water flooding is one of the EOR mechanisms where injected water with certain specification was used to maintain reservoir pressure and improve reservoir sweep efficiency. In order to improve the oil recovery and increasing production efficiency, water flooded was recently used in most of the reservoirs where it was initially used to support the reservoir pressure and maintain it above the oil bubble pressure and take the advantage of the viscous forces into displacement of the oil by the water. Injection of water with different composition within the initial formation, water can disturb the established reservoir chemical equilibrium and results in improved oil recovery. To evaluate the usage of smart water as EOR fluid a well chemical understanding of the most important parameters dictating the wetting properties of oil reservoir is highly needed. Low salinity water injection (LSWI) is one of the emerging improved oil recovery techniques for boosting the oil recovery from waterfloods [3]. The LSWI technique received increasing attention in the oil industry and has the potential of being cost effective compared to other EOR techniques because of its simplicity, higher oil recovery performance, and environmental friendly when compared with conventional high salinity water flooding and other EOR approaches. The main concerns with using LSWI are water sourcing and water disposal. The successful EOR approach using LSWI has been related to several factors linked to the composition of injected brine such as fine clay minerals migration, interfacial tension reduction, wettability alteration, controlling pH, and emulsion formation. The effect of low salinity water flooding has been shown at laboratory-scale and to a limited extent at field-scale using both sandstone and carbonate rocks. The modification of the injected brine composition was proved that it could improve the oil recovery factor of conventional water flooding up to 38% [4].

Several studies have been heighted that the salinity of the connate water is the main factor controlling the higher oil recovery where the salinity of the injected water must be much lower than the salinity of the connate water. Investigator found that an improved in oil recovery by using low salinity water injection only happen when clay sandstone minerals is existed along with the crude oil. The interpretation for this phenomenon was linked to permeability reduction caused by the fine migration leading to increasing in more pressure drop and higher oil production, since most observations reviled that pressure drop was strictly tied to incremental recovery [5]. Which is not the case with the carbonates rock since carbonate rock reservoir where it may contain clays minerals, but these minerals as regularly is trapped within the rock matrix and not considerably exaggerated by the injected fluid.

Furthermore, the interaction between the negatively polar crude oil and positively charged carbonates is much higher than the sandstone rocks [6] therefore, the wettability alteration mechanism in the case of carbonate rock is more complicated compare with the silica and clay due to the various contributions including fine migration, pH [7], and salinity effect [8, 9]. Change in wettability in carbonate rock from oil wet to water wet or mix wet can lead to an incremental improve in oil recovery. Høgnesen et al. [10] concluded that any modification in the composition of the injected water can altered the rock wettability. Additionally, the effect of sulfate ions within the injected water in wettability iteration have been investigated by Høgnesen et al. [10] where their results show that increasing the concentration of sulfate ion at high temperature modify the wettability in carbonates, and result in higher oil recovery. Moreover, Webb et al. [11] conducted a study that compared oil recovery from a North Sea carbonate core samples using sulfate-free brine, with seawater contains sulfate. Their results displayed that the seawater contains sulfate ions has ability to modify the wettability of the carbonate system to water wet state.

Generally, few studies regarding the effect of wettability alteration for carbonate rocks have been established so far and this was mainly due to incomplete understanding of complex chemical interactions between rock, oil and brine. Experimental and modeling studies are still progressing to gain more insight into the mechanism underlying the effect of LSWI on oil recovery. Therefore, this work presents a numerical simulation study to evaluate production potential of naturally fractured reservoirs under low salinity water flooding. The simulation model based on hybrid approach of using permeability tensor and discrete fracture approaches. The governing equations use for the simulation are expanded using finite element technique.

2. Background

Low salinity water flooding is one of the incremental oil recovery techniques which inject low salinity water to alter the wettability or assist at fine migration from the reservoir to increase the oil production and reduce the residual oil saturation. The mobility and migration of the reservoir fines mainly depends on the salinity of the water flood where injection of the low salinity water enhances the fines migration. Migrate particles can clog pore throats and act as fluid flow hindrance and results in reduction of the permeability in the clogged throats. Where in the oil and gas recovery practices these phenomena is intentionally avoided as it damages the permeability and hence reduction in permeability translates into reduction in productivity from reservoir wells. Meanwhile, by inducing damage to the permeability of the reservoir in water swept zones; less permeable water zone to a high permeability oil zone will be occuring. The process of decreasing the fluid mobility in one zone is known as mobility control process. As result of low salinity water flooding; fine mobilization and its accumulation on pore throats can be identified by an increase in the pressure drop.

Studies have confirmed a pronounced effect of LSWI on oil recovery in both secondary and tertiary modes of injection. The importance of LWSI enhanced because it can be integrated with other EOR method such as chemical or miscible gas flooding. Bernard [12] the relative effectiveness of fresh and salt waters in flooding oil from cores containing clays investigated the effect of Dang et al. [13] evaluated the merits of combining CO_2 with LSW injection (CO_2-LSWAG) on the EOR through modeling, optimization and uncertainty assessment. They found that CO_2-LSWAG.

Different mechanisms have been suggested regarding the effect of LSWI on sandstone rocks including fines migration, pH, multi-ion exchange (MIE), salting-in, and wettability alteration [14]. The wettability alteration process underlies the low salinity effect as the decrease in salinity increases the size of the double layer between the clay and the oil interface, which leads to organic material release [15].

Furthermore, the behavior of low salinity waterflooding can be summarized as follows [5]: (1) more oil can be produced if the brine salinity of the injected water is lower than the initial brine salinity in the core (2) reducing the initial brine salinity of the injected water during the stage enhance the pressure drop, (3) pressure drops reduced due to reducing of the core permeability due to presence of fine migration [16], (4) existence of connate water is beneficial to enhance the effect of low salinity water [17].

In addition, the effect of the salinity of the cannot water on the oil recovery have been presented Sharma et al. [18] where they concluded that oil recovery increased significantly with the salinity of the connate brine. In addition, the performance of waterfloods is strongly affected by the composition of the crude oil and its ability to wet the rock surfaces, the salinity of the connate brine in the reservoir, and the height above the oil/water contact. This contribution was also supported by McGuire et al. [4].

Although the mechanism for the effect of law salinity water flooding is benign different from each other the main concept is to improved oil recovery.

In this study, we simulate low salinity water flooding under geomechnical effects. The simulation based on finite element technique, hybrid approach of permeability tensor and discrete fracture. The low salinity effect is taken under wettability changes during the simulation process.

3. Reservoir model development

This study presents a novel three dimensional poroelastic numerical model for simulation multiphase fluid flow in naturally fractured basement reservoirs. A hybrid approach is used in the simulation of fluid flow. In this approach small fractures are considered as part of matrix and flow is simulated by using single continuum approach. While flow through long fractures is simulated using a discrete fracture approach. The reservoir is divided into a number of grid blocks. Grid based full permeability tensors for short to medium fractures are calculated using Darcy's diffusivity equation. Cubic law is to simulate flow through long fractures that cuts through different grid blocks.

4. Methodology

4.1. Grid based full permeability tensors

A fractured porous medium composed of matrix with nonzero permeability and fractures with high permeability based on the fracture aperture [19]. In this medium, the fluid flows through matrix and fractures with transfer between these two structures. In which, each point in the matrix can be assigned a bulk permeability km, while each point in the fractures can be assigned k_{eff}.

In order to calculate the effective permeability tensors which represents an average permeability for the two structures, 3D cube is used to represent the matrix and fractures porous media (**Figure 1**) [17].

The fractured porous media is bounded in an impermeable cover with boundary conditions for pressures (P1 and P2).

The boundary conditions are: $p(x=0)=p_1, p(x=L)=p_2, J.\ n=0\ and\ v=0\ on\ s1$

The seepage velocity calculated based on the flow rate integration over fracture surfaces and matrix porous media and by using total volume of the block.

$$v = -\frac{k_m}{\mu} \nabla p \tag{1}$$

Where μ the fluid viscosity and Δp is the pressure change. The continuity equation for local seepage velocity in the matrix read as:

$$\nabla . v = 0 \tag{2}$$

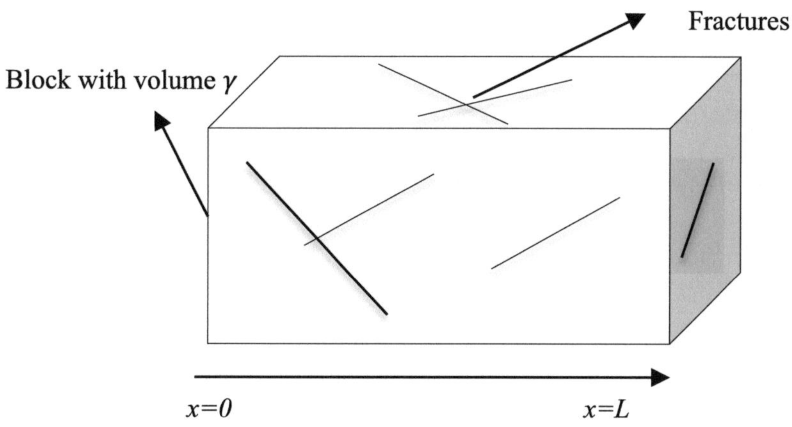

Figure 1. 3D cube used for permeability tensor calculations.

The hydraulic properties of fracture can be can be characterized by fracture transmissivity (aperture) and main flow rate is set parallel and normal to fracture plane. The flow rate J in fractures is usually defined by unit width of fracture and can be expressed by:

$$J = -\frac{k_{eff}}{\mu} \nabla_s p \tag{3}$$

In case of the flow is parallel to fracture plane, the seepage velocity normal to the fracture induces a pressure drop expressed by:

$$v = -\frac{1}{\mu} \nabla p \tag{4}$$

The effective fracture permeability of fracture can be describes by its aperture b as (in case the fractures are empty):

$$k_{eff} = -\frac{b^3}{12} \tag{5}$$

The mass conservation equation for the flow in a fracture is:

$$\nabla_s . J = -(\vec{v} + \vec{-v}) . n \tag{6}$$

Where n the unit vector is normal to fracture plane, $\overline{v}^{+}$ is the seepage velocity in the matrix on the side of n and $\overline{v}$ is the seepage velocity on the opposite side.

This transport equation is implemented with the above-mentioned boundary conditions to calculate the permeability tensors.

Therefore, the total seepage velocity over the block is obtained by integrating the flow rates over fracture surfaces and matrix porous media. Then the results divided by the total block volume to calculate the block effective permeability tensor.

$$v_x = \frac{1}{\gamma}\left\{\int_{\gamma m}^{-} v_x \, dv + \int_{sf}^{-} J_x \, ds\right\} = \frac{-k_{eff}}{\mu}\frac{\partial p}{\partial x} \tag{7}$$

Where, sf is the surface for all fractures and γ is the matrix volume.

4.2. Simulation of low salinity water flooding

Darcy's law and continuity equations are governed two phase fluid flow system through fractures and matrix porous media [18].

The Darcy's law is expressed as:

$$\phi s_{\pi} u^{\pi s} = \frac{k_{ij} k_{r\pi}}{\mu_{\pi}}[-p_{\pi} + p_{\pi} g_i]\pi = w, nw \tag{8}$$

Continuity equation for wetting phase incorporating the effective overburden, maximum, and minimum stresses can be expressed as follow:

$$-\nabla^T\left[\frac{k_{ij}k_{rw}}{\mu_w\beta_w}\nabla(p_w+\rho_w gh)\right]+\phi\frac{\partial}{\partial t}\left[\frac{\rho_w s_w}{\beta_w}\right]+\rho_w\frac{s_w}{\beta_w}\left[\left(1-\frac{D}{3K_m}\right)\frac{\partial\varepsilon}{\partial t}+\frac{Dc}{3K_m}+\left(\frac{1-\phi}{K_m}-\frac{D}{(3K_m)^2}\right)\frac{\partial p}{\partial t}\right]$$

$$+\rho_w Q_w = 0 \tag{9}$$

Continuity equation for non-wetting phase can be written as follow:

$$-\nabla^T\left[\frac{k_{ij}k_{ro}}{\mu_o\beta_o}\nabla(p_o+\rho_o gh)\right]+\varphi\frac{\partial}{\partial t}\left[\frac{\rho_o s_o}{\beta_o}\right]+\rho_o\frac{s_o}{\beta_o}\left[\left(1-\frac{D}{3K_m}\right)\frac{\partial\varepsilon}{\partial t}+\frac{Dc}{3K_m}+\left(\frac{1-\varphi}{K_m}-\frac{D}{(3K_m)^2}\right)\frac{\partial p}{\partial t}\right]$$

$$+\rho_o Q_o = 0 \tag{10}$$

Where, φ is the porosity of the media, s_π is the saturation for each phase, $u^{\pi s}$ is the relative velocity vector between fluid phase and solid phase, k_{ij} is the permeability tensor, $k_{r\pi}$ is the relative permeability for each fluid phase π, μ_π, ρ_π and ,p_π are dynamic viscosity, density of fluid, and fluid pressure for each phase respectively, g_i is the gravity acceleration vector, β_π is the fluid formation volume factor, K_m is the bulk modulus of solid grain, D is the elastic stiffness matrix, and Q_π represents external sources or sinks.

The effect of low salinity has been taken on wettability changes which affect the relative permeability curve. Therefore, the salt concentration at each grid block has been calculated using Eq. (10) and the calculated saturations from Eqs. (9) and (10).

$$\frac{\partial C S_w}{\partial t}+Q_t\frac{\partial C f_w(S_w, C)}{\partial x} = 0 \tag{11}$$

Rearranging Eq. (11):

$$\frac{\partial}{\partial t}(S_w C)+Q_t\frac{\partial}{\partial x}(C) = 0 \tag{12}$$

Where C is concentration of a relevant solute in water, S_w is water saturation, fw is water fractional flow, and Q_t is total flow rate.

The changes of oil and water relative permeability and capillary pressure are calculated using Eqs. (11)–(13).

$$k_{rw} = F_1 k^L_{rw}+(1-F_1)k^H_{rw} \tag{13}$$

$$k_{ro} = F_1 k^L_{ro}+(1-F_1)k^H_{ro} \tag{14}$$

$$P_{cow} = F_2 P^L_{cow}+(1-F_2)P^H_{cow} \tag{15}$$

Where F_1 and F_2 are functions of the salt concentration, k_{rw} is the water relative permeability and k_{ro} is the oil relative permeability, P_{cow} is oil–water capillary pressure. The subscripts H and L are standing for high salinity and low salinity respectively.

The end point saturations are calculated by using the following equations:

$$S_{wco} = F_1 S^L_{wco} + (1 - F_1) S^H_{wco} \tag{16}$$

$$S_{wcr} = F_1 S^L_{wcr} + (1 - F_1) S^H_{wcr} \tag{17}$$

$$S_{wmax} = F_1 S^L_{wmax} + (1 - F_1) S^H_{wmax} \tag{18}$$

$$S_{owcr} = F_1 S^L_{owcr} + (1 - F_1) S^H_{owcr} \tag{19}$$

Where F_1 and F_2 are functions of the salt concentration, S_{wco} is the connate water saturation, S_{wcr} is the critical water saturation, S_{wmax} is the maximum water saturation and S_{owcr} is the critical oil saturation in water. The subscripts H and L are standing for high salinity and low salinity curves, respectively.

5. Validation of the model using 3D reservoir with a single fracture

In this case, a simple 3D reservoir with one fracture is used for validating the numerical model. The reservoir is built by using a commercial black oil reservoir simulator (CMG IMEX). The model dimensions are (10 m × 10 m × 5 m). The reservoir model includes 15 horizontal layers. Of these 15 horizontal layers, the 8th layer from the top is embedded as a fracture layer with different rock properties. The mesh is refined around the fracture layer

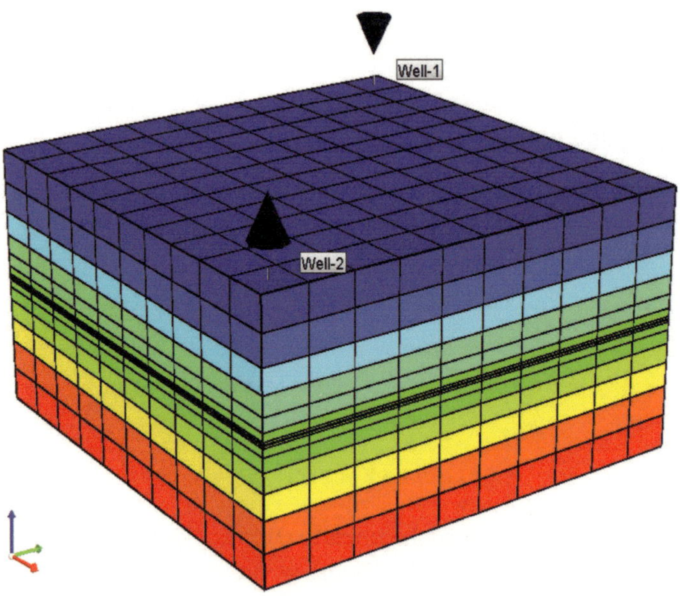

Figure 2. Reservoir fracture model used in CMG to validate the developed two phase fluid flow numerical model.

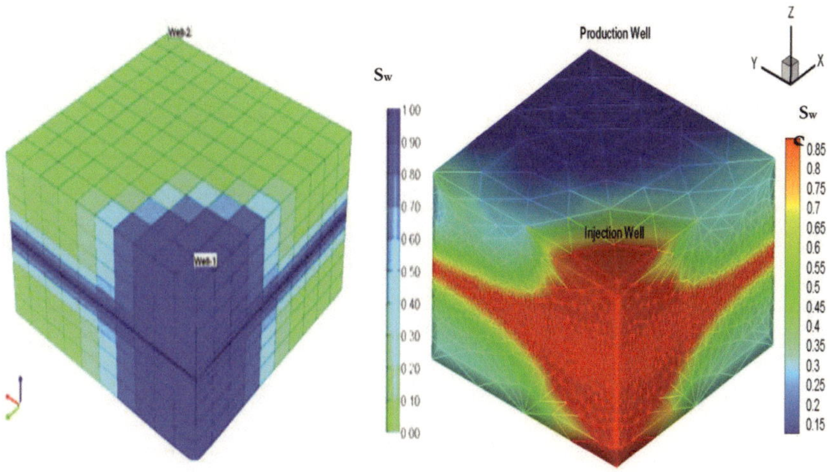

Figure 3. Water saturation profile after one day of water injection (a) CMG 3D model with one fracture introduced at eighth layer. (b) 3D model with one fracture introduced at the middle of the reservoir model used with the developed multiphase numerical model.

as shown in **Figure 2**. The matrix and fracture permeability are set at 0.01 md and 2 Darcy respectively. Linear relative permeability curve, one injector and one producer are used for the simulation of fluid flow.

Figure 3a and **b** show the water saturation profile obtained from CMG and two phase finite element numerical model, respectively. As can be seen from the figures, at the beginning of the injection, water moves fast inside the fracture due to its high permeability compared to the matrix permeability. With the pass of time (approximately a day), the water saturation starts to increase gradually inside the matrix elements. These figures show that the finite element numerical model is able to predict the same water saturation profile as that of CMG black oil reservoir simulator but with greater accuracy.

6. Fracture characterization

In this study, object based simulation technique is used to generate subsurface discrete fracture maps [20]. In this model, fractures are treated as objects and placed in the domain stochastically. The number of generated fractures is controlled by fracture intensity ($0.1\ m^{-1}$) and fractal dimension parameters. The fractures are treated as objects with varying in radius, dip and azimuth angles.

6.1. Fracture intensity

The fracture intensity is an important parameter to give an indication about the probability of fractures occurrence in a discrete fracture model [21]. The fracture intensity is defined as the number of fractures per unit bulk volume. To calculate the fracture intensity, the reservoir is divided into a number of grid blocks and fractures that cut each block are defined. Then,

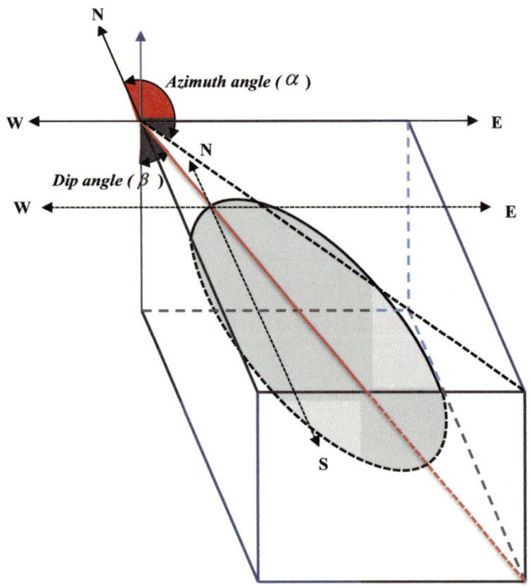

Figure 4. Description of the plane in which the fracture lies. The ellipse represents a fracture with (α is the fracture azimuth angle and β is the dip angle).

the number of these fractures is divided by the bulk volume of the corresponding grid block. Fracture intensity map is extracted from geological interpretations of reservoirs. Fracture intensity is expressed as:

$$\text{Fracture Intensity} = \frac{\sum_{i=1}^{N} Area}{Volume} \tag{20}$$

Where, N is the total number of fractures that intersect the corresponding grid block.

6.2. Fracture dip and azimuth

Each fracture in a discrete fracture model is defined by its properties which include fracture azimuth angle, dip angle, center point (x, y and z) and radius. Fracture azimuth is defined by the angle formed between the fractures plan and the geographic north. These angles are inferred from core and Fullbore Formation Micro imager (FMI) data. Fracture dip is the angle between the fracture plane and horizontal plane (see **Figure 4**) and inferred from the geological interpretations (e.g. dip of the geological formation). In object based simulation, the fractures azimuth angles are characterized by the Gaussian distribution while dip angles are assumed to be followed by normal distribution.

7. Results and discussion

The case study has been taken from a fractured reservoir located in United Arab Emirates. The studied reservoir (4500 m x 4500 m x 300 m) contains more than 14,000 long to short fractures. The reservoir has been divided into grid blocks (100 m × 100 m × 50 m) and fractured intersected with each block are determined. The fractures with length (l < 40 m) have

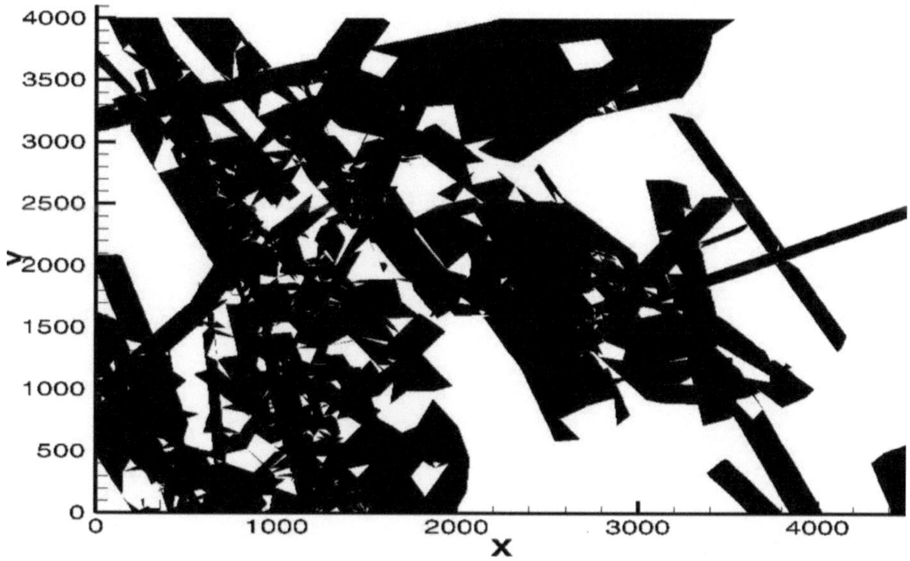

Figure 5. 2D view of fracture network for the studied reservoir.

Figure 6. 3D view of fracture network for the studied reservoir.

been accounted for effective permeability tensors and long fractures (l > 40 m) are discretized explicitly in the reservoir domain for fluid flow simulation (**Figures 5** and **6**).

The area of the reservoir is 20 km^2 and the average porosity is 5% and OOIP is 1000 MM barrel. The reservoir PVT is collected from PVT report and the initial oil formation volume factor that has been used during the simulation runs was 1.25 rb/STB. The relative permeability curve and capillary pressure used are presented in **Figure 7**.

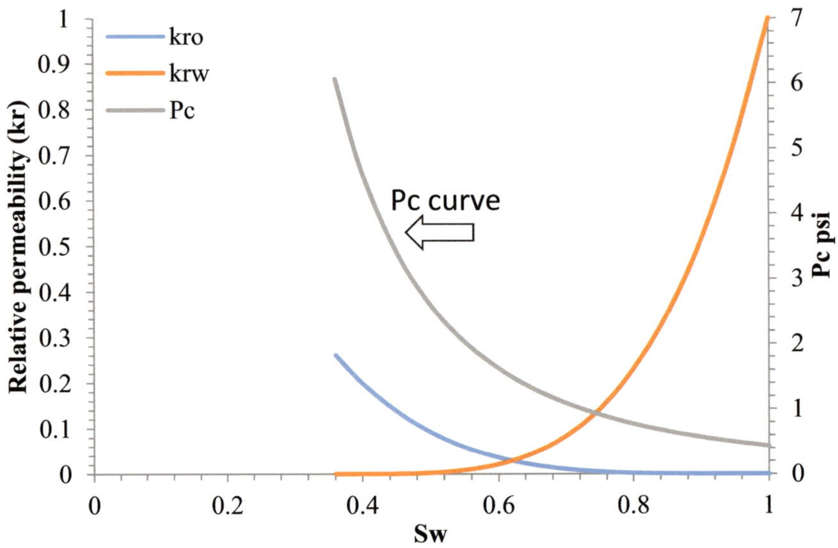

Figure 7. Relative permeability and capillary pressure curves used in simulation of fluid flow.

Parameter	Value
Oil viscosity (cp)	3
Initial Oil formation volume factor rb/stb	1.05
Oil density (lb/ft3)	52

Table 1. Oil physical properties used in simulation model.

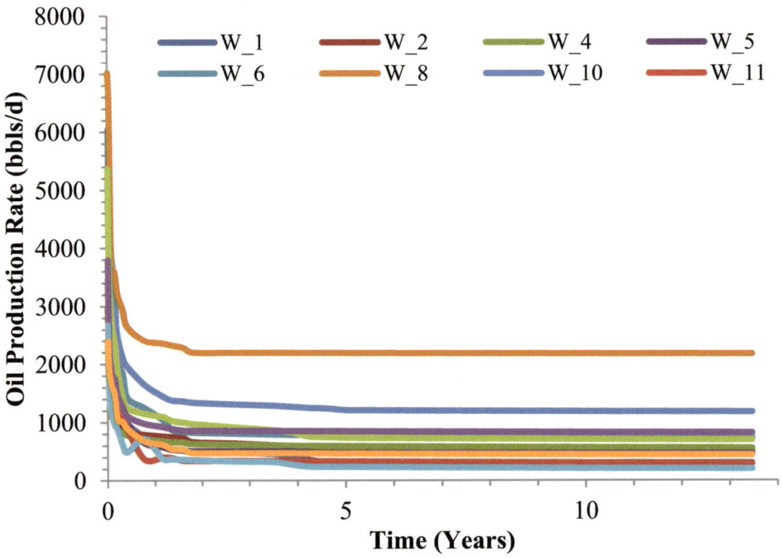

Figure 8. Oil production rate for wells in the studied fractured reservoir.

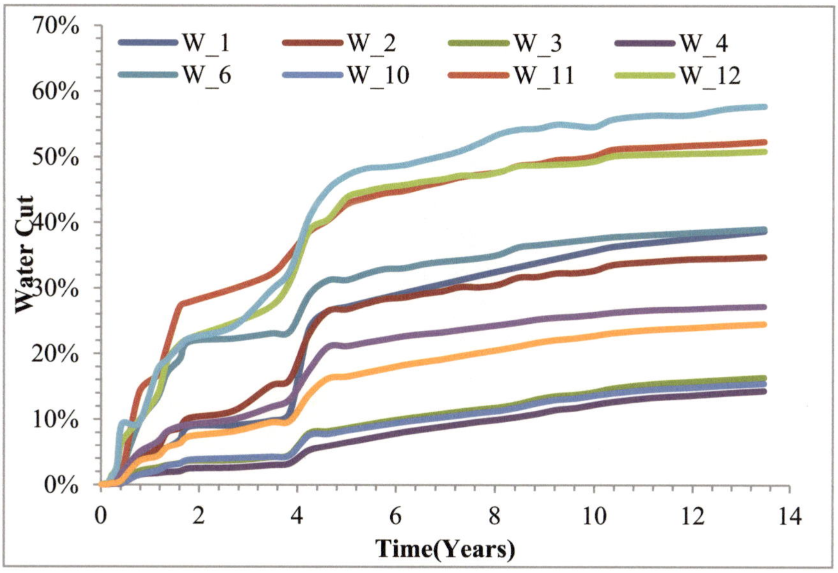

Figure 9. Water cut for wells in the studied fractured reservoir.

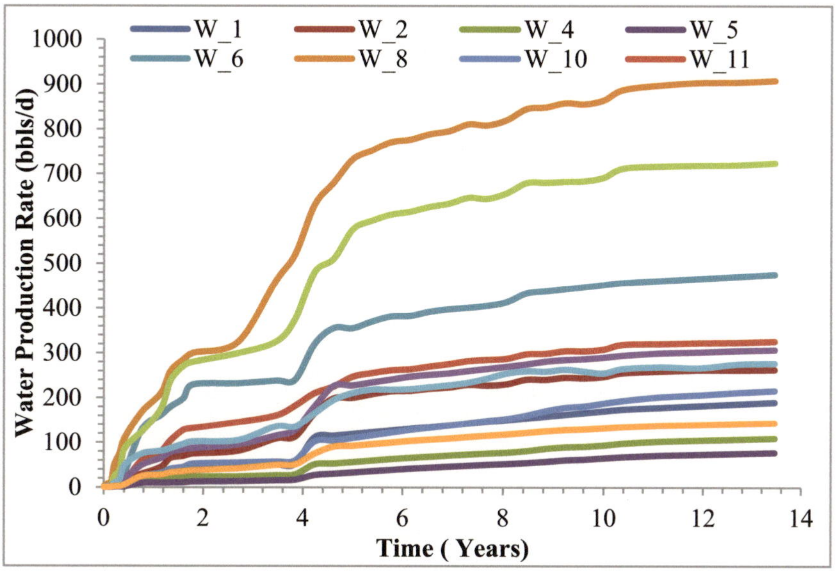

Figure 10. Water production rate for wells in the studied fractured reservoir.

Low salinity water flooding is applied on the studied reservoir to maintain the reservoir pressure and for testing wells production profile under water injection scenario. The fractured reservoir thickness is divided into three zones. The first zone is from 800 to 1100 m is the main production zone. The second zone is from 1100 to 1200 m is a separation between the injection and production zones and the third zone is from 1200 to 1400 m is the injection zone.

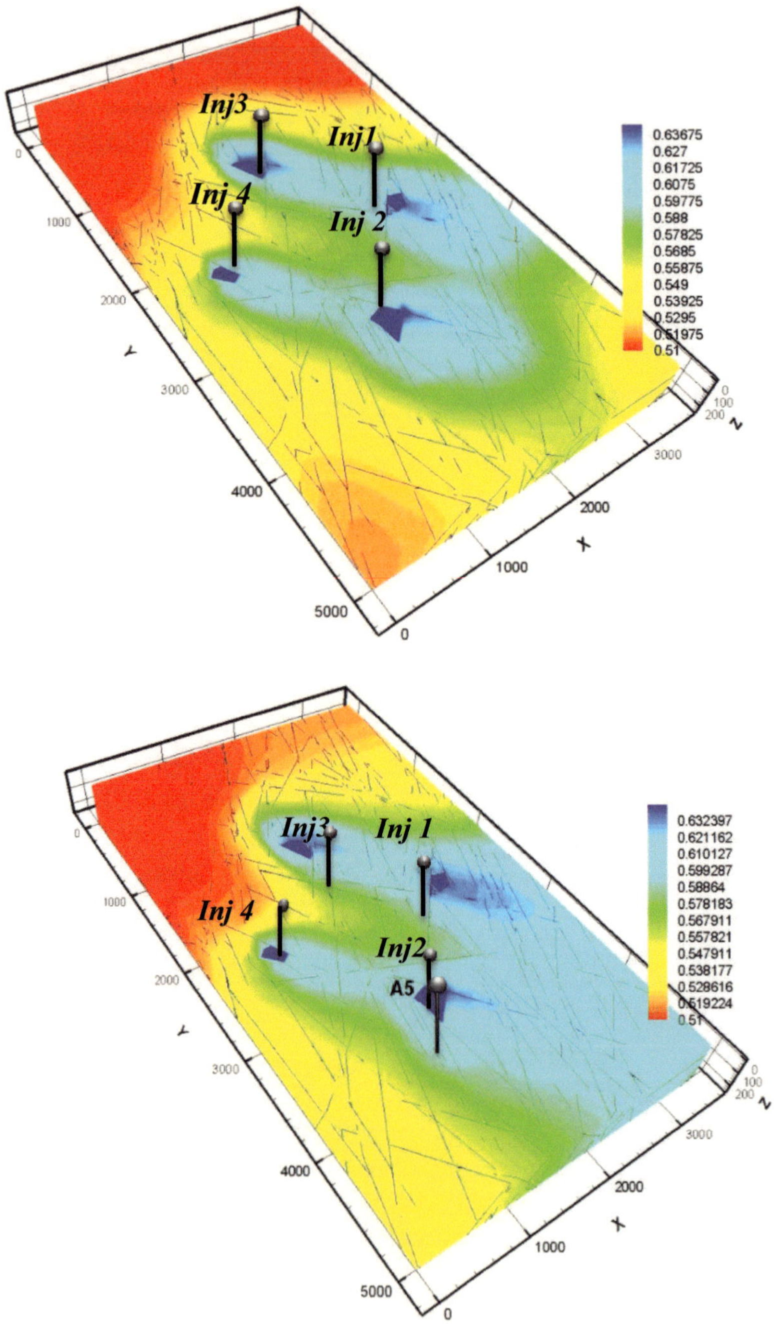

Figure 11. Water saturation profile after five months and 5 years of water injection respectively.

Four injectors were selected in the studied reservoir and are distributed as five spots to displace oil around 12 producers. All the injection wells are vertical and the amount of water injected in each well was 4500 bbls/d. water is injected into the bottom region while the oil was produced from the basement top section.

http://dx.doi.org/10.5772/intechopen.71299

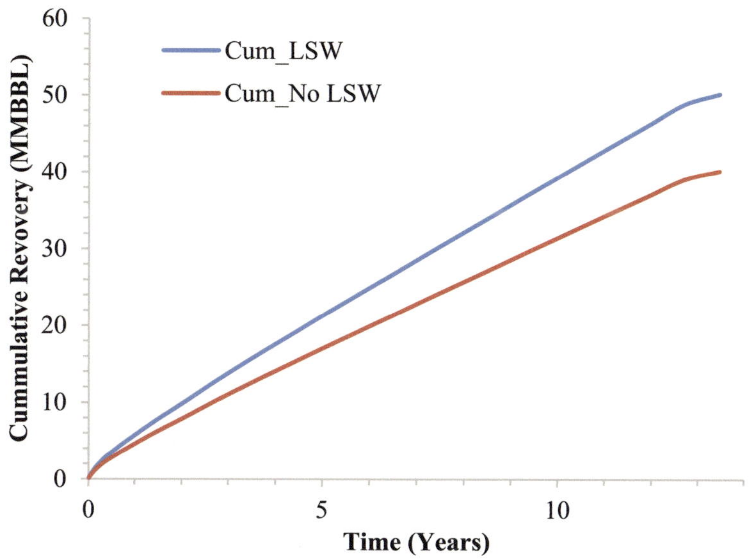

Figure 12. Cumulative oil production rate under conventional and low salinity water flooding.

The reservoir model was a two phase model, containing only oil and water for simplifications. The initial connate water salinity was set to 750 kg/m³ total dissolved salts (TDS) approximately the same salinity as regular seawater, and Brine salt content is 0.25 fraction (Ca^{2+}, Mg^{2+}, Na^{+}, and SO_4^{2-} concentration was observed.).

Table 1 shows oil physical properties.

Figure 8 shows wells water production rate. Well #W_8 has the highest water production rate because of the location of this well is bounded by the four injectors. Well #W_5 has the lowest water production rate because of the location of this well is far from the inaction area. **Figure 9** is showing the water cut at the producers, while **Figure 10** is showing the water production rate for all producers. As shown from **Figure 7**, well #W_8 has highest oil production rate owing to the sweep efficiency in the area surrounded by well #W_8 is very high. **Figure 11** shows the water saturation profile after 5 months and 5 years of water injection respectively.

Figure 12 shows the comparison between the cumulative oil production under conventional and low salinity water flooding for the same fractured reservoir. The presented results at **Figure 12** show that low salinity water flooding process increases he recovery factor after 13 years of water injection by 1.4% which proves that this type of fractures reservoir requires a comprehensive study on how to increase the recovery factor by understanding the mechanisms behind the production process.

8. Conclusion

In this study, we made a comprehensive review on low salinity water flooding and the factors controlling the oil recovery. In addition, simulation of low salinity water flooding performed on a fractured reservoir located in Arab region. The simulation was based on a finite element technique a hybrid mode of using permeability tensors for low and short fractures and discretization of long fractures explicitly inside the reservoir domain. In which, the presented technique eliminate the use of dual porosity model as this technique does not consider fluid distribution within the matrix blocks during the simulation period and only can be applied for small number of large scale interconnected fractures. The simulation of low salinity water flooding in this study strongly based on wettability change. The results show that the recovery increases significantly by using injected brine volume.

Author details

Reda Abdel Azim[1*], Sara Faiz[1], Shaik Rahman[2], Ahmed Elbagir[1] and Nour Al Obaidi[1]

*Address all correspondence to: reda.abdelazim@aurak.ac.ae

1 Chemical and Petroleum Engineering Department, American University of Ras al-Khaimah, Ras al-Khaimah, UAE

2 New South Wales University, Australia

References

[1] Lu T, Li Z, Li S, Wang P, Wang Z. Enhanced heavy oil recovery after solution gas drive by water flooding. Journal of Petroleum Science and Engineering. 2016;**137**:113-124

[2] Sheng J. Critical review of low-salinity waterflooding. Journal of Petroleum Science and Engineering. 2014;**120**:216-224

[3] Al-Shalabi EW, Sepehrnoori K, Delshad M, Pope G. A novel method to model low-salinity-water injection in carbonate oil reservoirs. SPE Journal. 2015;**20**(05):1154-1166

[4] McGuire PL, Chatham JR, Paskvan FK, Sommer DM, and Carini FH, Low salinity oil recovery: An exciting new EOR opportunity for Alaska's North Slope. In SPE Western Regional Meeting. Society of Petroleum Engineers, 2005

[5] Tang G-Q, Morrow NR. Influence of brine composition and fines migration on crude oil/brine/rock interactions and oil recovery. Journal of Petroleum Science and Engineering. 1999;**24**(2):99-111

[6] Thomas C, Duvall ML, Robertson EP, Barrett KB, Bala GA. Surfactant-based EOR mediated by naturally occurring microorganisms. SPE Reservoir Engineering. 1993;**8**(04):285-291

[7] Piñerez T, Austad T, Skule S, Puntervold T, Wrobel S, Hamon G. Linking low salinity EOR effects in sandstone to pH, mineral properties and water composition. In: SPE Improved Oil Recovery Conference. Society of Petroleum Engineers; Tulsa, Oklahoma, USA. 11-13 April; 2016

[8] Yousef AA, Al-Salehsalah SH, Al-Jawfi MS. New recovery method for carbonate reservoirs through tuning the injection water salinity: Smart waterflooding. In: SPE EUROPEC/EAGE Annual Conference and Exhibition. Society of Petroleum Engineers; Vienna, Austria. 23-26 May; 2011

[9] Tang G, Morrow NR. Injection of Dilute Brine and Crude Oil/Brine/Rock Interactions, in Environmental Mechanics: Water, Mass and Energy Transfer in the Biosphere: The Philip Volume (eds P. A.C. Raats, D. Smiles and A. W. Warrick), American Geophysical Union, Washington D.C. 2002. DOI: 10.1029/129GM16

[10] Strand S, Høgnesen EJ, Austad T. Wettability alteration of carbonates – Effects of potential determining ions (Ca_2^+ and SO_4^-) and temperature. Colloids and Surfaces A: Physicochemical and Engineering Aspects. 2006;**275**(1):1-10

[11] Webb K, Black C, Edmonds I. Low salinity oil recovery – The role of reservoir condition corefloods. In: IOR 2005-13th European Symposium on Improved Oil Recovery. Budapest, Hungary. 2005

[12] Bernard GG. Effect of floodwater salinity on recovery of oil from cores containing clays. In: SPE California Regional Meeting. Society of Petroleum Engineers; Los Angeles, California. 1967

[13] Dang C, Nghiem L, Nguyen L, Chen Z, Nguyen Q. Evaluation of CO_2 low salinity water-alternating-gas for enhanced oil recovery. Journal of Natural Gas Science and Engineering. 2016;**35**:237-258

[14] Rezaei Doust A, Puntervold T, Strand S, Austad T. Smart water as wettability modifier in carbonate and sandstone: A discussion of similarities/differences in the chemical mechanisms. Energy & fuels. 2009;**23**(9):4479-4485

[15] Ligthelm DJ, Gronsveld J, Hofman J, Brussee N, Marcelis F. Novel Waterflooding strategy by manipulation of injection brine composition. In: EUROPEC/EAGE Conference and Exhibition. Society of Petroleum Engineers; Amsterdam, The Netherlands. 2009

[16] Zhang Y, Morrow NR. Comparison of secondary and tertiary recovery with change in injection brine composition for crude-oil/sandstone combinations. In: SPE/DOE Symposium on Improved Oil Recovery. Society of Petroleum Engineers; Tulsa, Oklahoma, USA. 2006

[17] Zhang Y, Xie X, Morrow NR. Waterflood performance by injection of brine with different salinity for reservoir cores. In: SPE Annual Technical Conference and Exhibition. Society of Petroleum Engineers; Anaheim, California, USA. 2007

[18] Sharma M, Filoco P. Effect of brine salinity and crude-oil properties on oil recovery and residual saturations. SPE Journal. 2000;**5**(03):293-300

[19] Abdelazim R, Rahman SS. Estimation of permeability of naturally fractured reservoirs by pressure transient analysis: An innovative reservoir characterization and flow simulation. Journal of Petroleum Science and Engineering. 2016;**145**:404-422

[20] Doonechaly NG, Rahman S, Cinar Y. A new finite-element numerical model for analyzing transient pressure response of naturally-fractured reservoirs. In: SPE Annual Technical Conference and Exhibition. Society of Petroleum Engineers. September 2013

[21] Tran NH, Chen Z, Rahman SS. Integrated conditional global optimisation for discrete fracture network modelling. Computers & Geosciences. 2006;**32**(1):17-27

Chapter 11

Experimental Study of the Effect of Composite Solvent and Asphaltenes Contents on Efficiency of Heavy Oil Recovery Processes at Injection of Light Hydrocarbons

Dmitry N. Borisov, Dmitry V. Milordov,
Svetlana G. Yakubova and Makhmut R. Yakubov

Additional information is available at the end of the chapter

http://dx.doi.org/10.5772/intechopen.72673

Abstract

The current state of research in the field of solvent injection techniques for increase of heavy oil production efficiency is discussed in the chapter. As a result of a series of experiments on the physical modeling of oil displacement processes in a porous medium in large-sized model, features of asphaltene precipitation and the formation of fixed residual oil upon injection of solvent based on light alkanes are revealed. The oil displacement by n-hexane was studied and the difference in the composition of residual oil in the zones of dispersion and diffusion has been shown. The influence of the composition of asphaltenes peculiarities on the dynamics of oil recovery and on the accumulated oil recovery during the injection of n-hexane, as well as the composition and quantity of asphaltenes precipitated in the porous medium, has been estimated. The effect of toluene and nonylphenol additives on the proportion of asphaltenes in the residual oil and cumulative oil recovery has been evaluated using the Ashalchinskoye field oil as an example of heavy oil in the physical modeling of injection of n-hexane as the base solvent.

Keywords: heavy oil, composition of asphaltenes, injection of solvent, reservoir modeling, enhanced oil recovery

1. Introduction

The results of the pilot projects on combined steam and solvent injection for increasing heavy oil (HO) and bitumen recovery [1] have confirmed the viability of their use. The main volume of heavy oil recovery by downhole methods is provided by using steam; for example, in

Canada, these are mainly cyclic steam stimulation (CSS) and steam assisted gravity drainage (SAGD). The main problems of thermal steam effect are mainly associated with energy costs and requirement of deep water purification for steam generators. The main criterion in thermal stem methods is the availability of fuel resources used for a steam generation: in order to recover 1 ton of heavy oil one is required to burn averagely about 300 kg fuel, and natural gas is often used as fuel. The combined use of solvents and steam significantly increases energy efficiency due to the decrease in the amount of heat required in order to decrease the bitumen viscosity. As a result of this, the operational costs are reduced as well as carbon dioxide emissions which are important from the ecological point of view. Since 2000, various projects have been realized in Canada for combined steam and solvent injection [2–8]: Solvent aided process (SAP), solvent-assisted SAGD (SA SAGD), liquid addition to steam to enhance recovery (LASER), expanding solvent SAGD (ES SAGD), and steam alternating solvent (SAS). Mostly light alkanes or their mixtures (propane, butane, pentane and hexane) and gas condensate were used as a hydrocarbon solvent, which were injected into paired horizontal, single horizontal and vertical wells.

Currently, increased attention is paid to providing methods of recovering heavy oil from thin reservoirs, where the variants of using SAGD are not applicable. For example, in western Canada (Saskatchewan province), where about 2/3 of all Canadian heavy oil resources are located (5.4 billion m^3 of proven and probable ones), 55% of the deposits are contained in the reservoirs of less than 5 m thick. In Russia, the Tatneft Company has such reservoirs [9]. In this connection, several variants of heavy oil recovery are considered using pure solvent: the cyclic solvent process (CSP) and cyclic solvent injection (CSI) technologies, which are designed for single wells. It has been experimentally established that the main technical limitations of common CSI process are the untimely restoration of the oil viscosity due to solvent release and loss of oil mobility during recovery associated with it. In order to overcome the said technical drawbacks in the recent years some non-thermal process variants are proposed: enhanced CSP (ECSP) [10], cyclic production with continuous solvent injection (CP-CSI) [11] and gas flooding-assisted cyclic solvent injection (GA-CSI) [12]. For example, in the ECSP method two components are used as a solvent: methane and propane. Methane provides for enveloping the reservoir by the effect and propane reduces the oil viscosity. Analysis of the obtained results indicates limitations of the propane use for maintaining the required oil viscosity which implies a necessity of using hydrocarbons and oil fractions with higher boiling points. In this connection in the pilot CSP project, the propane–butane mixture is used as a solvent and other solvents are further assessed.

The Nsolv process involves the injection of a pure, heated solvent vapor into a bitumen reservoir. Its main advantage is a possibility of realization in the reservoirs of small thickness. Moreover, the capital investments have been reduced by more than 40% compared to the SAGD technology, mainly due to the absence of necessity of using steam generators and water treatment plants. It is also stated [13] that the oil viscosity, coking ability and metal contents are decreased in the Nsolv process due to partial upgrading of oil in reservoir. The technology enhanced solvent extraction incorporated electromagnetic heating (ESEIEH) it comprises solvent injection with preliminary electromagnetic heating of the reservoir using low-frequency radio waves is carried out. The emitters in the wellbores generate electromagnetic waves and after the heating the solvent is introduced. At use of the pure solvent, the development of productive deposits of small thickness is possible. No water is required in this case, and capital investments, operational costs and total energy costs are significantly reduced (up to 85%). The experience shows that the dissolving mixtures injected into the deposits should be mostly gaseous in order to fill the free pore volume formed as a result of development using primary methods. Moreover, the dissolving mixtures should have good oil solubility and should be readily available and relatively not expensive.

At use of light alkane hydrocarbons as solvents one must take into the account the data about asphaltenes precipitated in the oil-saturated reservoir and should develop methods of inhibiting the precipitation [14]. The possible methods of preventing asphaltene precipitation are the addition of aromatic compounds or use of asphaltene stabilizers (inhibitors). In various periods many investigators have analyzed the in vitro effect of the various chemical compounds on precipitation of asphaltenes in oils and model asphaltene systems [15–20]. For example, in the paper [21] the effect of three groups of additives on asphaltene precipitation was studied: aromatic compounds, heteroatom compounds (O, S, N) and hydrogen donor solvents (tetraline, decaline). It was established that all additives except nitrogen-containing compounds have a trend to inhibit asphaltene precipitation. In the toluene-naphthalene-phenanthrene the molecular weight increases additive efficiency. The isomers of toluene and xylene dissolve asphaltenes, xylene being more effective than toluene. Nonylphenol is more effective than toluene and xylene isomers due to its alkyl chain and polar hydroxyl group. The paper [22] demonstrates that asphaltenes do not precipitate when the ratio of alkanes to aromatic compounds is below 7 and if this value exceeds 8 there is a possibility of asphaltene precipitation.

Depending on chemical and structural characteristics of asphaltene precipitation inhibitors and oil characteristics (resin, asphaltenes and aromatics contents) their efficiency may be different [23]. It is suggested to use both natural inhibitors such as oil resins and deasphalted oils [24], and synthetic amphiphiles based on alkyl phenols, alkyl benzene and alkyl sulfonic acids [21, 25, 26] as asphaltene inhibitors.

Apart from the contents of the solvent and collector properties, these processes should take into the account various factors such as a ratio of resins and asphaltenes in the oil [27–29], contents of the asphaltenes themselves [30], the presence of water, clay, etc. [31–35].

In order to understand the asphaltene precipitation processes in the reservoir upon injection of light alkane-based solvents, it is necessary to carry out physical modeling using specifically selected methodological approaches [36–38].

In this chapter, we used a special stand for physical modeling of the heavy oil displacement process from the reservoir. An experimental evaluation of light alkanes solvents injection options in a reservoir model for heavy oil extraction has been carried out. We compared amount and composition of displaced and residual oil and investigated the distribution of precipitated asphaltenes in the sandstone after oil displacement. Additives to prevent asphaltene precipitation in a porous medium were selected and their minimum required amount in the composition of the displacement solvent was determined. Using the increased size model of reservoir allows to obtain more accurate data on the dynamics of oil production during the experiment.

2. Materials and methods

2.1. Physical modeling of heavy oil displacement from the porous medium using the large oil reservoir model

To model oil displacement processes, we used Ashalchinskoye field HOs. Density, viscosity and composition are presented in **Table 1**. Asphaltenes were precipitated from petroleum by 20-fold volume excess of n-hexane. After 24 h, the obtained precipitate was filtered and washed with boiling n-hexane in Soxhlet apparatus up to decolorization of flowing solvent to remove as much maltenes as possible. Solvent from maltenes solution was removed up to a constant weight using rotary evaporator. Maltenes were separated into Saturates, Aromatics and Resins through the adsorption method on silica using n-hexane for Saturates, Aromatics and isopropanol:benzene (1:1 v/v) for Resins according to the known procedure (ASTM, 1993) [39].

The physical modeling of the oil displacement processes by hydrocarbon solvents was carried out using a laboratory model which was a transparent cell made out of an acrylic glass with the inner chamber dimensions of 450 × 450 × 20 mm [40, 41].

A mixture of heavy oil (**Table 1**) and sand (fraction 0.250–0.315 mm) was prepared prior to each experiment. The displacement process was carried out at 20°C by injecting the solvent into the cell via the upper opening at the constant excess pressure of 1563.1 Pa (0.0154 atm) provided by stable liquid level in the metering vessel. n-Hexane as a model basic solvent (one of the main component of distillate fractions, use of which is possible in oil refineries) and also additives of toluene and nonylphenol (product of PJSC "Nizhnekamskneftekhim") were used. The samples of the displaced HO mixed with solvent were sampled via the bottom opening. The test samples (HO and solvent mixtures) were sampled starting from ≈5.5 hours after the solvent was injected into the reservoir model; further, the fluid sampling was carried out with 1-hour intervals. Duration of each experiment was 16 hours. For all of the solvents the accumulated oil recovery, oil percentage in the recovered HO + solvent mixture, amount of residual oil in the reservoir model after displacement and asphaltenes contents in it were assessed. After each of the experiments, the residual oil was removed from the model and separated into nine identical sectors (**Figure 3**). In order to remove the residual oil the sand from each of the sectors was separately washed by benzene several times with subsequent solvent evaporation till constant weight. The asphaltenes were precipitated from the oil by the 20-fold excess volume of n-hexane according to the aforesaid technique.

Density, g/cm^3, 20°C	Viscosity, mm^2 /s, 20°C	Composition, wt%			
		Asphaltenes	Saturates + Aromatics	Resins	Light fractions <200°C
0.9540	3083	6.3	61.1	3.8	1.8

Table 1. Density, viscosity and composition of Ashalchinskoye field HOs.

2.2. Preparation of model oils with various asphaltene contents and physical modeling of their displacement by solvents from a porous medium

Asphaltenes (A0) of the Ashalchinskoye field HOs obtained at the separation of deasphalted oil were fractioned into two fractions, A1 and A2, by selected toluene/hexane mixture at a ratio of 35/65% v/v [30, 42]. After this, the "common" asphaltenes A0, and asphaltenes of the A1 and A2 fractions were mixed with deasphalted oil in the same ratio they were present in the natural oil. Therefore, three artificial model oils were obtained differing only in contents and structure of asphaltenes (**Table 2**).

The comparative analysis of chemical contents and structure of A0, A1 and A2 isolated from the HO was carried out using various investigation methods.

Investigation of the molecular weights of the asphaltenes was carried out using the UltraFlex III mass spectrometer (Bruker, Germany) with time-of-flight (TOF) detector using the Matrix-activated laser desorption and ionization (MALDI) method, using 1,8,9-trihydroxyanthracene as a matrix.

IT spectra of the asphaltenes were recorded in the 4000–400 cm^{-1} range using the Tensor-27 Fourier IT spectrometer (Bruker, Germany) with the optical resolution of 4 cm^{-1}. The asphaltenes were first ground in a mortar with KBr powder, and a pellet was made from the obtained mixture using a hand press. The obtained values of the intensities allowed for calculating the aromaticity spectral coefficients (Arom = $D_{1600}/D_{720\,+\,1380}$ – reflects the percentage of C=C bonds in aromatic fragments relative to the C-H bonds in the aliphatic structures), aliphaticity (Al = D_{1450}/D_{1600} – reflects the percentage of C-H bonds in aliphatic fragments relative to the aromatic C=C bonds); condensation (Cond = $D_{1600}/D_{740\,+\,860}$ – reflects the percentage of C-H bonds in the aromatic structures); oxidation (Ox = D_{1700}/D_{1600} reflects the personage of carbonyl groups R-C=O at the presence of OH group) [43].

Thermal stability and possible phase transitions of the starting asphaltenes A0 and also of the isolated A1 and A2 fractions were assessed using combined thermogravimetry (TG) and differential scanning calorimetry (DSC) methods using the STA 449 CJupiter thermal analyzer (Netzsch, Germany) in the argon atmosphere with total flow of 50 ml/min in the temperature range of 30–800°C. The heating rate was 10°C/min. The weight of the sample was 6–7 mg.

Measurements of the stable free radical (SFR) and vanadyl complexes (VC) contents were carried out using the SE/X-2544 EPR spectrometer (RadioPan, Poland) under the same conditions at room temperature. At determination of SFR and VC contents in the studied samples

Sample	Composition	Viscosity, mm^2/s, 50°C
Model oil A0	Deasphalted oil + asphaltenes A0	966.8
Model oil A1	Deasphalted oil + asphaltenes A1	1054.7
Model oil A2	Deasphalted oil + asphaltenes A2	647.6

Table 2. Composition and properties of model oils.

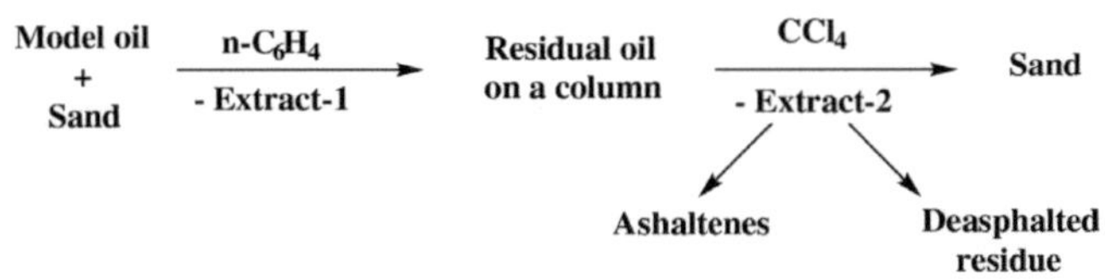

Figure 1. The general scheme of experiments on the displacement of model oils.

ampoules with 3 mm inner diameter out of molybdenum glass that did not produce a visible signal in the EPR spectrum were used. The ampoules were densely filled with asphaltenes 13–14 mm in height. The intensity of the stable free radical line was determined using the amplitude of the single line in the EPR spectrum center (g = 2.003). In order to calculate the intensity of the vanadyl complexes lines the amplitude of the +1/2 line that is near the SFR line in the weaker magnetic field area was determined. The intensity of the SFR and VC signals of the studied samples was discounted to the sample weights.

Vanadium and nickel concentrations in oils and asphaltenes were measured by means of direct flame atomic absorption spectrometry using AAS-1 N spectrophotometer (Carl Zeiss Jena, Germany), with approved standard samples of metal concentration in oil products used as blank solutions.

The experimental displacements of the model oils A0, A1 and A2 in the porous medium by n-hexane were carried out using a 3 × 100 cm glass column wherein the oil + sand mixture (fraction 0.25–0.35 mm, oil saturation 10%). Further, the 3 pore volumes of hexane at 20°C were passed by injecting the solvent through the upper opening of the tube at the constant excess pressure of 1563.1 Pa (0.0154 atm) provided by stable liquid level in the metering vessel, and 5 cm^3 samples were sampled. After 3 pore volumes of hexane were passed, additional displacement of the residual oil from the model was carried out by hexane until full disappearance of coloration of the solvent passing through the column (Extract 1). The asphaltene concentrates precipitated in the porous volume were extracted by washing (extraction) from the column by tetrachloromethane (Extract 2). Then the solvent was evaporated from Extract 2, the remainder was dried in vacuo until constant weight and was dissolved in small amount of toluene for further isolation of asphaltenes by 20-fold excess n-hexane. The general scheme of experiments on the displacement of model oils is outlined in **Figure 1**.

3. Results and discussion

3.1. Physical modeling of the process of HO displacement from the reservoir by n-hexane

At modeling the HO recovery processes from the reservoir using solvents, the experiment results were represented in the form of diagrams of plots of Cumulative oil recovery and oil percentage in the recovered HO + solvent mixture vs. time. The experiments were carried out under the 10% oil saturation of the sand (**Figure 2**).

http://dx.doi.org/10.5772/intechopen.72673

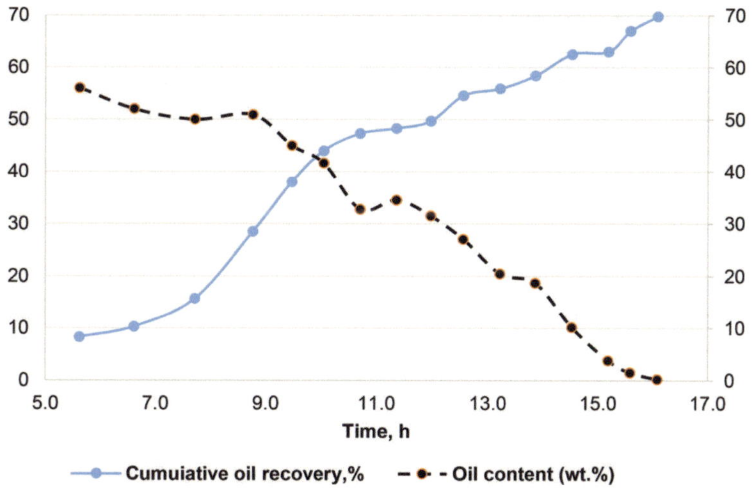

Figure 2. Cumulative oil recovery and oil content in the recovered HO + solvent mixture at carrying out the displacement experiments by hexane.

The results of experiments indicate the following supposed mechanism of the oil recovery by solvent based on n-hexane. At the first step, increased oil yield was noted from the model with a maximum rate which stabilizes over some time. The stabilization of the oil yield from the model occurs at the transition of the displacement mode from the dispersion (affected by gravity force) to diffusion (in a peripheral area). Regarding the content of asphaltenes in the residual oil, for all 9 samples from various model sectors higher values are indicated compared to the starting bitumen (**Figure 3**, **Table 1**).

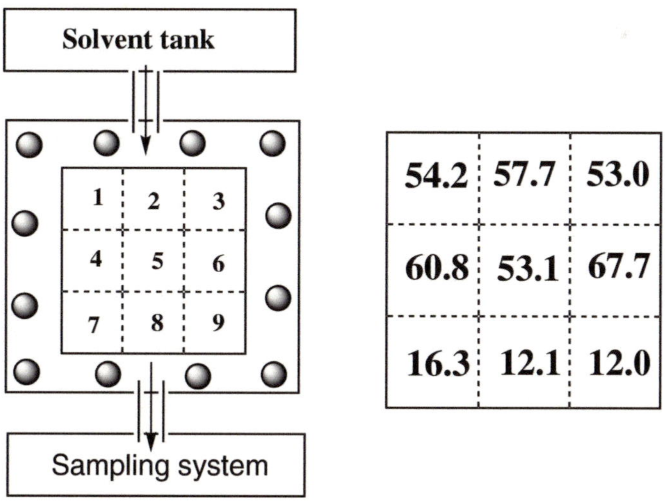

Figure 3. Scheme of an experimental setup of modeling the physical and chemical effect of solvents on the oil-saturated reservoir and the asphaltene amount (wt%) in the residual oil by cell sectors.

It was found that the highest asphaltene concentration in the residual oil is detected in upper sectors where the starting step of the oil displacement occurs and where the oil contacts the pure solvent (**Figure 3**). Hexane alters the solubility of asphaltenes in oil because light saturated hydrocarbons affect the interaction of asphaltenes and resins destabilizing the asphaltenes [28, 29]. Introduction of aromatic compounds in the system may positively affect the stability of asphaltenes in the system [20, 23–26] because the asphaltenes have a polyaromatic structure.

3.2. Investigation of the effect of fractional asphaltene contents in the HO on the recovery process by n-hexane

In order to identify the features of supermolecular structure formation in the porous medium depending on the fraction contents of the asphaltenes, the approach was used that allows discovering the content and a number of other oil components. The natural object, heavy oil of the Ashalchinskoye field, was separated into the deasphalted oil (maltenes) and A0 asphaltenes, from which by fractioning the two fractions were isolated, the so-called A1 fraction ("island" type), the part of the asphaltenes characterized by highest condensation, and the A2 fraction ("archipelago" type), which is regarded to be as less condensed, comprising larger number of heteroatoms and alkyl substituents [42, 44, 45]. Then the obtained "common" A0 asphaltenes, the A1 and A2 asphaltene fractions were mixed with deasphalted oil at the same ratio as they are present in the natural oil (**Table 2**).

In order to enable a possibility of quantitative production of the asphaltene fractions and their characterization by physical and chemical investigation methods and also in order to obtain a sufficient amount of model oils and carrying out modeling of the oil displacement from porous medium by n-hexane, it was necessary for the A0 asphaltenes to be separated into fractions in the ~ 50:50% ratio. Experimentally in the 25 to 40% range, the ratio of the solvent (toluene) to precipitant (hexane) was found equal to 35%. As a result of fractioning the A1 and A2 asphaltene fractions were obtained at the ratio of 51:49%. The A1 asphaltene fraction isolated at various toluene percentage in the precipitant was characterized by a number of physical and chemical methods: EPR with determination of SFR and VC contents, spectrophotometry with determination of specific light absorption coefficient (K_{la}), atomic absorption spectroscopy with determination of vanadium and nickel contents (**Table 3**) and also IR spectroscopy (**Table 4**) and elemental analysis (**Table 5**).

In accordance with the obtained results, at increase in toluene percentage from 25 to 40% in the mixture with hexane the decrease in the A1 fraction yield occurs from 62 to 39.5 wt%, correspondingly (**Table 3**). Thereby, as the toluene percentage in the hexane mixture increases, the gradual increase of vanadium, nickel, SFR, VC and specific K_{la} occurs in contents of the A1 asphaltenes. According to FTIR data (**Table 4**), the A1 asphaltenes at their precipitation from the mixture of hexane with toluene are characterized by the increase in the aromaticity and condensation coefficients, decrease in the aliphaticity and oxidation coefficients. Elemental contents data demonstrate that as the toluene percentage increases, the carbon content insignificantly increases in the asphaltenes while hydrogen content decreases. As a whole,

Amount of toluene in the mixture, % vol	Yield A1, wt%	V, wt%	Ni, wt%	VC*, ×10^18 rel.sp./g	SFR**, ×10^18 rel.sp./g	SFR/VC	K_{la}***
25	62.0	0.220	0.0278	33.8	122	3.61	76.26
30	52.1	0.225	0.0278	34.4	126	3.66	77.20
35	51.0	0.235	0.0285	35.0	124	3.54	78.91
40	39.5	0.240	0.0291	35.1	132	3.76	79.30

*VC=Vanadyl complexes.
**SFR = Stable free radicals.
***K_{la} = light absorption coefficient.

Table 3. Fraction yield, contents of metals, paramagnetic centers, VC, K_{la} value for A1 asphaltenes isolated at various toluene content in hexane.

Amount of toluene in the mixture, % vol	Arom	Cond	Al	Ox
25	0.43	1.15	2.39	0.18
30	0.46	1.19	2.37	0.17
35	0.47	1.25	2.33	0.17
40	0.50	1.26	2.34	0.15

Table 4. Structural-group composition of A1 according to FTIR data with different toluene content in hexane.

Amount of toluene in the mixture, % vol	C,%	H, %	N, %	S, %	C/H
25	72.69	7.24	2.46	5.28	10.04
30	75.44	7.45	2.66	5.16	10.13
35	75.86	7.46	2.92	5.01	10.16
40	77.58	7.61	3.11	5.63	10.19

Table 5. Elemental composition of A1 with different toluene content in hexane.

differences between the asphaltenes A1 obtained by using 25 and 40% toluene are quite significant in all elemental analysis parameters (**Table 5**).

In order to find out the main structural differences of the A0 asphaltenes and A1 and A2 fractions, the comparative analysis of data of some physical and chemical investigation methods was carried out (**Tables 6–9**).

Analysis of EPR, spectrophotometry and atomic absorption spectrometry data allowed for identifying a number of principal differences in the asphaltene fractions (**Table 6**). Thus, from the A1 fraction higher values of all parameters, K_{la}, vanadium, nickel, SFR and VC contents were found. The least difference in the values is found in the VC content which does not exceed 10%. In all other parameters, the observed value difference is 1.6 to 1.9 fold.

Sample	V, wt%	Ni, wt%	VC, $\times 10^{18}$ rel.sp./g	SFR, $\times 10^{18}$ rel.sp./g	SFR/VC	K_{la}
A0	0.185	0.0232	33.8	100	2.96	60.28
A1	0.235	0.0285	35.0	124	3.54	78.91
A2	0.120	0.0161	30.8	69	2.24	49.79

Table 6. Results of determining the contents of metals, SFR, VC and also K_{la} for the asphaltenes.

At comparison of the molecular weight of the asphaltenes the value of the average molecular weight is used that, according to the MALDI mass spectra, is equal to 1600 and 1200 amu for the asphaltenes of the A1 and A2 fractions, correspondingly, and for the A0 asphaltenes the value of 1400 amu is found.

Thermal stability and possible phase transitions for A0 asphaltenes and isolated A1 and A2 fractions were assessed by the combined thermogravimetry (TG) and differential scanning calorimetry (DSC) methods (**Table 7**). The threshold temperature of the intimation of intensive thermal destruction of A1 and A2 fraction asphaltenes in significant degree are determined by their structural features: ratio of naphthene-aromatic and fused aromatic structures. For the A1 fraction, the higher condensation degree is characteristics than for the A2 fraction and contents of the alkyl substituents in the A2 asphaltene fraction molecules is higher than in the A1 asphaltene fraction molecules (**Table 8**). This factor mostly determines thermal stability of the asphaltenes. Intensive weight loss begins in the A1 asphaltene fraction at 436°C and in the A2 fraction at 360°C. The temperature of half weight loss for the A1 and A2 asphaltene fraction is 501.6 and 470.2°C, correspondingly, and for the A0 asphaltene fraction it has an intermediate value of 493.0°C. Also, the characteristic feature of the A1 asphaltene fraction compared to A2 is a higher yield of residue (coke) after heating to 800°C.

Most of the parameters allow for identifying the asphaltene fractions reflecting specifics of their structural features. Thus, from the A2 asphaltene fraction higher values of aliphaticity

Sample	Loss of mass in the temperature range (°C),%								The temperature of half weight loss,°C	Yield of residue, %
	30–100	100–200	200–300	300–400	400–500	500–600	600–700	700–800		
A0	0.36	1.12	1.94	5.99	41.98	4.92	4.96	4.99	493.0	33.74
A1	1.15	3.86	0.98	4.74	39.03	4.17	1.59	1.02	501.6	43.46
A2	—	2.79	6.25	9.57	40.58	3.62	1.10	0.63	470.2	35.46

Table 7. Results of thermal analysis for the asphaltenes and for their A1 and A2 fractions.

Sample	Arom	Cond	Al	Ox
A0	0.46	1.33	2.44	0.23
A1	0.53	1.73	2.32	0.16
A2	0.44	1.24	2.76	0.25

Table 8. Structural-group composition of samples according to FTIR data.

Sample	C,%	H, %	N, %	S, %	C/H
A0	78.42	8.13	2.08	4.43	9.65
A1	75.41	7.75	2.91	3.45	9.79
A2	72.48	7.64	1.79	6.05	9.49

Table 9. Elemental analysis data for the asphaltenes and for their A1 and A2 fractions.

and oxidation are found. At the same time for the A1 asphaltene fraction higher condensation and aromaticity can be noted (**Tables 8** and **9**).

As a result of performed experiments, it was found that at the displacement of the A1 model oil with n-hexane the number of supermolecular asphaltene-resin structures formed in the porous medium is more than 2-fold higher compared to the A2 model oil. The asphaltene content in such asphaltene-resin concentrates differs insignificantly for the A1 and A2 oils, 74.5 and 61.1 wt% correspondingly. Thereby in the case with model A1 oil, the main weight (59%) of the asphaltenes present in the oil remains in contents of the asphaltene concentrates formed in the porous medium. For the A2 oil, this parameter is much lower, just 18.9% (**Table 10**).

The identified features of asphaltene precipitate formation in the porous medium affect dynamics and the oil recovery rate. Thus, for the A2 model oil maximum oil recovery rate is observed at displacement with hexane (**Figure 4**).

As a whole, at the comparison of the A1 and A2 asphaltenes obtained after oil displacement the analogous picture is observed as for the starting asphaltenes A1 and A2. For the A1 asphaltenes higher contents of vanadium, nickel, vanadyl complexes and also K_{la} are observed; however, the SFR content in the A2 asphaltenes is significantly higher (**Table 11**). In most cases, the values of the analyzed parameters remarkably differ at the comparison of the starting asphaltene fractions and those formed after oil displacement. Therefore, contents and

Sample	Asphaltenes in Extract-2, wt %	Asphaltenes percentage in the Extract-2 vs. potential content in the starting oil, wt%
Model oil A0	70.9	45.7
Model oil A1	74.5	59.0
Model oil A2	61.1	18.9

Table 10. Results of displacement of model oils with n-hexane from the porous medium (asphaltenes percentage in the extract 2 vs. potential content in the starting oil, wt%).

Sample	V, wt%	Ni, wt%	VC, $\times 10^{18}$ rel.sp./g	SFR, $\times 10^{18}$ rel.sp./g	K_{la}
A0	0.185/0.160	0.0232/0.020	33.8/22.5	100/116	60.28/56.04
A1	0.235/0.175	0.0285/0.025	35.0/31.4	124/116	78.91/70.87
A2	0.120/0.165	0.0161/0.021	30.8/30.9	69.0/163	49.79/60.38

Table 11. Contents of metals, paramagnetic centers, VC, K_{la} values for the asphaltenes isolated from the starting model oils and sediments formed after the oil displacement process with hexane (model oil/extract 2).

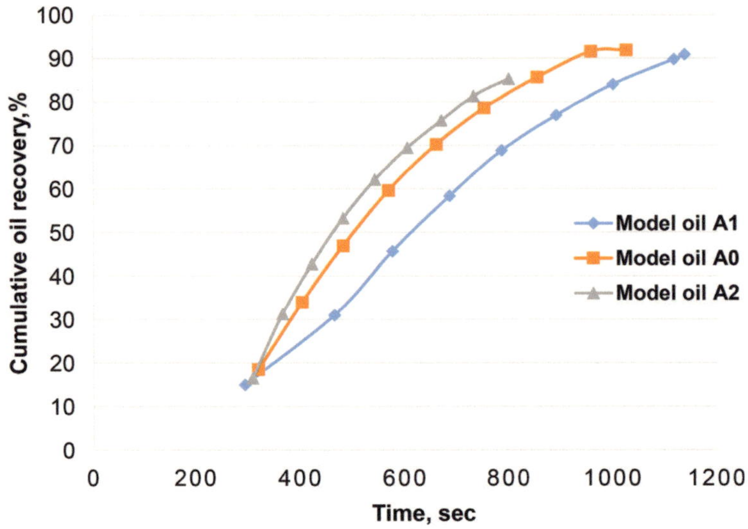

Figure 4. Cumulative oil recovery at the displacement of model oils with n-hexane from the porous medium.

properties of the asphaltenes formed during displacement of heavy oil in the porous medium by n-alkanes differ from the starting asphaltenes. Thereby, from the high molecular weight asphaltene fraction, A1, this trend has insignificant changes whereas for the A2 fraction significant increase is observed. This fact allows for supposing that during oil displacement process in case of the A2 asphaltenes their further fractioning occurs in the porous medium, as a result of which higher molecular weight asphaltene concentrates are formed compared to the starting asphaltenes.

3.3. Physical modeling of the HO displacement process with n-hexane with additives inhibiting precipitation of asphaltenes on a large oil reservoir model

The primary function of asphaltene stabilizers in the oils is played by resins forming solvate shells that prevent asphaltene precipitation. When the solvate shells are dissolved by saturated hydrocarbons, the asphaltenes precipitate. Therefore, in order to stabilize the oil + solvent mixtures one can use the additives of oil resins, synthetic alkyl phenols or aromatic compounds which are asphaltene solvents.

In order to increase the colloidal stability of the asphaltene components of the Ashalchinskoye HO field during its displacement with n-hexane, an experimental selection of chemical additives was carried out (sand fractions 0.250–0.35 mm, oil saturation 10%, temperature 20°C).

Originally, toluene and nonylphenol were used as an additive for the main solvent to decrease the amount of asphaltene residue. Moreover, the preliminary injection of free toluene was tested with subsequent injection of n-hexane. In this case, the toluene to n-hexane ratio was 5 and 95% correspondingly. As a result of experiments, the cumulative oil recovery was assessed (in percents to the oil contained in the model cell) over time (**Figure 5**).

http://dx.doi.org/10.5772/intechopen.72673

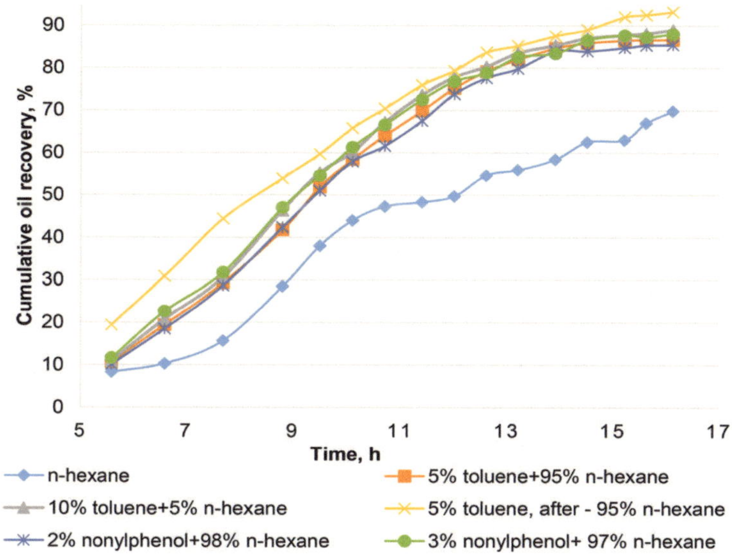

Figure 5. Cumulative oil recovery at carrying out the experiments on HO displacement with various solvents.

Use of toluene and nonylphenol mixed with n-hexane at displacement significantly affects alteration of total oil recovery and oil content in the extracted products. The cumulative oil recovery at HO displacement only with n-hexane is 69.8% whereas the addition of 5 and 10% toluene to n-hexane increases the accumulated oil recovery to 86.6 and 89.0% correspondingly. Use of nonylphenol gives the values of cumulative oil recovery comparable with those with toluene use of 85.4 and 88.0% for 2 and 3% nonylphenol in n-hexane, correspondingly. The highest oil recovery (93.2%) from the reservoir model is achieved when toluene is injected prior to n-hexane at 5:95 toluene–hexane ratio.

After each experiment, the residual oil was extracted from the model separately in each of the nine sectors by benzene extraction. An amount of asphaltenes in the residual oil samples was measured (**Figure 6**).

At use of n-hexane for the oil displacement, in the residual oil on average 50 to 60 wt% asphaltenes is contained. The asphaltene content is significantly lower in the lower part of the model cell, which is on average about 12 to 16 wt%. The results show that use of the compositions based on n-hexane with toluene and nonylphenol allows for significantly decreasing the asphaltene contents in the residual oil compared to the using of n-hexane only. For the 10 wt% toluene or 3 wt% nonylphenol in the mixture with n-hexane the asphaltene contents in the residual oil in the lower part of the model cell is on average less than 10 wt% what insignificantly exceeds their contents in the starting oil. Use of such method as pre-injection of toluene prior to injecting the base solvent allows for achieving even higher inhibition of asphaltenes in the residual oil. In this case, the amount of toluene used is only 5% of the total solvent amounts, which finally allows for significantly decreasing its consumption and, as a consequence, improves the project economy.

54.2	57.7	53.0
60.8	53.1	67.7
16.3	12.1	12.0

a

30.7	37.7	33.0
30.6	44.6	43.7
8.9	10.5	8.6

b

29.5	34.9	28.4
29.0	26.3	19.6
7.9	11.3	7.3

c

10.4	11.6	14.0
14.1	14.0	15.4
7.0	8.7	7.4

d

38.7	43.7	43.0
36.6	49.6	48.7
10.9	12.5	12.6

e

26.9	33.3	27.4
25.2	28.1	26.5
8.9	10.3	8.6

f

Figure 6. Changes in the amount of asphaltenes in residual oil for different solvent used. (a) n-hexane, (b) 5% toluene +95% n-hexane, (c) 10% toluene +90% n-hexane, (d) 5%toluene, after – 95% n-hexane, (e) 2% nonylphenol +98% n-hexane, (f) 3% nonylphenol +97% n-hexane.

The increased asphaltene contents observed in all of the performed experiments at transition from the upper part of the model to the lower ones can be explained by the fact that at the beginning of each experiment upon the oil contact with solvent asphaltene precipitation occurs and then the solvent is enriched by deasphalted oil that facilitates decrease in the precipitate amount in the residual oil.

If one refers to the distribution of the asphaltene contents in horizontal, at the use of pure n-hexane, in the side parts of the model, where the slow diffusion process occurs, an elevated content of asphaltenes is observed compared to the central part of the model where the contact of the oil with n-hexane occurs faster. Upon addition of nonylphenol and toluene to n-hexane aligning of asphaltene contents in the residual oil in the neighboring model segments is observed and achievement of better encompassing of the model porous medium by a solvent.

Natural inhibitors acting as solvating agents are also effective; they, in contrast to the aromatic solvents are compliant with ecological safety regulations. These are oil resins and deasphalted oil [23, 46, 47].

Earlier in our paper [48] we have carried out a comparative analysis of various methods for assessing the efficiency of asphaltene precipitation stabilizers for Ashalchinskoye field HOs. According to the results of performed studies, it was established that the deasphalted oil of Ashalchinskoye field HOs is quite an effective stabilizer of asphaltene precipitation, which is not inferior to the aromatic hydrocarbons. The oil resins as asphaltene precipitation inhibitors differ insignificantly in efficiency from expensive synthetic products. It was established that the percentage of benzene resins of not above 6.5% per hexane volume allows for fully providing asphaltene stabilization during heavy oil displacement process. Also, it was found that alcohol-benzene resins at normal conditions are not the effective asphaltene precipitation inhibitors.

4. Conclusions

In modeling, the oil displacement process by injecting n-hexane, the influence of the molecular weight and condensation of asphaltenes on parameters of their deposition in a porous medium is shown. It was revealed that features of structural and group composition of heavy oil asphaltenes affect the dynamics and completeness of oil extraction. Under the same conditions, the process of displacement of model oil with low molecular weight asphaltenes makes it possible to achieve almost 5% more accumulated oil recovery than that for high molecular weight asphaltenes.

In a separate series of experiments, it was shown that the composition and properties of asphaltenes formed during the displacement of heavy oil in a porous medium by n-alkanes differ from that of the initial condition of asphaltenes. Herewith for the fraction of high-molecular-weight asphaltenes A1 this trend is weak, while for fraction A2 it is pronounced significantly. This fact suggests that in the process of oil displacement the less condensed asphaltenes A2 undergo additional fractionation in a porous medium, resulting in higher molecular weight asphaltene concentrates in comparison with the initial asphaltenes.

As a result of several experiments for the physical modeling of heavy oil displacement in porous medium in a large-sized model, the opportunities of using various composite solvents based on n-hexane with toluene and nonylphenol additives were assessed.

By taking the example of the Ashalchinskoye field heavy oil, we have shown that accumulated oil recovery by n-hexane only appears at the level of 69.8%, whereas adding the toluene in 5–10 or 2–3% nonylphenol amount rises the recovery to 85–89%. The largest oil recovery (93.2%) from the reservoir model is achieved when toluene is injected prior to n-hexane (5% of toluene relative to the volume of n-hexane).

Author details

Dmitry N. Borisov*, Dmitry V. Milordov, Svetlana G. Yakubova and Makhmut R. Yakubov

*Address all correspondence to: boriku@gmail.com

A.E. Arbuzov Institute of Organic and Physical Chemistry, Kazan Scientific Center, Russian Academy of Sciences

References

[1] Rassenfoss S. A tricky Tradeoff - can adding a little solvent yield a lot more heavy crude? Journal of Petroleum Technology. 2012;**6**:58-64. DOI: 10.2118/0612-0058-JPT

[2] Orr B. ES-SAGD: Past, Present, and Future. SPE 129518. 2009. DOI: 10.2118/129518-STU

[3] Boone T. An Integrated Technology Development Plan for Solvent-Based Recovery of Heavy Oil. SPE 150706. 2011. DOI: 10.2118/150706-MS

[4] Stark S. Increasing Cold Lake Recovery by Adapting Steam flood Principles to a Bitumen Reservoir. SPE 145052. 2011. DOI: 10.2118/145052-MS

[5] Gupta S, Gittins S. Semi analytical Approach for Estimating Optimal Solvent Use in Solvent Aided SAGD Process. SPE 146671. 2011. DOI: 10.2118/146671-MS

[6] Edmunds N, Maini B, Peterson J. Advanced solvent-additive processes by genetic optimization. Journal of Canadian Petroleum Technology. 2010;**49**:34-41. DOI: 10.2118/140659-PA

[7] Sharma J, Gates I. Dynamics of Steam-Solvent Coupling at the Edge of an ES-SAGD Chamber. SPE 128045. 2010. DOI: 10.2118/128045-MS

[8] Gates I. Design of the Injection Strategy in Expanding-Solvent Steam Assisted Gravity Drainage. In: Paper at the Second CDEN International Conference on Design Education, Innovation, and Practice Kananaskis; 18–20 July; Alberta, Canada; 2005. DOI: 10.24908/pceea.v0i0.3888

[9] Yakubov M, Amerkhanov M, Khisamov R, Khanipova Y. World experience of solvents injection for extraction of heavy oil and solvent-based processes potential in TATNEFT PJSC heavy oil fields. Oil Industry. 2017;**2**:78-81. DOI: 10.24887/0028-2448-2017-2-78-81.

[10] Jamaloei B, Dong M, Mahinpey N, Maini B. Enhanced cyclic solvent process (ECSP) for heavy oil and bitumen recovery in thin reservoirs. Energy and Fuels. 2012;**26**(5):2865-2874. DOI: 10.1021/ef300152b

[11] Jiang T, Zeng F, Jia X, Gu Y. A new solvent-based enhanced heavy oil recovery method: Cyclic production with continuous solvent injection. Fuel. 2014;**115**:426-433. DOI: 10.1016/j.fuel.2013.07.043

[12] Jia X, Zeng F, Gu Y. Gasflooding-assisted cyclic solvent injection (GA-CSI) for enhancing heavy oil recovery. 2015;**140**:344-353 DOI: 10.1016/j.fuel.2014.09.066

[13] Krawchuk P. Development and Operation of the Nsolv BEST Demonstration. Paper WHOC16–135 presented at the World Heavy Oil Congress 2016, Calgary, Alberta, 6–9 September, 2016

[14] Yakubov M, Yakubova S, Borisov D, Gennady Romanov G, Yakubson K. Asphaltene Precipitation Inhibitors and Phase Behavior Control for Bitumen Recovery by Solvent Injection. Canada. Society of Petroleum Engineers; 2014. p. 1913-1918. DOI: 10.2118/170165-MS

[15] Yarranton H, Masliyah J. Molar mass distribution and solubility modeling of asphaltenes. Alche journal. 1996;**42**(12):3533-3543. DOI: 10.1002/aic.690421222

[16] Tojima M, Suhara S, Imamura M, Furuta A. Effect of heavy asphaltene on stability of residual oil. Catalysis Today. 1998;**43**(3–4):347-351. DOI: 10.1016/S0920-5861(98)00163-1

[17] Rogel E. Theoretical approach to the stability of visbroken residues. Energy and Fuels. 1998;**12**(5):875-880. DOI: 10.1021/ef970222r

[18] Schabron J. The solubility and three-dimensional structure of asphaltenes. Petroleum Science and Technology. 1998;**16**:361-375. DOI: 10.1080/10916469808949788

[19] Koots J, Speight J. Relation of petroleum resins to asphaltenes. Fuel. 1975;**54**(3):179-184. DOI: 10.1016/0016-2361(75)90007-1

[20] Miadonye A, Evans L. The solubility of asphaltenes in different hydrocarbon liquids. Petroleum Science and Technology. 2010;**28**(14):1407-1414. DOI: 10.1080/10916460902936960

[21] Ghloum E, Al-Qahtani M, Al-Rashid A. Effect of inhibitors on asphaltene precipitation for Marrat Kuwaiti reservoirs. Journal of Petroleum Science and Engineering. 2010; **70**(1–2):99-106. DOI: 10.1016/j.petrol.2009.10.003

[22] Rogel E. Effect of inhibitors on asphaltene aggregation: A theoretical framework. Energy and Fuels. 2011;**25**:472-481. DOI: 10.1021/ef100912b

[23] Clarke P, Pruden B. Asphaltene precipitation: Detection using heat transfer analysis, and inhibition using chemical additives. Fuel. 1997;**76**(7):607-614. DOI: 10.1016/S0016-2361(97)00052-5

[24] Chang C, Fogler H. Asphaltene stabilization in alkyl solvents using oil-soluble amphiphiles. In: SPE International symposium on Oilfield Chemistry;New Orleans, Louisiana. Society of Petroleum; 2–5 March, 1993. p. 339-345. DOI: 10.2118/25185-MS

[25] González G, Middea A. Peptization of asphaltenes by various oil soluble amphiphiles. Colloids and Surfaces. 1991;**52**:207-217. DOI: 10.1007/s12182-0160098-1

[26] Kyeongseok O, Milind D. Effect of organic additives on the onset of asphaltene precipitation. Energy and Fuels. 2002;**16**:694-699. DOI: 10.1021/ef010223q

[27] Guzmán R, Ancheyt J, Trejo F, Rodríguez S. Methods for determining asphaltene stability in crude oils. Fuel. 2017;**188**:530-543. DOI: 10.1016/j.fuel.2016.10.012

[28] Carnahan N, Salager J, Antón R, Dávila A. Properties of resins extracted from Boscan crude oil and their effect on the stability of asphaltenes in Boscan and Hamaca crude oils. Energy and Fuels. 1999;**13**(2):309-314. DOI: 10.1021/ef980218v

[29] Khvostichenko D, Andersen S. Electrodeposition of asphaltenes. 2. Effect of resins and additives. Energy and Fuels. 2010;**24**(4):2327-2336. DOI: 10.1021/ef900970j

[30] Petrova L, Abbakumova N, Borisov D, Yakubov M, Zaidullin I, Romanov G. Interrelation of flocculation, precipitation, and structure of asphaltene fractions. Chemistry and Technology of Fuels and Oils. 2013;**49**(1):25-31. DOI: 10.1007/s10553-013-0407-y

[31] Mukhametshina A, Kar T, Hascakir B. Asphaltene precipitation during bitumen extraction with expanding solvent steam assisted gravity drainage (ES-SAGD): Effects on pore-scale displacement. SPE Journal. 2016;**21**(2):380-392. DOI: 10.2118/170013-PA

[32] Mohammadzadeh O, Rezaei N, Chatzis I. Production characteristics of the steam-assisted gravity drainage (SAGD) and solvent-aided SAGD (SA-SAGD) processes using a 2-D macroscale physical model. Energy and Fuels. 2012;**26**(7):4346-4365. DOI: 10.1021/ef300354j

[33] Kar T, Ovalle C, Rogel E, Vien J, Hascakir B. The residual oil saturation determination for steam assisted gravity drainage (SAGD) and solvent-SAGD. Fuel. 2016;**172**:187-195. DOI: 10.1016/j.fuel.2016.01.029

[34] Kar T, Hascakir B. The role of resins, asphaltenes, and water in water–oil emulsion breaking with microwave heating. Energy and Fuels. 2015;**29**:3684-3690. DOI: 10.1021/acs.energyfuels.5b00662

[35] Kar T, Mukhametshina A, Unal Y, Hascakir B. The effect of clay type on steam assisted gravity drainage performance. Journal of Canadian Petroleum Technology. 2015;**54**(6): 412-423. DOI: 10.2118/173795-PA

[36] Mohammadzadeh O, Rezaei N, Chatzis I. Pore-level investigation of heavy oil and bitumen recovery using solvent – aided steam assisted gravity drainage (SA-SAGD) process. Energy and Fuels. 2010;**24**:6327-6345. DOI: 10.1021/ef100621s

[37] Ahmadloo F, Asghari K, Henni A, Freitag N. Experimental investigation of capillarity and drainage height roles in the vapor extraction process. Petroleum Science and Technology. 2013;**31**:204-214. DOI: 10.1080/10916466.2010.525580

[38] Yazdani A, Maini B. Modeling of the VAPEX process in a very large physical model. Energy and Fuels. 2008;**22**:535-544. DOI: 10.1021/ef700429h

[39] ASTM D2007–93. Standard test method for characteristic groups in rubber extender and processing oils by the clay-gel adsorption chromatographic method. ASTM, 1993

[40] Yakubov M, Borisov D, Sinyashin K, Amerkhanov M, Khisamov R. Physical modeling of ultraviscous oil displacement by using solvent on a large model of oil reservoir. Journal of Petroleum Science and Engineering. 2017;**154**:457-461. DOI: 10.1016/j.petrol.2017.02.024

[41] Yakubov M, Borisov D, Sh R, Amerkhanov M, Khisamov R. Effect of solvent composition on heavy oil displacement while modeling. Oil Industry. 2014;**10**:106-109

[42] Trejo F, Centeno G, Ancheyta J. Precipitation, fractionation and characterization of asphaltenes from heavy and light crude oils. Fuel. 2004;**83**:2169-2217. DOI: 10.1016/j.fuel.2004.06.008

[43] Petrova L, Abbakumova N, Zaidullin I, Borisov D. Polar-solvent fractionation of asphaltenes from heavy oil and their characterization. Petroleum Chemistry. 2013;**53**(2):81-86. DOI: 10.1134/S0965544113020084

[44] Acevedo S, Castro A, Negrin J, Fernández A, Escobar G, Piscitelli V, Dessalces G. Relations between asphaltene structures and their physical and chemical properties: The rosary-type structure. Energy and Fuels. 2007;**21**(4):2165-2175. DOI: 10.1021/ef070089v

[45] Ganeeva Y, Yusupova T, Romanov G. Asphaltenenano-aggregates: Structure, phase transitions and effect on petroleum systems. Russian Chemical Reviews. 2011;**80**(10):993-1008. DOI: 10.1070/RC2011v080n10ABEH004174

[46] León O, Contreras E, Rogel E, Dambakli G, Acevedo S, Carbognani L, Espidel J. Adsorption of native resins on asphaltene particles: A correlation between adsorption and activity. Langmuir. 2002;**18**(13):5106-5112. DOI: 10.1021/la011394q

[47] Al-Sahhaf T, Fahim M, Elkilani A. Retardation of asphaltene precipitation by addition of toluene, resins, deasphalted oil and surfactants. Fluid Phase Equilibria. 2002;**194–197**: 1045-1057. DOI: 10.1016/S0378-3812(01)00702-6

[48] Yakubov M, Yakubova S, Borisov D, Milordov D, Usmanova G, Gryznov P, Romanov G. Change in the composition and properties of asphaltene for physical modeling of processes of replacement of heavy oil-based solvent of n-alkanes. Herald of the Kazan Technological University. 2013;**16**(22):277-280

Section 2

Environmental Solutions

Chapter 12

Total Acid Number Reduction of Naphthenic Acids Using Supercritical Fluid and Ionic Liquids

Pradip Chandra Mandal and Mitsuru Sasaki

Additional information is available at the end of the chapter

http://dx.doi.org/10.5772/intechopen.71812

Abstract

Naphthenic acids (NAs) are complex mixture of predominately alkyl-substituted cyclo-aliphatic carboxylic acids and small amount of acyclic acids present in crude oil, heavy oil and in oil sands bitumen. They are toxic components in refinery wastewater and in oil sand extraction water and lead to corrosion problems within the oil refineries. Therefore, the amount of NAs needs to suppress in petroleum oils and wastewater came from petroleum industry. This paper reviews the supercritical fluids (SCFs)- and ionic liquids (ILs)-based acidity reduction process from heavy oils by reviewing open literature. The potential benefits of SCFs- and ILs-based acidity reduction process of heavy oils are also explored. The reviewed articles reveal that total acid number (TAN) removal increase with increasing reaction time and temperature by the action of SCF. Supercritical methanol (SC-MeOH) has higher potentiality for removing acidity of NAs than supercritical water (SCW) without deposition of coke. TAN removal from NAs using SCF follows first order kinetics on TAN removal. ILs can reduce acidity of heavy oil either forming zwitterionic species or building cage structure around NAs through specific chemical bonds. Thus, non-catalytic SCF- and ILs-based TAN reduction process can open a new window to reduce acidity of heavy oils.

Keywords: supercritical fluid, naphthenic acid, total acid number, ionic liquid

1. Introduction

Heavy oil is a type of unconventional oil deposit (global estimated reserve of heavy oil and bitumen is 6.2 trillion barrels) all over the world. Such type of oil reservoirs are vastly undeveloped due to the difficulties of production, transportation and refining efficiently and inexpensively. Heavy oils contain significant quantity of metals, sulfur and nitrogen hetero-atom-containing compounds and naphthenic acids (NAs) as contaminants. NAs are a mixture of saturated aliphatic and alicyclic carboxylic acids. They are accountable for acidity of crude oils or heavy oils. NA content of heavy oil can be figured out in terms of total acid number (TAN), mg of KOH per g of oil. Crudes containing high TAN, specifically TAN $\geq$ 2, have lower demand [1, 2]. Moreover, their existence in crude oil creates multiple problems: they are corrosive [3–6] in nature, lessen quality of product and make environmental disposal problems [7–9]. Therefore, their removal from heavy oils is considered as a pressing issue for petroleum industries.

A number of deacidification processes have been disclosed in open literature to deacidify acidic petroleum oil, such as non-catalytic non-destructive methods, non-catalytic destructive methods and catalytic destructive methods.

High TAN containing crudes are usually mixed off with low TAN containing crudes to eliminate excessive corrosion in refinery equipment. This process is obviously expensive and non-selective [7]. In a different way, polymeric compounds including adequate basicity may be used to entrap or neutralize NAs. This process, however, involves the use of costly neutralizing agents, which are difficult to recover from deacidified products. Another method, extractive separation, usually requires a multistage wash using various solvent to single NA compounds or their derivatives such as water-oil emulsion and salt [10–12]. But valuable hydrocarbon compounds are removed to do so making the process costly.

The metal oxide (such as magnesium oxide and aluminum oxide) solid solutions show high adsorption capacity and can readily remove at least 95% of the NAs present in a liquid hydrocarbon feedstock at temperatures of 30–80°C [12, 13]. Recent adsorption technology is only attainable for low temperature distillate fractions. The prior process requires a highly alkaline environment, a strong base such as alkali metal hydroxides (for example, caustic soda). Unfortunately, the caustic does not simply provide an alkaline environment but in time is neutralized by acidic components of the hydrocarbon stream. Therefore, this process requires its continuous replacement and replenishment. It is alarming that disposal of spent caustic solutions is itself an environmental problem [13].

Corrosion inhibitors such as organic poly sulfide or phosphites or phosphoric acid have been introduced to passivate metal surfaces by building a defensive layer on them. But, this process requires maintaining adequate thickness of the protective layer that enforces the costs of multiple inhibitor injections.

Esterification is a well-known process for removing NA compounds from petroleum oils. In this process, NAs reacts with alcohol in the presence or absence of catalyst to form ester [14]. Metal carboxylates [15] and metal oxides [16] are fruitfully applied in esterification at reaction temperatures of 250–350°C over enhanced reaction times [12].

Non-catalytic NA destruction methods, such as thermal decomposition [7, 17], have a limited success for acidity reduction of crude oil. High temperature, in excess of 400°C, is mostly required which can offer thermal cracking of the crude resulting coke formation [2]. In this regard, catalytic decarboxylation process [18] has come out as an alternative and economic process of NAs decomposition.

The success of zeolites [12] and metal oxides [19] for deacidification of crude oil is high. But the process still requires moderately high operating temperatures (300–400°C). Moreover, many of these applications require hydrogen [7] to increase deacidification rate. But, hydrogen is not usually available at a production site. Hydrotreating in the presence of alumina supported metal oxides such as cobalt, molybdenum or nickel oxides has been successfully used in this regard [20, 21]. Metal oxides saturated with copper [22] and nickels [23] in the liquid phase at moderate temperatures (greater than 200°C) have also been utilized for the decarboxylation of fatty acids. Catalytic process, however, would be less feasible in crude oil systems due to catalyst poisoning by aromatics compounds, nitrogen containing species, metals and sulfur containing species of crude oil.

Supported palladium and platinum catalysts are active to destruct NAs via decarboxylation [24] and decarbonylation [25] in the liquid or gas phase. It is seen that transition metals supported catalysts can efficiently decarboxylate a series of fatty acids at temperatures of 80–100°C even without hydrogen [26]. In addition, the same titania-based metal catalysts can resist sulfur significantly for fine chemical synthesis [27]. Though their activity and chemical stability is high, platinum group metal-based catalysts are cost-effective process for large scale operation.

A common industrial method of NAs removal from the kerosene/diesel fractions is a dilute caustic wash of them using alkali/alkaline earth metals. In this process, NAs is separated from oil fraction as calcium or sodium naphthenates, which is dissolved in water soluble fraction. Then, it is possible to recover the water insoluble NAs by acidifying calcium or sodium naphthenates with a mineral acid [12]. This process is an uneconomical one; as it is considered inexpensive, incomplete removal of all NAs due to its poor solubility coupled with the formation of emulsion [12]. Treating of petroleum products for removing of NAs via a dilute caustic wash is a common industrial method. As an alternative of this process, the researchers are using organic bases such as glycolic solutions [28], monoethanolamine or 2-(Dibutylamino) ethanol [29] or imidazole derivatives [30] as alcoholic solutions. However, excess amount of organic base is required to reach good extraction process. This large excess of base resulted in carryover of amine into the petroleum phase and can create problem in the downstream catalyst units used for upgrading the oil [31].

Recently, specific ionic liquids (ILs) have been used in different applications including acidity removal from petroleum industry. For example, sulfur-containing acids are successfully extracted from crude oil using ILs containing a cation with a connected amino group [32]. In addition, amino acid ionic liquids (AAILs) can neutralize NAs by the reaction where one amino group of the amino acid-anion can react with one carboxylic acid group to form zwitterionic species. In this regard, lysine-derived AAILs indicated higher extraction capability compared to other AAILs because AAILs derived from lysine contain two amino groups [12]. Mandal et al. [2, 33] have disclosed a new supercritical fluid-based deacidification process at high temperatures and pressures.

This review paper examines the open literatures of petroleum oil deacidification process using environmental benign solvent, supercritical fluids and ionic liquids.

2. Naphthenic acid

NAs are complex mixtures of alkyl-substituted acyclic and cycloaliphatic carboxylic acids found in hydrocarbon deposits (petroleum, oil sands bitumen and crude oils) [34–36]. They are originally come from aerobic microbial degradation of petroleum hydrocarbon [34]. They are described by the general chemical formula $C_nH_{2n+z}O_2$, where n indicates the number of carbon and z is either zero or a negative, even integer that secludes lack of hydrogen resulting from ring formation. The numbers of ring present in the compounds are obtained by dividing the absolute value of z by 2. The rings of NAS can be fused or bridged. **Figure 1** depicts examples of typical structures of NAs, which have a different number of rings [36].

NAs are non-volatile, chemically stable and perform as surfactants. The polarity and non-volatility of NAs increase with increasing molecular weight delivering specific compounds with different physical, chemical, and toxicological properties [34, 36–39]. NAs act as typical carboxylic acids having acid strength analogous with the higher fatty acids. NAs are little weaker acid than low molecular weight carboxylic acids like acetic acid [36, 40–42].

Commercial NAs, produced via extraction of petroleum distillates, have industrial applications as shown in **Table 1**. Sodium salt of NAs is soluble in water.

Steel alloys that can resist corrosion by sulfur-containing compounds can be suitable for corrosion by NAs [43, 44]. The process of NAs corrosion is obscure till now though it can chelate with the metal ion by the carboxylate forming hydrogen gas [45]. The amount of NA passing

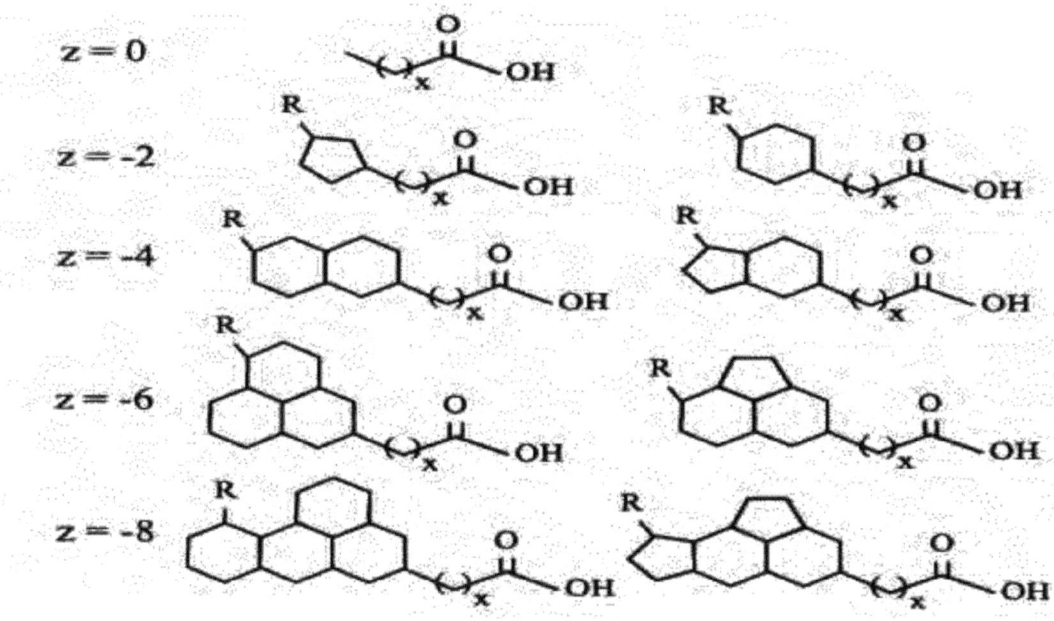

Figure 1. General structure of naphthenic acid.

NA metal salt	Industrial applications
Na salt	Emulsifying agent for agricultural insecticides Additive for cutting oils emulsion breaker in oil industry
Ca naphthenate	Additive for lubricating oil
Fe and Mn naphthenates	Fuel additives for improved combustion and reduced corrosion
Pb and Ba salts	Catalyst for oil-based paints
Cu and Zn naphthenate	Wood preservatives
Co naphthenate	Curing agent in rubbers and resins Adhesion promoter of steel cord to rubber
Mn, Pb, Co, and Ca soap	Oxidative catalysts

Table 1. Industrial usage of NAs [36].

through the system together with operating temperature of 220–400°C favor corrosion [43, 46]. NAs decompose at temperatures above 400°C creating a protective film over the alloy [43, 46]. Corrosivity increases with increasing total acid number (TAN) of crude oil. Yet, the spread of corrosion by NAs rely not only on TAN but also on the availability of the carboxylic acid group and the type of compounds [43, 44, 46].

Very few reports have been disclosed discussing NAs in refinery wastewater due to their analytical difficulties although discharged wastewater from petroleum refineries. Low molecular weight NAs (typically <500 Daltons) are readily dissolved as their carboxylates in water at neutral and alkaline pH [34]. Bitumen extraction from the Athabasca oil sands deposit in northeastern Alberta, Canada releases NAs into tailings pond water [44]. Holowenko et al. [47] discovered that tailings pond water contain 20–120 mg NAs per liter, which is sufficient to create environmental pollution. Dorn [48] concluded that NAs concentrations greater than 2.5–5 mg per liter in refinery effluent would be toxic to fish. Rogers et al. [49] supposed that the liver was the primary target of NAs toxicity. The low molecular weight NAs are more capable to biodegradation than the high molecular weight NAs [35, 44]. Scott et al. [35] disclosed that the commercial NAs are more readily biodegradable compounds than the NAs present in the oil sand tailings water.

3. History of SCF and IL

3.1. History of SCF

Baron Charles Cagniard de la Tour in 1822 revealed supercritical phenomena [50] while carrying out experiments in a sealed cannon barrel filled with various fluids at various temperatures. G. Gore in 1861 showed that camphor and naphthalene dissolved in liquid carbon dioxide, whereas many carboxylic acids did not dissolved in liquid carbon dioxide. The term, critical point, was eventually coined by Thomas Andrew in 1869 [50] as the end of the vapor pressure curve in a phase diagram. The supercritical fluid (SCF) phenomena were first described by Hannay and Hogarth (1879) [50], who reported on the solubility of cobalt chloride in supercritical ethanol. Villard in 1896 [50] on his review of SCF solubility phenomena described the ability of methane, ethylene, carbon dioxide and nitrous oxide to dissolve a number of liquids and solid hydrocarbons such as carbon disulfide, camphor, stearic acid and paraffin wax. A few years later, E. H. Buchner (1906) [50] reviewed the literature and also made significant additions to the experimental database of high pressure SCF-solute mixtures. He became the first in a long line of researches to measure the solubility of a model compound, naphthalene, in supercritical carbon dioxide (sc-CO_2). Chemists continued to experiment with SCF for almost a century, although they remained something of a curiosity until the 1970s, when rising energy costs led chemists to consider SCF as a cheaper alternative to liquid extraction and distillation. Industrial use of SCFs as solvent began in 1950s. Nowadays the uses of supercritical fluid extraction (SFE) are largely apparent in the decaffeination of green coffee, food, nutraceuticals, perfumes and cosmetics, pharmaceuticals, textile, electronics, aerogels, ceramic and innovative materials, oil industry, laundry dry cleaning, analytical processes, like, supercritical fluid chromatography, nanoparticle and microparticle formation, generation of co-crystals in pharmaceutical processes, biodiesel production and several other new and upcoming applications [51].

Supercritical water (SCW) is water at a temperature and pressure above its critical points (374°C and 22.1 MPa). It is known as an environmentally benign solvent for a comprehensive variety of chemical reactions; it takes part in a reaction not only as a solvent but also as a reactant [52, 53]; thus, SCW is a logical choice to obstruct the precipitation of AS at the elevated cracking temperature beyond 400°C [54]. In addition, aromatic bonds—ethers and esters—and aliphatic bonds are freely broken in SCW [52, 55, 56]. Thermal reactions of heavy oil have been studied in SCW, showing that two major reactions occur under supercritical conditions: oxidation and hydrolysis [57–59]. These reactions allow tar to be decomposed successfully into useful chemical compounds in SCW [60, 61]. Hu et al. [62] treated oil shale using SCW showing a higher conversion and a larger oil recovery than toluene extraction process. SCW also delivered easier decomposition of the polar components in oil shale likened with supercritical toluene [63]. Park and Son [64] studied extraction of Athabasca oil sand bitumen in a microreactor using SCW, achieving a maximum 24% conversion at a temperature of 380°C, pressure of 30 MPa and a reaction time of 90 min. These results show that SCW can be used as an effective green solvent for the extraction and decomposition of heavy hydrocarbons.

Methanol's critical property data are Tc = 239.63°C, ρ_c = 277.49 kg/m^3 and Pc = 8.08 MPa [65], becoming supercritical with disappearing phase boundaries between liquid and gas at these conditions. Yadav and Chandra [66] discovered that the hydrogen bonding between methanol molecules decreases significantly with moving to the supercritical state from the ambient one. Supercritical methanol (SC-MeOH) can degrade polyethylene terephthalate (PET), which can be used for recycling waste plastics [67, 68]. SC-MeOH can also be used to produce biodiesel [69–70]. SC-MeOH can convert cellulose into methylated cellotriose, methylated cellobiose, methyl α- and β-D-glucosides, levoglucosan and 5-hydroxymethylfurfural [71]. Wang et al. [72] have studied the kinetics of esterification reaction between low-concentration NAs and methanol with or without using catalyst at temperatures of 180–280°C in an autoclave reactor. Acidity reduction of NAs using SC-MeOH without the addition of catalyst has been disclosed in open literature by Mandal et al. [33]. They have proved that approximately 100% TAN was reduced at a temperature of 350°C, a methanol partial pressure (MPP) of 10 MPa and a reaction time of 60 min. In addition, TAN reduction kinetics were followed first-order kinetics with Arrhenius parameters of activation energy 5.78 kcal/mol and a pre-exponential factor 1.56 s^{-1}.

The SCF processes are the currently most interesting environmental benign process [73]. The sc-CO_2 and the SCW have proved to be very efficient as a green extracting agent [74] and they are efficiently replacing harmful organic solvents nowadays. In addition, recent scientific reports have utilized sc-CO_2 for the synthesis of graphene-related materials. Moreover, continuous hydrothermal systems offer the ability to carry out synthesis in a high-throughput mode, enabling the discovery of new materials [75]. Furthermore, SCW can remove heavy metals from metalloporphyrin compounds effectively and efficiently in an environmentally friendly way [76]. Thus, the sc-CO_2 and the SCW have attracted tremendous interest among academia and industry in this modern era as a green process.

3.2. Ionic liquid (IL) and its history

An ionic liquid (IL) is ionic, slat-like materials. The ions in ILs are poorly coordinated resulting low melting point (below 100°C, or even at room temperature). An IL contains charged cation, a bulk organic structure with low symmetry and anion held together by Coulombic interactions. It is hard to break these bonds and make ILs evaporate. The extensively used cations are based on ammonium, sulfonium, phosphonium, imidazolium, pyridinium, picolinium, pyrrolidium, etc. with different substitutes. The anions of ILs may be halogen (first-generation ILs), organic ([CH3COO]$^-$, $[CF_3CO_2]^-$, [PhCOO]$^-$, etc.) or inorganic ($[BF_4]^-$, $[PF_6]^-$, $[AlCl_4]^-$, etc.). The properties of ILs depend on mutual fit of cation and anion, size, geometry and charge distribution. Usually, the cation has an impact on the hydrophobicity or hydrogen bonding ability and the anion controls the water miscibility [77, 78]. The structures of some commonly used ionic liquid systems are tabulated in **Figure 2** [78, 79].

The field of ionic liquid (IL) is not new, but their application as solvents in chemical processes for synthesis and catalysis has recently become important. This interesting field began in 1914 after preparation of ethylammonium nitrate ($[C_2H_5NH_3][NO_3]$) with the melting point (mp) of 12°C by Paul Walden [78–80]. This compound is recognized as the first IL by many researchers [80] but due to its high reactivity has not really found a use. After two decades of silence, IL appears in an US patent no. 1943176 in 1934. The writer of this patent claimed that the cellulose has dissolved in halide salts of nitrogen-containing bases, such as 1-benzylpyridinium chloride, 1-ethylpyridinium chloride, etc. at the temperature above 100°C. After another declined phase in the history of IL, it re-emerged in the period just after World War II in 1948 as an another US patent [79]. It was basically appeared in the open literature in 1951 [79] as the

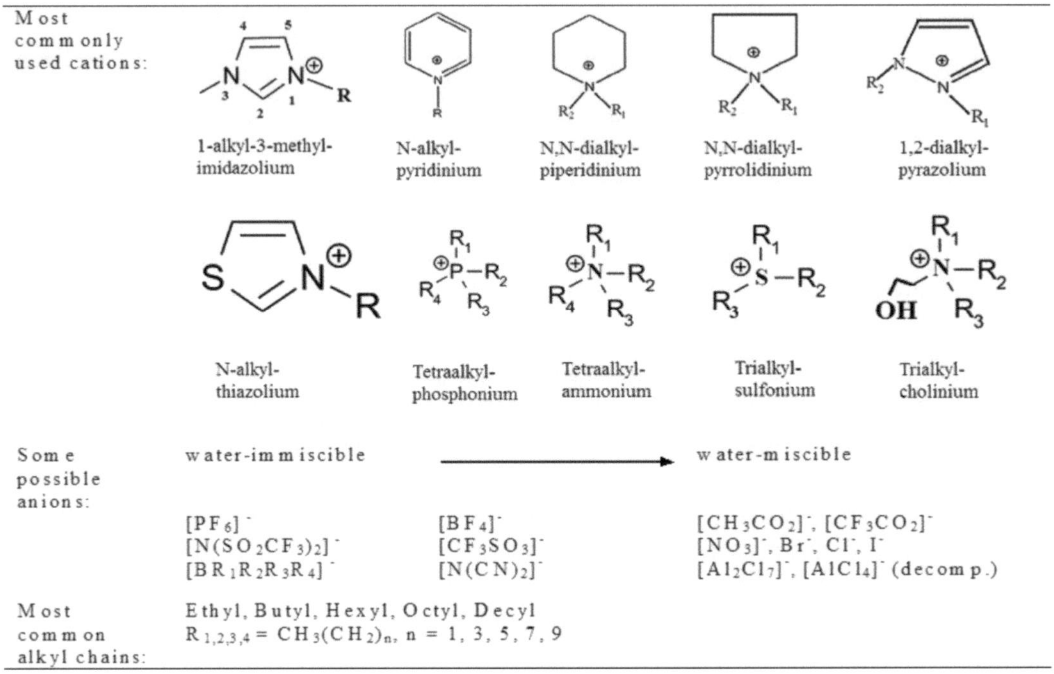

Figure 2. Some commonly used IL systems.

application of mixture of aluminum (III) chloride and 1-ethylpyridinium bromide to the electrodeposition of aluminum. But they have failed to investigate the IL at that time due to the complicated mixture of bromide and chloride salts. Later the physical and chemical properties of this IL were studied by Osteryoung group in 1975, aided by Bernard Gilbert [79]. The IL was created international attention in 1970s as aluminum chloride-based molten salts were utilized in 1940s for preparation of nuclear warheads batteries. Wilkes in 1970 attempted to develop better batteries for nuclear warheads and space probes that needed molten salts to operate [78]. The chemists of that period searched new salts that can remain liquid at lower temperatures as the molten salts were sufficiently hot to damage the neighboring materials. Wilkes and his co-workers carried on improving their ILs for using as battery electrolytes and in the long run they identified one low melting point IL [78]. Wilkes and Hussey discovered the $[C_2mim]Cl-AlCl_3$ IL system, liquid at room temperature, which was the first genuine example of IL system [79]. ILs became one of the most promising chemicals as solvents in the late 1990s.

ILs are versatile compounds due to their interesting properties. The properties of ILs are variable and can be in theory adjusted for any application, i.e., they are not all non-corrosive, but they can be designed to be non-corrosive. In general, they are characterized to have low vapor pressure, thermal and chemical stability and non-flammability. The research areas on ILs are growing very rapidly and it has numerous potential applications, such as, microwave-assisted organic synthesis, catalysis, biocatalysis, separation, extraction, electrochemistry, nanomaterials synthesis, polymerization reactions and corrosion inhibitors [81]. ILs can also offer surprising efforts to reassess and optimize existing technologies and processes related to the petroleum industry. A large number of research studies on the assessment and utilization of ILs in pollutant removal including sulfur-, nitrogen- and flour-containing compounds; aromatics; naphthenic acids and asphaltene of refinery feeds are disclosed in open literature.

4. TAN removal using SCF

4.1. Effect of temperature on TAN reduction using SCF

The initial TAN of NA, used raw materials for potential experiments by Mandal et al. [2, 33], was 241.55 (mg of KOH/g of NA). TAN reduction takes place at high temperatures and pressures in the presence of SCW. But, SC-MeOH can reduce TAN at lower temperature and pressure than SCW. Temperature has a considerable effect on TAN reduction using SCF. For exploring temperature effect on TAN reduction using SCW, Mandal et al. [2] carried out a lot of experiments at temperatures of 400–490°C and water partial pressure (WPP) of 25–45 MPa. In addition, SC-MeOH effect on TAN reduction was explored at temperatures of 300–350°C and MPP of 10 MPa [33].

Figure 3 illustrates the high TAN removal which increased with increasing reaction time and temperature. Type of SCF was also an important factor on TAN reduction process. Mandal et al. [2] in 2012 disclosed that approximately 89% TAN is reduced at reaction temperature of 490°C and reaction time of 60 min in treatment without water depositing approximately 4.05% solid which was nothing but lean hydrogen content high molecular weight hydrocarbons. Solid formation is the indication of loss of valuable compounds and thus potential

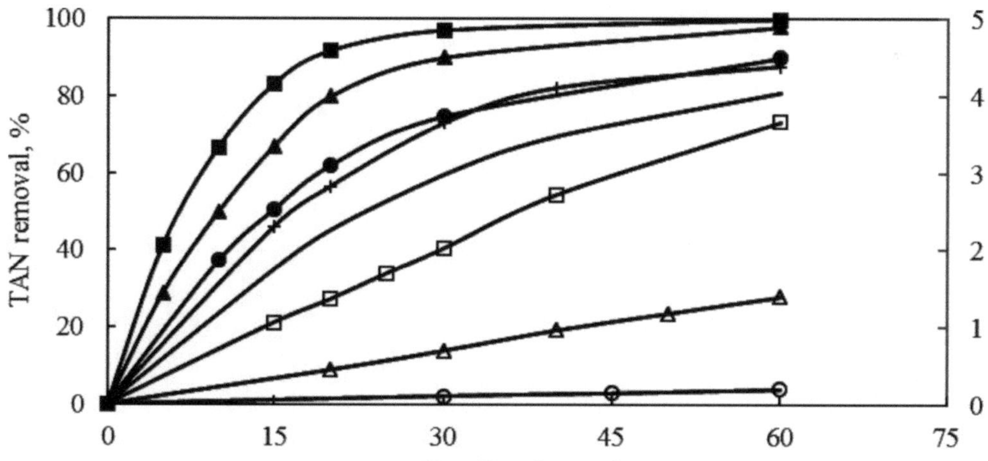

Figure 3. Variation of TAN as a function of reaction times, temperatures and water or methanol partial pressures (symbols: O, 400°C, WPP of 45 MPa; Δ, 450°C, WPP of 45 MPa; □, 490°C, WPP of 45 MPa; +, 490°C, no water; –, 490°C, no water (solid); •, 300°C, MPP of 10 MPa; ▲, 325°C, MPP of 10 MPa; ■, 350°C, MPP of 10 MPa).

revenue. Thus, it is desirable to reduce solid deposition from the economical point of view. SCW have the capability to reduce solid formation. Mandal et al. [2] do not observe solid deposition during their experiments using SCW at temperatures of 400–450°C. Solid deposition is detected at a temperature of 490°C, but depends on WPP and reaction time. The amount of solid deposition falls to 1.79 and 0.10% at WPPs of 30 and 45 MPa, respectively, indicating the drastic reduction of solid deposition at a WPP of 45 MPa. At extreme operating conditions (at temperature of 490°C, reaction time of 60 min and WPP of 45 MPa) of SCW treatment, the TAN reduction is reduced to approximately 73% which is less than the value (89%) obtained during pyrolysis without water.

4.2. Reaction kinetic study and mechanism

In 2012, Mandal et al. [2] conducted pioneering study on TAN removal using SCW. Their study revealed that TAN removed from NAs using SCW followed first-order kinetics. Mandal et al. [33] on another study showed that TAN removal from NAs using SC-MeOH also followed first-order kinetics. Wang et al. showed that the esterification reaction between NAs and subcritical methanol follows second-order kinetics with or without SnO catalyst in their study. The feedstock used by Wang et al. was the second vacuum fraction processed by a Chinese corporation having TAN content 3.69 (mg of KOH/g of feed stock), which was a mixture of NAs and higher hydrocarbons. The feedstock used by Mandal et al. [2, 33] was basically a blend of carboxylic acids having a TAN content 241.55 (mg of KOH/g of NA). The variation of properties of feed stock and reaction operating conditions were responsible for the variation of reaction order. **Figure 4** shows the Arrhenius type dependency of NAs removal rates on temperature. Based on such type of plot, Mandal et al. [2, 33] discovered the

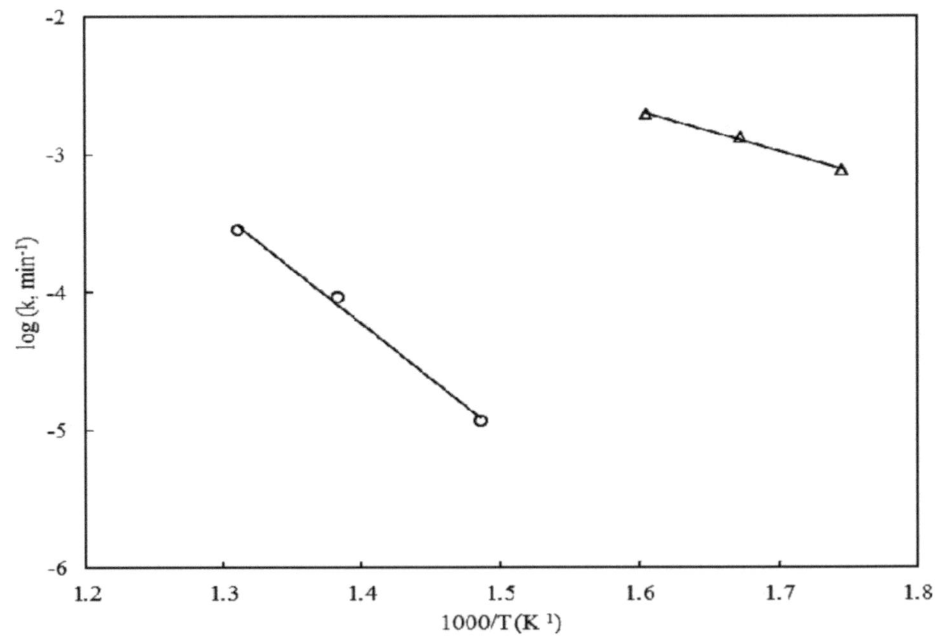

Figure 4. Arrhenius plot for reaction between NAs and SCF (symbols: O, SCW; Δ, SC-MeOH).

activation energy and pre-exponential factor. Thus, the temperature-dependent rate constant can be expressed by the following Arrhenius equations:

$$k_T(T) = 1.43 \times 10^5 e^{\frac{-66.24}{RT}}, \text{ for SCW} \quad (1)$$

$$k_T(T) = 1.56 e^{\frac{-5.78}{RT}}, \text{ for SC-MeOH} \quad (2)$$

The respective activation energies of the reaction of NAs with SCW and SC-MeOH are 66.24 and 5.78 kcal per mol. The activation energy of NAs for reacting with SCW is 91.27% higher than the activation energy of NAs for reacting with SC-MeOH at lower pressure. This indicates that 91.27% higher energy is required for the reaction of NAs with SCW.

According to the study of Mandal et al. [2, 33], SCW and SC-MeOH reduce acidity of NAs in different reactions and pathways (**Figure 5**). SCW acts as a hydrogen donor [82] and it can remove metal from metalloporphyrin where hydrogenation reaction is involved. NAs react with SCW to form stable compounds [2] and gaseous product (CO_2, CO, petroleum gas) where hydrogenation reaction is happening to occur. Based on the analysis using GC/MS and MALDI TOF/MS, Mandal et al. in 2012 [2] showed that hydrocarbons are formed through decarboxylation of NA followed by long-chain scission and formation of high molecular weight hydrocarbons. It is also observed that SCW can produce low molecular weight hydrocarbons suppressing char formation and production of high molecular weight hydrocarbons from alkylbenzene and NAs [2, 82]. NA is not hindered the quality of petroleum oil except corrosion problem.

On the other hand, SC-MeOH without a distinct catalyst can reduce acidity promptly by forming ester (like nonanoic acid, methyl ester; decanoic acid, methyl ester and so on), saturated alkylbenzene (like heptylbenzene) and alkane (like decosane). SC-MeOH as an excellent solvent can dissolve feedstock so that reactant molecules are in mutual close proximity to react readily. At the first step of reaction, SC-MeOH reacts with NAs to form ester and water. The thermal stability of ester depends on its carbon skeleton [33, 83]. Mandal et al. in 2013 [33] observed that concentration of saturated alkylbenzene and alkane is increasing with increasing reaction time. In the second step of reaction, the thermal effect of SC-MeOH can

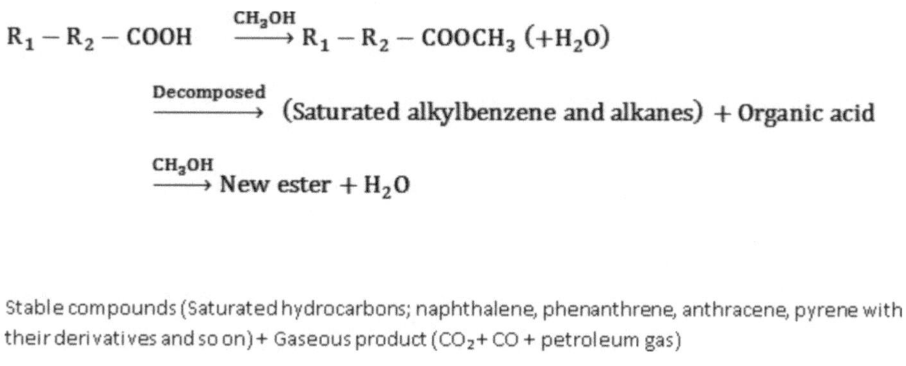

Figure 5. Reaction mechanism of NAs with SCF.

decompose unstable ester to form saturated alkylbenzene and alkane and produce new ester. In the third step of reaction, produced esters can react with SC-MeOH to form new ester and water.

5. TAN reduction using IL

Literature study revealed that NAs are successfully removed from crude fractions by aqueous base washing [84, 85]. But serious emulsion problems are experienced during operation.

Shi et al. [86] has proposed a fancy method to isolate NAs from highly acidic crude oils by forming ILs. In this method, the researchers prepared BrØnsted ILs by acid-base reaction between basic imidazole and NAs to form naphthenates ILs according to the following reaction:

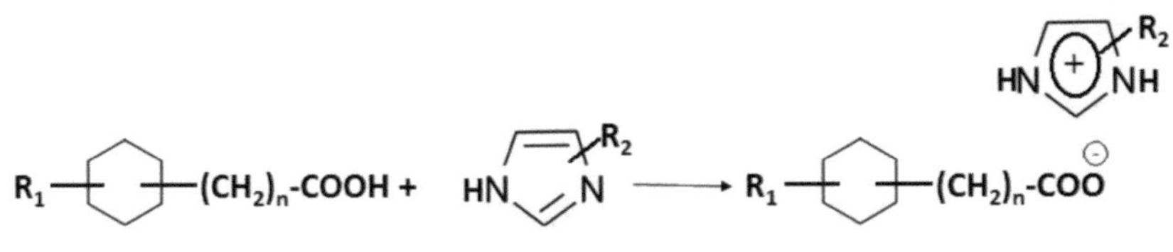

(3)

The acid removal rate has been affected by the 2-methylimidazole content, reagent/oil ratio, reaction time and reaction temperature. All of them had a positive effect on the acid removal rate except the reagent/oil ratio which had a negative effect on the oil yield rate. The same approach was employed by Baden Aniline and Sida Factory (BASF), the largest Ludwigshafen, Germany based chemical company in the world to develop the first successful large-scale industrial application of ILs in 2003 [81] through the process biphasic acid scavenging ILs (BASIL). In this process, they are using N-alkylimidazole derivative to scavenge acid that is formed in the alkoxyphenylphosphines production process as a by-product forming ILs. The reaction results in the formation of the IL N-alkylimidazolium chloride having a melting point of 75°C. The IL separates as a clear liquid phase from the pure product and is reused.

Crude oil can be deacidified with a basic IL [87]. The basic IL is a liquid salt having the general formula $[C^+][A^-]$, where C^+ is a cation having a basic moiety, preferably represented by a formula $[Cat^+\text{-}(Z\text{-}Bas)_n]$, and/or A^- is an anion having a basic moiety, preferably represented by $[An^-\text{-}(Z\text{-}Bas)]$. Here, Cat^+ is a positively charged moiety, An^- is an anionic moiety, Bas is a basic moiety, Z is a covalent bond joining Cat^+/An^- and Bas or is a divalent linking group and n is an integer from 1 to 3. The Cat^+ moiety comprises a heterocyclic ring structure selected from imidazolium, pyridinium, pyrazolium and so on. The An^- moiety comprises a CO_2^- or SO_3^- group. Base comprises at least one basic nitrogen, phosphorus, sulfur or oxygen atom and preferably selected from $-N(R_1)(R_2)$, $-P(R_1)(R_2)$, $-S(R_3)$, $-O(R_3)$, where R_1 and R_2 are independently selected from hydrogen, linear or branched alkyl, cycloalkyl, aryl and substituted ring and R_3 is selected from linear or branched alkyl, cycloalkyl, aryl and substituted aryl. Z comprises divalent organic radical and preferably selected from a divalent alkylene radical, alkyleneoxyalkylene radical, polyoxyethylene radical and alkylenearylene or alkylenearylenealkylene radical.

Sun and Shi in 2012 [88] revealed that the ILs of imidazole anion having strong alkalinity had excellent performance on deacidification. They discovered that the lower the agent/oil mass ratio, the higher the NAs conversion and high temperature could decrease the reagent/oil ratio to achieve maximum deacidification rate with each IL. When [AMIm]Im is used, the performance of deacidification succeeded the order [OMIm]Im > [HMIm]Im > [BMIm]Im > [EMIm] Im. Duan et al. in 2013 [89] have studied on pyridinium, imidazolium and imidazolide-based ILs at different alkalinity. They showed that the stronger alkalinity of ILs was, the higher deacidification would be and the deacidification succeeded the order [AMIm]Im > [AMIm] Br > [APy]Br. In addition, when the length of alkyl chain increased, each kind of ILs would have more excellent performance on deacidification.

Sun et al. [88, 89] explained NAs removal mechanism using ILs with the help of techniques proposed by Holbrey et al. [90]. Holbrey et al. [90] revealed that IL molecules themselves could form cage structure through specific chemical bonds. In this cage, the target molecules would be captured through the formation of liquid clathrate due to the π-π interaction between ILs and target molecules. This phenomenon would be helpful to explain the following facts: (1) the larger the π interaction between ILs and NAs, the stronger the IL alkalinity which means of great electron density, the better the effect of deacidification would be. (2) For the same anion containing ILs, the longer alkyl chain of ILs, the higher the deacidification activity as well as alkalinity, the more inclusive for NAs by increasing the space of the cage. Thus, NAs conversion increased with the larger of alkyl chain of the ILs. (3) Unregenerated ILs indicated good performance on deacidification in the following iterative reactions because each ILs molecules could adapt more than one NA molecules. The NAs could not be withdrawn completely from low NA containing oil as the surrounding alkane molecules made NA molecule difficult to penetrate into the clathrate structure of ILs. When fresh acidic oil added into this system, the surrounding alkane structure was broken and ILs showed good performance again. (4) The NA molecules could easily escape from the cage structure at high temperature due to unfavorable adsorption process at this condition. So, the reuse of IL would be decreased at such temperature condition.

Andersen et al. [12] have studied a number of tetraalkylammonium and tetraalkylphosphonium amino acid-based ILs to remove NAs from crude oil. A model oil having TAN 4.00 mg KOH g_{oil}^{-1} has been charged in a glass vessel and heated the same at a temperature of 25°C for 1 h. An exact amount of ILs has then been mixed with the oil for reaction for 20 min. The percentage of NA removal obtained from the experiments is summarized in **Table 2**.

Table 2 revealed that all AAILs have showed excellent and swift acid removal after 20 min except $[NTf_2]^-$-based ILs. The low extraction efficiency of $[NTf_2]^-$-based ILs implies its weak interaction between the IL and the NA. There is a little effect of the cation type on acid removal

IL anion	NA Removal (%)	
	Tributylmethylammonium $[N_{4441}]^+$	Tetrabutylphosphonium $[P_{4444}]^+$
Bis (trifluoromethanesulfonyl)imide ($[NTf_2]^-$)	2	3
Serinate ($[Ser]^-$)	31	32
Threonate ($[Thr]^-$)	35	37
Valinate ($[Val]^-$)	21	31
Cystinate ($[Cys]^-$)	26	27
Prolinate ($[Pro]^-$)	36	38
Lysinate ($[Lys]^-$)	46	41
Taurinate ($[Tau]^-$)	37	32

Table 2. Percentage of NA removal using a range of AAILs at a temperature of 25°C [12].

although $[P_{4444}]^+$ have the pioneering extraction efficiency due to their lower viscosity than the other. It is noticeable that $[N_{4441}]$ [Val] have the lowest acid removal capability than $[P_{4444}]$[Val] due to the highest melting point of $[N_{4441}]$ [Val] compared to $[P_{4444}]$[Val]. $[Lys]^-$-based ILs have two amine groups per molecule of IL that can extend its acid removal capability. The other functional groups present on the anion such as hydroxyl group (–OH) in $[Ser]^-$ and $[Thr]^-$ and thiol (–SH) group in $[Cys]^-$-based ILs have a little impact on the corresponding extraction efficiency.

The following two reactions are the possible reactions for removing NA using AAIL:

$$[Cation]\left[H_2 - C(R) - COO\right] + R' - COOH \leftrightharpoons [Cation]\left[OOC - C(R) - NH_3\right][OOC - R'] \tag{4}$$

$$[Cation]\left[H_2 - C(R) - COO\right] + R' - COOH \leftrightharpoons [Cation][OOC - R'] + NH_2 - C(R) - COOH \tag{5}$$

In the first reaction, the amino acid anion constructs IL-NA complex known as zwitterionic species by protonating the amino group and forming the carboxylate anion. In the second reaction, the protonation of amino acid anion and ion exchange with NA form the naphthenate salt releasing the amino acid. Andersen et al. [12] have confirmed based on infrared spectroscopy that the reaction mechanism involves a stoichiometric reaction between one amino group and one carboxylic acid group forming a zwitterionic species. Lysine anion contains two amino groups. Therefore, lysine-derived AAILs showed higher extraction capability compared to other AAILs.

Based on current studies, SCF process is more effective than IL-based process for reducing TAN of NAs. SC-MeOH without the addition of catalyst can reduce approximately 100% TAN at a temperature of 350°C, a methanol partial pressure (MPP) of 10 MPa and a reaction time of 60 min [33]. On the other hand, the IL, tributylmethylammonium lysinate, can reduce approximately 46% NA at a temperature of 25°C [12]. A lot of ILs are available in open market or possibility of production in laboratory that can be used to reduce TAN of NAs at room temperatures at high extent. The benefit of using ILs is needed room or low temperature to reduce TAN of NAs. In addition, it is possible to separate ILs after treatment reducing cost of process. Therefore, IL-based processes are attracting the researchers nowadays.

6. Conclusion

A number of TAN reduction methods have been developed for reducing the TAN content of crude oil to maintain TAN content of crude oil below the integer value 2. The SCF process and the IL-based process, known as environmental benign processes, are new processes in this area. SCF without the addition of catalyst can reduce TAN of NAs and substantial amounts of NAs are converted into non-acidic products. The reaction kinetics is consistent with first-order dependency on TAN removal. Like SCF, ILs can also reduce acidity of heavy oils either by forming zwitterionic species or creating cage structure around NAs through specific chemical bonds. In comparison, SCF process is more efficient than IL-based process for reducing TAN of NAs. Therefore, IL-based processes are attracting the researcher's attention due to their availability in laboratory or in open market and separability tendency after using.

Acknowledgements

This work was supported by the Short-Term Internal Research Fund (STIRF), Universiti Teknologi PETRONAS (UTP), Malaysia.

Author details

Pradip Chandra Mandal[1] and Mitsuru Sasaki[2*]

*Address all correspondence to: msasaki@kumamoto-u.ac.jp

1 Department of Pipeline Construction, Titas Gas T & D Co. Ltd., Dhaka, Bangladesh

2 Institute of Pulsed Power Science, Kumamoto University, Kumamoto, Japan

References

[1] Bienstock MG, Matragrano JG, Patel RD, Bearden Jr. R. Thermal process for reducing total acid number of crude oil. US Pat. No.: 6,086,751; 2000

[2] Mandal PC, Wahyudiono, Sasaki M, Goto M. Reduction of total acid number (TAN) of naphthenic acid (NA) using supercritical water for reducing corrosion problems of oil refineries. Fuel. 2012;**92**:288

[3] Hau J. Predicting sulfidic and naphthenic acid corrosion. Corrosion. 2009;**65**(12):831-844

[4] Blum SC, Olmstead WN, Bearden R. Thermal decomposition of naphthenic acids, US Pat No.: 5,820,750; 1998

[5] Smith DF, Rodgers RP, Rahimi P, Teclemariam A, Marshall AG. Effect of thermal treatment on acidic organic species from Athabasca bitumen heavy vacuum gas oil, analyzed by negative-ion electrospray Fourier transform ion cyclotron resonance (FT-ICR) mass spectrometry. Energy & Fuels. 2009;**23**:314-319

[6] Petkova NB, Angelova MK, Tasheva JTS. Possibilities for substitution of the applied inhibiting system including copper-ammonia carbonate complex for blocking the corrosive components in the crude oil processing. Oxidation Communications. 2010;**33**(4):926-930

[7] Blum SC, Olmstead WN, Bearden R. Thermal decomposition of naphthenic acids. US Pat. No.: 5,820,750; Oct., 1998

[8] de O Gomes A, Guimaraes RCL, Batista IP. Total acid number distribution in a Brazilian refinery oil sample after exposure to different levels of heating in the process units, preprints-American Chemical Society. Division of Petroleum Chemistry. 2009;**54**(1):26-29

[9] Petkova N, Angelova M, Petkov P. Establishing the reasons and type of the enhanced corrosion in the crude oil atmospheric distillation unit. Petroleum and Coal. 2009;**51** (4):286-292

[10] Braccknell SND, Osborne CG. Naphthenic acid corrosion in oil refineries. U.S. Patent 6,464,859 (Oct. 15, 2002)

[11] Duncum SN, Osborne CG. Process for deacidifying a crude oil system. US 6,464,859 B1 (Oct. 15, 2002)

[12] Andersen K, Goodrich P, Hardacre C, Hussain A, Rooney DW, Wassell D. Removal of naphthenic acids from crude oil using amino acid ionic liquids. Fuel. 2013;**108**:715-722

[13] Gillespie RD, Arena BJ. Naphthenic acid removal as an adjunct to liquid hydrocarbons sweeting, U.S. Patent 5,389,240; 1995

[14] Huagn Y-Z, Zhu J-H, Wang Y-Y, Wu B-C. Removal of naphthenic acids from high acid crude oil through catalytic esterification on zinc-aluminum hydrotalcite-like compounds. Journal of China University of Petroleum (Edition of Natural Science). 2010;**34**(6) 147-150+156

[15] Sartori G, Savage DW, Blum SC, Dalrymple DC, Wales WE. Metal compounds as accelerators for petroleum acid esterification, U.S. Patent 5,948,238; Sept. 7, 1999

[16] Wang YZ, Liu YP, Liu CG. Removal of naphthenic acids of a second vacuum fraction by catalytic esterification. Petroleum Science and Technology. 2008;**26**:1424-1432

[17] Lazar A, Evans JM. Process of treating hydrocarbon oils. U.S. Patent 1,953,353; Apr. 3, 1934

[18] Takemura Y, Nakamura A, Taguchi H, Uchi K. Catalytic decarboxylation of benzoic acid. Industrial and Engineering Chemistry Product Research and Development. 1985;**24**:213-215

[19] Zhang A, Ma Q, Wang K, Liu X, Shuler P, Tang Y. Naphthenic acid removal from crude oil through catalytic decarboxylation on magnesium oxide. Applied Catalysis A. 2006; **303**:103-109

[20] Halbert TR, Riley KL, Trachte KL, Vannauker DL. Process for reduction of total acid number in crude oil. U.S. Patent 5,910,242; 1999

[21] Thomas R, Kenneth LR, Kenneth LT, David L. Contacting with a hydrotreating catalyst in the presence of hydrogen treat gas containing hydrogen sulfide. U.S. Patent 5,910,242 A; Jun 8, 1999

[22] Natal Santiago MA, Sanchez-Castillo MA, Cortright RD, Dumesic JA. Catalytic reduction of acetic acid, methyl acetate, and ethyl acetate over silica-supported copper. Journal of Catalysis. 2000;**193**:16-28

[23] Maier W, Roth W, Thies I, Van Rague Schleyer P. Gas phase decarboxylation of carboxylic acids. Chemische Berichte. 1982;**115**:808-812

[24] Kubickova I, Snare M, Eranen K, Maki-Arvela P, Murzin DY. Hydrocarbons for diesel fuel via decarboxylation of vegetable oils. Catalysis Today. 2005;**106**:197-200

[25] Snare M, Kubickova I, Maki-Arvela P, Eranen K, Murzin DY. Heterogeneous catalytic deoxygenation of stearic acid for production of biodiesel. Industrial and Engineering Chemistry Research. 2006;**45**:5708-5715

[26] Manyar HG, Paun C, Pilus R, Rooney DW, Thompson JM, Hardacre C. Highly selective and efficient hydrogenation of carboxylic acids to alcohols using titania supported Pt catalysts, Chemical Communications 46 (2010) 6279-6281

[27] Burch R, Paun C, Cao X-M, Crawford P, Goodrich P, Hardacre C, et al. Catalytic bimetallic Pt/re-based catalysts. Journal of Catalysis. 2011;**283**:89-97

[28] Wang Y, Chu Z, Qiu B, Liu C, Zhang Y. Removal of naphthenic acids from a vacuum fraction oil with an ammonia solution of ethylene glycol. Fuel. 2006;(17-18):2489-2493

[29] Lu R, Xu X, Yang J, Gao J. Reduction of total acid number of crude oil and distillate. Energy Sources, Part A: Recovery, Utilization, and Environmental Effects. 2007;(1):47-57

[30] Shi L, Shen B, Wang G. Removal of naphthenic acids from Beijing crude oil by forming ionic liquids. Energy & Fuels. 2008;(6):4177-4181

[31] Laredo GC, Altamirano E, De los Reyes JA. Inhibition effects of nitrogen compounds on the hydrodesulfurization of dibenzothiophene: Part 2. Applied Catalaysis A. 2003; **243**:207-214

[32] Collins IR, Plechkova N, Anderson K. Process for removing sulfur-containing acids from crude oil. WO Patent 138307:A2; 2007

[33] Mandal PC, Wahyudiono, Sasaki M, Goto M, Non-catalytic reduction of total acid number (TAN) of naphthenic acids (NAs) using supercritical methanol. Fuel Processing Technology. 2013;**106**:641-644

[34] Clemente JS, Prasad NGN, MacKinnon MD, Fedorak PM. A statistical comparison of naphthenic acids characterized by gas chromatography-mass spectrometry. Chemosphere. 2003;**50**:1265-1274

[35] Scott AC, MacKinnon MD, Fedork PM. Naphthenic acids in Athabasca oil sands tailing waters are less biodegradable than commercial naphthenic acids. Environmental Science & Technology. 2005;**39**:8388-8394

[36] Headley JV, McMartin DW. A review of the occurrence and fate on naphthenic acids in aquatic environments. Journal of Environmental Science and Health. 2004;**39**(8):1989-2010

[37] Brient JA, Wessner PJ, Doyle MN. Naphthenic acids. In: 4th E, Kroschwitz JI, editors. Kirk-Othmer Encyclopaedia of Chemical Technology. New York, NY: John Wiley and Sons; 1995. pp. 1017-1029

[38] Herman DC, Fedorak PM, Costerton JW. Biodegradation of cycloalkane carboxylic acids in oil sands tailings. Canadian Journal of Microbiology. 1993;**39**:576-580

[39] Headley JV, Peru KM, McMartin DW, Winkler M. Determination of dissolved naphthenic acids in natural waters using negative-ion electrospray mass spectrometry. Journal of AOAC International. 2002;**85**:182-187

[40] Whelan JK, Farrington JW. Organic Matter: Productivity, Accumulation and Preservation in Recent and Ancient Sediments. New York, NY: Columbia University Press; 1992

[41] Tissot BP, Welte DH. Petroleum Formation and Occurrence. Berlin, Germany: Springer-Verlag; 1984

[42] Snowdon L, Lloyd R, Powell TG. Immature oil and condensate — Modification of hydrocarbon generation model for terrestrial organic matter. AAPG Bulletin. 1982;**66**(6):755-788

[43] Kane RD, Cayard MS. Understanding critical factors that influence refinery crude corrosiveness. Materials Performance. 1999;**38**(7):48-54

[44] Clemente JS, Fedorak PM. A review of the occurrence, analyses, toxicity, and biodegradation of naphthenic acids. Chemosphere. 2005;**60**:585-600

[45] Slavcheva E, Shone R, Turnbull A. Review of naphthenic acid corrosion in oil refining. British Corrosion Journal. 1999;**34**:125-131

[46] Turnbull A, Slavcheva E, Shone B. Factors controlling naphthenic acid corrosion. Corrosion. 1998;**54**:922-930

[47] Holowenko FM, MacKinnon MD, Fedorak PM. Characterization of naphthenic acids in oil sands wastewaters by gas chromatography-mass spectrometry. Water Research. 2002;**36**:2843-2855

[48] Dorn PB. Case histories – The petroleum industry. In: Toxicity reduction: Evaluation and control. In: Ford DL, editor. Water Quality Management Library. Vol. 3. Lancaster, PA: Technomic Publishing Company; 1992. pp. 183-223

[49] Rogers VV, Wickstrom M, Liber K, MacKinnon MD. Acute and subchronic toxicity of naphthenic acids from oil sands tailings. Toxicological Sciences. 2002;**66**:347-355

[50] Berche B, Henkel M, Kenna R. Critical phenomena: 150 years since Cagniard de la Tour. Journal of Physical Studies. 2009;**13**(3) 3001(4 p.)

[51] Marchetto F, Aziz AA. A review of supercritical fluid extraction technology and application. Journal Teknologi. 2014;**69**(6):27-31

[52] Savage PE. Organic chemical reactions in supercritical water. Chemical Reviews. 1999;**99**:603-622

[53] Akiya N, Savage PE. Roles of water for chemical reactions in high-temperature water. Chemical Reviews. 2002;**102**:2725-2750

[54] Cheng ZM, Ding Y, Zhao LQ, Yuan PQ, Yuan WK. Effect of supercritical water in vacuum residue upgrading. Energy & Fuels. 2009;**23**:3178-3183

[55] Tagaya H, Suzuki Y, Kadokawa J, Karasu M, Chiba K. Decomposition of model compounds of phenol resin waste with supercritical water. Chemistry Letters. 1997;**26**:47-48

[56] Lachance R, Paschkewitz J, DiNaro J, Tester JW. Thiodiglycol hydrolysis and oxidation in sub- and supercritical water. Journal of Supercritical Fluids. 1999;**16**:133-147

[57] Goto M, Nada T, Ogata A, Kodama A, Hirose H. Supercritical water oxidation for the destruction of municipal excess sludge and alcohol distillery wastewater of molasses. Journal of Supercritical Fluids. 1998;**13**:277-282

[58] Sasaki M, Kabyemela B, Malaluan R, Hirose S, Takeda N, Adschiri T. Cellulose hydrolysis in subcritical and supercritical water. Journal of Supercritical Fluids. 1998;**13**: 261-268

[59] Adschiri T, Shibata R, Arai K. Phenol recovery by BPA tar hydrolysis in supercritical water. Sekiyu Gakkaishi. 1997;**40**:291-297

[60] Wahyudiono, Fujinaga S, Sasaki M, Goto M. Recovery of phenol through the decomposition of tar under hydrothermal alkaline conditions. Chemical Engineering and Technology. 2006;**29**:882-889

[61] Wahyudiono MS, Goto M. Non-catalytic liquefaction of tar with low temperature hydrothermal treatment. Journal of Material Cycles and Waste Management. 2007;**9**:173-181

[62] Hu H, Zhang J, Guo S, Chen G. Extraction of Huadian oil shale with water in sub- and supercritical states. Fuel. 1999;**78**:645-651

[63] Funazukuri T, Yokoi S, Wakao N. Supercritical fluid extraction of Chinese Maoming oil shale with water and toluene. Fuel. 1988;**67**:10-14

[64] Park JH, Son SH. Extraction of bitumen with sub- and supercritical water. Korean Journal of Chemical Engineering. 2011;**28**(2):455-460

[65] Abdulagatov IM, Polikhronidi NG, Abdurashidova A, Kiselev SB, Ely JF. Thermodynamic properties of methanol in the critical and supercritical regions. International Journal of Thermophysics. 2005;**26**(5):1327-1368

[66] Yadav VK, Chandra A. Dynamics of supercritical methanol of varying density from first principles simulations: Hydrogen bond fluctuations, vibrational spectral diffusion, and orientational relaxation. The Journal of Chemical Physics. 2013 Jun 14;**138**(22)

[67] Goto M, Koyamoto H, Kodama A, Hirose T. Degradation kinetics of polyethylene terephthalate in supercritical methanol. AICHE Journal. 2002;**48**(1):136-144

[68] Yang Y, Lu Y, Xiang H, Xu Y, Li Y. Study on methanolytic depolymerization of PET with supercritical methanol for chemical recycling. Polymer Degradation and Stability. 2002;**75**:185-191

[69] Demirbas A. Biodiesel production from vegetable oil by supercritical methanol. Journal of Scientific & Industrial Research. 2005;**64**:858-965

[70] Wei C-Y, Huang T-C, Chen H-H. Biodiesel production using supercritical methanol with carbon dioxide and acetic acid. Journal of Chemistry. 2013;**2013**:1-6

[71] Ishikawa Y, Saka S. Chemical conversion of cellulose as treated in supercritical methanol. Cellulose. 2001;**8**(3):189-195

[72] Wang Y-Z, Liu Y-P, Liu C-G. Kinetics of the esterification of low-concentration naphthenic acids and methanol in oils with or without SnO as a catalyst. Energy & Fuels. 2008;**22**:2203-2206

[73] Knez Ž, Markočič E, Leitgeb M, Primožič M, Knez Hrnčič M, Škerget M. Industrial applications of supercritical fluids: A review. Energy. 2014;**77**:235-243. DOI: 10.1016/j.energy.2014.07.044

[74] Andrade CKZ, Dar AR. Applying green processes and techniques to simplify reaction work-ups. Tetrahedron. 2016;**72**(47):7375-7391. DOI: 10.1016/j.tet.2016.09.055

[75] Patel D, Kellici S, Saha B. Green process engineering as the key to future processes. PRO. 2014;**2**:311-332. DOI: 10.3390/pr2010311

[76] Mandal PC, Goto M, Sasaki M. Removal of nickel and vanadium from heavy oils using supercritical water. J. Jpn. Petrol. Inst. 2014;**57**(1):18-28

[77] Huddleston JG, Visser AE, Reichert WM, Willauer HD, Broker GA, Rogers RD.Characterization and comparison of hydrophilic and hydrophobic room temperature ionic liquids incorporating the imidazolium cation. Green Chemistry. 2001;**3**:156-164

[78] Keskin S, Kayrak-Talay D, Akman U, Hortacsu O. A review of ionic liquids towards supercritical fluid applications. Journal of Supercritical Fluids. 2007;**43**:150-180

[79] Plechkova NV, Seddon KR. Application of ionic liquid in the chemical industry. Chemical Society Reviews. 2008;**37**:123-150

[80] Koel M. Ionic liquid in chemical analysis. CRC Press (ISBN 1-4200-4646-2); 2008. p. XXVII

[81] Martinez-Palou R, Luque R. Applications of ionic liquids in the removal of contaminants from refinery feedstocks: An industrial perspective. Energy & Environmental Science. 2014;**7**:2414-2447

[82] Mandal PC, Shiraishi T, Wahyudiono, Sasaki M, Goto M. Heptylbenzene decomposition in supercritical water: A simulation study. International Journal of Engineering & Technology (IJET). 2010;**7**(4):689-699

[83] Shopova N, Milkova T. Thermal decomposition of cholesteryl esters of cinnamic acid derivatives and their effect on the a-tetralylhydroperoxide free-radical-induced decomposition. Thermochimica Acta. 2000;**356**:101-107

[84] Varadaraj R, Savage DW. Process for neutralization of petroleum acids, U.S. Patent 6,030,523, 2000

[85] Sartori G, Savage DW, Balinger BH. Process for decreased the acidity of crudes using crosslinked polymeric amines. U.S. Patent 6,121,411; 2000

[86] Shi LJ, Shen BX, Wang GQ. Removal of naphthenic acids from Beijiang crude oil by forming ionic liquids. Energy & Fuels. 2008;**22**:4177-4181

[87] Hamer CK et al. Process of deacidifying crude oil. EP patent 1, 911, 829A1, Bulletin 2008/16. pp. 1-20

[88] Sun Y, Shi L. Basic ionic liquids with imidazole anion: New reagents to remove naphthenic acids from crude oil with high total acid number. Fuel. 2012;**33**:83-87

[89] Duan J, Sun Y, Shi L. Three different types of heterocycle of nitrogen-containing alkaline ionic liquids treatment of acid oil to remove naphthenic acids. Catalysis Today. 2013;**212**:180-185

[90] Holbrey JD, Reichert WM, Nieuwenhuyzen M, Sheppard O, Hardacre C, Rogers RD. Liquid clathrate formation in ionic liquid-aromatic mixtures. Chemical Communications. 2003:476-477

Chapter 13

Conducting Polymers Films Deposited on Carbon Steel and Their Interaction with Crude Oil

Oscar E. Vázquez-Noriega, Javier Guzmán,
Nohra V. Gallardo-Rivas,
Reinaldo David Martínez Orozco,
Ana M. Mendoza-Martínez,
María Yolanda Chávez Cinco,
Luciano Aguilera Vázquez and Ulises Páramo-García

Additional information is available at the end of the chapter

http://dx.doi.org/10.5772/intechopen.70091

Abstract

The formation of scale/solids deposits inside the pipelines is a frequent problem in the petrochemical industry. These scales can be organic as the asphaltenes and inorganic as the accumulations of salts, which apart from blocking the inside of the pipes can also cause a change in the integrity of the steel. Therefore, it is necessary to avoid the conditions where deposition occurs, together with chemical and mechanical methods of remediation to mitigate the deposition. In this work we intend to use conductive polymers in order to inhibit the deposition of asphaltenes on carbon steel surfaces, by using polypyrrole (PPy) as material capable of conducting electrical current. The electrodeposition of PPy on carbon steel were performed by cyclic voltammetry (CV) and chronoamperometry (CA). The results showed that under certain experimental conditions it is possible to make a PPy film with adequate characteristics. Important factors were the grip and electrochemical stability of the formed film on steel, which depends on the electrosynthesis technique and in some cases favoured by a pre-treatment with a 10% HNO3 solution applied to the steel prior to electropolymerization. The PPy films deposited with pre-treatment completely covered the steel surface and showed better stability, adherence and generated a hydrophobic material.

Keywords: conducting polymers, corrosion protections, carbon steel, polypyrrole, asphaltenes, crude oil

1. Introduction

The formation of solid deposits is a frequent problem in the hydrocarbons transport due to asphaltenes precipitation where the crude oils flow. The asphaltenes are the fraction of higher molecular weight and polarity of the crude oil and they are responsible for these solid deposits, which usually give rise to several problems in the transport and processing of crude oil, mainly by the obstruction of the pipelines [1–6].

Common oilfield scales are crystalline deposits, resulting from the precipitation of mineral compound such as carbonates and sulfates present in the injected water, as method for enhanced oil recovery. These scale deposits may appear as a thick layer adhered to the inner walls of the pipes, they are often several centimeters thick with particles sizes up to 1 cm or more. One of the primary problems of scale formation in pipes is the reduction in flow rate by increasing the inner surface roughness of the pipe and reducing the flow ability area. This produces an increase in the pressure drop and consequently production decreases. By increasing the growth of deposited minerals, it becomes impossible to access deeper pipe sections, and finally the scale deposits end up blocking the flow of production. On the other hand, organic scales formation results from asphaltenes and waxes deposition. Usually, the localized corrosion can be developed beneath or around these deposits present in the steel surfaces, due to the presence of bacteria or sulfurous gas, which reduces the integrity of the metal.

The petroleum in its natural state is a mixture of organic compounds of varied structures and different molecular weights. In general, it is possible to group the oil constituents into four well-defined organic components: saturated, aromatic, resin, and asphaltenes [7], known as SARA. The study of the heavy oil fraction (asphaltenes) has increased in recent years due to the problems that they represent in the production and conversion processes. In general, the structure of asphaltenes is considered to consist of a condensed aromatic nucleus with side alkyl chains and heteroatoms incorporated into many of the cyclic structures. The condensed aromatic system may contain from 4 to 20 benzene rings [8]. Nowadays, there is a considerable debate about the structure of asphaltenes, particularly in the size of aromatic groups and how they are linked to other groups in the structure. The type and amount of deposits of heavy organic compounds varies depending on the hydrocarbons present, and the relative amount of each organic family involved. In general, asphaltenes deposition can be explained in detail on the basis of four effects or mechanisms: (1) effect of polydispersity, (2) colloidal steric effect, (3) aggregation effect, and (4) electrokinetic effect.

On the other hand, during crude oil transportation through pipes, there may be sludge deposits along the line and these are also called sediments. The deposits at the bottom of the storage tanks or crude transport lines are water, salts, sand, and heavy hydrocarbons, and their average concentration is about 25% water, 5% inorganic compounds, and 70% vol. of hydrocarbons [9]. In general, asphaltenes deposits cause problems in five points: extraction, transportation, processing, economic benefit of crude oil, and environmental pollution. There are several methods to prevent and/or remove asphaltenes deposits: mechanical methods [10], chemical cleaning, pressure [11, 12], temperature and flow rate manipulations [13], additives, and chemical inhibitors [14, 15]. The conductive polymers applications are highly diverse and several studies have pointed out the different variables that affect the properties and performance of polypyrrole (PPy) electrodeposition on the steel surface such as the conductivity, stability, roughness, adhesion, film thickness, and so on [16–21].

When an electroactive conducting polymer coats a metal surface, it can act as a membrane-like selective permeable layer, allowing the diffusion of certain ions and rejects others, depending on their chemical affinity, electronic structure, and inter- and intra-molecular arrangement. In this context, some works have been reported on the inhibition of corrosion and stability of conductive polymer films on steel substrates [22–24].

1.1. Conductive polymers

Conductive polymers are those synthetic polymers that are capable of conducting electrical current. These polymers may owe their conductivity to intrinsic properties of the material or modifications. Conductive polymers have a wide range of applications due to their physicochemical characteristics; many of these properties include electrical conductivity, electroactivity, electrochromism, environmental stability, chemical stability, among others [25, 26]. The scientific and technological development has given rise to two types of electronic conductive materials with polymer matrix: the intrinsically conductive polymers and the extrinsically conductive polymers [27].

Several reviews in the literature focus on the corrosion protection by conductive polymers such as polyaniline (PANi), polypyrrole (PPy) and polythiophene (PTh). Although a number of possible protection mechanisms are proposed, the possible passivation of the metal by polymers such as polypyrrole or polyaniline is frequently indicated. In this work, we propose the use of PPy films to inhibit asphaltenes depositions. It is shown that the efficacy of conducting polymers for corrosion protection depends on the application modus and the experimental conditions, that is, depending on the suitable conditions, a conductive polymer may have excellent protection capacity or may lead to negative response of the coating. An important part of the deposition of PPy lies in the surface treatment of the electrode surface.

Intrinsically conductive polymers are organic polymer in which the electrical conductivity originates from the extended electrons conjugation along the polymer chain. The most common conductive polymers (polyacetylene, polyparaphenylene, polypyrrole, polythiophene, and polyaniline) have carbon atoms in the backbone with sp^2 hybridization. This hybridization creates covalent σ bonds between the carbons of the main chain and those of the branched chains. The sp^2 hybridization leaves an unbonded orbital p (usually pz); these orbitals overlap and form a bond, with a distribution of $C = C$ double bonds alternating with single carbon-carbon bonds along the chain.

Extrinsically conductive polymers are those that owe their conductivity to the inclusion of conductive materials such as metals, graphite, or charge transfer complexes in the polymer matrix, generally thermoplastic. In this case, above the percolation concentration, the conductive paths along the material give it electronic conductivity, while the polymer matrix allows the material to be processed in industrial operations to achieve different types of products and finishes [28]. The conductive polymers are formed from suitable monomers, and can be obtained either by chemical synthesis or by electrochemical methods (**Table 1**).

Conducting polymers	General comments	Refs
Polypyrrole (PPy)	PPy is chemically and thermally stable. The conductivity of PPy strongly depends on the preparation technique.	[19, 20]
Polyaniline (PANi)	PANi is easy to synthesize and used as corrosion inhibitor and relatively inexpensive.	[29, 30]
Polythiophenes (PTh)	PTh poses high charge-carrier mobility due to their relative structural order, is soluble and demonstrates solvatochromism and thermochromism effect.	[31, 32]
Polyacetylene	Despite its discovery started the development of conductive polymers, to date, polyacetylene has no commercial applications.	[33]
Polyphenylenes	Thermally stable up to 500–600°C but is quite insoluble in most solvent, with potential applications as light-emitting diode due to their electroluminescence properties.	[34]

Table 1. Typical conducting polymers and their common applications.

1.2. Electrochemical synthesis of conductive polymers

The electrochemical synthesis of any conductive polymer has its particularities; however, there are a number of common factors in its synthesis that have been exposed and that must be taken into account at the time of its preparation. In principle, it is assumed that, except for the initiation step, the electrochemical polymerization will proceed by a similar mechanism as the thermal polymerization for the same monomer in a comparable environment.

Experimentation usually begins with the evaluation of the potential window in the selected electrolyte medium, which will allow defining the electrochemical parameters for the subsequent electrolysis. In the evaluation stage as in the synthesis itself, strict control of electrochemical variables (electrode potential, electrode nature, current density, solution conductivity, electric field, etc.) and aspects such as medium, presence or not of protonating agents, oxygen, inert atmosphere, and so on [35] are observed.

From a scientific point of view, the flow of an anodic current through an electrochemical system, formed by a monomer, a solvent, and an electrolyte, can initiate reactions such as the formation of an oxide layer on the electrode, oxidation of monomer on the metal, oxidation of the solvent, and oxidation of the electrolyte. But in this process, the polymerization develops an electrode coating the chemical nature of the electrode changes after a few seconds of polymerization from a metal electrode to a polymer electrode. So, the above reactions will occur at different potentials on the new electrode, and a new metal-polymer interface appears in addition to that of the growth-dissolution polymer and then new reactions will occur, oxidation of the polymer and degradation thereof [27].

1.3. Applications of conducting polymers electrochemically synthesized

Conductive polymers have been widely used in the fields of electrochemistry, electroanalysis, electrocatalysis, batteries and capacitors, and so on [36]. In these applications, the electrochemical activity and the conductivity are two important properties of the conducting polymers, because they play fundamental roles. In addition, small ions and molecules are able to diffuse into matrices of conducting polymers, providing other advantages over conventional electrode materials. This intrinsic property allows electrochemical reactions to take place along the matrices of the conducting polymers and thus increase the active sites for the electrochemical processes by using a 3D electrode. However, in order to efficiently utilize all active sites and improve mass transport during the electrode process, the film thickness of the conductive polymer should be reduced to facilitate diffusion of the ion in the polymeric matrix. Considering these factors, conductive polymer nanomaterials show different characteristics that can provide advantages over other materials. In addition, nanostructures can produce conductive polymers with new surface properties and new functions [36]. Different applications of conducting polymers prepared nanomaterials by electrochemical techniques as been reported: sensors [37–39], electrochemical capacitors [40], fuel cell electrodes [41], batteries [42, 43], electrochromic devices [44], and electrochemical actuators [45–47].

1.4. Inhibitors of asphaltenes deposition by using conducting polymer coatings

Polypyrrole (PPy) and polythiophene (PTh) are conductive polymers, which have been used as corrosion protection and have the characteristic that they can coat the steel [48, 49]. This process can be carried out by electropolymerization in aqueous phase; in addition, PPy exhibits good mechanical properties, thermal stability, and high conductivity [49]. The performance of corrosion protection by bi-layered PPy coatings was investigated by Kowalski et al.; they used an inner PPy layer doped with molybdophosphate ions to protect steels from corrosion. The polymer layer maintained the passive state of the steel in an acid solution and a neutral NaCl solution for several days [22].

Warren et al. [16] reported that anion dopants containing sulfonates, such as dodecylbenzenesulfonic acid (DBSA), can be used to form PPy films with high conductivities, good stability, mechanical properties, and apparent order. However, when there is a chemical-physical mismatch at the metal-polymer interface, the adhesion between the PPy layer and the metal substrate is generally poor [50].

Beck and Michaelis [51] described anodic electrodeposition of black PPy films on steel electrodes from aqueous electrolytes containing the monomers and oxalic acid, with strong adherence and low surface roughness. Su and Iroh [52] investigated the electrodeposition mechanism of PPy coatings on steel substrates from aqueous oxalate solutions. Their results revealed the formation of a passive layer on the steel substrate before reaching the electropolymerization potential of pyrrole.

Tüken et al. [53] prepared multilayer coating of polypyrrole/polyphenol on mild steel by cyclic voltammetry (CV). The corrosion performance of this multilayer coating was investigated by electrochemical impedance in sulfate solution. Another study carried out by the same authors consisted of the coating of copper with a mixture of polymers (PPy/PTh) [31]. They observed that the electrochemical synthesis of stable and homogeneous PTh films could not be achieved on copper electrode by direct oxidation of thiophene in acetonitrile-$LiClO_4$, due to insufficient surface passivation, and the copper dissolution in monomer oxidation potential region.

On the other hand, Pekmez et al. [32] informed that the electrochemical synthesis of an anticorrosive polybithiophene (PBTh) on stainless steel is feasible, and the obtained PBTh coatings strongly adhered to the working substrate.

Rocha et al. [14] carried out a study on the inhibition of asphaltenes precipitation in Brazilian crude oils using amphiphiles substances such as low molecular mass ethoxylated nonylphenols, vegetable oils (coconut essential oil, sweet almond, andiroba, and sandalwood oil), and organic acids (linoleic, caprylic, and palmytic). These compounds showed great efficiency in the asphaltenes precipitation inhibition, through a mechanism of asphaltenes stabilization as a function of its surfactant capacity.

Castellano et al. [15] performed a theoretical investigation of σ-π and π-π interactions on benzene, pyridine, and thiophene dimers; they observed the influence of these interactions with asphaltene stability on crude oil and concluded that chemical interactions between species with opposite polarities lead to an intermolecular association in the asphaltenes, which are responsible for the phenomenon of aggregation. This study was aimed to understand why these compounds tend to aggregate and then flocculate in oil operation processes.

The applications of the conducting polymers are highly diverse and rely on the final properties from the synthesis conditions [23, 24]. Therefore, conducting electrochemical studies to understand the effect of the synthesis variables that affect electrodeposition of conductive polymers is necessary, in order to attain the required conditions in each specific application, as it is the case for application in corrosion protection [54]. When the electroactive conducting polymer is coated to the electrode surface, it may work as a selective permeable layer, which allows certain ions and molecules to pass, according to the degree of cross-linking of the films and supported molecules on the monomer. So the study of the surface properties of polymeric materials is justified, and the way in which the species present in the system can cause the material to deteriorate. In this context and considering the studies reported in the literature, the inhibitors of the asphaltenes deposit using PPy and PTh electrodeposited in carbon steel, is a novel application of this type of materials, in the literature there are few reports in this line of research.

1.5. Interfacial interaction (contact angle)

The deposition of a coating on a solid generates new interfaces between dissimilar materials and involves considerations of wettability, spreading, interface evolution, and adhesion. The interaction at solid-liquid interface is determined by a balance between the adhesive and cohesive forces. Adhesive forces between a liquid and a solid cause a liquid drop to spread on the surface. The cohesive forces within the liquid cause the drop to maintain a stable position and avoid contact with the surface. This solid-liquid interaction at the interphase is called wettability. The wettability of electrochemically deposited conductive polymer films depends to a large extent on several parameters, such as deposition conditions (applied voltage, transferred charge, etc.), dopant, and working electrode roughness [55]. Mecerreyes et al. [56] carried out a study where they obtained a hydrophobic PPy film (water contact angle of >90°) using a perfluorinated dopant anion, and a hydrophilic film using a ClO_4^- doping anion. Controlling the wettability of a solid surface is important in many applications, for example, in self-cleaning surfaces, liquid lenses, smart fabrics, and in biomedicine [57–59].

It has been reported that the roughness of hydrophobic solid increases its hydrophobicity due to two different ways: roughness increases the surface area of the solid, which geometrically enhances hydrophobicity and due to the air that can remain under the drop [60, 61]. It is important to note that a roughness at two or more length scales has been implicated as the cause of imitating the "lotus effect," which is the characteristic of a lotus leaf to promote water repellency and self-cleaning [62]. One method to surface wettability control is by oxidizing or reducing the polymer film by modifying thereby the surface morphology. Several research groups have produced films of superhydrophobic conductive polymers, creating micro- and nanostructured surfaces by tempering methods [63–66]. However, these methods have a disadvantage due to the complexity of manufacturing processes [67].

For both improved oil recovery and crude oil transport, it is necessary to develop chemical additives that modify the wetting behavior of reservoir rock (also known as a core), in order to facilitate the crude oil extraction, or to prevent it from wetting the pipe's inner wall and allow crude oil to easily flow through pipelines [68]. The oilfield scale formations are associated to interface activity of polar components of the crude oil. Asphaltenes are the most polar fraction of the crude and contain large amounts of active species [69] and because of this, asphaltenes are reported as the major fractions responsible for altering surface wettability through the interaction of polar functional groups with polar sites of solid surface [70, 71]. A study by Kaminsky and Radke [72] indicates that low solubility asphaltenes can diffuse through water films to arrive at rock surfaces without significant wettability alteration; the rupture of the water film followed by direct deposition of crude oil onto rock allows explaining the wettability reservoir rock.

2. Experimental

2.1. Reagents and chemicals

The electrolytes used were 0.1 M aqueous solutions of $KNO_3$and KCl (both from J. T. Baker, reagent grade). Also a 0.1 M[1] of pyrrole (Py, Sigma Aldrich) solution was prepared with previous purification in a bed column packed with silica and activated carbon. All the solutions were prepared with deoxygenated water (Millipore, 18.2 MΩ) for 15 min with an atmosphere of pure nitrogen (Praxair, 99.99%) before each experiment. The film stability was evaluated in a KCl electrolyte according to a procedure reported [73].

The crude oil used in this work originates from the Gulf of Mexico and presents the following features: 15°API, 25% weight of asphaltenes, density 0.9647 g cm^{-3}, and a kinematic viscosity of 1.697 mPa s.

2.2. Materials

A conventional three-electrode cell was used for the electrodeposition of PPy over carbon steel (CS-1018) as the working electrode, graphite electrode as counter-electrode, and aqueous saturated calomel electrode, SCE (Tacussel), as reference electrode. Initially, the CS-1018 electrode was polished with different grain sandpaper to achieve a defined surface and subjected to ultrasonic baths (Branson 2510) for 5 min to remove contaminants from the surface. To improve the adhesion of the polymer to the steel surface, it was necessary to carry out a pretreatment of the electrode with a mordant layer to increase the roughness. For this purpose, some authors [23, 24] immerse the metal electrode in acidic solutions of HCl or HNO_3. In this work, the electrode was treated with acid solutions of 10% HNO_3 with an immersion time of 2 min. The treated surfaces were designated as treated in HNO_3(T) and only polished surfaces without acidic treatment were designated as untreated in HNO_3(NT).

2.3. Equipment

The polypyrrole electrosynthesis was performed by cyclic voltammetry and chronoamperometry (CA) techniques by using a Gamry Reference-600 potentiostat, in a three-electrode cell at room temperature (20±3°C). The electrodeposition by cyclic voltammetry was carried out at a scan rate of 100 mVs^{-1}, in a potential window of −0.8 to 1.0 V/SCE for 40 cycles of polymerization. The conditions for the chronoamperometry were at constant potential of 1.0 V/SCE, for 300 s. For the characterization of the polymeric films, a scanning electron microscope, SEM (Jeol, JSM-6390LV) coupled with energy dispersive spectroscopy (EDS) analysis (Oxford Instruments, INCAx-sight), an atomic force microscope, AFM (Veeco, Innova Scanning Probe Microscope), and an equipment to measure contact angle (Chem Instruments, CAM-plus) were used. The contact angle measurements of polymer surfaces were conducted with deionized water and crude oil, analyzing two areas on the sample and considering three measurements in each zone taken every 2 min for 10 min.

3. Results and discussion

3.1. CS-1018 electrode characterization

The SEM and AFM micrographs of the CS-1018 substrates are shown in **Figure 1**, where the lines attributed to the mechanical polishing are observed. The images show that the treated substrates, CS-1018 T, present a rougher surface due to chemical attack with HNO_3. The observed roughness measurements were of Rq = 0.0509 μm and Rq = 0.739 μm, for the untreated and treated surfaces, respectively. This difference in the roughness value will be important for the adhesion effect of the polymer material to be synthesized on it.

http://dx.doi.org/10.5772/intechopen.70091

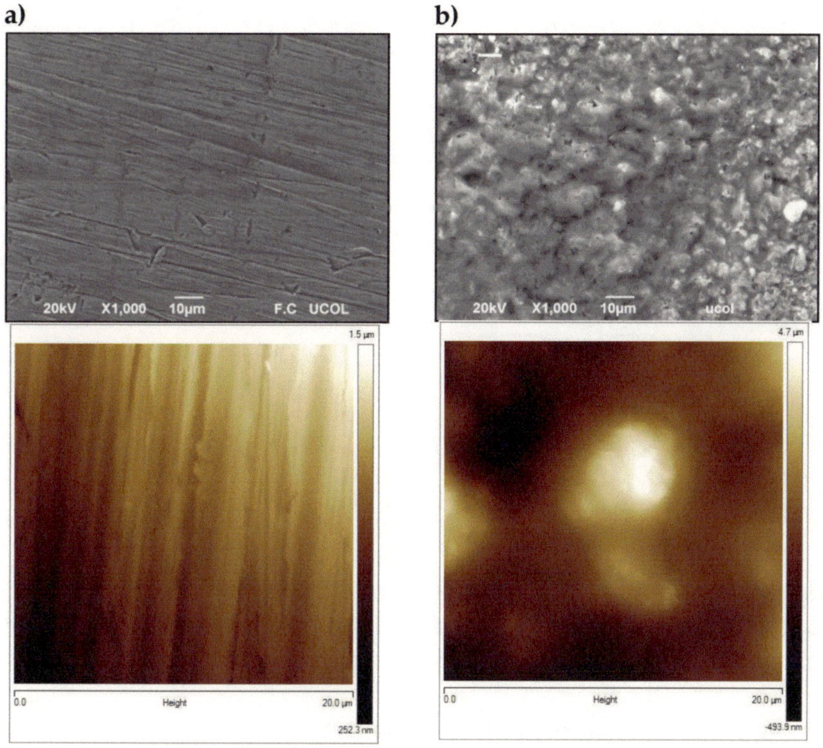

Figure 1. SEM and AFM micrographs of CS-1018: (a) untreated (NT) and (b) treated (T) in 10% HNO_3.

3.2. PPy film electrodeposition

In this work, the results obtained by two electrochemical techniques are presented: CV and CA. The electrodeposits were made in four different electrolytes, in order to find the best characteristics of the film. **Figure 2a** shows an example with the cyclic voltammograms of the polymerization of polypyrrole in electrolyte of KNO_3. The voltammograms show the characteristic signals [74–76]. **Figure 2b** shows the signal of PPy deposited with applied constant potential method [77].

3.3. PPy morphology on CS-1018

Figure 3 shows the deposited PPy films on the CS-1018 with and without treatment in the four different electrolytes, electrodeposited by CA technique. When KI and KF electrolytes were used, deposit films were not obtained, but when K_2SO_4 and KNO_3 were used as electrolytes, the formation of PPy films was obtained.

Figure 4a shows the SEM image of PPy deposited on CS-1018 NT, in which a homogeneous film is observed that completely covers the surface of the steel. However, it is possible to observe lines in said film due to the polishing process of the steel, indicating that the deposited film is thin. By contrast, in the PPy film deposited on AC-1018 T, non-uniform circular

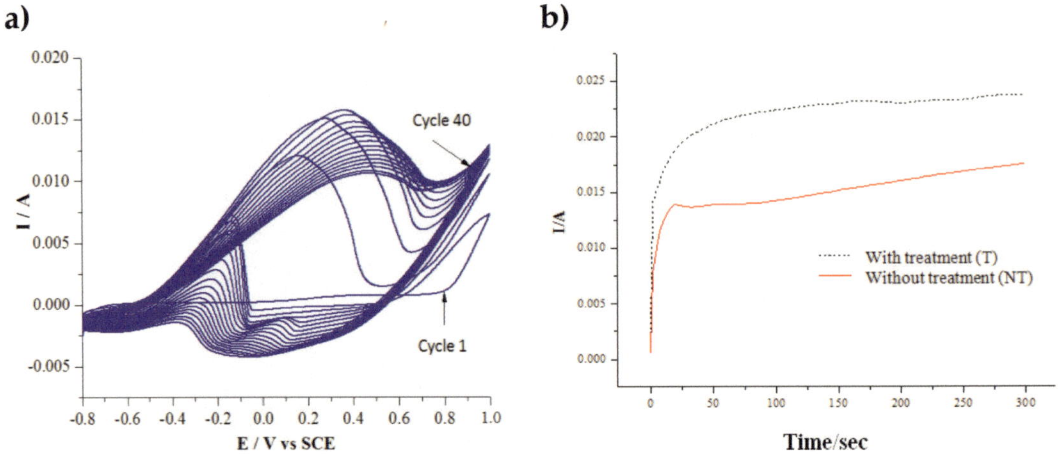

Figure 2. Electrochemical formation of PPy on CS-1018 with (a) CV and (b) CA.

agglomerates with needle-shaped scales are observed, as can be seen in **Figure 4b**. This suggests that acid treatment to the steel affected the morphology of the deposited polymer. This observed morphologies show a characteristic topography of conductive polymers [78].

3.4. Contact angle characterization of the PPy surface

In order to analyze the surface of deposited PPy films, contact angle measurements were performed with and without water and crude oil on selected samples, and steel substrates with and without acid treatment. The measurements were performed in two areas of the sample

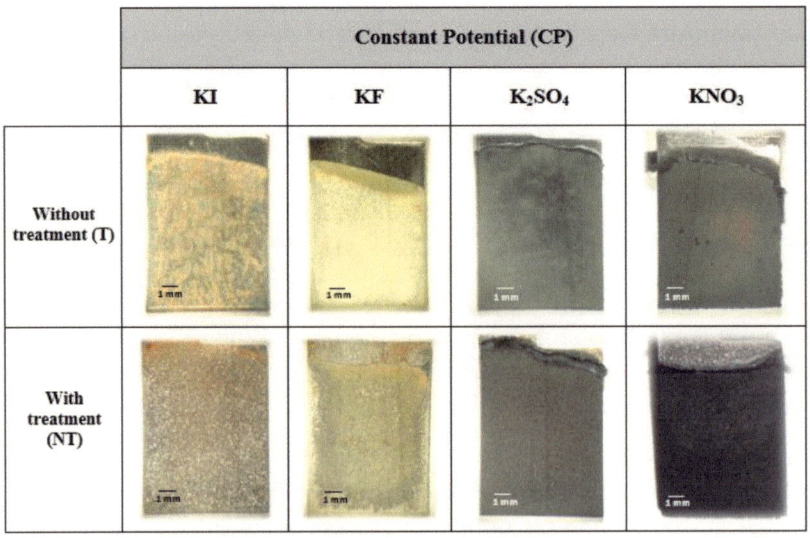

Figure 3. PPy films electrodeposited on CS-1018 substrates with and without treatment in four electrolyte media by CA technique.

http://dx.doi.org/10.5772/intechopen.70091

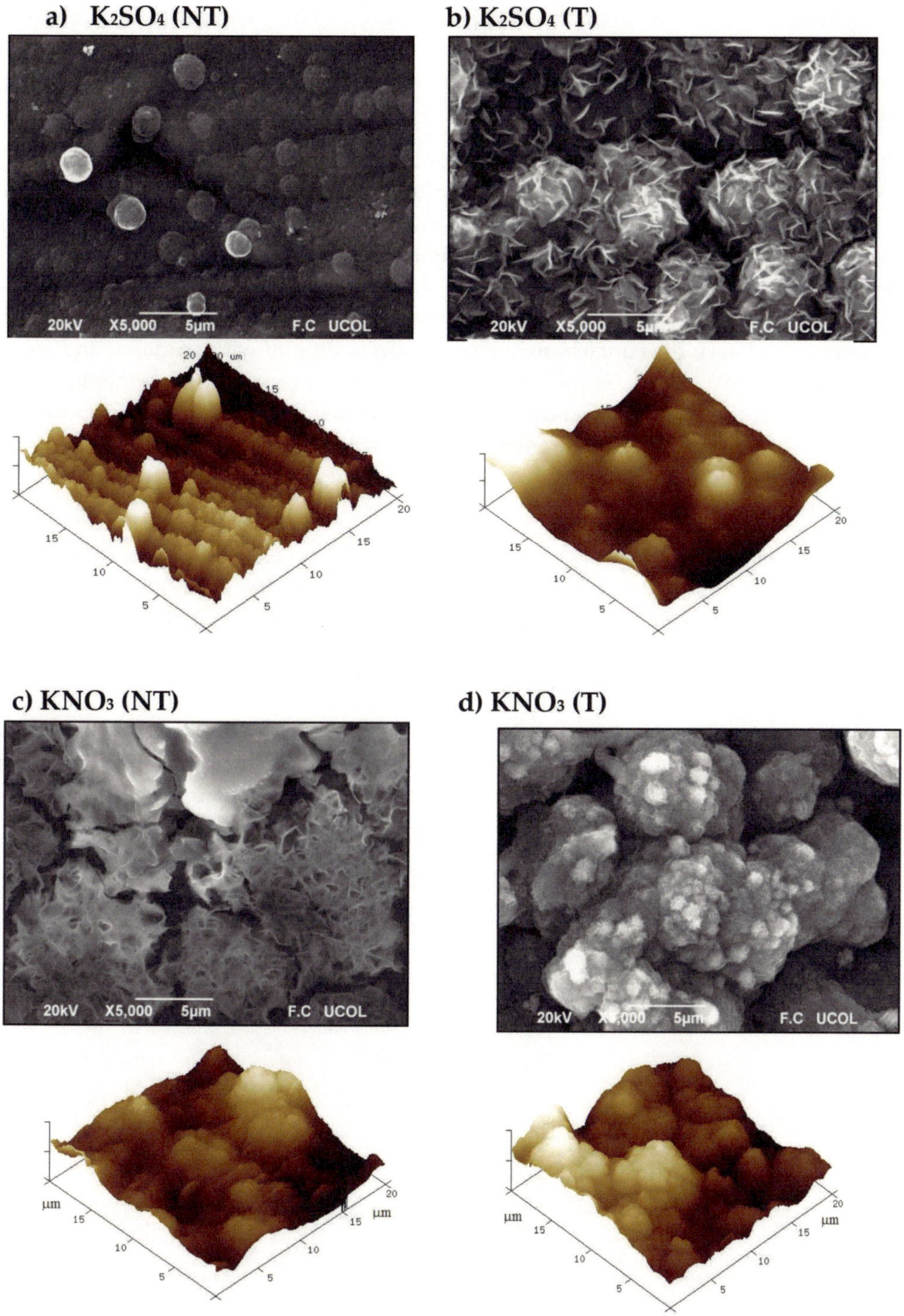

Figure 4. SEM and AFM micrographs of the PPy deposited by the chronoamperometry technique on untreated (NT) and treated (T) CS-1018 substrates: (a) in K_2SO_4 electrolyte untreated substrate, (b) K_2SO_4 electrolyte treated substrate, (c) KNO_3 at 0.1 mol L-1 concentration on untreated and (d) treated substrates with 10% HNO_3.

and in each zone, three measurements were taken every 2 min. The contact angle values were obtained from the averages of the measurements. **Figure 5** shows the experimental setup, wherein a drop of water makes contact with the substrate CS-1018 NT and how its image is projected to measure the contact angle.

Table 2 shows the values of the contact angle measurements with water and crude oil of the treated and untreated electrodes, and the respective formed polypyrrole films. It is noted that the contact angle of the CS-1018 NT surface is about 70° with water and crude oil. In other words, the untreated metal surface has the same affinity for both liquids. When the steel surface is subjected to acid treatment, CS-1018 T, the contact angle with water increases to 110°. This result indicates that nitric acid treatment induces a slightly hydrophobic behavior. On the other hand, the contact angle also increases slightly with crude oil and remains at a value close to 90°, that is, at the boundary between the oleophilic/oleophobic balances. Contact angle results show that the acid treatment to the metal surface provides hydrophobic properties to the surface; this measurement is in line with AFM results, showing greater surface roughness for electrodes exposed to acid medium. These results are consistent with those reported in the literature [52, 60, 74, 79].

According to this result, the polypyrrole films deposited on CS-1018 T presented rougher surfaces than those synthesized on CS-1018 NT. In both cases, after the synthesis and coating with PPy of the electrodes, the contact angle with water exhibited an increase as the surface roughness increases. PPy deposition on CS-1018 NT has a slightly hydrophilic behavior contrary to the case when the polymer is synthesized on a treated surface.

The contact angle values with crude follow the same trend as the water. This contact angle value is lower in the untreated surface, with apparently less roughness. However, these contact

Figure 5. Experimental setup for contact angle measurement.

Sample	Contact angle/(°)	
	Water	Oil
CS-1018 NT	70	70
CS-1018 T	110	85
PPy-KNO_3 on CS-1018 NT by CV	80	83
PPy-KNO_3 on CS-1018 T by CV	125	90
PPy-K_2SO_4 on CS-1018 NT by CA	70	70
PPy-K_2SO_4 on CS-1018 T by CA	90	80

Table 2. Contact angle values with water and oil of the electrodeposited polymers.

angle values in the presence of the polymer coating are close to 90°, the threshold value for determining the oleophilic nature of the electrode surface.

Contact angle measurements show that the surface CS-1018 T PPy-KNO_3 forms a contact angle greater than the surface of CS-1018 NT PPy-KNO_3. That is, that the greater the porosity of the metal surface, a higher roughness of the polymer deposited is obtained, and consequently the polypyrrole-oil interaction decreases.

The roughness plays an important role for the hydrophobicity of the polymer deposit. According to the literature [80], hydrophobicity increases as the roughness of the material grows. **Figure 6** shows the average values of contact angle with oil as a function of roughness of the deposited

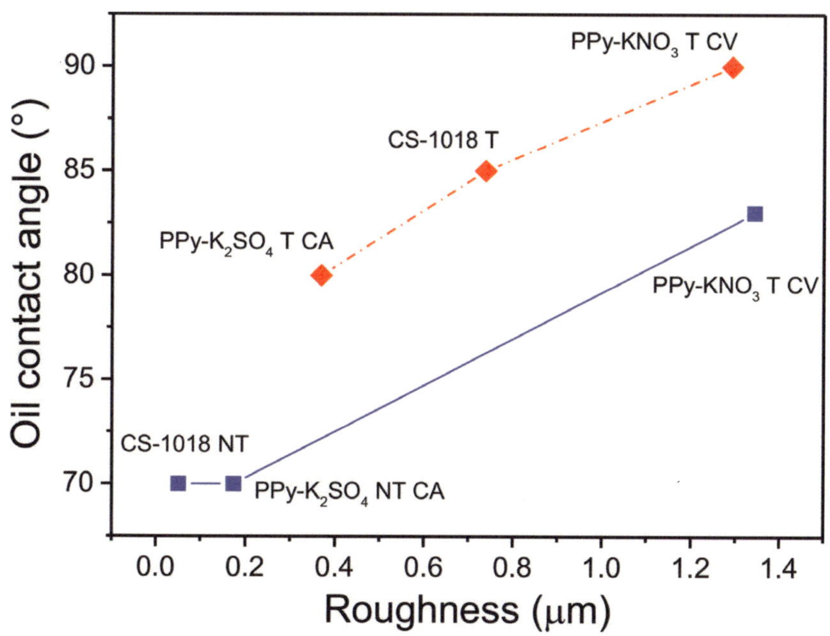

Figure 6. Relation between oil contact angle with the roughness of the polymer films.

polymer films. The materials with greater roughness presented greater contact angle than those with low roughness. It is observed that the contact angle of the PPy-KNO_3 with acid treatment by CV increased around 20° with respect to CS-1018 NT. Therefore, this result shows that such polymer under certain conditions is efficient to reduce contact of asphaltenes with CS-1018.

4. Conclusions

Acid pretreatment modifies the roughness of the CS-1018 substrate generating an oxide layer that influences both the morphology and the stability of electrodeposited PPy films, creating surfaces with different arrangements, which depends on the electropolymerization technique employed. The roughness difference is directly related to the stability of the polymer film formed and its surface properties (wettability). CV electrodeposition is the most appropriate method for this type of application. The PPy materials deposited with KNO_3 by CV, with and without treatment in an acid medium, presented a greater homogeneity and roughness. The roughness is directly proportional to the hydrophobicity of the PPy film, which was evidenced with an increase in contact angle values (lower affinity to crude oil). The inhibition of the asphaltene deposition is evidenced by obtaining contact angles of more than 90°. The results indicate that this methodology is cost-effective, versatile, and scalable for the synthesis of electrodes for applications to inhibit asphaltenes deposition.

Acknowledgements

O.E. Vázquez-Noriega acknowledges the scholarship granted by CONACYT-Mexico. We acknowledge the funding through CONACyT-Mexico projects No. 177480 and the Fis. D. Pozas of the University of Colima for the study of SEM-EDS microscopy. RD Martínez-Orozco appreciates the postdoctoral scholarship by Fondo Sectorial-CONACYT-SENER grant No. 2138.

Author details

Oscar E. Vázquez-Noriega[1], Javier Guzmán[2], Nohra V. Gallardo-Rivas[1], Reinaldo David Martínez Orozco[1], Ana M. Mendoza-Martínez[1], María Yolanda Chávez Cinco[1], Luciano Aguilera Vázquez[1] and Ulises Páramo-García[1]*

*Address all correspondence to: uparamo@itcm.edu.mx

1 Centro de Investigación en Petroquímica, División de Estudios de Posgrado e Investigación, Instituto Tecnológico de Cd. Madero, Prol. Bahía de Aldhair y Av. De las Bahías, Parque de la Pequeña y Mediana Industria, Altamira, Tamaulipas, Mexico

2 Instituto Mexicano del Petróleo, Eje Central Lázaro Cárdenas, Col. San Bartolo Atepehuacán, Mexico D.F., Mexico

References

[1] McSween HY, Richardson SM, Uhle ME. Geochemistry: Pathways and Processes, 2 ed., Columbia University Press, 2003. 432 p. ISBN: 9780231509039

[2] Luo P, Wang X, Gu Y, Zhang H, Moghadam S. Asphaltene precipitation and its effects on the vapour extraction (VAPEX) heavy oil recovery process. In, International Thermal Operations and Heavy Oil Symposium, 20-23 October, Calgary, Alberta, Canada, SPE/PS/CHOA 117527, 2008. p. 1-9

[3] Al-Sahhaf TA, Fahim MA, Elkilani AS. Retardation of asphaltene precipitation by addition of toluene, resins, deasphalted oil and surfactants. Fluid Phase Equilibria. 2002;**194-197**:1045-1057

[4] Leontaritis KJ, Ali Mansoori G. Asphaltene deposition: A survey of field experiences and research approaches. Journal of Petroleum Science and Engineering. 1988;**1**:229-239

[5] Rogel E, Miao T, Vien J, Roye M. Comparing asphaltenes: Deposit versus crude oil. Fuel. 2015;**147**:155-160

[6] Panuganti SR, Vargas FM, Gonzalez DL, Kurup AS, Chapman WG. PC-SAFT characterization of crude oils and modeling of asphaltene phase behavior. Fuel. 2012;**93**:658-669

[7] Maia Filho DC, Ramalho JBVS, Spinelli LS, Lucas EF. Aging of water-in-crude oil emulsions: Effect on water content, droplet size distribution, dynamic viscosity and stability. Colloids and Surfaces A: Physicochemical and Engineering Aspects. 2012;**396**:208-212

[8] Speight JG. The Chemistry and Technology of Petroleum, 4th ed. CRC Press, 2006. 953 p. ISB: 9781439873892

[9] Hu G, Li J, Zeng G. Recent development in the treatment of oily sludge from petroleum industry: A review. Journal of Hazardous Materials. 2013;**261**:470-490

[10] Afghoul AC, Amaravadi S, Boumali A, Calmeto JCN, Lima J, Lovell J, Tinkham S, Zemlak K, Staal T. Tubería flexible: La próxima generación. Oilfield Review. 2004;**16**:38-57

[11] Santhana K, Van Gisbergen SJ, Harris J, Ferdiansyah E, Brady ME, Al-Harthy S, Pandey A. Eliminating multiple interventions using a single rig-up coiled-tubing solution. SPE Production & Operations. 2008;**23**(02):119-124

[12] Frenier WN, Ziauddin M, Venkatesan R. Organic deposits in oil and gas production, Society of Petroleum Engineers, 2010. 362 p. ISBN: 978-1-55563-291-5

[13] Park SJ, Ali Mansoori G. Aggregation and deposition of heavy organics in petroleum crudes. Energy Sources. 1988;**10**:109-125

[14] Rocha Junior LC, Ferreira MS, da Silva Ramos AC. Inhibition of asphaltene precipitation in Brazilian crude oils using new oil soluble amphiphiles. Journal of Petroleum Science and Engineering. 2006;**51**:26-36

[15] Castellano O, Gimon R, Soscun H. Theoretical study of the σ–π and π–π interactions in heteroaromatic monocyclic molecular complexes of benzene, pyridine, and thiophene dimers: Implications on the resin–asphaltene stability in crude oil. Energy & Fuels. 2011;**25**: 2526-2541

[16] Warren LF, Anderson DP. Polypyrrole films from aqueous electrolytes. The effect of anions upon order. Journal of the Electrochemical Society. 1987;**134**:101-105

[17] Su W, Iroh JO. Electrodeposition mechanism, adhesion and corrosion performance of polypyrrole and poly(N-methylpyrrole) coatings on steel substrates. Synthetic Metals. 2000;**114**:225-234

[18] Patois T, Lakard B, Monney S, Roizard X, Fievet P. Characterization of the surface properties of polypyrrole films: Influence of electrodeposition parameters. Synthetic Metals. 2011;**161**:2498-2505

[19] González MB, Saidman SB. Electrodeposition of bilayered polypyrrole on 316 L stainless steel for corrosion prevention. Progress in Organic Coatings. 2015;**78**:21-27

[20] Su W, Iroh JO. Morphology and structure of the passive interphase formed during aqueous electrodeposition of polypyrrole coatings on steel. Electrochimica Acta. 1999;**44**: 4655-4665

[21] Mollahosseini A, Noroozian E. Electrodeposition of a highly adherent and thermally stable polypyrrole coating on steel from aqueous polyphosphate solution. Synthetic Metals. 2009;**159**:1247-1254

[22] Kowalski D, Ueda M, Ohtsuka T. The effect of counter anions on corrosion resistance of steel covered by bi-layered polypyrrole film. Corrosion Science. 2007;**49**:3442-3452

[23] Tallman DE, Spinks G, Dominis A, Wallace GG. Electroactive conducting polymers for corrosion control. Journal of Solid State Electrochemistry. 2002;**6**:73-84

[24] Spinks GM, Dominis AJ, Wallace GG, Tallman DE. Electroactive conducting polymers for corrosion control. Journal of Solid State Electrochemistry. 2002;**6**:85-100

[25] Rajagopalan R, Iroh JO. Characterization of polyaniline–polypyrrole composite coatings on low carbon steel: A XPS and infrared spectroscopy study. Applied Surface Science. 2003;**218**:58-69

[26] Tat'yana VV, Oleg NE. Polypyrrole: A conducting polymer; its synthesis, properties and applications. Russian Chemical Reviews. 1997;**66**:443

[27] Otero TF. Polímeros conductores: Síntesis, propiedades y aplicaciones electroquímicas. Revista Iberoamericana de Polímeros. 2003;**4**:1-32

[28] Arias J. Síntesis y caracterización de polímeros conductores basados en anilinas sustituidas y su aplicación en electrocatálisis [Thesis doctoral]. Universidad de Alicante, Spain; 2007

[29] Deshpande PP, Jadhav NG, Gelling VJ, Sazou D. Conducting polymers for corrosion protection: A review. Journal of Coatings Technology and Research. 2014;**11**:473

[30] Ohtsuka T. Corrosion protection of steels by conducting polymer coating. International Journal of Corrosion. 2012;**2012**:7, Article ID 915090

[31] Tüken T, Yazıcı B, Erbil M. Polypyrrole/polythiophene coating for copper protection. Progress in Organic Coatings. 2005;**53**:38-45

[32] Pekmez NÖ, Abacı E, Cınkıllı K, Yağan A. Polybithiophene and its bilayers with polyaniline coatings on stainless steel by electropolymerization in aqueous medium. Progress in Organic Coatings. 2009;**65**:462-468

[33] Basescu N, Liu Z-X, Moses D, Heeger AJ, Naarmann H,TheophilouN. High electrical conductivity in doped polyacetylene. Nature. 1987;**327**:403-405

[34] Leising G, Tasch S, Brandstätter C, Graupner W, Hampel S, List EJW, Meghdadi F, Zenz C, Schlichting P, Rohr U, Geerts Y, Scherf U, Müllen K. Efficient full-colour electroluminescence and stimulated emission with polyphenylenes. Synthetic Metals. 1997;**91**:41-47

[35] Gravert DJ, Janda KD. Organic synthesis on soluble polymer supports: Liquid-phase methodologies. Chemical Reviews. 1997;**97**:489-510

[36] Li C, Bai H, Shi G. Conducting polymer nanomaterials: Electrosynthesis and applications. Chemical Society Reviews. 2009;**38**:2397-2409

[37] Dai L, Soundarrajan P, Kim T. Sensors and sensor arrays based on conjugated polymers and carbon nanotubes. Pure and Applied Chemistry. 2002;**74**:1753-1772

[38] Barlett PN, Cooper JM. A review of the immobilization of enzymes in electropolymerized films. Journal of Electroanalytical Chemistry. 1993;**362**:1-12

[39] Peng H, Zhang L, Soeller C, Travas-Sejdic J. Conducting polymers for electrochemical DNA sensing. Biomaterials. 2009;**30**:2132-2148

[40] Frackowiak E, Béguin F. Carbon materials for the electrochemical storage of energy in capacitors. Carbon. 2001;**39**:937-950

[41] Liu H, Song C, Zhang L, Zhang J, Wang H, Wilkinson DP. A review of anode catalysis in the direct methanol fuel cell. Journal of Power Sources. 2006;**155**:95-110

[42] Sivakkumar SR, Kim D-W. Polyaniline/carbon nanotube composite cathode for rechargeable lithium polymer batteries assembled with gel polymer electrolyte. Journal of The Electrochemical Society. 2007;**154**:A134-A139

[43] Huang XH, Tu JP, Xia XH, Wang XL, Xiang JY. Nickel foam-supported porous NiO/polyaniline film as anode for lithium ion batteries, Electrochemistry Communications. 2008;**10**:1288-1290

[44] Cho SI, Lee SB. Fast electrochemistry of conductive polymer nanotubes: Synthesis, mechanism, and application. Accounts of Chemical Research. 2008;**41**:699-707

[45] Smela E. Conjugated polymer actuators for biomedical applications. Advanced Materials. 2003;**15**:481-494

[46] Bar-Cohen Y. Artificial muscles based on electroactive polymers as an enabling tool in biomimetics. Proceedings of the Institution of Mechanical Engineers, Part C: Journal of Mechanical Engineering Science. 2007;**221**:1149-1156

[47] Otero TF. Soft, wet, and reactive polymers. Sensing artificial muscles and conformational energy. Journal of Materials Chemistry. 2009;**19**:681-689

[48] Hu J, Zhu H, Ma Y, Yi T, Mao X, Lin A, Gan F. Corrosion protection of stainless steel by separate polypyrrole electrode in acid solutions. Materials and Corrosion. 2011;**62**:68-73

[49] Kowalski D, Ueda M, Ohtsuka T. Corrosion protection of steel by bi-layered polypyrrole doped with molybdophosphate and naphthalenedisulfonate anions. Corrosion Science. 2007;**49**:1635-1644

[50] Prissanaroon W, Brack N, Pigram PJ, Liesegang J. Co-doped polypyrrole coatings for stainless steel protection. Surface Review and Letters. 2006;**13**:319

[51] Beck F, Michaelis R. Strongly adherent, smooth coatings of polypyrrole oxalate on iron. The Journal of Coatings Technology. 1992;**64**:59

[52] Wencheng S, Iroh JO. Electrodeposition mechanism of polypyrrole coatings on steel substrates from aqueous oxalate solutions. Electrochimica Acta. 2000;**46**:1-8

[53] Tüken T, Yazıcı B, Erbil M. A new multilayer coating for mild steel protection. Progress in Organic Coatings. 2004;**50**:115-122

[54] Rohwerder M, Duc LM, Michalik A. In situ investigation of corrosion localised at the buried interface between metal and conducting polymer based composite coatings. Electrochimica Acta. 2009;**54**:6075-6081.

[55] Chang JH, Hunter IW. Characterization and control of the wettability of conducting polymer thin films. Materials Research Society. 2010;**1228**:7-12

[56] Mecerreyes D, Alvaro V, Cantero I, Bengoetxea M, Calvo PA, Grande H, Rodriguez J, Pomposo JA. Low surface energy conducting polypyrrole doped with a fluorinated counterion. Advanced Materials. 2002;**14**:1521-4095

[57] Russell TP. Surface-responsive materials. Science. 2002;**297**:964-967

[58] Langer R, Tirrell DA. Designing materials for biology and medicine. Nature. 2004;**428**: 487-492

[59] Luzinov I, Minko S, Tsukruk VV. Adaptive and responsive surfaces through controlled reorganization of interfacial polymer layers. Progress in Polymer Science. 2004;**29**:635-698

[60] Lafuma A, Quere D. Superhydrophobic states. Nature Materials. 2003;**2**:457-460

[61] Feng L, Li S, Li Y, Li H, Zhang L, Zhai J, Song Y, Liu B, Jiang L, Zhu D. Super-hydrophobic surfaces: From natural to artificial. Advanced Materials. 2002;**14**:1857-1860

[62] Gao L, McCarthy TJ. The "lotus effect" explained: two reasons why two length scales of topography are important. Langmuir. 2006;**22**:2966-2967

[63] Wang X, Berggren M, Inganäs O. Dynamic control of surface energy and topography of microstructured conducting polymer films. Langmuir. 2008;**24**:5942-5948

[64] Xiangjun W, Kristofer T, Olle I. Single- and bilayer submicron arrays of fluorescent polymer on conducting polymer surface with surface energy controlled dewetting. Nanotechnology. 2005;**16**:437

[65] Lee W, Jin M-K, Yoo W-C, Lee J-K. Nanostructuring of a polymeric substrate with well-defined nanometer-scale topography and tailored surface wettability. Langmuir. 2004;**20**: 7665-7669

[66] Xu L, Wang J, Song Y, Jiang L. Electrically tunable polypyrrole inverse opals with switchable stopband, conductivity, and wettability. Chemistry of Materials. 2008;**20**:3554-3556

[67] Xu L, Chen W, Mulchandani A, Yan Y. Reversible conversion of conducting polymer films from superhydrophobic to superhydrophilic. Angewandte Chemie International Edition. 2005;**44**:6009-6012

[68] Vargas G, Santillán J, Rincón R, Trejo A, Romero A. Determinación experimental del efecto de la temperatura sobre el ángulo de contacto de un crudo de 15° API sobre un núcleo de yacimiento, Memoria en extenso. In: XXVI Congreso Nacional de Termodinámica, 19-23 September 2011, México, D.F. pp. 751-759

[69] Andersen SI, Christensen SD. The critical micelle concentration of asphaltenes as measured by calorimetry. Energy & Fuels. 2000;**14**:38-42

[70] Liu L, Buckley JS. Alteration of wetting of mica surfaces. Journal of Petroleum Science and Engineering. 1999;**24**:75-83

[71] Standal S, Haavik J, Blokhus AM, Skauge A. Effect of polar organic components on wettability as studied by adsorption and contact angles. Journal of Petroleum Science and Engineering. 1999;**24**:131-144

[72] Kaminsky R, Radke CJ. Asphaltenes, water films, and wettability reversal. Society of Petroleum Engineers Journal. 1997;**2**:485-493

[73] John R, Wallace GG. The use of microelectrodes to probe the electropolymerization mechanism of heterocyclic conducting polymers. Journal of Electroanalytical Chemistry and Interfacial Electrochemistry. 1991;**306**:157-167

[74] Wang Y, Northwood DO. An investigation into the nucleation and growth of an electropolymerized polypyrrole coating on a 316L stainless steel surface. Thin Solid Films. 2008;**516**:7427-7432

[75] Wang Y, Rajeshwar K. Electrocatalytic reduction of Cr(VI) by polypyrrole-modified glassy carbon electrodes. Journal of Electroanalytical Chemistry. 1997;**425**:183-189

[76] Lehr IL, Saidman SB. Morphology and properties of polypyrrole electrosynthesized onto iron from a surfactant solution. Synthetic Metals. 2009;**159**:1522-1528

[77] Vázquez-Noriega OE, Guzmán J, Gallardo-Rivas NV, Reyes-Gómez J, Mendoza-Martínez AM, Rivera-Armenta JL, Páramo-García U. Polypyrrole films deposited on

carbon-steel CS-1018 and its interaction with Mexican crude oil. International Journal of Electrochemical Science. 2015;**10**:6378-6391

[78] Ansari Khalkhali R, Price WE, Wallace GG. Quartz crystal microbalance studies of the effect of solution temperature on the ion-exchange properties of polypyrrole conducting electroactive polymers. Reactive and Functional Polymers. 2003;**56**:141-146

[79] dos Santos RG, Mohamed RS, Bannwart AC, Loh W. Contact angle measurements and wetting behavior of inner surfaces of pipelines exposed to heavy crude oil and water. Journal of Petroleum Science and Engineering. 2006;**51**:9-16

[80] Gennes P-Gd, Brochard-Wyart FO, Quéré D. Capillarity and Wetting Phenomena: Drops, Bubbles, Pearls, Waves. New York, NY: Springer; 2004

Chapter 14

Desulfurization of Dibenzothiophene by *Pseudomonas fluorescens* (UCP 1514) Leading to the Production of Biphenyl

Thayse A.L. Silva, Manfred Schwartz,
Patrícia M. Souza, Ian Garrard,
Galba M. Campos-Takaki and Elias B. Tambourgi

Additional information is available at the end of the chapter

http://dx.doi.org/10.5772/intechopen.70430

Abstract

Dibenzothiophene (DBT) is a typical recalcitrant thiophenic sulfur component of fuels, and its desulphurization has been a model reaction in the treatment of these compounds. Based on this information, the potential of *Pseudomonas fluorescens* (UCP 1514) on the desulfurization of dibenzothiphene was studied, in order to use it for reducing the sulfur content of diesel oil in compliance with environmental regulations. The result of biodegradation by the bacteria was determined by undertaking high-performance liquid chromatography of the metabolites produced. These can also be identified by gas chromatography with a mass spectrometry detector, and doing so revealed a sulfur-free product, biphenyl, as the final product of the degradation process. The results showed a decrease of 73% in dibenzothiophene content, which means that *P. fluorescens* removes sulfur from dibenzothiophene with a good selectivity to form biphenyl. These promising results indicate that *P. fluorescens* has an interesting potential to degrade sulfur-containing compounds in diesel oil and thereby could help in removing sulfur content from diesel oil. The process of microbial desulfurization described herein can be used particularly after carrying out hydrodesulfurization. Consequently, the sulfur content could be reduced even further. Applying *P. fluorescens* UCP 1514 in dibenzothiophene could help to understand the nature of the biodegradation process and to achieve the regulatory standards for sulfur level in fossil fuels.

Keywords: biodegradation, biodesulfurization, dibenzothiophene, diesel, *Pseudomonas fluorescens*

1. Introduction

Direct combustion of fossil fuels leads to sulfur oxide emissions that contribute to acid rain and air pollution. Dibenzothiophene (DBT) and its derivatives are the main organic sulfur compounds found in petroleum. These compounds are present due to the pathway by which petroleum is formed [1].

Various types of heterocycles containing oxygen, nitrogen, and sulfur are found in the environment, originating from anthropogenic or natural sources. Dibenzofurans, dibenzodioxins, and dibenzothiophene are among the most important environmental pollutants and are well covered in the literature [2].

As these heterocyclic compounds are relatively stable, the traditional processes, such as acid alkali treatment, are inefficient at removing them from petroleum.

Hydrogenation has been widely used for petroleum desulfurization. Because of the rigorous operating conditions (high pressure and temperature in the presence of a catalyst) and the consumption of hydrogen, this technique can be expensive both in the capital investment needed and in its operating costs [3].

Several microorganisms are used to remove sulfur compounds which are recalcitrant to HDS, such as dibenzothiophene (DBT) and its alkylated derivatives [4]. Degradation aided by microbes is a viable bioremediation technology for organic pollutants. There is a wide range of microorganisms which could be involved in the breaking of chemical bonds. So, bioremediation uses the metabolic versatility of microorganisms to degrade hazardous pollutants [5].

Biodesulfurization (BDS) is one of the most promising technologies that is used together with the traditional hydrodesulfurization (HDS) to reduce the sulfur content in fossil fuels [6]. The limited efficiency of the HDS method leads to residues which may contain thiophenic compounds. Most of the sulfur remaining in a fuel after a HDS process is in the form of thiophenic compounds, of which dibenzothiophene (DBT) is a typical recalcitrant. Therefore desulfurizing DBT has been used as a model reaction when treating fossil fuels [7].

Research on biodesulfurization using DBT has resulted in two different biochemical pathways, named Kodama [8] after their author and 4S [9]. The Kodama pathway is considered unsuitable for treating fuel because water-soluble sulfur compounds are produced, which are thus unavailable for burning and so reduce the caloric value of the fuel.

On using the 4S pathway, DBT is transformed into 2-hydroxybiphenyl (2-HBP) and sulfite. In this pathway, the carbon skeleton of DBT is released intact under the formation of 2-HBP; therefore, its calorific value is not lost [10].

We report experiments of bioremediation of DBT by *Pseudomonas fluorescens*. This microorganism was found to degrade DBT via the sulfur-selective pathway and hence was applied so as to desulfurize diesel oil. The bacteria succeeded in appreciably decreasing the sulfur in DBT, thus indicating their potential for use as biocatalysts in desulfurizing fossil fuels.

http://dx.doi.org/10.5772/intechopen.70430

2. Materials and methods

2.1. Microorganism and culture conditions

The experiments were carried out using a pure culture of bacterium *P. fluorescens* (UCP 1514) belonging to the Nucleus for Research in Environmental Sciences (NPCIAMB), of the Catholic University of Pernambuco. The bacterial culture was maintained in solid medium nutrient agar consisting of meat extract (5.0 g); peptone (10.0 g); NaCl (5.0 g); agar (15.0 g/l of distilled water), to which was added by syringe; and 2 mM DBT dissolved in dimethylfomamide (DMF), at 4°C as a stock of bacterial inoculum.

2.2. Fermentation conditions

The culture was grown in Luria-Bertani (LB) medium containing tryptone (10.0 g), yeast extract (5.0 g), NaCl (10.0 g), and glucose (5.0 g/l) as supplement, as per Konishi et al. [11]. The following DBT concentration and solvent was used as stock solution: 1 M in dimethylfomamide (DMF) [12]. The solution was sterilized in a Millipore® filter.

The Erlenmeyer flasks were autoclaved at 120°C for 30 min. Batch kinetic experiments were carried out in 250 mL Erlenmeyer flasks by adding 2 mM DBT solution to 100 mL LB liquid culture medium.

Another culture was prepared as inoculum just before batch experiment, and while the bacteria were in the exponential phase, a volume of 5 mL consisting of 108 cells/mL was transferred to 100 mL of liquid medium. All experiments were made in triplicate followed by control trials without inoculum, performed in a rotary shaker operating at 150 rpm at a constant temperature of 37°C for 144 h. Aliquots were obtained every 24 h.

Bacterial growth was monitored through biomass and pH using a pH meter (Orion 310). The specific growth rate (μ_{esp}) and generation time (T_G) were determined according to [13]. For the specific growth rate, Eq. (1) was used:

$$\begin{aligned} &\mu_{esp} = \left(ln_x - ln_{x_0} / T - T_0\right) \\ &\text{Where,} \\ &X = \text{Final biomass} \\ &X_0 = \text{Initial biomass} \\ &T = \text{Final time} \\ &T_0 = \text{Initial time} \\ &\text{The generation time was determined by:} \\ &T_G = ln_2 / \mu_{esp} \end{aligned} \tag{1}$$

After cultivation, the samples were centrifuged at 5000 × g for 15 min at 5°C to separate the biomass and metabolic liquid. After separation the biomass was lyophilized to determine the dry weight of biomass per unit volume.

2.3. HPLC and GC-MS analysis

In order to determine the intermediates of the pathway of DBT biodesulfurization by microorganisms, the bacterial culture was grown at 37°C in a rotary shaker at 150 rpm for different periods and was then centrifuged at 5000 × g for 15 min at 5°C to separate the biomass and metabolic liquid.

The supernatant was extracted with an equal volume of ethyl acetate. The organic layer was removed, and the aqueous layer was acidified with 5 N HCl to pH 2.0 and again extracted with an equal volume of ethyl acetate.

The two extracts were pooled, evaporated to dryness under vacuum in a rotavapor system and resuspended in 1 mL of ethyl acetate. DBT was quantified by HPLC analysis, for which the organism was grown in LB containing 2 mM DBT for different time periods, mixed with an equal volume of ethyl acetate and analyzed by using an HPLC Varian UV-VIS detector, model 320. Separation was carried out with a C18 RP column (4.6 × 250 mm), solvent liberation system, model 210 Varian Star®, with the following separation conditions: mobile phase acetonitrile 75% and phosphate buffer 10 mM (pH 6.0). The eluate was detected at 232 nm [14].

Gas chromatography (GC) analysis was performed in a Varian Star 3600CX with a DB-1 (100% dimethylpolysiloxane) fused capillary column (30 m × 0.25 mm); column temperatures were programmed from 50°C for 5 min, raised to 10°C/min, and then increased to 250°C for 5 min, with a total time of 30 min for integrating purposes.

Injector and detector temperatures were 250°C. 3.0 μL of a solution of about 2 mM DBT in ethyl acetate was injected. DBT analysis was carried out using a Varian Star CX 3600 equipped with Varian Saturn 2000. The carrier gas was helium 5.0 WHITE MARTINS, MS split dibenzothiophene WAX. Mass spectra were taken at 70 eV. Scanning speed was 1.5 scans/s from m/z 40 to 500.

The samples containing DBT were also analyzed by GC-MS; identification was made by the computerized matching of the mass spectra obtained with those stored in the MAINLIB library of the GC-MS data system and by the Spectral Database for Organic Compounds SDBS library.

3. Results

3.1. Effects of DBT on the growth of *P. fluorescens*

P. fluorescens UCP 1514 grew in the control as well as in DBT medium (**Figure 1**). The beginning of the exponential phase was observed between 24 and 72 h of cultivation. The growth rate of the control was 0.65 h^{-1}, and in experiments with DBT, it was 0.2 h^{-1}, with a generation time of 0.027 and 0.006 h in the control and in the DBT culture, respectively (**Table 1**).

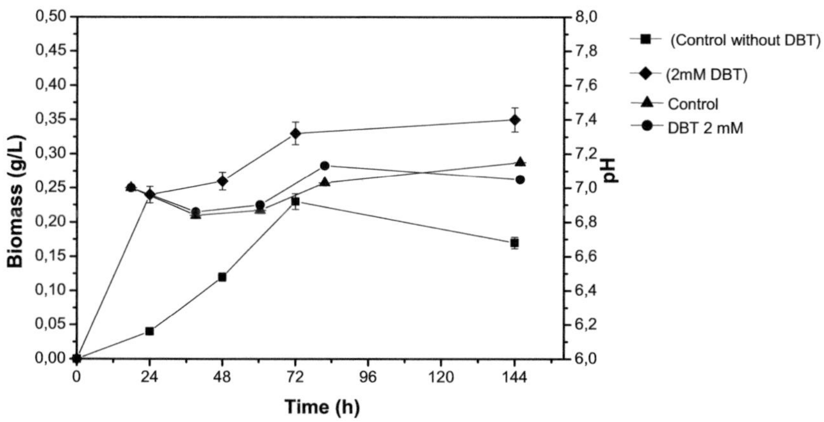

Figure 1. Growth and pH of *Pseudomonas fluorescens* cultured in Luria-Bertani medium with 2 mM dibenzothiophene (DBT) as sole sulfur source. (•) pH with DBT, (■) pH control, (•) biomass with DBT, and (■) biomass control.

The pH did not decrease during cultivation. With the control, growth was observed until 72 h of cultivation with 0.23 g dry cell/L. After this period, the decline phase of microorganisms began, with 0.17 g dry cell/L at 144 h. However, in experiments with DBT, the opposite occurred; after adaptation time, the population of the microorganism continued to increase, presenting 0.35 g dry cell/L at 144 h as well as maintaining it, probably due to using hydrocarbon and its metabolites as the main energy source [15].

3.2. The use of DBT by *P. fluorescens*

To study the ability of the bacteria to use DBT in liquid cultures, the amount of DBT in the cultures was measured by HPLC as a function of time. DBT utilization and its decrease over time by *P. fluorescens* are shown in **Figure 2**. Biodegradation started after 48 h of culture, presenting a decrease on the initial DBT concentration from 2 to 1.98 mM and at the end of fermentation fell to 0.54 mM, having metabolized about 73% of DBT.

Desulfurizing DBT by *P. fluorescens* and determining pathway intermediates

In order to determine the pathway intermediates of DBT degradation by *P. fluorescens*, GC-MS of ethyl acetate extracts of cultures grown over different periods were performed and revealed several peaks (**Figures 3–5**). Besides the solvent peaks, three peaks were further analyzed by their mass spectra to deduce the structures.

Experiments	μ_{esp}	T_G
Control	0.65 h^{-1}	0.027 h
Medium with 2 mM DBT	0.32 h^{-1}	0.006 h

Table 1. Values of μ_{esp} and T_G during kinetics growth of *P. fluorescens*.

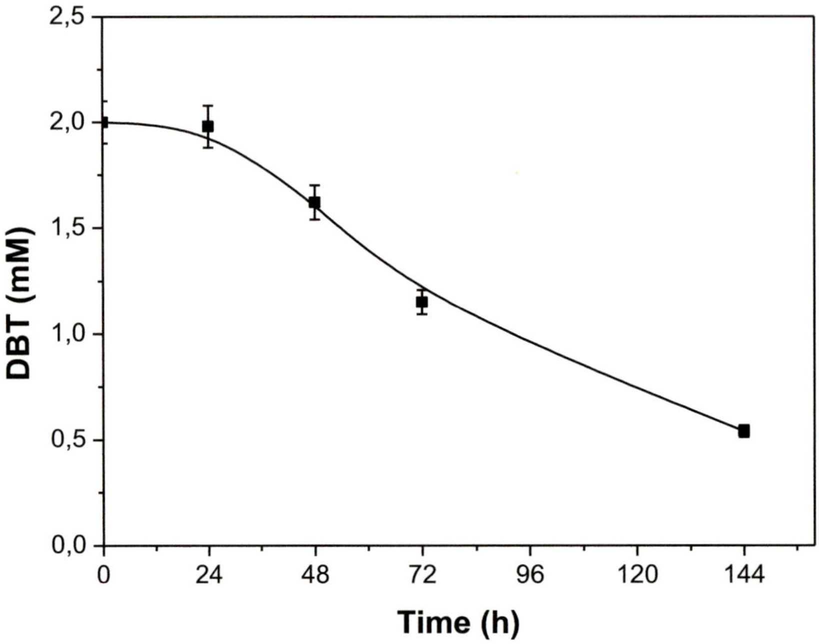

Figure 2. Time course of dibenzothiophene (DBT) utilization by *Pseudomonas fluorescens* (UCP 1514A) cultured in Luria-Bertani medium with 2 mM DBT as sole sulfur source.

As shown in **Figures 3–5**, three compounds were detected as metabolites of DBT (16.7 min, m/z = 184); one metabolite was identified as benzoic acid (11.9 min, m/z = 105), the second as methanecarbothiolic acid (0.66 min, m/z = 42), and the third as biphenyl (12.3 min, m/z = 154). Based on the structures of the metabolites (**Figure 6**) and the DBT-desulfurizing pathway [16], we suggest that DBT (1) degradation by *P. fluorescens* should begin by identifying dibenzothiophene sulfone with degradation to benzoic acid and changes from methanecarbothiolic acid to 2-mercaptobenzoic acid (2).

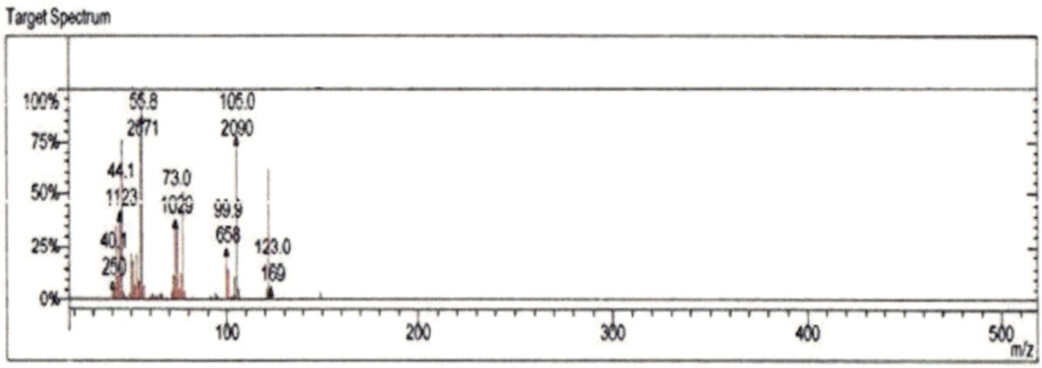

Figure 3. Representative mass spectra of dibenzothiophene in *Pseudomonas fluorescens*. (A) Benzoic acid, (A.1) benzoic acid (Spectral Database for Organic Compounds SDBS library).

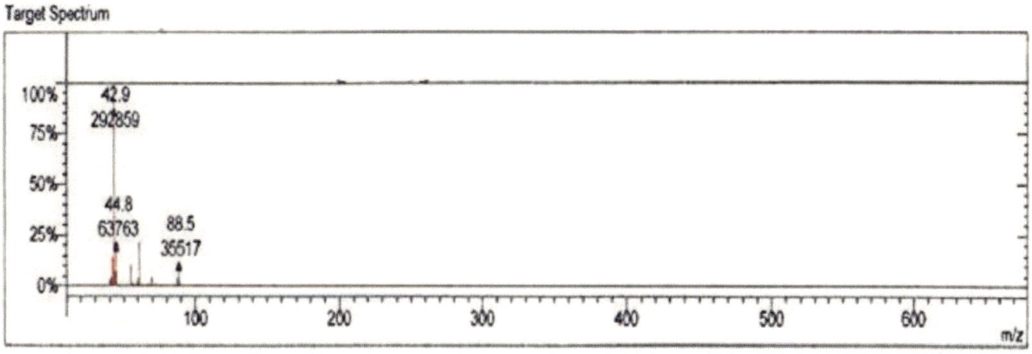

Figure 4. Representative mass spectra of dibenzothiophene in *Pseudomonas fluorescens*. (B) Methanecarbothiolic acid, (B.1) methanecarbothiolic acid (Spectral Database for Organic Compounds SDBS library).

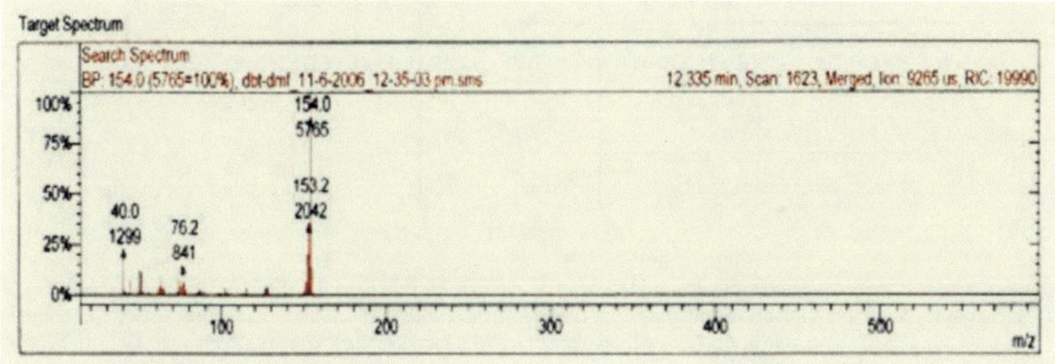

Figure 5. Representative mass spectra of dibenzothiophene in *Pseudomonas fluorescens*. (C) Biphenyl, (C.1) biphenyl (Spectral Database for Organic Compounds SDBS library).

According to [17], metabolite (3) had a molecular ion at m/z 216 and fragment ions at m/z 200 and 184 produced by two sequential losses of O (M+ −16 and −32) and thus was identified as dibenzothiophene sulfone (**Table 2, Figure 6**).

The detection of dibenzothiophene sulfone demonstrates the presence of the step of sulfur oxidation. According to [11], metabolite (4) was assumed to be dibenzo[c] [1,2]oxatiin S-oxide (m/z = 166). The abundance of fragment ions at (m/z = 182) may be due to loss of S = O from the molecular ion. Metabolite (5) was deemed to be dibenzo[c] [1,2]oxatiin S,S-dioxide. DBT was transformed to DBT sulfone [18], and a ring cleavage product 4-methoxybenzoic acid (6) was detected.

This fragmentation could be explained by the loss of oxygen after a loss of carbonyl. According to [1] metabolite (7) was assumed to be 2-HBP (m/z = 170); metabolite (8) was assigned as 2-methoxybiphenyl (2-MBP) (m/z = 184), and the last metabolite was designated as biphenyl (m/z = 154) (9), corresponding to loss of hydrogen and phenol group formation.

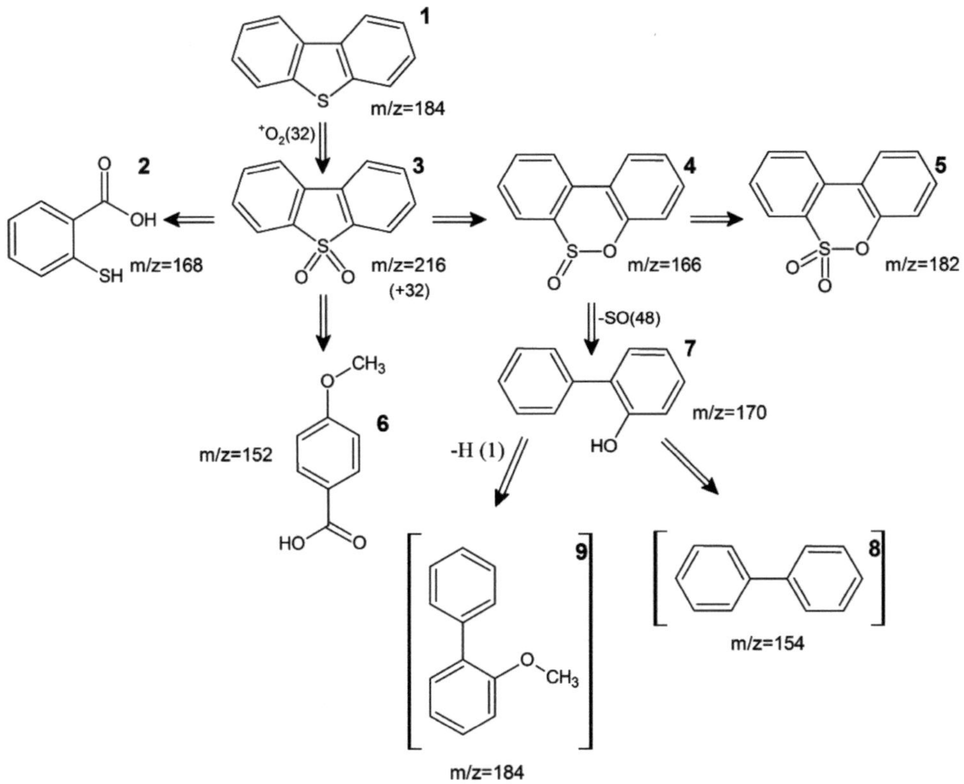

Figure 6. Proposed catabolic pathways of dibenzothiophene by *Pseudomonas fluorescens* (UCP 1514A). (1) Dibenzothiophene, (2) 2-mercaptobenzoic acid (Me), (3) dibenzothiophene 5,5′-dioxide, (4) dibenzo[e][1,2]oxathiin S-oxide, (5) dibenzo[c][1,2]oxathiin S,S-dioxide, (6) 4-methoxybenzoic acid, (7) 2-hydroxybiphenyl, (8) biphenyl, and (9) 2-methoxybiphenyl. The structures in brackets were the metabolites detected; the other structures are proposed intermediates or metabolites, but not detected.

ID	Fragment ion *m/z* (relative abundance, %)	Compound	Molecular ion *m/z*
1	185 (14.6), 184 (100), 139 (10.8), 92 (7.8)	Dibenzothiophene	184
2	165 (36), 136 (100), 108 (35)	2-Mercaptobenzoic acid (Me)	168
3	216 (14), 200 (17), 184 (100)	Dibenzothiophene sulfone	216
4	118 (100), 90 (47), 63 (33), 39 (20), 48 (19), 166 (9), 77 (5), 110 (4), 138 (3)	Dibenzo[c][1,2]oxatiin S-oxide	166
5	89 (100), 118 (99), 182 (73), 63 (65), 43 (38), 134 (25), 51 (22), 71 (18), 97 (7), 109 (5), 150 (4)	Dibenzo[c][1,2]oxatiin S,S-dioxide	182
6	152 (74.6), 135 (100), 107 (14.5), 92 (10), 77 (20.5)	4-Methoxybenzoic acid	152
7	171 (12.8), 170 (100), 141 (23.8), 169 (46.5), 115 (16.7)	2-Hydroxybiphenyl	170
8	186 (1.2), 185 (14.2), 184 (100), 169 (42.8), 141 (25.7), 139 (8.6)	2-Methoxybiphenyl	184
9	154 (100), 152 (18), 151 (5.1), 76.2 (10.3)	Biphenyl	154

Table 2. Mass spectral characteristics of proposed degradation products.

4. Discussion

P. fluorescens UCP 1514 remained viable in the presence of DBT, probably due to this element being considered essential for the formation of amino acids such as cysteine, and methionine, some vitamins, and other compounds that are important for the survival of microorganisms [19].

The development of BDS has been prompted by increasingly stringent regulations regarding limiting the sulfur content in transportation fuels. BDS is based on 4S pathway. When the pathway is used, DBT is not degraded, but it is transformed into 2-HBP which partitions in the oil phase, resulting in no loss of caloric value of the fuel [10]. DBT-containing hydrocarbon was desulfurized by growing cells of *Pseudomonas* sp. [20], as well as by growing and resting cells of *Gordonia* sp. at 30°C [21]. Martinez et al. [22] studied the biodesulfurization of dibenzothiophene by resting cells of *Pseudomonas putida* CECT5279, where complete conversion into HBP was reached in 90 min, using a gas flow of 2 vvm, thus showing that the 4S route is highly sensitive to oxygen availability working under oxygen-limiting conditions.

Several authors stated that the initial degradations of anthracene and other compounds do not depend on lignolytic activity and suggested that cytochrome P-450 monooxygenase could be responsible for this initial step [23].

The catabolic pathway of DBT was proposed, based on the metabolites tentatively identified by their mass spectra (**Figure 6**). The analysis of DBT metabolites suggested that *P. fluorescens* UCP 1514 can decompose DBT by multiple pathways. Ref. [24] reported dibenzothiophene 5-5′-dioxide, which is involved in the sulfur-specific process of biodesulfurization named as 4S pathway [7]. In this study, DBT degradation by *P. fluorescens* started from dibenzothiophene sulfone which was probably further metabolized to 2-mercaptobenzoic acid, according to [24]. Ref. [25] detected several DBT metabolites including benzo[b]thiophene-2,3-diol, 2-mercaptobenzoic acid, and 2,2′-dithiosalicylic acid from *P. fluorescens* 17 and 26 cultures.

The dibenzothiophene was transformed to DBT sulfone and from this compound to dibenzo[c][1,2]oxathiin S-oxide and dibenzo[c][1,2]oxathiin S,S-dioxide or metabolized to 4-methoxybenzoic acid, a ring cleavage product [26]. The conversion of DBT S,S-dioxide into dibenzo[c][1,2]oxathiin S-oxide could be explained by the oxidative cleavage of one of the two C-S bonds in DBT S,S-dioxide and circularization of the cleavage product under acidic conditions. However, it is likely that DBT sulfone is also an intermediate compound in the bacterial DBT degradation pathway. No peak corresponding to DBT sulfone was detected in our analysis. Benzo[c][1,2]oxathiin S-oxide was also found as the metabolite from benzothiophene in *Gordonia* sp. 213E strain [27]. This can be formed from (Z)-2-(2′-hydroxyphenyl)ethene1-sulfinate, the thiophene ring-open form, under acidic extraction conditions [11]. As shown in **Figure 6**, two possible metabolites, not including any sulfur in their molecular structures, were presented at the end of the biodegradation pathway of DBT.

Based on the deduced structures such as 2-MBP and biphenyl, it is interesting to note that *P. fluorescens* (UCP 1514) can produce two extra sulfur-free metabolites. In addition, according to the molecular structure of the two metabolites, it is presumed that they might be further produced from 2-HBP or from the intermediate metabolite HPBS in parallel with 2-HBP.

It was also supposed that 2-MBP was formed by the methylation of the hydroxyl group of 2-HBP [28]. These results suggest that producing 2-MBP and biphenyl has the advantage of partially eliminating the enzyme inhibitory effect of 2-HBP, thereby prolonging desulfurizing activities. Therefore, further studies investigating the enzymatic and genetic properties of UCP 1514 are being carried out to explore the possibilities of inducing this strain to form more 2-MBP and biphenyl. As reported, 2-HBP was toxic to bacterial cells, and once the concentration of 2-HBP was above 0.2 mM, the biodesulfurization of DBT was inhibited [29].

The inhibitory effects of 2-HBP, the end product of the desulfurization of dibenzothiophene (DBT) by the 4S pathway, were noted early in the study of biodesulfurization [30–32]. However, product inhibition has not been fully appreciated as the main impediment to developing improved biocatalysts, and effective means of overcoming the inhibition caused by 2-HBP are just beginning to be developed [33]. Only about half of the DBT consumed can be accounted for as 2-HBP in some biodesulfurization studies, implying that 2-HBP accumulates intracellularly [34–37]. This results in some increase in the activity of the 4S pathway. Moreover, the high enzymatic activities reported for the compartmentalization of the reaction steps of the 4S pathway illustrate that the inhibition by 2-HBP of desulfurization enzymes and the metabolism of cells strongly influence the desulfurization activity of biocatalysts [33]. Because 2-HPB strongly inhibits the 4S pathway, there is advantage in selecting any culture that alleviates/overcomes this inhibition. Among the numerous biodesulfurization cultures isolated, there are several that possess an extended 4S pathway such that 2-HPB is not the end product of DBT desulfurization, and instead 2-methoxybiphenyl (2-MBP) [38–40], biphenyl [41], or 2,2-dihydroxybiphenyl [42] are produced. The toxicity of 2-MBP is reported to be only slightly less than 2-HBP [43], but this study examined the inhibition of DBT biodesulfurization by externally added 2-MBP or 2-HBP and did not address the intracellular accumulation of 2-HBP [44].

In biodesulfurization, DBT is converted to 2-HBP, which increases the possibility of environmental pollution [45]. In contrast, the production of 2-MBP and biphenyl may partially eliminate the inhibitory effect of products and pollution from diesel oil combustion. In this study, intermediate metabolites such as DBTO, DBTO2, and HPBS were not detected by GC-MS analysis, which could be explained by the lability of desulfurized DBT metabolites [16] or the existence of an additional degradation pathway for DBT [46].

The capabilities of the isolated bacteria to survive and desulfurize a wide range of S compounds present in crude oil are desirable traits for the development of a robust BDS biocatalyst to upgrade crude oils and refinery streams [47]. Therefore, UCP 1514 desulfurizes DBT using a sulfur-specific degradation pathway, with selective cleavage of the C-S bonds, and is considered to be novel and different from other pathways.

Further work is required to find out the actual mechanism of DBT metabolism. There are some technologies that oil refineries use to remove sulfur, such as oxidative desulfurization (ODS), photocatalytic desulfurization, and HDS. HDS involves catalytic treatment of fuel at high temperatures (>300°C) and pressures (>100 atm) to remove the bulk of inorganic sulfur. Simple organic sulfur can be removed by HDS, but this process is inadequate for producing low-sulfur fuels as it is unable to remove the complex polycyclic sulfur compounds present in petroleum and coal.

Oxidative desulphurization (ODS) enables ultra-low-sulfur content to be attained in diesel fuels by oxidizing refractory sulfur compounds that are difficult to remove with hydrodesulphurization when the sulfur content to be attained needs to be below 10 mg/kg [48]. On using ODS, sulfur-containing hydrocarbons can be only oxidized to sulfoxides and sulfones using H2O2 as oxidants in acetonitrile as solvents [49].

Desulfurization can be performed efficiently by irradiation combined with a chemical catalyst. But the consumption of chemicals seemed to be excessive, which would result in a higher cost, according to [3].

Until now, conventional refining processes have been performed at much higher temperatures. Therefore thermophilic biodesulfurization is desirable without cooling the stock to 30°C [50]. Moreover, thermophilic biodesulfurization also reduces the viscosity of crude oil, which makes the process more practicable [51].

5. Conclusion

P. fluorescens UCP 1514 desulfurizes DBT through a sulfur-specific degradation pathway, with selective cleavage of the C-S bonds, and is considered to be novel and different from other pathways. Further work is required to find out the actual mechanism of DBT metabolism. The results indicate that *P. fluorescens* could have a good potential to be used in biocatalytic desulfurization of fossil fuels. Further work aimed at developing the strain is underway.

Acknowledgements

This research study was supported by the Brazilian National Council for Technological and Scientific Development (CNPq) and Financial Agency of Studies and Projects (FINEP) to whom the authors are grateful.

Author details

Thayse A.L. Silva[1,3], Manfred Schwartz[2], Patrícia M. Souza[1], Ian Garrard[3],
Galba M. Campos-Takaki[1]* and Elias B. Tambourgi[4]

*Address all correspondence to: galba_takaki@yahoo.com.br

1 UNICAP, Nucleus of Research in Environmental Sciences (NPCIAMB), Recife, PE, Brazil

2 SI Group Crios Resinas S.A., Rio Claro, SP, Brazil

3 Brunel University, Institute for Bioengineering, West London, United Kingdom

4 School of Chemical Engineering (FEQ), Department of Chemical Engineering,
State University of Campinas (UNICAMP), SP, Brazil

References

[1] Li F, Zhang Z, Feng J, Cai X, Xu P. Biodesulfurization of DBT in tetradecane and crude oil by a facultative thermophilic bacterium *Mycobacterium goodie* X7B. Journal of Biotechnology. 2007;**127**:222-228. DOI: 10.1016/j.jbiotec.2006.07.002

[2] Nojiri H, Omori T. Molecular bases of aerobic bacterial degradation of dioxins: Involvement of angular dioxygenation. Bioscience, Biotechnology, and Biochemistry. 2002;**66**:2001-2016. DOI: 10.1271/bbb.66.2001

[3] Qu Z, Yan N, Zhao Y, Jia J, Wu D. Removal of dibenzothiophene from the simulated petroleum by γ-irradiation induced reaction. Energy and Fuels. 2006;**20**:142-147. DOI: 10.1021/ef050171v

[4] Kilbane JJ. Biodesulfurization: How to make it work? Arabian Journal for Science and Engineering. 2017;**42**:1-9. DOI: 10.1007/s13369-016-2269-1

[5] Jong-Su S, Young-Soo K, Qing XL. Bacterial degradation of aromatic compounds. International Journal of Environmental Research. 2009;**6**:278-309. DOI: 10.3390/ijerph6010278

[6] Monticello DJ. Biodesulfurization and the upgrading of petroleum distillates. Current Opinion in Biotechnology. 2000;**11**:540-546. DOI: 10.1016/s0958-1669(00)00154-3

[7] Oldfield C, Pogrebinsky O, Simmonds J, Olson E, Kulpa CF. Elucidation of the metabolic pathway for dibenzothiophene desulfurization by *Rhodococcus* sp. strain IGTS8 (ATCC 53968). Microbiology. 1997;**143**:2961-2973. DOI: 10.1099/00221287-143-9-2961

[8] Kodama K, Umehara K, Shimazu K, Nakatani S, Minoda Y, Yamada K. Identification of microbial products from dibenzothiophene and its proposed oxidation pathway. Agricultural and Biological Chemistry. 1973;**37**:45-50. DOI: 10.1271/bbb1961.37.45

[9] Kertesz M, Wieteck C. Desulfurization and desulfonation: Application of sulfur-controlled gene expression in bacteria. Applied Microbiology and Biotechnology. 2001;**57**:460-466. DOI: 10.1007/s002530100800

[10] Mohebali G, Ball AS, Rasekh B, Kaytash A. Biodesulfurization potential of a newly isolated bacterium *Gordonia alkanivorans* RIPI90A. Enzyme and Microbial Technology. 2007;**40**:578-584. DOI: 10.1016/j.enzmictec.2006.05.012

[11] Konishi J, Onaka T, Ishii Y, Suzuki M. Demonstration of carbon sulfur bond targeted desulfurization of dibenzothiophene by thermophilic *Paenibacillus* sp. strain A11-2 capable of desulfurizing dibenzotiophene. FEMS Microbiology Letters. 2007;**187**:151-154. DOI: 10.1016/s0378-1097(00)00186-5

[12] Setti L, Rossi M, Lanzarini G, Pifferri PG. The effect of n-alkanes in the degradation of dibenzothiophene and of organic sulfur compounds in heavy oil by *Pseudomonas* sp. Biotechnology Letters. 1994;**14**:515-520. DOI: 10.1007/bf01023178

[13] Pirt SJ. Principles of Microbe and Cell Cultivation. London: Blackwell Scientific Publications; 1975. p. 214-215. DOI: 10.1002/aic.690220342

[14] Abbad-Andaloussi S, Lagnel C, Warzywoda M, Monot F. Multi-criteria comparison of resting cell activities of bacterial strains selected for biodesulfurization of petroleum compounds. Enzyme and Microbial Technology. 2003;**32**:446-454. DOI: 10.1016/s0141-0229(02)00320-4

[15] Kropp KG, Andersson JT, Fedorak PM. Bacterial transformation of 1, 2, 3, 4-tetrahidrodibenzothiophene and dibenzothiophene. Applied and Environmental Microbiology. 1997;**63**:3032-3042. DOI: 10.1021/es960869a

[16] Furuya T, Kirimura T, Kuno K, Usami S. Thermophilic biodesulfurization of dibenzothiophene and its derivatives by *Mycobacterium phlei* WU-F1. FEMS Microbiology Letters. 2001;**204**:129-133. DOI: 10.1016/s0378-1097(01)00397-4

[17] Jong-Su S, Young-Soo K, Il KC, Qing XL. Degradation of dibenzothiophene and carbazole by *Arthrobacter* sp. P1-1. International Biodeterioration and Biodegradation. 2006;**58**:36-43. DOI: 10.1016/j.ibiod.2006.04.005

[18] Ichinose M, Saito M, Wada H, Kitano A, Kondo N, Nishiyasu T. Modulation of arterial baroreflex dynamic response during muscle metaboreflex activation in humans. The Journal of Physiology. 2002;**544**:939-948. DOI: 10.1113/JPHYSIOL.2002.024794

[19] Kirimura K, Furuya T, Ishii Y, Kino K, Usami S. Biodesulfurization of dibenzothiophene and its derivatives through the selective cleavage of carbon-sulfur bonds by moderately thermophilic bacterium *Bacillus subtilis* WU-S2B. Journal of Bioscience and Bioengineering. 2001;**91**:262-266. DOI: 10.1263/jbb.91.262

[20] Setti L, Lanzarini G, Pifferi PG. Dibenzothiophene biodegradation by a *Pseudomonas* sp. in model solutions. Process Biochemistry. 1995;**30**:721-728. DOI: https://doi.org/10.1016/0032-9592(94)00040-9

[21] Chang JH, Chang YK, Ryu HW, Chang Ho N. Desulfurization of light gas oil in inmobilized-cell systems of *Gordona* sp. CYKS1 and *Nocardia* sp. CYKS2. FEMS Microbiology Letters. 2000;**182**:309-312. DOI: 10.1016/s0378-1097(99)00604-7

[22] Martinez I, Santos VE, Gomez E, Garcia-Ochoa F. Biodesulfurization of dibenzothiophene by resting cells of *Pseudomonas putida* CECT5279: Influence of the oxygen transfer rate in the scale-up from shaken flask to stirred tank reactor. Journal of Chemical Technology and Biotechnology. 2016;**91**:184-189. DOI: 10.1002/jctb.4559

[23] Verdin A, Sahraoui ALH, Durand R. Degradation of benzo[a]pyrene by mitosporic fungi and extracellular oxidative enzymes. International Biodeterioration and Biodegradation 2004;**53**:65-70. DOI: https://doi.org/10.1016/j.ibiod.2003.12.001

[24] Seo JS, Keum YS, Cho I, Li QX. Degradation of dibenzothiophene and carbazole by *Arthrobacter* sp. P1-1. International Biodeterioration and Biodegradation. 2006;**58**:36-43. DOI: 10.1016/j.ibiod.2006.04.005

[25] Finkelstein ZI, Baskunov BP, Vavilova LN, Golovleva LA. Microbial transformation of dibenzothiophene and 4.6-dimethyldibenzothiophene. Microbiology. 1997;**66**:402-407. DOI: 10.1007/s11021-005-0067-y

[26] Eibes G, Cajthaml T, Moreira MT, Feijoo G, Lema JM. Enzymatic degradation of anthracene, dibenzothiophene and pyrene by manganese peroxidase in media containing acetone. Chemosphere. 2006;**64**:408-414. DOI: 10.1016/j.chemosphere.2005.11.075

[27] Gilbert SC, Morton J, Buchanan S, Oldfield C, McRoberts A. Isolation of a unique benzothiophene desulfurizing bacterium: *Gordona* sp. strain 213E (NCIMB 40816) and characterization of the desulfurization pathway. Microbiology. 1998;**144**:2545-2553. DOI: 10.1099/00221287-144-9-2545

[28] Li W, Zhang Y, Wang MD, Shi Y. Biodesulfurization of dibenzothiophene and other organic sulfur compounds by a newly isolated *Microbacterium* strain ZD-M2. FEMS Microbiology Letters. 2005;**247**:45-50. DOI: 10.1016/j.femsle.2005.04.025

[29] Ohshiro T, Suzuki K, Izumi Y. Regulation of di- benzothiophene degrading enzyme activity of *Rhodococcus erythropolis* D-1. Journal of Fermentation and Bioengineering. 1996;**81**:121-124. DOI: 10.1016/0922-338x(96)87588-x

[30] Bordoloi NK, Rai SK, Chaudhuri MK, Mukherjee AK. Deep-desulfurization of dibenzothiophene and its derivatives present in diesel oil by a newly isolated bacterium *Achromobacter* sp. to reduce the environmental pollution from fossil fuel combustion. Fuel Processing Technology. 2014;**119**:236-244. DOI: doi.org/10.1016/j.fuproc.2013.10.014

[31] Matsui T, Noda KI, Tanaka Y, Maruhashi K, Kurane R. Recombinant *Rhodococcus* sp. strain T09 can desulfurize DBT in the presence of inorganic sulfate. Current Microbiology. 2002;**45**:240-244. DOI: 10.1007/s00284-002-3739-0

[32] Martinez I, Mohamed ME, Rozas D, Garcia JL, Diaz E. Engineering synthetic bacterial consortia for enhanced desulfurization and revalorization of oil sulfur compounds. Metabolic Engineering. 2016;**35**:46-54. DOI: 10.1016/j.ymben.2016.01.005

[33] Guobin S, Huaiying Z, Weiquan C, Jianmin X, Huizhou L. Improvement of biodesulfurization rate by assembling nanosorbents on the surfaces of microbial cells. Biophysical Journal. 2005;**89**:L58-L60. DOI: 10.1529/biophysj.105.073718

[34] Wang P, Krawiec S. Kinetic analysis of desulfurization of dibenzothiophene by *Rhodococcus erythropolis* in batch and fed-batch cultures. Applied and Environmental Microbiology. 1996;**62**:1670-1675 PMID: 16535315 PMCID: PMC1388853

[35] Pan J, Wu F, Wang J, Xu L, Khayyat NH, Stark BC, Kilbane II JJ. Enhancement of desulfurization activity by enzymes of the *Rhodococcus dsz* operon through coexpression of a high sulfur peptide and directed evolution. Fuel. 2013;**112**:385-390. DOI: https://doi.org/10.1016/j.fuel.2013.04.065

[36] Nekodzuka S, Nakajima-Kambe T, Nomura N, Lu J, Nakahara T. Specific desulfurization of dibenzothiophene by *Mycobacterium* sp. strain G3. Biocatalysis and Biotransformation. 1997;**15**:17-27. DOI: http://dx.doi.org/10.3109/10242429709003607

[37] McFarland BL. Biodesulfurization. Current Opinion in Microbiology. 1999;**2**:257-264. DOI: 10.1016/S1369-5274(99)80045-9

[38] Ismail W, El-Said W, Raheem ASA, Mohamed ME, El Nayal AM. Biocatalytic desulfurization capabilities of a mixed culture during non-destructive utilization of recalcitrant organosulfur compounds. Frontiers in Microbiology. 2016;**7**:1-14. DOI: 10.3389/fmicb.2016.00266

[39] Li FL, Xu P, Ma CQ, Luo LL, Wang XS. Deep desulfurization of a hydrodesulfurization-treated diesel oil by a facultative thermophilic *Mycobacterium* sp. X7B. FEMS Microbiology Letters. 2003;**223**:301-307. PMID: 12829302

[40] Yu B, Tao F, Li F, Hou J, Tang H, Ma C, Xu P. Complete genome sequence of *Mycobacterium goodie* X7B, a facultative thermophilic biodesulfurization bacterium with industrial potential. Journal of Biotechnology. 2015;**212**:56-57. DOI: 10.1128/AEM.00081-06

[41] Akhtar N, Ghauri MA, Anwar MA, Akhtar K. Analysis of the dibenzothiophene metabolic pathway in a newly isolated *Rhodococcus* spp. FEMS Microbiology Letters. 2009;**301**:95-102. DOI: 10.1111/j.1574-6968.2009.01797.x

[42] Oldfield C, Wood NT, Gilbert SC, Murray FD, Favre FR. Desulfurization of benzothiophene and dibenzothiophene by actinomycete organisms belonging to the genus *Rhodococcus* and related taxa. Antonie van Leeuwenhoek. 1998;**74**:119-132. DOI: 10.1023/A:1001724516342

[43] Chen H, Cai YB, Zhang WJ, Li W. Methoxylation pathway in biodesulfurization of model organic compounds with *Mycobacterium* sp. Bioresource Technology. 2009;**100**:2085-2087. DOI: 10.1016/j.femsle.2005.04.025

[44] Kilbane JJ II. Biodesulfurization: How to make it work? Arabian Journal for Science and Engineering. 2016;**42**:1-9. DOI: 10.1007/s13369-016-2269-1

[45] Ichinose H, Wariishi H, Tanaka H. Bioconversion of recalcitrant 4-methyldibenzothiophene to water-extractable products using lignin-degrading basidiomycete *Coriolus versicolor*. Biotechnology Progress. 1999;**15**:706-714. DOI: 10.1021/bp990082z

[46] Kropp KG, Fedorak PM. A review of the occurrence, toxicity, and biodegradation of condensed thiophenes found in petroleum. Canadian Journal of Microbiology 1998;**44**:605-622. DOI: https://doi.org/10.1139/cjm-44-7-605

[47] Mohamed ME-S, Al-Yacoub ZH, Vedakumar JV. Biocatalytic desulfurization of thiophenic compounds and crude oil by newly isolated bacteria. Frontiers in Microbiology. 2015;**6**:112. DOI: 10.3389/fmicb.2015.00112

[48] Palaić N, Sertić-Bionda K, Margeta D, Podolski Š. Oxidative desulphurization of diesel fuels. Chemical and Biochemical Engineering Quarterly. 2015;**29**:323-327. DOI: 10.15255/CABEQ.2015.2184

[49] Chica A, Corma A, Domine ME. Catalytic oxidative desulfurization (ODS) of diesel fuel on a continuous fixed-bed reactor. Journal of Catalysis. 2006;**242**:299-308. DOI: 10.1016/j.jcat.2006.06.013

[50] Konishi J, Ishii Y, Onaka T, Okumura K, Suzuki M. Thermophilic carbon-sulfur-bond targeted biodesulfurization. Applied and Environmental Microbiology. 1997;**63**:3164-3169. DOI: 10.1111/j.1574-6968.2000.tb09152.x

[51] Borgne SL, Quintro R. Biotechnological processes for the refining of petroleum. Fuel Processing Technology. 2003;**81**:155-169. DOI: 10.1016/s0378-3820(03)00007-9

Chapter 15

Reduction in the Sulfur Content of Fossil Fuels by *Cunninghamella elegans* (UCP 0596) to Dibenzothiophene Compound

Patrícia Mendes de Souza,
Thayse Alves de Lima e Silva,
Marcos Antonio Barbosa Lima,
Luciana de Oliveira Franco, Manfred Schwartz,
Paulo Henrique da Silva, Lúcia Roberta Barbosa,
Aline Elesbão do Nascimento, Kaoru Okada and
Galba Maria de Campos-Takaki

Additional information is available at the end of the chapter

http://dx.doi.org/10.5772/intechopen.70432

Abstract

Biodesulfurization (BDS) is one of the most promising technologies used together with traditional hydrodesulfurization (HDS) to reduce the sulfur content of fossil fuels. In this research study, a strain of *Cunninghamella elegans* (UCP 596) was isolated from mangrove sediments to metabolize an organosulfur dibenzothiophene (DBT) compound in the concentrations of 0.5 and 1 mM and transform to DBT sulfone (DBT-5-dioxide), followed by dibenzothiophene 5,5-dioxide and 2-hydroxybiphenyl metabolites, thus suggesting the use of the "4S" metabolic pathway. The fungus also degraded the DBT completely in the first 24 h of growth on a 2.0 mM DBT concentration by angular deoxygenation, which suggests that a new second metabolic pathway was used. The DBT was consumed as the carbon source, and the sulfur was removed in the form of sulfite ion. A new product, benzoic acid, was formed at the end of the catabolism of DBT by *C. elegans* using an angular route.

Keywords: *Cunninghamella elegans*, biodesulfurization, dibenzothiophene, angular deoxygenation

1. Introduction

Environmental pollution, acid rain, and health problems are caused when sulfur dioxide is emitted into the atmosphere as the result of the combustion of petroleum fractions. To solve these problems, regulations are increasingly stringent in order to minimize the levels of sulfur emitted into the atmosphere. Hydrodesulfurization (HDS) is a conventional technology used to remove sulfur from fossil fuels. This is achieved by using metal catalysts and hydrogen gas, but despite the extremely high pressures and temperatures, the process does not eliminate heterocyclic organosulfur compounds, especially those such as dibenzothiophene (DBT) [1–3]. Hydrodesulfurization is a very effective technique for removing thiols, sulfides, and disulfides, but it is not suitable for removing thiophene compounds. Therefore, petrochemical industries have sought techniques that enable the sulfur to be removed from these heterocyclic compounds [4].

An alternative is biodesulfurization (BDS), a more efficient and low-cost process, which uses microorganisms to desulfurise these compounds, by promoting selective metabolism of the sulfur (attacking C-S) without degrading the carbon skeleton (CC), thus keeping the energy source of the molecule intact. Dibenzothiophene is considered a model compound for studying the biological desulfurization of fossil fuels and persistent compounds such as S-heterocycles in the environment [4, 5].

Several microorganisms have been studied with a view to using them to remove sulfur biochemically from DBT. Prokaryotic organisms that desulfurise organosulfur compounds without metabolizing the carbon skeleton are uncommon and are usually used in ways that seek to oxidize sulfur selectively [5–12].

Eukaryotic organisms, such as *Cunninghamella elegans*, grow on DBT by forming DBT-5-oxide and DBT-5-dioxide, but biphenyl is not formed [13]. *Trichosporon* sp. uses phenol and dibenzothiophene as the only source of carbon and sulfur, respectively, and studies on it have shown that it has the capacity to transform biphenyl and dibenzothiophene, and therefore could be used to remove these toxic compounds [14].

Trametes trogii UAMH 8156, *Trametes hirsuta* UAMH 8165, *Phanerochaete chrysosporium* ATCC 24725, *Trametes versicolor* IFO 30340 (formerly, *Coriolus* sp.), and *Tyromyces palustris* IFO 30339 all oxidized DBS to dibenzyl sulfoxide prior to oxidation to dibenzyl sulfone [15].

The fungus *Paecylomyces* sp. specifically removes sulfur by oxidizing DBT and produces 2,2′-dihidroxibifenil [16, 17]. Biological oxidation of sulfur is primarily catalyzed by two enzymatic systems, P450 monooxygenases and flavin-containing mono oxygenases (FMO). However, in most organisms, the enzymes responsible for oxygenating sulfur (S-oxygenation) have not been clearly identified. *C. elegans* catalyzes the S-oxygenation of DBT; consequently, oxygenases were identified as being responsible for the S-oxygenation of DBT in *C. elegans* [18].

Dibenzothiophene is a heterocycle compound that is regarded as the most potent environmental pollutant. Microbial degradation of this pollutant is attractive, and the bioprocesses for DBT biodegradation are environment friendly. This study focuses on investigating the biotechnological potential of *C. elegans* UCP 0596 to degrade DBT and the products formed.

2. Materials and methods

2.1. Preserving the microorganism

A microorganism was isolated from mangrove sediment of the Rio Formoso, Pernambuco, Brazil. The fungus was identified as *C. elegans* (UCP-596), deposited in the Culture Collection of the Catholic University of Pernambuco, and registered in the World Federation for Culture Collection-WFCC. The fungus was maintained on potato dextrose agar (PDA) medium at 5°C and transferred to a new medium every 4 months.

2.2. Chemicals

The DBT was purchased from Aldrich, cat: D3, 220-2 and a stock solution prepared in NN-dimethylformamide at a concentration of 1 M (w/v). The solution was sterilized in a Millipore® filter, as described by Araújo et al. [22]. All other chemicals were of analytical grade. All organic solvents were of HPLC grade (E Merck).

2.3. Inoculum and culture conditions

Inoculum from these actively growing cultures was used. *C. elegans* was grown in Potato Dextrose Agar medium (PDA) at 28°C during 6 days, until sporulation, and after this period, spore suspension was prepared containing 107 sporangioles/mL. For the assay to biodegrade dibenzothiophene in *C. elegans*, 500 mL Erlenmeyer flasks containing 200 mL of Sabouraud liquid medium (control) were used. The medium was treated with dibenzothiophene in concentrations of 0.50, 1.0, and 2 mM and inoculated with 5% of sporangiole suspension, after which the mixture was incubated in an orbital shaker at 150 rpm, at 28°C, and aliquots were collected every 24 h until 96 h of growth. Thereafter, a metabolic liquid was obtained by centrifugation at 5000 g for 15 min at 5°C to separate the biomass.

2.4. Extraction of the metabolite produced

The metabolic liquid was extracted by chromatography, using the method described by Labana et al. [19]. To determine the intermediate compounds in the metabolic pathway used by *C. elegans* to degrade DBT, the supernatants were extracted with an equal volume of ethyl acetate (60 :60 mL). The organic layer was removed, and the aqueous layer was acidified to pH 2.0 with 5 N HCl solution, and extracted with an equal volume of ethyl acetate. The extracts were dried using a rotor, evaporated in a vacuum, and re-suspended with 2 mL of ethyl acetate and analyzed by gas chromatography-mass spectrometry (GC-MS).

2.5. Analysis using gas chromatography-mass spectrometry (GC-MS)

The analysis was conducted by using a Varian Star 3600 CX Gas Chromatograph, coupled to a Varian Saturation 2000 Mass Spectrometer, with a CP-WAX 58 FFAP-CB column, 50 m, 0.32 mm ID, DF = 0.2 mm. The carrier gas used was White Martins helium 5.0, Pressure 8 PSI. The programmed temperature was 50°C for 5 min, which was increased by 10°C min^{-1}, until the temperature reached 250°C for 5 min, with a total time of 30 min. The temperature of the injector and detector was 250°C. Approximately 3.0 μL of each solution extracted with ethyl acetate was injected into the chromatograph. Spectrometry was performed at 70 eV. The scan speed was 1.5 scans s^{-1} at 40–500 m/z. The samples containing DBT were also analyzed using the GC/MS. DBT was identified by comparing the mass spectro obtained in the MAINLIB library of the GC/MS system and the Spectral Database for Organic Compounds SDBS library.

2.6. Determining the removal of DBT

The level of removal of DBT from a solution of 1–10 mM DBT was determined by using a UV Sensor to visualize a curve with at a wavelength of 250 nm running through the metabolic liquid for different concentrations of DBT (0.5, 1.0, and 2.0 mM) at a constant temperature of 28°C and shaken at 150 rpm for 144 h. Measurements were made every 24 h [20].

3. Results and discussion

3.1. Biodegradation of DBT by *C. elegans*

The remarkable ability of fungi to survive in different niches is a consequence of the evolution of enzyme systems, which have coexisted for billions of years with an enormous variety of natural substances of different origins. This diversity of substrates, which have the potential for microbial growth in hydrophobic sources, induced the production of enzymes that are suitable for transforming organic molecules with very different structures. Enzyme "arsenals" have even been able to act on synthetic chemical substances that are derived from human activities. When there are hydrophobic sources, there is no doubting that the response of the metabolism of certain microorganisms gives some additional advantages to microbial cells. This includes exploiting new ecological niches as energy sources [14]. Among fungi, *C. elegans* has been reported as having the ability to oxidize and degrade several Polycyclic Aromatic Hydrocarbons (PAHs) such as anthracene, acenaphthene, benzo(a)anthracene, benzo(a)pyrene, phenanthrene, fluoranthene, and naphthalene, as well as nitrated hydrocarbons, which are considered to be mutagenic and carcinogenic agents [15].

In this study, a strain of *C. elegans* (UCP 0596) grown on Sabouraud culture medium supplemented with 0.50 mM DBT was able to degrade about 70% of DBT in 24 h of growth. In the end, about 97% of DBT of the growing cell of *C. elegans* was removed after 96 h of cultivation, and the residual DBT was determined as being 0.015 mM (**Figure 1**). The results showed that a DBT degradation of 0.50 and 1.0 mM occurred due to the function of the concentrations and the elapsed time. On the other hand, in *C. elegans* grown on Sabouraud medium containing only 1.0 mM DBT as source, the concentration decreased to 44% in the 24 h of growth. At the end of cultivation, the content of DBT was reduced to 0.019 mM, which corresponds to biodegradation of 98.1%. The life of the DBT 2 mM by *C. elegans* records a decrease in the amount of DBT, 100% of which has been removed at 24 h, for the concentrations studied (**Figure 1**).

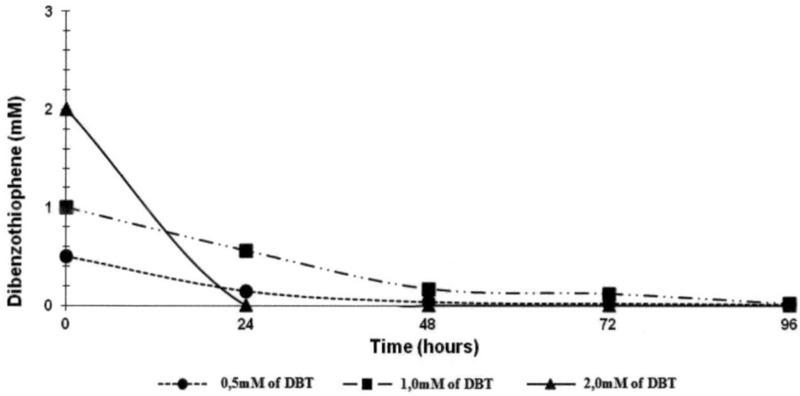

Figure 1. Degradation of dibenzothiophene by *Cunninghamella elegans* (UCP 596) over time in liquid Sabouraud culture medium, containing 0.50, 1.0, and 2.0 mM of DBT, at 28°C, for 96 h, at 150 rpm.

However, the environmental problems with DBT degradation have received much more attention from researchers worldwide. Organic compounds containing sulfur are a small but important fraction of some fuels and due to it being difficult to biodegrade them, they are considered to be recalcitrant compounds. The presence of sulfur is undesirable because it contributes to the corrosion of equipment in the refinery, and also to the emission of sulfur oxides (SOx) into the atmosphere by the combustion of oil, thus causing environmental problems, such as air pollution and being a potential cause of acid rain [17].

The metabolites produced by biodegradation of *C. elegans* were identified as dibenzothiophene 5-oxide, dibenzothiophene 5,5-dioxide in the first 48 h of growth and 2-hydroxybiphenyl within 72 h of growth. These results suggest the fungus used the "4S" metabolic pathway. The results are in agreement with studies by Schlenk et al. [13].

However, although these studies detected several products because DBT was biodegraded, desulfurization with the formation of 2-hydroxybiphenyl was not observed. In our results, dibenzothiophene dioxide 5-5, a compound found in the "4S" pathway, was found.

This pathway is one strategy for reducing these emissions, namely to remove sulfur from mineral carbon, petroleum and its derivatives before combustion. Currently, physical and chemical processes deemed to be hydrodesulfurization (HDS) ones are being used in refineries to remove inorganic sulfur (**Table 1**). These treatments incur very high costs since they involve using chemical catalysts under extreme conditions (200–425°C) and high pressures of from 150 to 205 psi [1]. Inorganic sulfur and organic sulfur can be removed by HDS, but this process is unsuitable for producing fuels with low sulfur content since this process is unable to remove sulfur compounds from complex polycyclical hydrocarbons, containing sulfur, which are present in petroleum and coal. Thus, tiophenic compounds represent a large amount of sulfur after treatment of HDS in fuels. Another strategy to reduce the sulfur content is to expose these subtracts to microorganisms that can specifically break the carbon-sulfur chain, thus releasing sulfur to a water-soluble portion, in an inorganic form. This process of microbial desulfurization or biodesulfurization (BDS) is an effective and low-cost technique [1, 2, 21].

Hydrodesulfurization (HDS)	Most commonly used method in the petroleum industry to reduce the sulfur content of crude oil. In most cases, HDS is performed by co-feeding oil and H_2 to a fixed-bed reactor packed with an appropriate HDS catalyst.
Extractive desulfurization	It is a liquid-liquid extraction process and the two liquid phases must be immiscible. It depends on the solubility of the organosulfur compounds in certain solvents.
Ionic liquid extraction	It is an interesting alternative to provide ultra clean diesel oils.
Adsorptive desulfurization	It depends on the ability of a solid sorbent to selectively adsorb organosulfur compounds from the oil.
Oxidative desulfurization (ODS)	It involves a chemical reaction between an oxidant and sulfur that facilitates desulfurization. ODS is a field of considerable interest at present.
Autoxidation	Refers to oxidation by atmospheric oxygen, i.e., oxygen in air.
Chemical oxidation	The use of a peroxide species avoids the initiation period associated with the slow in situ formation of hydroperoxides by autoxidation. The sulfur-containing compounds can directly be oxidized by the hydroperoxide to yield a sulfoxide and then a sulfone.
Catalytic oxidation	Reduce the energy barrier of oxidation by facilitating the oxidation reaction itself on the catalytically active surface; some materials serve as oxygen carriers and are more active oxidation agents than oxygen; some catalysts facilitate the decomposition of hydroperoxides, thereby accelerating the propagation step in the oxidation reaction.
Ultrasound oxidation	Provides energy for the oxidation process by ultrasound, but it does not affect the oxidation chemistry.
Photochemical oxidation	It has a high efficiency and requires mild reaction conditions. The method involves two steps: first, sulfur compounds are transferred from the oil into a polar solvent and then the transfer is followed by photooxidation or photodecomposition under UV irradiation.
Biodesulfurization (BDS)	Biodesulfurization takes place at low temperatures and pressure in the presence of microorganisms that are capable of metabolizing sulfur compounds. It is possible to desulfurize crude oil directly by selecting appropriate microbial species
Aerobic biodesulfurization	Aerobic BDS was proposed as an alternative to hydrodesulfurization of crude oil.
Anaerobic biodesulfurization	The main advantage of anaerobic desulfurization processes over aerobic desulfurization is that oxidation of hydrocarbons to undesired compounds, such as colored and gumforming products, is negligible.
Alkylation-based desulfurization	It has been tested with thiophenic sulfur compounds at small scale, and it is commercially applied for light oil at large scale as the olefinic alkylation of thiophenic sulfur (OATS) process developed by British Petroleum.
Chlorinolysis-based desulfurization	Chlorinolysis involves the scission of C–S and S–S bonds through the action of chlorine.
Supercritical water-based desulfurization	The effect of supercritical water (SCW) on desulfurization of oil is marginal. The purpose of using SCW (critical point of water: 374°C and 22.1 MPa) as reaction medium is to break C–S bonds.

Source: Javadli and Klerk [42].

Table 1. Technologies used for sulfur reduction in oil and gas industry.

3.2. Pathways for biodegradation of DBT by *C. elegans*

On the basis of these findings, the highest DBT biodegradation was observed in **Figure 2** which shows that the fungus *C. elegans* used the angular deoxygenating pathway. The fungus used DBT as a carbon source, removes the sulfur, and formed sulfite and sulfate, and at the end of biodegradation benzoic acid was produced. In this study, the DBT was catabolized by *C. elegans* for dibenzothiophene 5-oxide and dibenzothiophene 5,5-dioxide compounds in the first 48 h of growth, culminating in the formation of benzoic acid, within 72 h of growth.

The use of gas chromatography coupled to mass spectrometry identified the possible products obtained from metabolizing DBT by *C. elegans*. The catabolic pathway of DBT was proposed based on the metabolites shown in **Table 2** and the mass spectrograms in **Figure 3**. Analysis of the metabolites showed that *C. elegans* (UCP 596) can degrade DBT using two metabolic pathways.

Figure 3 confirmed the four compounds found were: dibenzothiophene 5-oxide (38.7 min, m/z = 200.2); dibenzothiophene 5,5-dioxide (29.1 min. m/z = 216.1) 2-hydroxybiphenyl (22.4 min., m/z = 170.2), and benzoic acid (22.3 min., m/z = 105.1), which were detected in the degradation of DBT (26.7 min, m/z = 184), by *C. elegans* (UCP 596).

The literature describes three well-known pathways in the process of degrading DBT by microorganisms: Kodama; angular dioxygenation; and the sulfoxide, sulfone, sulfonate, sulfate pathway deemed the "4S" [6, 9, 18–20, 23]. **Figure 4** shows the proposed routes of microbial degradation.

Dahlberg et al. [24] reported that DBT 5,5-dioxide is involved in the process of biodesulfurization by the 4S pathway. Degradation pathways of DBT by bacteria have been widely studied as in: Brevibacterium [7], Arthrobacter [24, 25], Mycobacterium [26–29]. *Pseudomonas delafieldii* and *P. putida* mineralized DBT completely, by using it as a source of carbon, sulfur, and energy, and formed as final compounds: benzoic acid, sulfite, and water by the pathway known as angular dioxygenation [30–32]. Under aerobic conditions, 3-hydroxy-2-formyl benzothiophene was formed by degrading the DBT with the bacterium *Pseudomonas* sp. [4], 1,2-dihydroxy-1,2-dihydrodibenzothiophene and dibenzothiophene 5-oxide were generated by Beijeninckia sp. [33]. Under anaerobic conditions, the biphenyl compound was formed by *Desulfovibrio desulfuricans* M6 [34]. *Corynebacterium* sp. SY1 [35] and *Rhodococcus rhodochrous* IGTS8 were only able to remove sulfur from DBT by converting the DBT into the compound 2-hydroxybiphenyl (2-HBP) [10, 36, 37]. These samples have the ability to remove organic sulfur selectively without degrading the carbon atoms [17]. The biological oxidation of sulfur is primarily catalyzed by two enzyme systems, namely P450 monooxygenases and flavin-containing monooxygenase (FMO). However, in most organisms, the enzymes responsible for oxygenating sulfur (S-oxygenation) have not been clearly identified [5].

A similar behavior was also observed with *C. elegans* (UCP 596), which was able to degrade DBT at the three concentrations studied (0.50 mM and 1 mM) after 24 and 48 h of growth. This showed the metabolites dibenzothiophene 5-oxide, 5,5-dibenzothiophene dioxide, and at the end of the experimental period of 96 h of growth, the compound 2-hydroxybiphenyl was detected. These results suggest the fungus uses the "4S" metabolic pathway

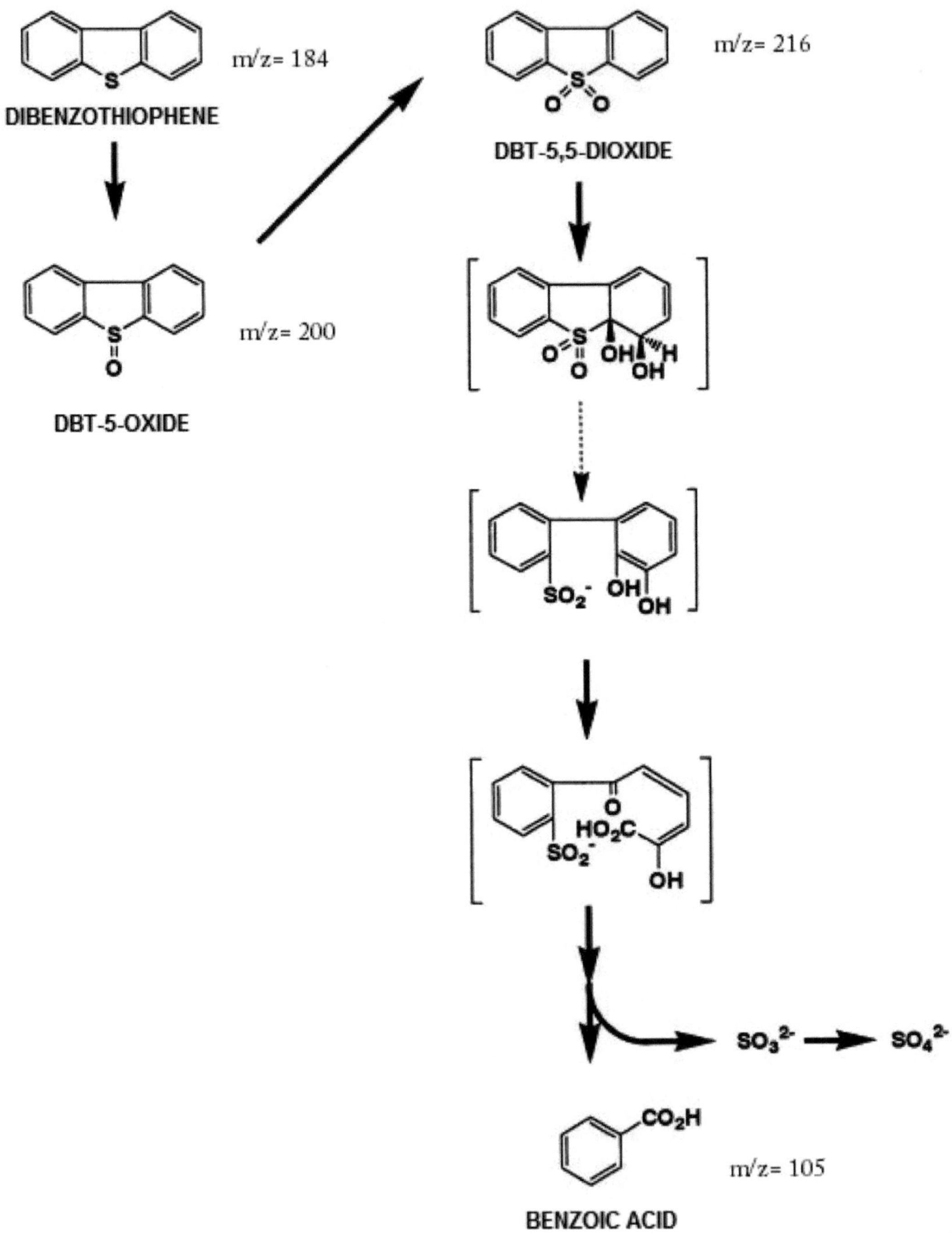

Figure 2. Suggestion for metabolic pathway used by *Cunninghamella elegans* (UCP 596) to remove the sulfur from the DBT inserted into the liquid Sabouraud culture medium, at 28°C, while being shaken at 150 rpm.

Molecular ion (m/z)	Probable compounds formed	Fragments of ions—m/z (relative to intensity %)
184	Dibenzothiophene	185 (13.37), 184 (100), 139.3 (12.14), 45.1 (10.43)
200.2	Dibenzothiophene 5-oxide	201.2 (19.05), 200.2 (100), 172.5 (43.06), 171.5 (73.15), 139.3 (12.42), 86.3 (5.95), 50.2 (14.25)
216.1	Dibenzothiophene 5,5-dioxide	216.1 (100), 187.2 (21.34), 144.3 (16.12), 104.3 (7.98), 50.2 (13.12)
170.2	2-Hydroxybiphenyl	227.3 (5.58), 170.2 (100), 140.2 (16.10), 114.2 (53.73), 85.2 (17.38), 41.2 (13.38)
105.1	Benzoic acid	118.0 (22.67), 105.1 (100), 58.2 (15.49), 45.1 (60.30)

Table 2. Characterization of the mass spectrum of the products from degrading DBT by *Cunninghamella elegans* (UCP 0596).

sulfoxide-sulfone-sulfonate-sulfate. As the specific pathway for removing sulfur atoms present in the DBT due to a progressive oxidative attack in the thiophene groups, this pathway involves a multi-enzyme system with three different activities The first enzyme is a monooxygenase of the DBT, which oxidizes DBT to 5,5-dioxide of DBT. The second enzyme is also a monooxygenase that converts 5,5-dioxide of DBT to 2-hydroxybiphenyl-2-sulfinate. Finally, a lyase that catalyzes the breaking of the C-S link, transforms the 2-hydroxybiphenyl-2-sulfinate into two end products, namely 2-hydroxybiphenyl (HBP) and sulfate. The microorganisms that use this pathway to metabolize DBT manage to remove the potentially toxic atom of the thiophenic compound, in the form of a sulfate compound, with only a slight loss of energy value occurring [16, 17, 20, 26]. Thus, the strains that use the "4S" pathway will be able to constitute a fundamental biological tool in the treatment, on a large scale, of fossil fuels, if biocatalysts in an industrial environment can be obtained [2, 36, 37].

In addition, benzoic acid was detected after 48 h of cell growth. This is a compound present in the Van Afferden or angular dioxygenation pathway that uses the DBT as a carbon source and removes the sulfur in the form of sulfite ion, the final product of which is benzoic acid. The metabolic pathway known as Van Afferden has no great interest in terms of biodesulfurization processes of fossil fuels, since the complete mineralization of the carbonaceous structure occurs, which will necessarily involve reducing the potential chemical energy of the fuels. However, microorganisms that use this metabolic pathway are potentially useful in formulating a mixed microbial inoculum for bioremediation processes of polyaromatic hydrocarbons containing sulfur released into the environment [7, 38, 39].

The results show that *C. elegans* (UCP 596) was able to degrade about 70% of DBT in 24 h of growth in Sabouraud culture medium treated with 0.50 mM of DBT, and that after 96 h of culture, the reduction was 97% of DBT from the initial content. On the other hand, the sample of *C. elegans* grown on Sabouraud treated with 1.0 mM of DBT, showed a reduction of 98.1%, after 96 h of growth, thus reducing the initial concentration to 0.019 mM of DBT.

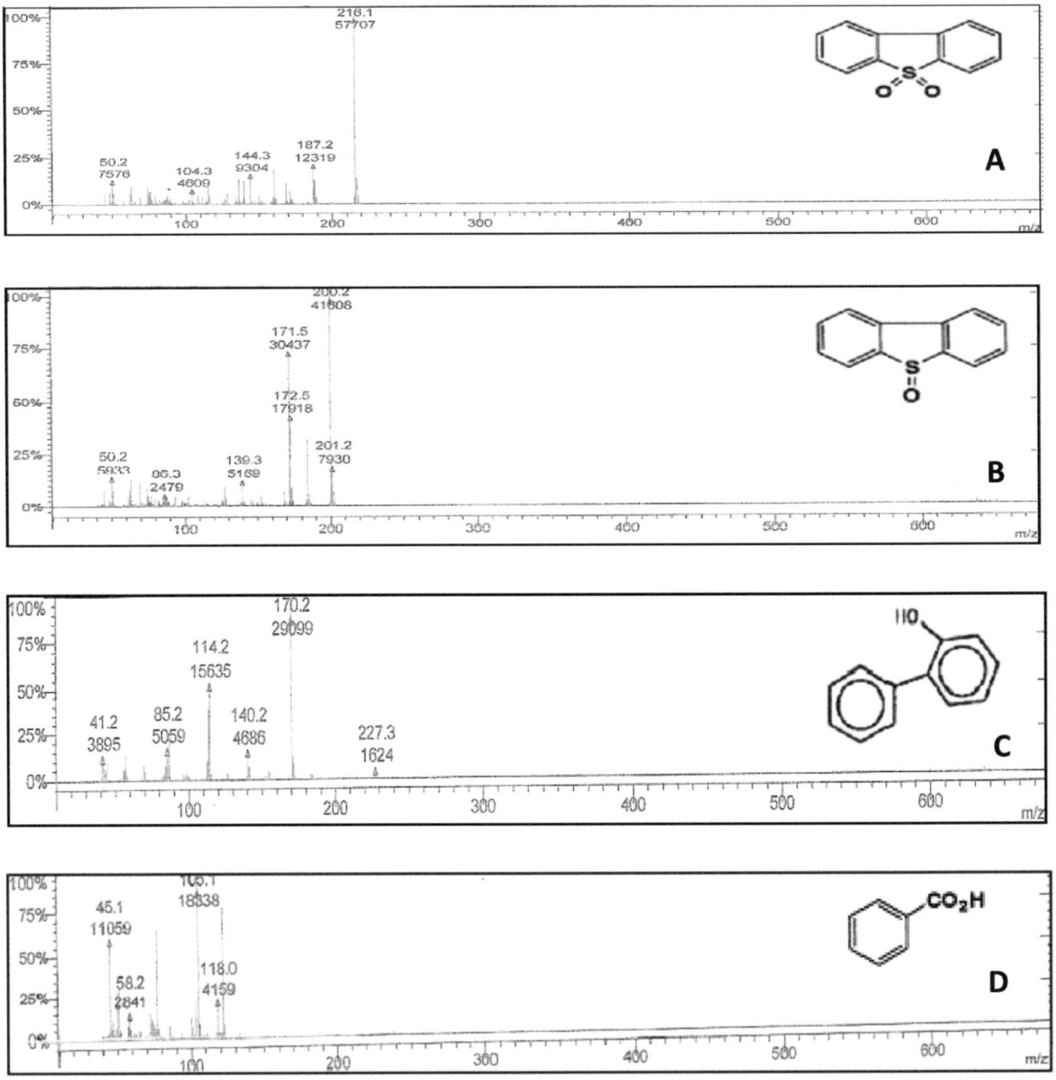

Figure 3. Mass spectrogram of dibenzothiophene 5-oxide (A), dibenzothiophene 5,5-dioxide (B), 2-hydroxybiphenyl (C), and benzoic acid (D). Metabolites formed on degrading dibenzothiophene by *Cunninghamella elegans* after 48 h of growth.

According to Schlenk et al. [13], the filamentous fungus *C. elegans* ATCC-36112, metabolized approximately 98% of dibenzothiophene at a concentration of 1.8 mg/mL for dibenzothiophene 5-oxide (86% of total metabolites) and dibenzothiophene 5-dioxide (14% of total metabolites), after incubation for 24 h. The fungus *Pleurotus ostreatus* grown in culture medium containing DBT metabolized about 84% of organosulfur in 21 days of growth. The results were analyzed by HPLC, and the main compounds formed were dibenzothiophene 5-oxide and dibenzothiophene 5,5-dioxide. The lignolitic mechanism of the fungus *P. ostreatus* may be

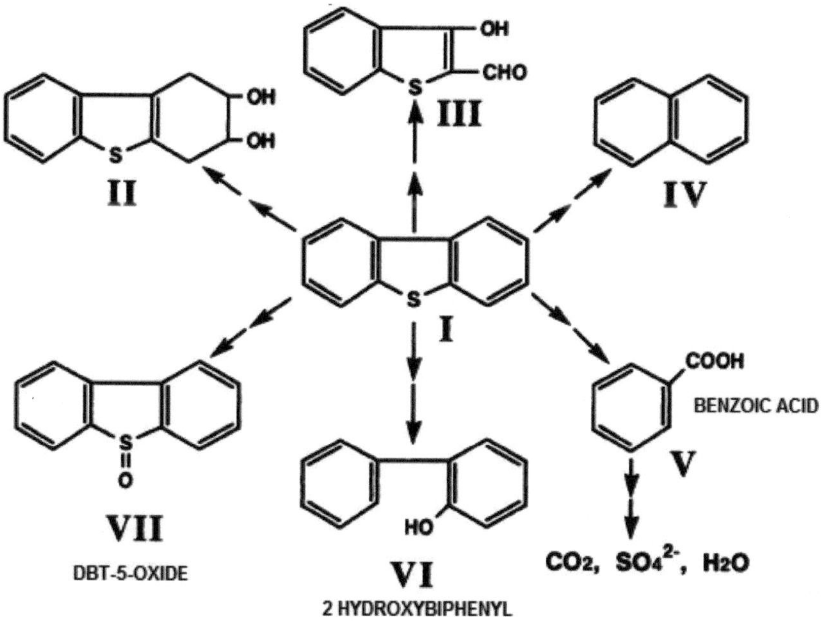

Figure 4. Pathways proposed for biodegrading dibenzothiophene by *Cunninghamella elegans*: (I) dibenzothiophene, (II) 1,2-dihydroxi-1,2-dihydrodibenzothiophene, (III) 3-hydroxy-2-formyl benzothiophene, (IV) biphenyl, (V) benzoic acid, (VI) 2-hydroxybiphenyl and (VII) dibenzothiophene 5-oxide.

involved in the metabolism of DBT [15, 40]. *Arthrobacter* sp. used the angular deoxygenating pathway for metabolizing 5,5-dioxide of DBT, a compound used as a source of carbon [41].

The data presented indicate the potential of *C. elegans* UCP 0596 using DBT as the sole source of sulfur and the greater efficiency of the fungus pathway at removing sulfur during the catabolism and that the products formed are of great importance. That activity pattern, which is shown by *C. elegans* UCP 0596 forming different biomolecules, gives information on their cellular characteristics, which undoubtedly express the potential abilities of the organism.

4. Conclusion

C. elegans UCP-596 has the ability to metabolize dibenzothiophene by an oxidation process which formed dibenzothiophene 5-oxide, dibenzothiophene 5,5-dioxide, and 2-hydroxybiphenyl metabolites. The greater efficiency of the fungus at removing sulfur suggests the use of the "4S" metabolic pathway. It can be considered for the upstream removal of sulfur.

The activity pattern of DBT degradation by *C. elegans* gives potential to benzoic acid production as the end of cellular catabolism. The fungus strain is able to desulfurize DBT in the use of angular dioxygenation pathway, as a second metabolic pathway. Its selective sulfur removal makes it proper for the industrial use of *C. elegans* UCP-596 strain.

Acknowledgements

This work was supported by Coordination Unit for Improving the Qualifications of Higher Education staff (CAPES), Foundation for Supporting Science and Technology of the State of Pernambuco (FACEPE), and National Council for Scientific and Technological Development (CNPq). The authors are grateful to the Catholic University of Pernambuco (Recife-PE, Brazil) for the use of its laboratories and the technician Severino Humberto de Almeida for his assistance.

Author details

Patrícia Mendes de Souza[1,5], Thayse Alves de Lima e Silva[1,5], Marcos Antonio Barbosa Lima[2], Luciana de Oliveira Franco[2], Manfred Schwartz[3], Paulo Henrique da Silva[5], Lúcia Roberta Barbosa[4], Aline Elesbão do Nascimento[5], Kaoru Okada[5] and Galba Maria de Campos-Takaki[5]*

*Address all correspondence to: galba_takaki@yahoo.com.br

1 Catholic University of Pernambuco – UNICAP, Recife, PE, Brazil

2 Department of Biology, Federal Rural University of Pernambuco, Recife, PE, Brazil

3 SI Group Crios Resinas S.A., Rio Claro, SP, Brazil

4 Foundation of Higher Education in Olinda – FUNESO, Olinda, PE, Brazil

5 Nucleus for Research in Environmental Sciences and Biotechnology, Catholic University of Pernambuco, Recife, PE, Brazil

References

[1] C-L Y, Zhai X-P, Zhao L-Y, Liu C-G. Mechanism of hydrodesulfurization of dibenzothiophenes on unsupported NiMoW catalyst. Journal of Fuel Chemistry and Technology. 2013;**41**:991-997. 8th ed. DOI: 10.1016/S1872-5813(13)60043-2

[2] Izumi Y, Ohshiro T, Ogino H, Hine Y, Shimao M. Selective desulfurization of dibenzothiophene by *Rhodococcus erythropolis* D-1. Applied and Environmental Microbiology. 1994;**60**:223-226. PMID: 16349153

[3] Hosseini SA, Yaghmaei S, Mousavi SM. Biodesulfurization of dibenzothiophene by a newly isolated thermophilic bacteria strain. Iranian Journal of Chemistry and Chemical Engineering Research Note. 2006;**25**:65-71. 3rd ed.

[4] Zhang SH, Chen H, Li W. Kinetic analysis of biodesulfurization of model oil containing multiple alkyl dibenzothiophenes. Applied Microbiology and Biotechnology. 2013;**97**:2193-2200. DOI: 10.1007/s00253-012-4048-6

[5] Amin GA. Integrated two-stage process for biodesulfurization of model oil by vertical rotating immobilized cell reactor with the bacterium *Rhodococcus erythropolis*. Journal of Petroleum & Environmental Biotechnology. 2011;**2**:1-4. DOI: 10.4172/2157-7463.1000107

[6] Bahuguna A, Lily MK, Munjal A, Singh RN, Dangwal K. Desulfurization of dibenzothiophene (DBT) by a novel strain *Lysinibacillus sphaericus* DMT-7 isolated from diesel contaminated soil. Journal of Environmental Sciences. 2011;**23**:975-982. DOI: 10.1016/S1001-0742(10)60504-9

[7] Van Afferden M, Schacht S, Klein J, Truper HG. Degradation of dibenzothiophene by *Brevibacterium* sp. Archives of Microbiology. 1990;**153**:324-328. DOI: 10.1007/BF00249000

[8] Bhatia S, Sharma DK. Thermophilic desulfurization of dibenzothiophene and different petroleum oils by *Klebsiella* sp. 13T. Environmental Science and Pollution Research International. 2012;**19**(8):3491-3497. 8th ed. DOI: 10.1007/s11356-012-0884-2

[9] Rath K, Mishra B, Vuppu S. Biodegrading ability of organo-sulphur compound of a newly isolated microbe *Bacillus* sp. KS1 from the oil contaminated soil. Archives of Applied Science Research. 2012;**4**:465-471. 1st ed.

[10] Kilbane JJ, Jackowski K. Biodesulfurization of water-soluble coal-derived material by *Rhodococcus rhodochrous* IGT S8. Biotechnology and Bioengineering. 1992;**40**:1107-1114. DOI: 10.1002/bit.260400915

[11] Boltes K, del Águila AR, García-Calvo E. Effect of mass transfer on biodesulfurization kinetics of alkylated forms of dibenzothiophene by *Pseudomonas putida* CECT5279. Journal of Chemical Technology and Biotechnology. 2013;**88**:422-431. DOI: 10.1002/jctb.3877

[12] Nassar HN, El-Gendy NS, Abo-State MA, Moustafa YM, Mahdy HM, El-Temtamy SA. Desulfurization of dibenzothiophene by a novel strain *Brevibacillus invocatus* C19 isolated from Egyptian coke. Biosciences, Biotechnology Research Asia. 2013;**10**:29-46. 1st ed. DOI: 10.13005/bbra/1090

[13] Schlenk D, Bevers RJ, Vertino AM, Cerniglia CE. P450 catalysed S-oxidation of dibenzothiophene by *Cunninghamella elegans*. Xenobiotica. 1994;**24**:1077-1083. DOI: 10.3109/00498259409038667

[14] Zahra E, Giti E, Sharareh P. Removal of dibenzothiophene, biphenyl and phenol from waste by *Trichosporon* sp. Scientific Research and Essays 1. 2006;**1**:72-76. 3rd ed.

[15] Van Hamme JD, Wong ET, Dettman H, Gray MR, Pickard MA. Dibenzyl sulfide metabolism by white rot fungi. Applied and Environmental Microbiology. 2003;**69**:1320-1324. DOI: 10.1128/AEM.69.2.1320-1324.2003

[16] Faison BD, Clark TM, Lewis SN, Ma CY, Sharkey DM, Woodward CA. Degradation of organic sulfur compounds by a coal-solubilizing fungus. Applied Biochemistry and Biotechnology. 1991;**28**:237-250. DOI: 10.1007/BF02922604

[17] Kilbane JJ, John J. Sulfur-specific microbial metabolism of organic compounds. Resources, Conservation and Recycling. 1990;**3**:69-79. DOI: 10.1016/0921-3449(90)90046-7

[18] Crawford D, Gupta RK. Oxidation of dibenzothiophene by *Cunninghamella elegans*. Current Microbiology. 1990;**21**:229-231. DOI: 10.1007/BF02092161

[19] Labana S, Pandey G, Jain RK. Desulphurization of dibenzothiophene and diesel oils by bacteria. Letters in Applied Microbiology. 2005;**40**:159-163. DOI: 10.1111/j.1472-765X.2004.01648.x

[20] Morais GS, Pesenti EC, Cestari MM, Navarro-Silva MA. Genotoxic effect of phenanthrene on *Chironomus sancticaroli* (Diptera: Chironomidae). Zoologia. 2014;**31**:323-328. 4th ed. DOI: 10.1590/S1984-46702014000400003

[21] Liu S, Suflita JM. Ecology and evolution of microbial populations for bioremediation. Trends in Biotechology. 1993;**11**:344-352. DOI: 10.1016/0167-7799(93)90157-5

[22] Araújo HWC, Freitas da Silva MC, Lins CM, Elesbão AN, Alves da Silva CA, Campos-Takaki GM. Oxidation of dibenzothiophene (DBT) by *Serratia marcescens* UCP 1549 formed biphenyl as final product. Biotechnology for Biofuels. 2012;**5**:1-9. DOI: 10.1186/1754-6834-5-33

[23] Calzada J, Alcon A, Santos VE, García-Ochoa F. Mixtures of *Pseudomonas putida* CECT 5279 cells of different ages: Optimization as biodesulfurization catalyst. Process Biochemistry. 2011;**46**:1323-1328. DOI: 10.1016/j.procbio.2011.02.025

[24] Dahlberg MD, Rohrer RL, Fauth DJ, Sprecher R, Olson GJ. Biodesulfurization of dibenzothiophene sulfone by *Arthrobacter* sp. and studies with oxidized Illinois no. 6 coal. Fuel. 1993;**72**:1645-1649. DOI: 10.1016/0016-2361(93)90349-7

[25] Seo JS, Keum YS, Cho IK, Li QX. Degradation of dibenzothiophene and carbazole by *Arthrobacter* sp. P1-1. International Biodeterioration and Biodegradation. 2006;**58**:36-43. DOI: 10.1016/j.ibiod.2006.04.005

[26] Furuya T, Kirimura T, Kuno K, Usami S. Thermophilic biodesulfurization of dibenzothiophene and its derivates by *Mycobacterium phlei* WU-F1. FEMS Microbiology Letters. 2001;**204**:129-133. DOI: 10.1016/S0378-1097(03)00169-1

[27] Okada H, Nomura N, Nakahara T, Maruhashi K. Analysis of dibenzothiophene metabolic pathway in mycobacterium strain G3. Journal of Bioscience and Bioengineering. 2002;**93**:491-497. DOI: 10.1016/S1389-1723(02)80097-4

[28] Li W, Zhang Y, Wang MD, Shi Y. Biodesulfurization of dibenzothiophene and other organic sulfur compounds by a newly isolated microbacterium strain ZD-M2. FEMS Microbiology Letters. 2005;**247**:45-50. DOI: 10.1016/j.femsle.2005.04.025

[29] Li F, Zhang Z, Feng J, Cai X, Xu P. Biodesulfurization of DBT in tetradecane and crude oil by a facultative thermophilic bacterium *Mycobacterium goodii* X7B. Journal of Biotechnology. 2007;**127**:222-228. DOI: 10.1016/j.jbiotec.2006.07.002

[30] Martin AB, Alcon A, Santos VE, García-Ochoa F. Production of a biocatalyst of *Pseudomonas putida* CECT5279 for DBT biodesulfurization: Influence of the operational conditions. Energy & Fuels. 2005;**19**:775-782. DOI: 10.1021/ef0400417

[31] Li F, Zhang Z, Feng J, Cai X, Xu P. Biodesulfurization of DBT in tetradecane and crude oil by a facultative thermophilic bacterium Mycobacterium goodii X7B. J. Biotechnol. 2007;**127**:222-228. DOI: 10.1016/j.jbiotec.2006.07.002

[32] Lin X, Liu J, Zhu F, Wei X, Li Q, Luo M. Enhancement of biodesulfurization by *Pseudomonas delafieldii* in a ceramic microsparging aeration system. Biotechnology Letters. 2012;**34**:1029-1032. DOI: 10.1007/s10529-012-0872-0

[33] Laborde AL, Gibson DT. Metabolism of dibenzothiophene by Beijeninckia species. Applied and Environmental Microbiology. 1977;**34**:783-790. PMID: 596875

[34] Setti L, Rossi M, Lanzarini G, Pifferi PG. Barrier and carrier effects of n-dodecane on the anaerobic degradation of benzothiophene by *Desulfovibrio desulfuricans*. Biotechnology Lettters. 1993;**15**:527-530. DOI: 10.1007/BF00129331

[35] Omori T, Monna L, Saiki Y, Kodama T. Desulfurization of dibenzothiophene by *Corynebacterium* sp. strain SY1. Applied and Environmental Microbiology. 1992;**58**:11-15. PMID: 1575493

[36] Denome SA, Olson ES, Young KD. Identification and cloning of genes involved in specific desulfurization of dibenzothiophene by *Rhodococcus* sp. strain IGTS8. Applied and Environmental Microbiology. 1993;**59**:283-2843. PMID: 16349035

[37] Oldfield C, Pogrebinsky O, Simmonds J, Olson E, Kulpa CF. Elucidation of the metabolic pathway for dibenzothiophene desulfurization by *Rhodococcus* sp. strain IGT S8 (ATCC 53968). Microbiology. 1997;**143**:2961-2973. DOI: 10.1099/00221287-143-9-2961

[38] Van Afferden M, Schacht S, Beyer M, Klein J. Microbial desulfurization of dibenzothiophene. American Chemicak Society Division of Fuel Chemistry. 1988;**33**:561-572

[39] Xiao P, Mori T, Kamei I, Kondo R. A novel metabolic pathway for biodegradation of DDT by the white rot fungi, *Phlebia lindtneri* and *Phlebia brevispora*. Biodegradation. 2011;**22**:859-867. 5th ed. DOI: 10.1007/s10532-010-9443-z

[40] Bezalel L, Hadar Y, PP F, Freeman JP, Cerniglia CE. Initial oxidation products in the metabolism of pyrene, anthracene, fluorene, and dibenzothiophene by the white rot fungus *Pleurotus ostreatus*. Applied and Environmental Microbiology. 1996;**62**:2554-2559. PMID: 16535361

[41] Nojiri H, Habe H, Omori T. Bacterial degradation of aromatic compounds via angular dioxygenation. The Journal of General and Applied Microbiology. 2001;**47**:279-305. 6th ed. PMID: 12483604

[42] Javadli R, Klerk A. Desulfurization of heavy oil. Applied Petrochemical Research. 2012;**1**:3-19. DOI: 10.1007/s13203-012-0006-6

Chapter 16

Microbial Bioremediation of Petroleum Hydrocarbon–Contaminated Marine Environments

Mouna Mahjoubi, Simone Cappello,
Yasmine Souissi, Atef Jaouani and Ameur Cherif

Additional information is available at the end of the chapter

http://dx.doi.org/10.5772/intechopen.72207

Abstract

Petroleum pollution has become a serious environmental problem, which can cause harmful damage to the environment and human health. This pollutant is introduced into the environment from both natural and anthropogenic sources. Various physicochemical and biological treatments were developed for the cleanup of contaminated environments. However, bioremediation is based on the metabolic capabilities of microorganisms, and it is considered as the most basic and reliable way to eliminate contaminants, particularly petroleum and its recalcitrant compounds. It is more effective alternative comparing to classical remediation techniques. A high diversity of potential hydrocarbon degrader's microorganisms was reported, and bacteria constitute the most abundant group, which has been well studied for hydrocarbon degradation. Several bioremediation approaches through bioaugmentation or/and biostimulation have been successfully applied. The interest on the optimizing of different parameters to achieve successful bioremediation technologies has been increased. In this chapter, we summarize the diversity and the hydrocarbon degradation potential of microorganism involved in the remediation of contaminated environments. We also present an overview of the efficient bioremediation strategies used for the decontamination of polluted marine environments.

Keywords: bioremediation, hydrocarbonoclastic, hydrocarbon degradation, toxicity, biosurfactant

1. Introduction

Oil is a primary and necessary energy source, which is widely used in different fields, including industry, transport and daily human activities [1]. However, the widespread distribution and the overexploitation of hydrocarbons become a serious problem causing harmful impacts on the environment and health due to their toxicity and their carcinogenic and mutagenic properties [2]. Hydrocarbon pollution may occur through several sources that may be natural or anthropogenic. We can distinguish numerous anthropogenic sources such as oil extraction and treatment field, transportation fuels accidents, leakage from underground storage tanks, petrochemical industry activities and release from oil refinery sites [3–5]. To reduce this contamination, different chemical, physical and biological treatment methods are considered. The choice of remediation methods depends on several parameters such as the type of pollutant and its characteristics (the physicochemical nature of the pollutant and its toxicity), the properties of contaminated site (the pollution source, the nature of the site) and the type of the pollution (old or recent). The treatment of a polluted site is carried out only after evaluating the type of pollutant, the environmental and human-associated risks and the treatment feasibility and predicted efficiency.

Depending on its location, we distinguish in situ and ex situ contaminated site remediation methods [6–9]. However, bioremediation is considered as one of the best environment-friendly way to remove hydrocarbons presenting several advantages compared to other methods [9–11]. In fact it is a natural, efficient and economic method. Moreover, it converts hydrocarbon onto less toxic compound through metabolic and enzymatic reactions [10, 12]. The biodegradation is mainly carried out by bacteria, yeast and fungi. Bacteria represent the major class of microorganisms involved in the degradation of hydrocarbons [2, 13]. There are two approaches for bioremediation: bioaugmentation and biostimulation [14–16]. Bioaugmentation consists in the addition of highly efficient oil-degrading bacteria to improve and enhance the degradation. While biostimulation consists in the modification of the environment conditions in order to stimulate indigenous bacteria activity.

The aim of this chapter is to provide an overview on the diversity of hydrocarbonoclastic bacteria and their hydrocarbon degradation potential as well as an update on the different bioremediation strategies adopted for the treatment of polluted marine environments particularly the Mediterranean basin. The mechanism by which hydrocarbonoclastic bacteria degrade hydrocarbons will also be discussed.

2. Hydrocarbon pollution in the Mediterranean Sea

The Mediterranean Sea is highly exposed to hydrocarbon pollution due to its geographical situation taking into account the location of most oil-producing and oil-consuming countries, placed, respectively, on the Southern and Northern sides of the basin [17]. It is an enclosed basin and is a very strategic maritime route that hosts about 20% of the global oil tanker traffic, and its coasts host about 80 petroleum harbors. Every year approximately 150,000 tons of hydrocarbons are transported in the Mediterranean Sea [18, 19].

The Mediterranean Sea is highly exposed to several sources of hydrocarbon pollution particularly linked to maritime traffic as well as extraction and refinery activities can affect human health and disturb ecological balance. Moreover, this pollution seriously threatens the marine life [20]. A very high microbial diversity was detected and estimated at 17,000 different species in the Mediterranean Sea constituting a habitat and a reservoir for novel microorganisms [20].

http://dx.doi.org/10.5772/intechopen.72207

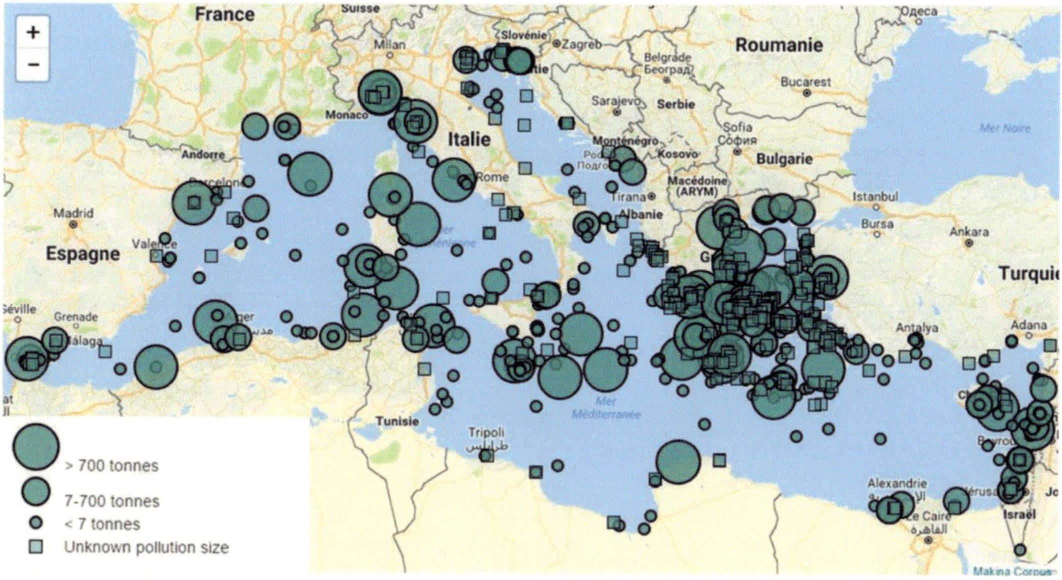

Figure 1. Accidents in the Mediterranean Sea, data from the REMPEC database from 1977 (http://www.rempec.org).

The Regional Marine Pollution Emergency Response Centre for the Mediterranean Sea (REMPEC) was established in 1989. It aims the protection of areas and biological diversity in the Mediterranean Sea against pollution by oil and other harmful substances. According to the REMPEC database on alerts and accidents in the Mediterranean Sea, a high number of marine accidents that caused pollution of the Mediterranean Sea by oil were recorded since 1977. It is estimated at 659 accidents (**Figure 1**) that have led to the spillage of approximately 310,000 tons of oil since 1977 and at least 120,000 tons of hazardous and noxious substances since 1988 [21].

3. Different classes of hydrocarbons

Crude oil is an extremely complex mixture of tens of thousands of hydrocarbons (aliphatics and aromatics) and nonhydrocarbons (containing sulfur, nitrogen, oxygen and various trace metals). The composition of oil varies depending on their origin. Moreover, in the same source, we can also observe a time-dependent variation. For that reason, it is difficult to give a detailed composition of oil [22].

In general, the different compounds of crude oil can be classified into three groups as described in **Figure 2**.

3.1. Aliphatic hydrocarbons

Aliphatic hydrocarbons are composed of hydrogen and carbon, which can be linear, branched or cyclic. Aliphatic compounds can be saturated or unsaturated. There are several types of aliphatic hydrocarbons, including alkanes, alkenes and alkynes. Alkanes are the most abundant constituents in crude oil and are the first component to be degraded.

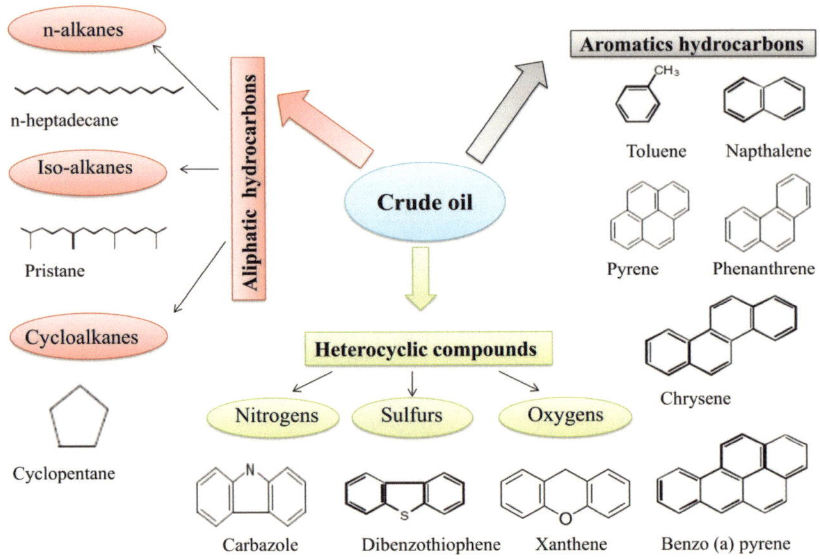

Figure 2. Different classes of hydrocarbon.

3.2. Aromatic hydrocarbons

Polycyclic aromatic hydrocarbons (PAHs) are a group of approximately 10,000 compounds that are atmospheric, water and soil pollutants containing one or more aromatic rings [23]. Among monoaromatic compounds, benzene, toluene and xylene were well reported and studied. Polycyclic aromatic hydrocarbons (PAHs), such as naphthalene, anthracene and phenanthrene, are also well documented. Due to their complex structure, PAHs are highly resistant to degradation and remain persistent in the ecosystem [24].

3.3. Heterocyclic compounds

Heterocyclic compounds are organic compounds containing at least one heterocyclic ring. They consist of compounds in which common heteroatoms are incorporated (oxygen, nitrogen and sulfur) into an organic ring structure in place of a carbon atom. They include polar compounds such as nitrogen (quinolines), sulfur (dibenzothiophenes) and oxygen (xanthene) atoms. Heterocyclic compounds are the most recalcitrant for degradation.

4. Risk hazards associated to hydrocarbon contamination

4.1. Impact on human health

Several human hazards are associated to the hydrocarbon dissemination. In fact, acute hydrocarbon exposure was reported to induce several pathologies including encephalopathy, arrhythmia, acidosis and dermatitis [25].

The most dangerous consequences of acute hydrocarbon ingestion likely arise from aspiration and result in pneumonitis [26]. Moreover, PAHs are well described for their toxic, mutagenic and/or carcinogenic properties [27–29]. Sixteen PAHs were identified by the US Environmental Protection Agency and the European Union as priority pollutant, and seven PAHs are regarded as carcinogens [30].

The American Association of Poison Control Centers Toxic Exposure Surveillance System studied 318,939 exposures to benzene, toluene/xylene, halogenated hydrocarbons, kerosene and lamp oil [31]. They demonstrated that hydrocarbons absorbed systemically and characterized by a low viscosity-induced higher hazard factors [31].

4.2. Hydrocarbon toxicity toward microorganisms

Considering the fact that hydrocarbon bioremediation is an important route for their environmental decontamination, these pollutants should not exert a toxic effect on the microbial community. However, several hydrocarbons and especially cyclic ones are known to be toxic to microbial cells. This toxicity is directly correlated to their partition coefficient between octanol and water (logP). The lipophilic compound accumulates into the lipid bilayers enhancing their availability to the cells. Due to hydrocarbon accumulation, the membrane loses its integrity and its permeability to protons and ions [32]. Due to those disturbances, dissipation of the proton motive force and impairment of intracellular pH homeostasis occur [32]. In order to overcome this toxicity and be able to degrade hydrocarbons, microbial strains should use a tolerance strategy toward those compounds. Sikkema and coworkers reported some adaptation mechanisms [32]. Microorganisms are thus able to increase structuring of the bilayer by changing the fatty acid conformation from cis to trans or by saturation of the fatty acid acyl chains. Another adaptation mechanism consists of the enhancement of the cross-linking degree between constituents of the cell wall and modifications of the cell wall hydrophobicity.

5. Hydrocarbon bioremediation: bioaugmentation, biostimulation and landfarming

Due to their high environmental and health-associated hazards, several physical and chemical hydrocarbon treatment methods have been investigated such as incineration, chlorination, UV oxidation and solvent extraction [24]. However, most of these approaches are associated to several drawbacks in terms of ecological impact and of removal efficiency.

In the meanwhile, the biological approach revealed to be of great interest for hydrocarbon removal. Several approaches were reported. We mainly distinguish (a) phytoremediation strategy that uses plants for decontamination purposes and (b) bioremediation that involves the use of microbial population for the cleanup of contaminated sites.

Bioremediation is considered as a tool for the decontamination of hydrocarbon polluted sites and as an extremely effective approach associated to eco-friendly features and relatively low cost. It is also considered as a safe approach. It is based on the catabolic activities of microorganisms and their ability to use pollutant as carbon and energy source by transforming it into less or non-toxic compounds.

Different bioremediation strategies are adopted depending on several factors such as site characteristics, type and concentration of pollutants [33]. Bioremediation can either be carried out ex situ or in situ. Ex situ methods involve the physical removal of the contaminated land for treatment process. However, in situ techniques involve treatment of the contaminated material in place [34].

The bioremediation efficiency depends mainly on the microbial structure, the sites to be decontaminated and the environmental conditions. These factors influence the degradation rate and the microbial activity. Moreover, bioremediation efficiency can be enhanced by three complementary approaches: bioaugmentation, biostimulation and landfarming [35].

5.1. Bioaugmentation

Bioaugmentation consists in the addition of selected microbial species, harboring specific catabolic abilities, into a contaminated environment. However, this technique involves the application of allochthonous or autochthonous microorganisms. The autochthonous bioaugmentation (ABA) consists on the exclusive introduction of indigenous microorganisms to the sites to be decontaminated [35, 36]. As described in **Figure** 3, the bioaugmentation treatment is detailed, where A is related to the cultivation of bacteria and/or consortium with high biodegradative capability in laboratory; B showed the addition on natural polluted environment; Cs explain the attack of bacteria and/or microbial consortium selected of pollutant (crude oil and/or hydrocarbons) and D showed the development of biodegradative process. The ABA approach consists on the addition of autochthonous microbial isolates or consortia that are adapted to the contaminated environments [37]. Different studies confirmed the efficiency of autochthonous microorganisms in the decontamination of hydrocarbon-polluted sites based on the fact that environmental conditions are suitable for their growth and metabolism [36, 38]. They significantly contribute to the biochemical activities in soil. Allochthonous microorganisms are defined as non-indigenous and originated from other site. When they are added into contaminated sites, they cannot easily integrate and adapt to the activities of the indigenous microbial community. Fodelianakis et al. [38] confirmed that bioaugmentation with allochthonous hydrocarbon-degrading bacterial consortia did not enhance the remediation process of hydrocarbon-contaminated sediments. Thus, the application of autochthonous microorganism is more effective [35]. Generally, Bioaugmentation is performed by using mixed consortia ([39]). Mrozik and Piotrowsk [40] showed the successful bioaugmentation of contaminated sites with a mixture of PAHs (naphthalene, phenanthrene, anthracene, pyrene, dibenzo[a] anthracene, benzo[a]pyrene) using a reconstructed microbial consortium. The addition of *Absidia cylindrospora* enhance the degradation of polycyclic aromatic hydrocarbons (more than 90% of fluorine was removed) more than of fluorine (a within 288 h) [41].

For a successful bioaugmentation treatment, environmental conditions, selected and indigenous microorganisms should be well controlled and characterized.

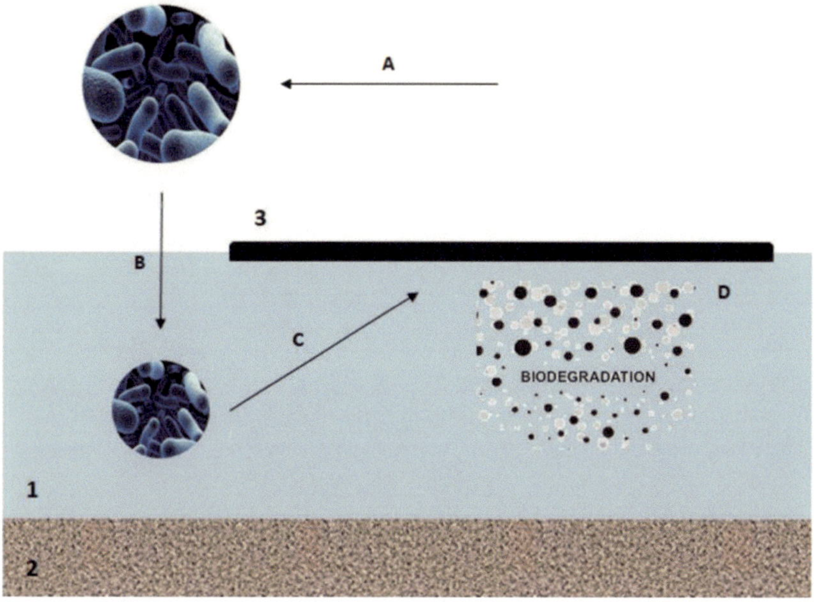

Figure 3. Schematic description of bioremediation (bioaugmentation) process. 1: Seawater; 2: marine sediment; 3: crude oil.

5.2. Biostimulation

Biostimulation consists in the addition of different substance to polluted environments to stimulate the natural biodegradation by indigenous degrading microorganisms. The biodegradation of hydrocarbons in marine environment is limited by low oxygen and nutriment availability, more specifically nitrogen and phosphorous [42]. The low bioavailability of pollutants is also considered as a limiting factor [42]. The biostimulation increases the growth of indigenous hydrocarbon degraders by the addition of growth-limiting nutrients. Moreover, the addition of specific substrate promotes and stimulates the degradation of the pollutant [43] (**Figure 4**). Adams et al. [44] showed the effectiveness of Biostimulation treatment with the addition of organic nutrients in contaminated soil containing more than 38,000 mg/kg TPH, and the result showed 100% removal of 2–3 ringed PAHs within the first 3 months.

5.3. Landfarming

Landfarming is a bioremediation approach, also known as land treatment or land application. This biotreatment process involves the spreading of excavated contaminated soils or sediments in a thin layer on a suitably prepared surface. In addition, the stimulation of the microbial activity in the soil is performed through the aeration and/or the addition of minerals and nutrients. Thus, the incorporated contaminated soil is periodically turned over or tilled to aerate the mixture. Various hydrocarbon contaminated soils were successfully

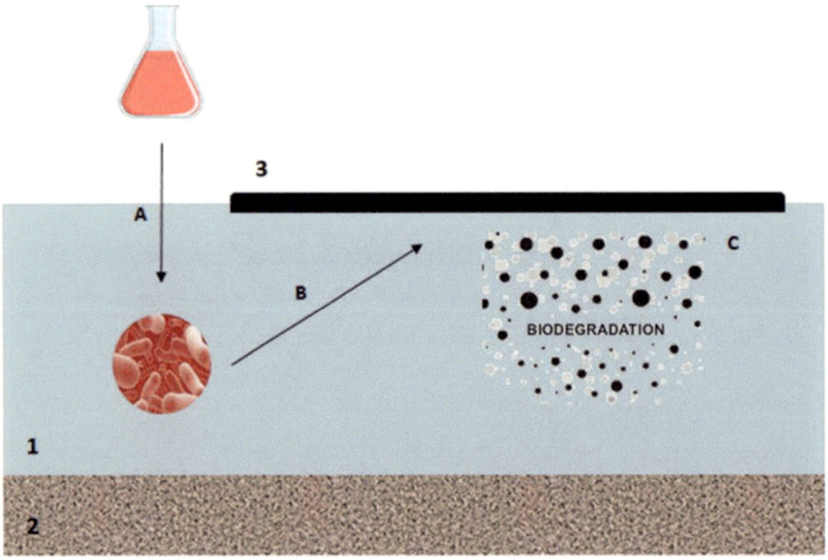

Figure 4. Schematic description of bioremediation (biostimulation) process. 1: Seawater; 2: marine sediment; 3: crude oil.

treated by landfarming [45]. Landfarming presents various advantages comparing to other treatment technologies. It has low cost, energy consumption and simple to implement. It has been successfully practiced for the remediation of hydrocarbon-contaminated soils [46]. The effectiveness of bioaugmentation in the bioremediation of oil-contaminated sediments, from the Mediterranean Sea, treated *ex situ* by landfarming was reported [38].

6. Factors affecting the bioremediation treatment

Various physical and chemical factors affect the rate of hydrocarbon degradation. Among many, we can cite the nature and amount of hydrocarbons, the type of polluted matrix (soil, sediment, water and effluents), the environmental conditions and the microbial community activity.

The environmental factors include the availability of nutrients, the temperature, the water content or humidity, oxygen, the bioavailability of the pollutant and the pH [47–49].

6.1. Nutrient availability

Inorganic sources such as nitrogen, phosphorus, potassium, hydrogen or oxygen are essential for microbial metabolism and affect the growth and the activity of microorganisms [48], whereas micronutrients, including zinc, manganese, iron, nickel, cobalt, molybdenum, copper and chlorine, are required but at a very low quantity. The ratios of carbon/nitrogen or carbon/phosphorus are considered as a determining factor of biodegradation rates and are high in hydrocarbon-contaminated sites which limit and affect the degradation rate [50].

Considerable effort has been devoted to the development of nutrient delivery systems that overcome the washout problems characteristic of open sea and intertidal environments; the use of slow release fertilizers and/or oleophilic nutrients can provide a continuous source of nutrients to polluted environment areas. Slow release fertilizers consist typically of inorganic nutrients in solid form coated with a hydrophobic compound like paraffin or vegetable oil. Successful alternative that overcomes the problem of quick dilution and wash out of water-soluble nutrients containing nitrogen and phosphorus is oleophilic biostimulants. The application of N and P sources in oleophilic form is considered to be the most effective nutrient application method, since oleophilic additives remain dissolved in the oil phase and thus are available at the oil-water or oil-sediment interface where they enhance bacterial growth and metabolism [43, 51].

6.2. Temperature

Temperature has a considerable effect on the ability of microorganisms to degrade hydrocarbons. At high temperatures, solubility, bioavailability, hydrocarbons distribution and diffusion rate increase, which promote the microbial biodegradation ability and enhance the biodegradation rate. On the other hand, very high temperature decreases oxygen solubility and limits the aerobic microbial biodegradation [52, 53].

Beside, Boopathy [34] confirmed that the degradation of pollutant in mesophilic temperatures is better and more efficient than the degradation at very low or high temperatures. However, it was reported that microorganisms are able to metabolize PAHs at extreme temperatures: for example, a high degradation rate of PAHs occurred in seawater at low temperatures (low as 0°C) and at 50°C [54]. Bargiela et al. [55] confirmed the correlation between the relative percentage of genes encoding enzymes participating in the biodegradation and temperature in oil-polluted sites along the coastlines of the Mediterranean Sea. The authors proved that low temperature increases bacterial richness with decreasing catabolic diversity. The relative percentage of gene sequences encoding enzymes for biodegradation ranged from 0.56 to 1.30% in low temperature (<3°C), whereas this number increased with site temperature 5.6-fold higher in the moderately warmer sites (from 13 to 26.5°C) [55].

6.3. Oxygen limitations

The initial steps in the catabolism of aliphatic, cyclic and aromatic hydrocarbons by bacteria and fungi involve the oxidation of the substrate by oxygenases for which molecular oxygen is required. Although anaerobic biodegradation has been shown to occur in different ecosystems including marine environments, its ecological significance has been generally considered to be minor and the biodegradation rate is rather low [24]. Conditions of oxygen limitation normally exist in aquatic sediments and soils. Oxygen depletion can occur in the presence of easily utilizable substrates that increase microbial oxygen consumption. In several instances, the concentration of dissolved oxygen can be close to zero leading to practically zero aerobic biodegradation rates. Although oxygen can be successfully delivered (in various forms) to hydrocarbon-contaminated soils and groundwater, enhancing biodegradation rates, this is not the case in marine environments as it is very difficult to implement such technologies in the field. Tilling is essentially the only option in aerating the top layers of contaminated sediments during low tide [34, 48]. Thus, oxygen represents a very significant and potentially factor which limit the rate of hydrocarbon degradation.

6.4. pH

pH varies less in aquatic environments and most bacteria and fungi capable of degrading hydrocarbon require a neutral pH [52, 53]. In general, microbial activity is influenced by extremely low or high pH [53]. According to Bamforth and Singleton [48], 40% of phenanthrene degradation was observed for *Burkholderia cocovenenans* at pH 5.5. However, the degradation at neutral pH in the same conditions was 80%. Additionally, Leahy and Colwell [53] reported that microbial degradation of naphthalene decreased at pH 5.0 compared to the highest rate of degradation observed at pH 7. Moreover, other reports described the efficiency of some microorganisms, such as *Pseudomonas,* to degrade hydrocarbon at alkaline pH [48]. A degradation of PAH in acidic contaminated environment (pH 2) by indigenous microorganisms was reported [54]. The adequate pH depends on microorganisms to be used for the bioremediation.

6.5. Bioavailability of hydrocarbon

The bioavailability is defined as the rate of substrate mass transfer into the microbial cells. It is considered as one of the most determinant parameters regarding hydrocarbon degradation rate. PAHs are characterized by a low bioavailability due to their low aqueous solubility. That is the reason why they are reported to be resistant to the degradation process and persistent in the environment [24]. Unsuccessful remediation of PAH-contaminated sites was reported due to the low bioavailability of PAHs [24]. It has been reported that bioavailability of hydrocarbon decreases with time [24]. Although the photo-oxidation increased the biodegradation of petroleum hydrocarbon by increasing its bioavailability which promote microbial activities [56]. Hydrocarbon and particularly PAHs become more bioavailable when they are dissolved or evaporated [27]. The bioavailability of pollutant in contaminated environment may be increased by the application of surfactants.

7. Distribution of hydrocarbon-degrading microorganisms

Oil spill in the marine environment can be highly toxic to the marine microbial communities. However, a part of these communities resist to this type of pollution and are able to metabolize the pollutant.

Several studies have revealed a huge diversity of microorganisms (bacteria, fungi, archaea and algae), which were isolated from contaminated environments and were able to degrade hydrocarbons. Recent research reported that 79 bacterial, 9 cyanobacterial, 103 fungal and 14 algal are able to use hydrocarbons as a sole source of carbon and energy [39]. Moreover, in the marine environment, at least 25 bacterial genera of hydrocarbon degrader were found [57].

Bacteria are the most important and abundant group known for hydrocarbon degradation.

However, there is a class of bacteria that uses exclusively hydrocarbons as substrates and has been characterized as obligate hydrocarbonoclastic bacteria (OHCB) [39, 58, 59]. This group is present at low or undetectable levels in marine environment before a contamination with hydrocarbon. Hydrocarbon-degrading microorganisms are widely distributed and ubiquitous in marine environment especially after a contamination with hydrocarbons. Their number and activity increase and can represent 100% of the total microbial community. However, they can be found in non-contaminated environment but in very small proportion which can represent 0.1% of the total microbial community [53, 60, 61].

Based on microbial isolation and cultivation from hydrocarbon-contaminated environment, several microorganisms were isolated and characterized. However, this method can detect only 1–2% of the most dominant microbial species and only a small fraction of those communities are able to be cultivated [62]. In addition, different molecular approaches were used to study the composition and the structure of the bacterial community in different environment such as denaturing gradient gel electrophoresis (DGGE), terminal restriction fragment length polymorphism (T-RFLP), automated ribosomal intergenic spacer analysis (ARISA) and 16SrRNA gene clone libraries [62–64].

With recent advances in genomics and sequencing technologies, more techniques based on high throughput sequencing of DNA (next generation sequencing technologies) are improved in metagenomic studies (**Figure 5**).

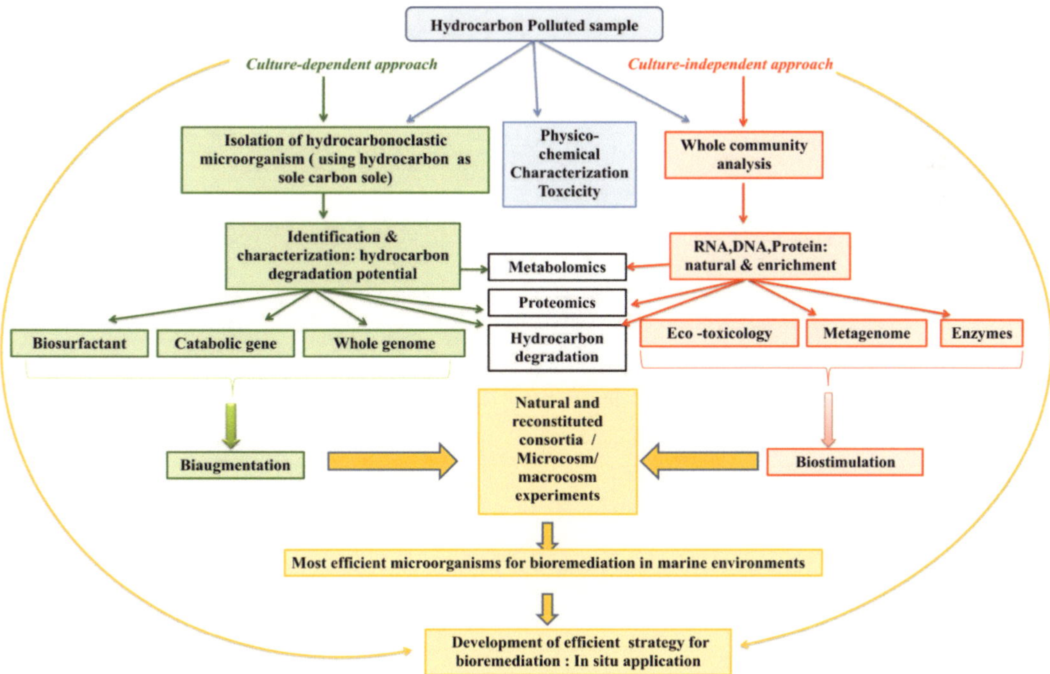

Figure 5. Schematic diagram explaining the different microbial hydrocarbon degradation strategy for bioremediation.

These hydrocarbonoclastic bacteria are of great importance since they belong to the key players in oil removal from contaminated marine sites. Among these OHCB, *Alcanivorax*, *Marinobacter*, *Thallassolituus*, *Cycloclasticus* and *Oleispira* are the most representative genera and they play a key role in the degradation of hydrocarbons [65]. As described in **Figure 6**, bacteria, which are able to degrade saturated and polycyclic aromatic hydrocarbons, were presented by red and blue, respectively, whereas the green organism is able to degrade dioxin compounds. The organisms shown in black color in **Figure 6** are able to degrade both saturated and PAH.

Among obligate hydrocarbonoclastic bacteria, *Alcanivorax borkumensis* is ubiquitous in hydrocarbon-polluted marine environment. This strain is able to metabolize linear and branched alkanes, but unable to degrade aromatic hydrocarbons [66, 67]. *Thallassolituus oleivorans* is highly specialized to degrade aliphatic hydrocarbons from C7 to C20 carbons [68]. Also, *Oleiphilus* and *Oleispira* are well known in the degradation of aliphatic hydrocarbons [58]. Whereas *Cycloclasticus* is reported to mineralize various PAHs such as naphthalene, phenanthrene, anthracene, pyrene and fluorene [69]. *Marinobacter* strains are able to degrade

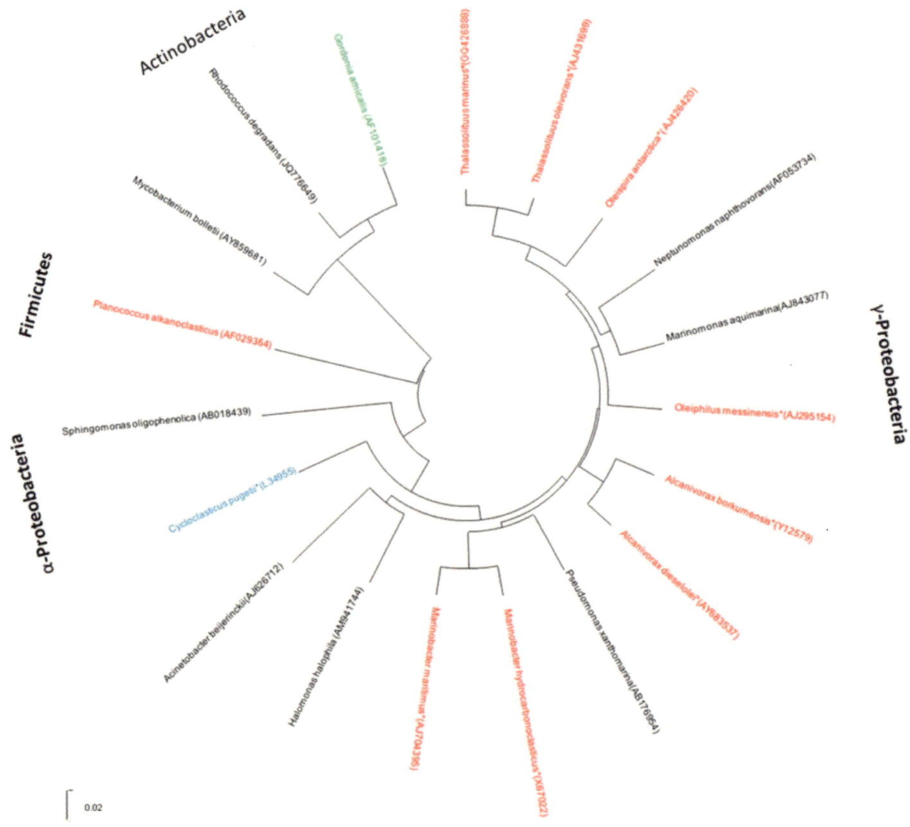

Figure 6. A phylogenetic tree of the distribution of hydrocarbon-degrading bacteria based on 16S rDNA partial sequences for type strains using MEGA 6.06. 16S rRNA sequence accession numbers of the reference type strains are indicated in parentheses. Microorganisms* showed obligate hydrocarbonoclastic bacteria.

hydrocarbon compounds efficiently [70]. Other non-obligate hydrocarbonoclastic bacteria have been described for their abilities to degrade different classes of hydrocarbons and were isolated from oil-contaminated environment. Among this group, able to degrade PAHs, we distinguish *Pseudomonas, Marinomonas, Halomonas* and *Micrococcus*. Moreover, *Acinetobacter* and *Sphingomonas* are well described in the degradation of aliphatic compounds [58].

Several studies reported that the efficiency of biodegradation ranged from: 6 to 82%, 0.13 to 50% and 0.003 to 100%, respectively, by fungi, soil bacteria and marine bacteria [71].

Among fungi, genera affiliated to *Amorphoteca, Neosartorya, Talaromyces, Aspergillus* and *Graphium* are able to degrade hydrocarbon. In addition, various studies reported the efficiency of hydrocarbon degradation by yeast belonging to *Candida lipolytica, Rhodotorula mucilaginosa, Geotrichum* sp. and *Trichosporon mucoides* [71].

Since crude oil is a complex mixture of hydrocarbon and individual microorganism is able to degrade only a limited number of crude oil components, several studies reported the efficiency of the degradation of crude oil by a whole consortium. It was defined as mixed cultures of hydrocarbon-degrading microorganism composed either by known hydrocarbon degrader microorganisms or by enrichment of indigenous microbial consortia. Several reports confirmed that the application of a consortium, with a broad enzymatic capacities to degrade a wide range of hydrocarbons, accelerate bioremediation treatment [33, 39, 44, 72]. A reconstructed consortium composed by a mixture of bacteria and fungi was able to achieve a high degradation of petroleum based on their complimentary metabolic potentials to metabolize the different components [72].

8. Mechanism of hydrocarbon degradation

The trend in biodegradation of petroleum hydrocarbons rate is performed in the following order: n-alkanes > branched alkanes > low molecular weight aromatics > cyclic alkanes. Resin and asphaltenes are the most recalcitrant compounds of the petroleum hydrocarbons [73].

8.1. Alkane-degrading bacteria

Alkanes are saturated hydrocarbons composed exclusively by carbon and hydrogen atoms. They can be n-alkanes (linear), cycloalkanes (cyclic) or isoalkanes (branched).

The major fraction of crude oil can be composed by alkanes which constitute more than 50% of crude oil depending on the type and source of oil.

Various aerobic and anaerobic microorganisms possess the metabolic capacity to use alkanes as sole carbon source and energy. Bacteria, fungi and yeasts were reported to degrade alkanes. The degradation of alkane by bacteria has been extensively studied [58]. The biodegradation of saturated hydrocarbons is essentially an aerobic process carried out by bacteria. Microorganism-degrading alkanes are widely distributed in the environment. The biodegradation of n-alkanes is better known than other classes of saturated hydrocarbons. Alkane hydroxylase is a key enzyme involved in alkane degradation [74].

Methane monooxygenase is involved in the degradation of short-chain length alkanes (C_2–C_4) [74]. The degradation of medium n-alkanes (C_5–C_{17}) was carried out via cytochrome P450 and monooxygenases including AlkB [73]. For the degradation of long n-alkanes > C_{18}, multiple alkane hydroxylases are reported [74].

Alcanivorax spp., initially undetectable in oil-contaminated sea water, were reported to constitute 90% of the whole bacterial community after nutrient addition in biostimulation treatment [39]. The detection of *alkB* genes, which encode the catalytic activity of alkane hydroxylase allowing the degradation of medium to long chain hydrocarbons, was also reported [74]. This gene was firstly discovered in *Pseudomonas putida* GPo1. Afterward, it was found in the majority of alkane-degrading bacteria. With advanced researches, 250 *alkB* gene homologs were reported in hydrocarbonoclastic bacteria [74].

8.2. Polycyclic aromatic hydrocarbons (PAHs)

PAHs represent a class of hazardous environmental pollutants that are accumulated in the environment from various origins and mainly from anthropogenic activities. PAHs are present in crude oil but in low concentrations. They are highly toxic and recalcitrant [60, 61].

Currently, there are many reports on aerobic catabolism of aromatics compounds and various catabolic pathways were investigated. In general, we distinguish two major strategies for PAHs degradation in the presence and absence of oxygen [74].

Usually, the degradation of PAHs by bacteria involves mono and dioxygenase since the first step is the hydroxylation of the aromatic ring *via* a dioxygenase which is a complex of enzymes composed by reductase, ferredoxin and terminal oxygenase subunits. Besides, several bacteria degrade PAHs through cytochrome P450 [24].

PAHs can be completely metabolized to carbon dioxide and water due to the presence of oxygen [61, 75].

The anaerobic biodegradation of PAHs is a slow process, and a few numbers of preliminary studies have reported the degradation pathways, catabolic genes and enzymes of this degradation. Among PAHs reported for their degradation in anoxic condition, we can cite naphthalene, anthracene, phenanthrene, fluorene, acenaphthene and fluoranthene [74].

9. Biosurfactant

The bioavailability of contaminants is considered as a major limit for the degradation and bioremediation of contaminated sites. This limit may be recovered by the use of biosurfactant produced by bacteria. Biosurfactant reduces the surface tension and facilitates the contact between microorganism and pollutants. Numerous studies reported the effect of biosurfactants on biodegradation of hydrocarbons and their industrial and commercial potentials [76, 77].

Biosurfactants are of great interest since these compounds are widely used in various fields, including industrial applications, agriculture, crude-oil recovery, cosmetics, pharmaceuticals, detergents, food processing and textile, and in various fields of biomedicine. Therefore, these compounds have advantages compared to chemical surfactant because these molecules are nontoxic, biodegradable and can be used under extreme conditions of temperature, pH and salinity [78–80].

Studies dealing with the biotechnological use of biosurfactants have significantly increased within the last years. Nowadays, with the research advance, the use of biosurfactants is being a potential alternative and promising tool for bioremediation.

Acinetobacter, Stenotrophomonas, Pseudomonas, Koccuria and *Bacillus,* isolated from hydrocarbon-contaminated sites in Tunisia, showed significant biosurfactant production and emulsification activities [2]. Several microorganism genera including *Corynebacterium, Rhodococcus, Pseudomonas, Bacillus, Achromobacter* and *Ochrobactrum* were identified as efficient biosurfactant producers (**Table** 1) [83]. Therefore, these bacteria may constitute potential candidates for bioremediation and can be useful for biotechnological applications. *Pseudomonas* has been widely studied for their ability to produce biosurfactants with a glycolipid structure. Rhamnolipids are the better characterized biosurfactants and enhanced oil recovery and bioremediation process [84–86]. *Bacillus subtilis* is known for its ability to produce lipopeptide biosurfactant known as surfactin, which is used in commercial application [82, 87]. Additionally, the ability of *Brevibacillus* sp. to produce biosurfactant during phenanthrene degradation and an association between biodegradation and production of biosurfactants was shown [88]. Other report describes the rhamnolipids produced by *Burkholderia plantarii* and their application in pharmaceutical and industries [89]. According to Bordoloi and Konwar [81], a higher solubilization of pyrene and fluorine due to the addition of biosurfactant was observed. *Rhodococcus erythropolis* produced Trehalose lipids, biosurfactant which stimulate the solubilization of phenanthrene to facilitate the degradation of this hydrocarbon [90].

Types	Biosurfactants	Microorganisms	Application	References
Glycolipids	Rhamnolipid	*Pseudomonas aeruginosa, Pseudomonas putida, Pseudomonas chlororaphis, Renibacterium salmoninarum*	Bioremediation Biocontrol agent	Bordoloi and Konwar [81] and Banat et al. [78]
	Trehalolipid	*Rhodococcus, Mycobacterium, Corynebacterium, Gordonia, Arthrobacter*	Bioremediation Biocontrol agent	Banat et al. [78]
	Sophorolipid	*Candida bombicola*	Bioremediation	Whang et al. [82]
		Candida apicola	Emulsion	
Lipopeptides	Surfactin	*Bacillus subtilis*	Biomedical application	Banat et al. [78]
	Viscosin	*Pseudomonas fluorescens*		Banat et al. [78]
	Rhodofactin	*Rhodococcus erythropolis*		
Fatty acid	Fatty acid	*Corynebacterium lepus*		
Phospholipids	Phospholipid	*Corynebacterium insidiosum*		

Table 1. Main types of biological surfactants, producing microorganisms and application.

10. Case studies of bioremediation strategies used for the decontamination of polluted marine environments

Numerous and different studies can be carried out, for example, as application of bioremediation processes in the Mediterranean Sea, however, is particularly interesting is the example that can be extracted from the work of Cappello et al. [91], Genovese et al. [65] and Fodelianakis et al. [38].

In the work of Cappello et al. [91] and Genovese et al. [65], a modular system aimed at optimizing the biodegradation process was implemented and tested for the remediation of hydrocarbon-contaminated sediments collected in Messina harbor (Italy). The system was designed to operate directly in the field but retains the advantages of controlled methods that do not impact the surrounding environment. The effects of the addition of air and/or slow-released fertilizer, temperature regulation and the addition of oil sorbents on the efficiency of the system were evaluated. The abundance of the indigenous microbial communities was monitored for 30 days of treatment. Measures of microbial activity, biochemical oxygen demand (BOD), screen of functional genes, quantitative of petroleum hydrocarbon degradation and ecotoxicological bioassays using the organism *Corophium orientale* were carried out.

The proposed system was designed according to the two needs highlighted by Nikolopoulou and Kalogerakis: (a) the development of low-cost oxygenation systems for aerobic bioremediation of contaminated anoxic sediments and (b) the development of novel biostimulant methods that are nontoxic to the marine environment (**Table 2**).

The results showed the effect of the air injection in the development of bacterial biomass with an increase of bacterial population by the DAPI and MPN counts with air insufflations comparing to the control (untreated sediments). The addition of nutrients and the regulation of temperature enhance the degradation rate to achieve a high degradation in the shortest possible time. In the mesocosm system, ~70% of the total oil and linear hydrocarbons were degraded with the development of mesophilic bacteria. Besides, the addition of fertilizers promoted the degradation of crude oil improving the degradation efficiency within the 30 days of the treatment when >90% of n-alkanes (C_{15}–C_{30}) and >60% of (alkyl)naphthalenes were degraded.

However, the authors found that biostimulation treatment through nutrient amendment to hydrocarbon-contaminated sediments enhanced the biodegradation activity of the indigenous microbial community and especially hydrocarbonoclastic bacteria such as *Alcanivorax*, *Marinobacter* and *Cycloclasticus*. In fact, sediment toxicity was evaluated by ecotoxicological method using *Corophium orientale*, an endemic and widespread Mediterranean amphipod, used as an indicator species in toxicity of harbor sediments [94]; a decrease of toxicity in treated sediments was observed (~30%) comparing to untreated system (95%). Thus, the success of the bioremediation treatment was explained by both reduction of the hydrocarbon in sediment and shift of toxicity.

As another example for bioremediation strategy to evaluate the effectiveness of bioaugmentation with allochthonous bacterial consortia in the bioremediation of oil-contaminated sediments collected from a coast adjacent to an oil refinery (Elefsina bay, Greece) treated ex

Type of bioremediation treatment	Microorganisms/nutrients added	Type of contaminant	Percentage removal (%)	Duration (days)	References
Bioaugmentation	*Acinetobacter baumannii*	Crude oil	43	35	Adams et al. [44]
	Candida tropicalis	Diesel	83	120	Adams et al. [44]
	Pseudomonas arthrobacter	Diesel	32	32	Adams et al. [44]
	Alcanivorax, Marinobacter Cycloclasticus and Thalassolituus	TERHC	79	14	Crisafi et al. [92]
	Acinetobacter	TPH	34	42	Wu et al. [93]
	Alcanivorax and *Thalassolituus*	Crude oil	70	20	Hassanshahian et al. [16]
	Absidia. Cylindrospora	Fluorene	90	12	Garon et al. [41]
	A bacterial consortia	Diesel	72	84	Adams et al. [44]
Biostimulation	Nitrogen and phosphorus	TPH	60	42	Wu et al. [93]
	Inorganic nutrients KH_2PO_4, NH_4Cl and $NaNO_3$	Crude oil	80	20	Hassanshahian et al. [16]
	Mixed of $(NH4)_2$, SO_4 and K_2HPO_4	Diesel	45	84	Adams et al. [44]
	Slow release fertilizers	Crude oil	90% of n-alkanes (C15–C30) and 60% of PAHs	30	Nikolopoulou and Kalogerakis [43]
	Inorganic nutrients (KNO_3 and K_2HPO_4)	Crude oil	97% of C12–C35 n-alkanes and 45% of PAH	45	Nikolopoulou and Kalogerakis [43]
	Organic nutrients (uric acid and lecithin) and biosurfactants (rhamnolipids)	n-Alkanes	97	45	Nikolopoulou et al. [36]
Bioaugmentation and biostimulation	*Rhodococcus* sp. and chemical surfactant and Tween 80	TPH	50	46	Adams et al. [44]
	Pseudomonas aeruginosa, Bacillus subtilis and fertilizer	TPH	75	70	Adams et al. [44]
	Inorganic nutrients + rhamnolipids and autochthonous bacterial consortia	Crude oil	80	30	Nikolopoulou et al. [36]

Table 2. Example of successful bioremediation strategies.

situ by landfarming. For the bioaugmentation treatment, a bacterial consortia composed by hydrocarbonoclastic strains isolated from different oil-contaminated sediments collected from North Africa (Morocco, Tunisia, Egypt) and Red Sea (Jordan) were used. Besides, the biostimulation was performed with the addition of N/P in the form of KNO_3/KH_2PO_4 to a final ratio of C:N:P (100:10:1). The experiments were carried out for 56 days. The result showed that biodegradation was mostly performed by the autochthonous degraders, while bioaugmentation did not enhance the remediation process.

11. Conclusion

Petroleum hydrocarbons are the most common pollutants in marine environment and the Mediterranean Sea is particularly endangered by hydrocarbon pollution. This pollution generates a great hazard to marine ecosystems. Bioremediation through bioaugmentation or biostimulation constitutes a promising strategy for the cleanup of oil contamination. It was based on the presence of catabolic genes and enzymes, which allow microorganisms to utilize hydrocarbons as carbon and energy source. This chapter presents an overview of different efficient strategies used for the decontamination of polluted marine environments. There are various factors that affect bioremediation efficiency such as oxygen, pH, temperature and nutrient availability. Thus, it has been confirmed that the combination of autochthonous bioaugmentation and biostimulation constitutes the best and efficient strategy for the decontamination of petroleum-contaminated marine environment. However, further researches are still required to answer multiple questions to develop a successful system for bioremediation.

Acknowledgements

The authors thank for financial support the European Union in the ambit of research projects: ULIXES (European Community's Seventh Framework Programme, FP7-KBBE-2010-4, CP-FP-SICA under grant agreement no. 266473) and MADFORWATER (European Union's Horizon 2020, WATER-5c-2015 under grant agreement no.688320) and the Tunisian Ministry of Higher Education and Scientific research in the ambit of the laboratory project LR11ES31.

Author details

Mouna Mahjoubi[1,2], Simone Cappello[3], Yasmine Souissi[1], Atef Jaouani[4] and Ameur Cherif[1]*

*Address all correspondence to: cherif.ameur@gmail.com

1 Univ. Manouba, ISBST, BVBGR-LR11ES31, Biotechpole Sidi Thabet, Ariana, Tunisia

2 Faculty of Science of Bizerte, University of Carthage, Bizerte, Tunisia

3 Istituto per l'Ambiente Marino Costiero (IAMC)-CNR of Messina, Messina, Italy

4 Laboratory of Microorganisms and Active Biomolecules, Faculty of Sciences of Tunis, University of Tunis El Manar, Tunis, Tunisia

References

[1] Turner NR, Renegar DA. Petroleum hydrocarbon toxicity to corals: A review. Marine Pollution Bulletin. 2017;**2**(119):1-16

[2] Mahjoubi M, Jaouani A, Guesmi A, Amor SB, Jouini A, Cherif H, Najjari A, Boudabous A, Koubaa N, Cherif A. Hydrocarbonoclastic bacteria isolated from petroleum contaminated sites in Tunisia: Isolation, identification and characterization of the biotechnological potential. New Biotechnology. 2013;**30**(6):723-733

[3] Cappello S, Caruso G, Zampino D, Monticelli LS, Maimone G, Denaro R, Tripodo B, Troussellier M, Yakimov MM, Giuliano L. Microbial community dynamics during assays of harbor oil spill bioremediation: A microscale simulation study. Journal of Applied Microbiology. 2007;**102**:184-194

[4] Cappello S, Santisi S, Calogero R, Hassanshahian M, Yakimov MM. Characterisation of oil-degrading bacteria isolated from bilge water. Water, Air, & Soil Pollution. 2012;**223**(6):3219-3226

[5] Jiang Z, Huang Y, Xu X, Liao Y, Shou L, Liu J, Zeng J. Advance in the toxic effects of petroleum water accommodated fraction on marine plankton. Acta Ecologica Sinica. 2010;**30**(1):8-15

[6] Costa AS, Romão LPC, Araújo BR, Lucas SCO, Maciel STA, Wisniewski A, Alexandre MDR. Environmental strategies to remove volatile aromatic fractions (BTEX) from petroleum industry wastewater using biomass. Bioresource Technology. 2012;**105**:31-39

[7] De Vasconcellos SP, Crespim E, da Cruz GF, Senatore DB, Simioni KCM, dos Santos Neto EV, Marsaioli AJ, de Oliveira VM. Isolation, biodegradation ability and molecular detection of hydrocarbon degrading bacteria in petroleum samples from a Brazilian offshore basin. Organic Geochemistry. 2009;**40**(5):574-588

[8] Greenwood PF, Wibrow S, George SJ, Tibbett M. Hydrocarbon biodegradation and soil microbial community response to repeated oil exposure. Organic Geochemistry. 2009;**40**(3):293-300

[9] Lin TC, Pan PT, Cheng SS. Ex situ bioremediation of oil-contaminated soil. Journal of Hazardous Materials. 2010;**176**(1):27-34

[10] Bacosa HP, Suto K, Inoue C. Bacterial community dynamics during the preferential degradation of aromatic hydrocarbons by a microbial consortium. International Biodeterioration & Biodegradation. 2012;**74**:109-115

[11] Liu PWG, Chang TC, Whang LM, Kao CH, Pan PT, Cheng SS. Bioremediation of petroleum hydrocarbon contaminated soil: Effects of strategies and microbial community shift. International Biodeterioration& Biodegradation. 2011;**65**(8):1119-1127

[12] Roy AS, Baruah R, Borah M, Singh AK, Boruah HPD, Saikia N, Deka M, Dutta N, Bora TC. Bioremediation potential of native hydrocarbon degrading bacterial strains in crude oil contaminated soil under microcosm study. International Biodeterioration& Biodegradation. 2014;**94**:79-89

[13] Dell'Anno A, Beolchini F, Rocchetti L, Luna GM, Danovaro R. High bacterial biodiversity increases degradation performance of hydrocarbons during bioremediation of contaminated harbor marine sediments. Environmental Pollution. 2012;**167**:85-92

[14] Bento F-M, Flávio AO, Camargo F-V-A, Okeke B-C, Frankenberger W-T. Comparative bioremediation of soils contaminated with diesel oil by natural attenuation, biostimulation and bioaugmentation. Applied Bioresource Technology. 2005a;**96**:1049-1055

[15] Liu PWG, Chang TC, Chen CH, Wang MZ, Hsu HW. Bioaugmentation efficiency investigation on soil organic matters and microbial community shift of diesel-contaminated soils. International Biodeterioration& Biodegradation. 2014;**95**:276-284

[16] Hassanshahian M, Emtiazi G, Caruso G, Cappello S. Bioremediation (bioaugmentation/biostimulation) trials of oil polluted seawater: A mesocosm simulation study. Marine Environmental Research. 2014;**95**:28-38

[17] Amer RA, Mapelli F, El Gendi HM, Barbato M, Goda DA, Corsini A, Lucia Cavalca L, Fusi M, Borin S, Daffonchio D, Abdel-Fattah YR. Bacterial diversity and bioremediation potential of the highly contaminated marine sediments at El-Max district (Egypt, Mediterranean Sea). BioMed Research International. 2015;**2015**:17

[18] Daffonchio D, Ferrer M, Mapelli F, Cherif A, Lafraya Á, Malkawi HI, Yakimov MM, Abdel-Fattah YR, Blaghenh M, Golyshini PN, Kalogerakis N, Boon N, Magagnini M, Fava F. Bioremediation of southern Mediterranean oil polluted sites comes of age. New Biotechnology. 2013;**30**(6):743-748

[19] Daffonchio D, Mapelli F, Cherif A, Malkawi HI, Yakimov MM, Abdel-Fattah YR, Boon N, Magagnini M, Fava F. ULIXES, unravelling and exploiting Mediterranean Sea microbial diversity and ecology for xenobiotics' and pollutants' clean up. Reviews in Environmental Science and Bio/Technology. 2012;**11**(3):207-211

[20] Coll M, Piroddi C, Steenbeek J, Kaschner K, Lasram FBR, Aguzzi J, Ballesteros E, Bianchi CN, Corbera J, Dailianis T, Danovaro R, Estrada M, Froglia C, Galil BS, Gasol JM, Gertwagen R, Gil J, Guilhaumon F, Kesner-Reyes K, Kitsos MS, Koukouras A, Lampadariou N, Laxamana E, Lopez-Fe de la Cuadra CM, Lotze HK, Martin D, Mouillot D, Oro D, Raicevich S, Rius-Barile J, Saiz-Salinas JI, San Vicente C, Somot S, Templado J, Turon X, Vafidis D, Villanueva R, Voultsiadou E. The biodiversity of the Mediterranean Sea: Estimates, patterns, and threats. PLoS One. 2010;**5**(8):e11842

[21] REMPEC: Regional Marine Pollution Emergency Response Centre for the Mediterranean Sea. http://www.rempec.org.

[22] Singh SN, Kumari B, Mishra S. Microbial degradation of alkanes. In: Microbial Degradation of Xenobiotics. Berlin Heidelberg: Springer; 2012. pp. 439-469

[23] Hassanshahian M, Abarian M, Cappello S. Biodegradation of Aromatic Compounds. In Biodegradation and Bioremediation of Polluted Systems-New Advances and Technologies. InTech; 2015

[24] Ghosal D, Ghosh S, Dutta TK, Ahn Y. Current state of knowledge in microbial degradation of polycyclic aromatic hydrocarbons (PAHs): A review. Frontiers in Microbiology. 2016;**7**:1369

[25] Tormoehlen LM, Tekulve KJ, Nanagas KA. Hydrocarbon toxicity: A review. Clinical Toxicology. 2014;**52**(5):479-489

[26] Mickiewicz M, Gomez HF. Hydrocarbon toxicity: General review and management guidelines. Air Medical Journal. 2006;**20**(3):8-11

[27] Abdel-Shafy HI, Mansour MS. A review on polycyclic aromatic hydrocarbons: Source, environmental impact, effect on human health and remediation. Egyptian Journal of Petroleum. 2016;**25**(1):107-123

[28] Rengarajan T, Rajendran P, Nandakumar N, Lokeshkumar B, Rajendran P, Nishigaki I. Exposure to polycyclic aromatic hydrocarbons with special focus on cancer. Asian Pacific Journal of Tropical Biomedicine. 2015;**5**(3):182-189

[29] Yu H. Environmental carcinogenic polycyclic aromatic hydrocarbons: Photochemistry and phototoxicity. Journal of Environmental Science and Health, Part C. 2002;**20**(2):149-183

[30] Sany SBT, Hashim R, Rezayi M, Salleh A, Rahman MA, Safari O, Sasekumar A. Human health risk of polycyclic aromatic hydrocarbons from consumption of blood cockle and exposure to contaminated sediments and water along the Klang Strait, Malaysia. Marine Pollution Bulletin. 2014;**84**(1):268-279

[31] Cobaugh DJ, Seger DL, Krenzelok EP. Hydrocarbon toxicity: An analysis of AAPCC TESS data. Przeglad Lekarski. 2006;**64**(4-5):194-196

[32] Sikkema J, De Bont JA, Poolman B. Mechanisms of membrane toxicity of hydrocarbons. Microbiological Reviews. 1995;**59**(2):201-222

[33] Santisi S, Cappello S, Catalfamo M, Mancini G, Hassanshahian M, Genovese M, Giuliano L, Yakimov MM. Biodegradation of crude oil by individual bacterial strains and a mixed bacterial consortium. Brazilian Journal of Microbiology. 2015;**46**(2):377-338

[34] Boopathy R. Factors limiting bioremediation technologies. Bioresource Technology. 2000;**74**(1):63-67

[35] Nikolopoulou M, Pasadakis N, Kalogerakis N. Evaluation of autochthonous bioaugmentation and biostimulation during microcosm-simulated oil spills. Marine Pollution Bulletin. 2013a;**72**(1):165-173

[36] Nikolopoulou M, Eickenbusch P, Pasadakis N, Venieri D, Kalogerakis N. Microcosm evaluation of autochthonous bioaugmentation to combat marine oil spills. New Biotechnology. 2013b;**30**(6):734-742

[37] Hosokawa R, Nagai M, Morikawa M, Okuyama H. Autochthonous bioaugmentation and its possible application to oil spills. World Journal of Microbiology and Biotechnology. 2009;**25**(9):1519-1528

[38] Fodelianakis S, Antoniou E, Mapelli F, Magagnini M, Nikolopoulou M, Marasco R, Barbato M, Tsiola A, Tsikopoulou I, Mahjoubi M, Jaouani A, Amer R, Hussein E, Al-Horani FA, Benzha F, Blaghen M, Malkawi HI, Abdel-Fattah Y, Cherif A, Daffonchio D, Kalogerakis N. Allochthonous bioaugmentation of crude oil-polluted sediments is inefficient in the presence of an effective degrading indigenous microbiome. Journal of Hazardous Materials. 2015;**287**:78-86

[39] Head IM, Jones DM, Röling WF. Marine microorganisms make a meal of oil. Nature Reviews. Microbiology. 2006;**4**(3):173

[40] Mrozik A, Piotrowska-Seget Z. Bioaugmentation as a strategy for cleaning up of soils contaminated with aromatic compounds. Microbiological Research. 2010;**165**(5):363-375

[41] Garon D, Sage L, Wouessidjewe D, Seigle-Murandi F. Enhanced degradation of fluorene in soil slurry by Absidia cylindrospora and maltosyl-cyclodextrin. Chemosphere. 2004;**56**(2):159-166

[42] Chandra S, Sharma R, Singh K, Sharma A. Application of bioremediation technology in the environment contaminated with petroleum hydrocarbon. Annals of Microbiology. 2013;**63**(2):417-431

[43] Nikolopoulou M, Kalogerakis N. Biostimulation strategies for fresh and chronically polluted marine environments with petroleum hydrocarbons. Journal of Chemical Technology and Biotechnology. 2009;**84**(6):802-807

[44] Adams GO, Fufeyin PT, Okoro SE, Ehinomen I. Bioremediation, biostimulation and bioaugmention: A review. International Journal of Environmental Bioremediation & Biodegradation. 2015;**3**(1):28-39

[45] Morais EBD, Tauk-Tornisielo SM. Biodegradation of oil refinery residues using mixed-culture of microorganisms isolated from a landfarming. Brazilian Archives of Biology and Technology. 2009;**52**(6):1571-1578

[46] Nikolopoulou M, Kalogerakis N. Biostimulation strategies for enhanced bioremediation of marine oil spills including chronic pollution. In: Handbook of Hydrocarbon and Lipid Microbiology. Berlin, Heidelberg: Springer. 2010;2521-2529

[47] Andreoni V, Gianfreda L. Bioremediation and monitoring of aromatic-polluted habitats. Applied Microbiology and Biotechnology. 2007;**76**(2):287-308

[48] Bamforth SM, Singleton I. Bioremediation of polycyclic aromatic hydrocarbons: Current knowledge and future directions. Journal of Chemical Technology and Biotechnology. 2005;**80**(7):723-736

[49] Haritash AK, Kaushik CP. Biodegradation aspects of polycyclic aromatic hydrocarbons (PAHs): A review. Journal of Hazardous Materials. 2009;**169**(1):1-15

[50] Breedveld GD, Sparrevik M. Nutrient-limited biodegradation of PAH in various soil strata at a creosote contaminated site. Biodegradation. 2000;**11**(6):391-399

[51] Silva-Castro GA, Uad I, Gónzalez-López J, Fandino CG, Toledo FL, Calvo C. Application of selected microbial consortia combined with inorganic and oleophilic fertilizers to

recuperate oil-polluted soil using land farming technology. Clean Technologies and Environmental Policy. 2012;**14**(4):719-726

[52] Margesin R, Schinner F. Biodegradation and bioremediation of hydrocarbons in extreme environments. Applied Microbiology and Biotechnology. 2001;**56**(5):650-663

[53] Leahy JG, Colwell RR. Microbial degradation of hydrocarbons in the environment. Microbiological Reviews. 1990;**54**:305-315

[54] Shallu S, Pathak H, Jaroli DP. Factors affecting the rate of biodegradation of polyaromatic hydrocarbons. International Journal of Pure and Applied Bioscience. 2014;**2**:185-202

[55] Bargiela F, Mapelli F, Rojo D, Chouaia B, Tornes J, Borin S, Richter DPM, Simone S, Christoph C, Genovese M, Denaro R, Martinez M, Fodelianakis S, Amer R, Bigazzi D, Han X, Chernikova J, Golsyhina O, Mahjoubi M, Jaouani A, Benzha F, Magagnini M, Hussein E, Al-Horani F, Cherif A, Blaghen M, Abdel-Fattah Y, Kalogerakis N, Barbas C, Malkawi H, Golyshin P, Yakimov M, Daffonchio D, Ferrer M. Bacterial population and biodegradation potential in chronically crude oil-contaminated marine sediments are strongly linked to temperature. Scientific Reports. 2015;**5**:11651

[56] Maki H, Hirayama N, Hiwatari T, Kohata K, Uchiyama H, Watanabe M, Yamasaki F, Furuki M. Crude oil bioremediation field experiment in the Sea of Japan. Marine Pollution Bulletin. 2003;**47**:74-77

[57] Kimes NE, Callaghan AV, Suflita JM, Morris PJ. Microbial transformation of the Deepwater horizon oil spill—Past, present, and future perspectives. Frontiers in Microbiology. 2014;**5**:603

[58] Harayama S, Kasai Y, Hara A. Microbial communities in oil-contaminated seawater. Current Opinion in Biotechnology. 2004;**15**(3):205-214

[59] Yakimov MM, Timmis KN, Golyshin PN. Obligate oil-degrading marine bacteria. Current Opinion in Biotechnology. 2007;**18**(3):257-266

[60] El Fantroussi S, Agathos SN. Is bioaugmentation a feasible strategy for pollutant removal and site remediation? Current Opinion in Microbiology. 2005;**8**(3):268-275

[61] Rodrigues EM, Kalks KH, Fernandes PL, Tótola MR. Bioremediation strategies of hydrocarbons and microbial diversity in the Trindade Island shoreline—Brazil. Marine Pollution Bulletin. 2015;**101**(2):517-525

[62] Zhang DC, Mörtelmaier C, Margesin R. Characterization of the bacterial archaeal diversity in hydrocarbon-contaminated soil. Science of the Total Environment. 2012;**421**:184-196

[63] Berry D, Gutierrez T. Evaluating the detection of hydrocarbon-degrading bacteria in 16S rRNA gene sequencing surveys. Frontiers in Microbiology. 2017;**8**:896

[64] Brightwell G, Boerema J, Mills J, Mowat E, Pulford D. Identifying the bacterial community on the surface of Intralox™ belting in a meat boning room by culture-dependent and culture-independent 16S rDNA sequence analysis. International Journal of Food Microbiology. 2006;**109**(1):47-53

[65] Genovese M, Crisafi F, Denaro R, Cappello S, Russo D, Calogero R, Santisi S, Catalfamo M, Modica A, Smedile F, Genovese L, Golyshin PN, Giuliano L, Yakimov MM. Effective bioremediation strategy for rapid in situ cleanup of anoxic marine sediment in mesocosm oil spill simulation. Frontiers in Microbiology. 2014;**5**:162

[66] Schneiker S, dos Santos VAM, Bartels D, Bekel T, Brecht M, Buhrmester J, Chernikova N, Denaro R, Ferrer M, Gertler C, Goesmann A, Golyshina OV, Kaminski F, Khachane AN, Lang S, Linke B, McHardy AC, Meyer F, Nechitaylo T, Pühler A, Regenhardt D, Rupp O, Sabirova JS, Selbitschka W, Yakimov MY, Timmis KN, Vorhölte F, Weidner S, Kaiser O, Golyshin PN. Genome sequence of the ubiquitous hydrocarbon-degrading marine bacterium *Alcanivorax borkumensis*. Nature Biotechnology. 2006;**24**(8):997-1004

[67] Yakimov MM, Golyshin PN, Lang S, Moore ER, Abraham WR, Lünsdorf H, Timmis KN. Alcanivoraxborkumensis gen. nov., sp. nov., a new, hydrocarbon-degrading and surfactant-producing marine bacterium. International Journal of Systematic and Evolutionary Microbiology. 1998;**48**(2):339-348

[68] Yakimov MM, Giuliano L, Denaro R, Crisafi E, Chernikova TN, Abraham W, Luensdorf H, Timmis KN, Golyshin PNR, Golyshin PN. *Thalassolituus oleivorans* gen. nov., sp. nov., a novel marine bacterium that obligately utilizes hydrocarbons. International Journal of Systematic and Evolutionary Microbiology. 2004a;**54**(1):141-148

[69] Kasai Y, Kishira H, Harayama S. Bacteria belonging to the genus Cycloclasticus play a primary role in the degradation of aromatic hydrocarbons released in a marine environment. Applied and Environmental Microbiology. 2002;**68**(11):5625-5633

[70] Yakimov MM, Gentile G, Bruni V, Cappello S, D'Auria G, Golyshin PN, Giuliano L. Crude oil-induced structural shift of coastal bacterial communities of Rod Bay (Terra Nova Bay, Ross Sea, Antarctica) and characterization of cultured cold-adapted hydrocarbonoclastic bacteria. FEMS Microbiology Ecology. 2004b;**49**(3):419-432

[71] Das N, Chandran P. Microbial degradation of petroleum hydrocarbon contaminants: An overview. Biotechnology Research International. 2010;**2011**:1-13

[72] Olajire AA, Essien JP. Aerobic degradation of petroleum components by microbial consortia. Journal of Petroleum & Environmental Biotechnology. 2014;**5**(5):1

[73] Rojo F. Degradation of alkanes by bacteria. Environmental Microbiology. 2009;**11**(10): 2477-2490

[74] Wang W, Shao Z. Enzymes and genes involved in aerobic alkane degradation. Frontiers in Microbiology. 2013;**4**:116

[75] Darmawan R, Nakata H, Ohta H, Niidome T, Takikawa K, Morimura S. Isolation and evaluation of PAH degrading bacteria. Journal of Bioremediation & Biodegradation. 2015;**6**(3):1

[76] Batista SB, Mounteer AH, Amorim FR, Totola MR. Isolation and characterization of biosurfactant/bioemulsifier-producing bacteria from petroleum contaminated sites. Bioresource Technology. 2006;**97**(6):868-875

[77] Bento FM, de Oliveira Camargo FA, Okeke BC, Frankenberger WT. Diversity of biosurfactant producing microorganisms isolated from soils contaminated with diesel oil. Microbiological Research. 2005b;**160**(3):249-255

[78] Banat IM, Makkar RS, Cameotra SS. Potential commercial applications of microbial surfactants. Applied Microbiology and Biotechnology. 2000;**53**(5):495-508

[79] Cameotra SS, Makkar RS. Recent applications of biosurfactants as biological and immunological molecules. Current Opinion in Microbiology. 2004;**7**(3):262-266

[80] Owsianiak M, Chrzanowski Ł, Szulc A, Staniewski J, Olszanowski A, Olejnik-Schmidt AK, Heipieper HJ. Biodegradation of diesel/biodiesel blends by a consortium of hydrocarbon degraders: Effect of the type of blend and the addition of biosurfactants. Bioresource Technology. 2009;**100**(3):1497-1500

[81] Bordoloi NK, Konwar BK. Bacterial biosurfactant in enhancing solubility and metabolism of petroleum hydrocarbons. Journal of Hazardous Materials. 2009;**170**(1):495-505

[82] Whang LM, Liu PWG, Ma CC, Cheng SS. Application of rhamnolipid and surfactin for enhanced diesel biodegradation—Effects of pH and ammonium addition. Journal of Hazardous Materials. 2009;**164**(2):1045-1050

[83] Huang XF, Guan W, Liu J, Lu LJ, Xu JC, Zhou Q. Characterization and phylogenetic analysis of bioemulsifier-producing bacteria. Bioresource Technology. 2010;**101**(1):317-323

[84] Aparna A, Srinikethan G, Smitha H. Production and characterization of biosurfactant produced by a novel pseudomonas sp. 2B. Colloids and Surfaces B: Biointerfaces. 2012;**95**:23-29

[85] Das K, Mukherjee AK. Crude petroleum-oil biodegradation efficiency of *Bacillus subtilis* and *Pseudomonas aeruginosa* strains isolated from a petroleum-oil contaminated soil from North-East India. Bioresource Technology. 2007;**98**(7):1339-1345

[86] Salihu A, Abdulkadir I, Almustapha MN. An investigation for potential development on biosurfactants. Biotechnology and Molecular Biology Reviews. 2009;**4**(5):111-117

[87] Desai JD, Banat IM. Microbial production of surfactants and their commercial potential. Microbiology and Molecular Biology Reviews. 1997;**61**(1):47-64

[88] Reddy MS, Naresh B, Leela T, Prashanthi M, Madhusudhan NC, Dhanasri G, Devi P. Biodegradation of phenanthrene with biosurfactant production by a new strain of Brevibacillus sp. Bioresource Technology. 2010;**101**:7980-7983

[89] Hörmann B, Müller MM, Syldatk C, Hausmann R. Rhamnolipid production by *Burkholderia plantarii* DSM 9509T. European Journal of Lipid Science and Technology. 2010;**112**(6):674-680

[90] Marqués AM, Pinazo A, Farfan M, Aranda FJ, Teruel JA, Ortiz A, Manresaa A, Espuny MJ. The physicochemical properties and chemical composition of trehalose lipids produced by Rhodococcus erythropolis 51T7. Chemistry and Physics of Lipids. 2009;**158**(2):110-117

[91] Cappello S, Calogero R, Santisi S, Genovese M, Denaro R, Genovese L, Giuliano L, Mancini G, Yakimov MM. Bioremediation of oil polluted marine sediments: A bio-engineering treatment. International Microbiology. 2015;**18**:127-134

[92] Crisafi F, Genovese M, Smedile F, Russo D, Catalfamo M, Yakimov ML, Denaro R. Bioremediation technologies for polluted seawater sampled after an oil-spill in Taranto Gulf (Italy): A comparison of biostimulation, bioaugmentation and use of a washing agent in microcosm studies. Marine Pollution Bulletin. 2016;**106**:119-126

[93] Wu M, Dick WA, Li W, Wang X, Yang Q, Wang T, Xu L, Zhang M, Chen L. Bioaugmentation and biostimulation of hydrocarbon degradation and the microbial community in a petroleum-contaminated soil. International Biodeterioration & Biodegradation. 2016;**107**:158-164

[94] American Public Health Association (A.P.H.A.). Standard Methods for the Examination of Water and Waste Water. 18th ed. Washington, DC: American Public Health Association; 1992

Chapter 17

Petroleum Degradation: Promising Biotechnological Tools for Bioremediation

Maddela Naga Raju and Laura Scalvenzi

Additional information is available at the end of the chapter

http://dx.doi.org/10.5772/intechopen.70109

Abstract

One of the most common chemicals involved in the soil contamination or soil pollution is petroleum hydrocarbons (PHs). As we know that PH-contaminated soil affects human health directly, such as (i) contact with soil, (ii) via inhalation of vaporized contaminants, and (iii) infiltration of soil contamination into groundwater aquifers used for human consumption. Microbiological processes play an important role in the removal of PHs and take advantage of the catabolic versatility of these organisms to degrade such compounds either partially or completely (mineralization). Thus, the present chapter moves around the relationship of microorganisms with PHs. Based on this concept, this chapter has been designed to address the following relevant issues: How to isolate PH-degrading microorganisms by co-enrichment and optimized enrichment methods? How to study the microbial community structure by high-throughput sequencing method? What are the metabolic versatilities of microorganisms for degrading PHs? How to treat the environmental problems through biological means? What are the available ecotoxicity studies for the analysis of residual PHs after the microbiological treatment at the PHs-contaminated sites? Thus, the aim of this chapter is to explain the importance of microorganisms in cleaning the oil-contaminated environments.

Keywords: biodegradation, bioremediation, ecotoxicology, microorganisms, petroleum hydrocarbons

1. Introduction

The most common contaminant in the environment is crude oil and its derivatives. Due to their wide spread occurrence and severe risks they pose to human health and water bodies (surface as well as ground), they require intense remediation practices at the contaminated sites. Strictly speaking, contamination is strongly correlated with the degree of industrialization and intensity of chemical usage. All hydrocarbon compounds derived from petroleum sources are generally described as total petroleum hydrocarbons (TPHs). Fuels such as petrol, diesel, kerosene, and lubricating oils/greases all come under the category of TPHs. Soil pollution by petroleum hydrocarbons (PHs) is mainly due to oil drilling, waste disposal (oil and fuel dumping), and accidental spilling as may occur during activities. Crude oil and refined fuel spills from tanker ship accidents have caused severe damages to ecosystems in many parts of the world. The quantity of oil spilled during accidents has ranged from a few million gallons to several hundred thousand gallons [1].

Chemically, hydrocarbons seem to be simple organic substances (comprising only carbon and hydrogen). However, there are many kinds of compounds with different chemical and physical nature. Analysis of the components at the spilled sites gives lots of information about their diversity. For instance, at the contaminated sites, TPH compounds that have an aliphatic structure (i.e., straight or branched chains of carbon molecules) will behave differently from aromatic compounds (ringed chains of carbons). Similarly, TPH compounds that have less carbon molecules (short-chain compounds) will also act differently. Solubility, volatility, and organic partitioning coefficients are greatly influenced by the number of carbon atoms. For example, compounds with less than 16 carbon atoms tend to be more mobile at the spilled sites due to their greater solubility, volatility, and lower organic partitioning coefficients. On the other hand, lightweight aromatic compounds, for example, benzene is highly toxic chemical. Generally, heavy weight TPHs have opposing properties, which are readily adsorb into the organic fractions of soil. Another important form of PHs is polycyclic aromatic hydrocarbons (PAHs), which are a class of chemicals that occur naturally in coal, crude oil, and gasoline. Basically, PAHs are neutral, nonpolar, heavy weight substances and composed of multiple aromatic rings. More importantly, PAHs have higher toxicity and are typically more persistent in the environment. They are also produced when coal, oil, gas, wood, garbage, and tobacco are burned. PAHs generated from these sources can bind to or form small particles in the air. Thus, PAHs making them of greater concern if they are released into the environment. Another important aspect of PHs is their forms or phases at the spilled site. This greatly depends on original composition of the source of spilled TPHs, geological and hydrogeological conditions at the spilled sites, and the age since the spillage occurred. More often, upon the spillage, the majority of TPHs mass will be partitioned within the soil phase. In certain instances, TPHs are able to float on the surface of the water table. In this form, TPHs are encountered as a phase-separated liquid and are also called light nonaqueous phase liquid (LNAPL), which is principally due to their buoyancy. Other two important forms of TPHs at the spilled site are dissolved and vapor forms. A percentage of TPHs will also be dissolved into the groundwater or trapped as a vapor within the soil "pore-space" in the unsaturated zone. Fate of crude oil at the spilled site is shown in **Figure 1**.

Human health effects from environmental exposure to PHs are vary, principally depends on type and quantity of PHs. For instance, large amounts of naphthalene in air can irritate eyes and breathing passages. Moreover, blood and liver abnormalities were observed in workers who have been exposed (either skin contact or inhalation of vapors) to large amounts of naphthalene [2]. Several PAHs (pure or mixture) are considered to be cancer-causing chemicals. In well-established animal model studies, PAHs were linked to skin, lung, bladder, liver, and stomach cancers [3]. Adult exposure to PAHs has been linked to cardiovascular disease [4].

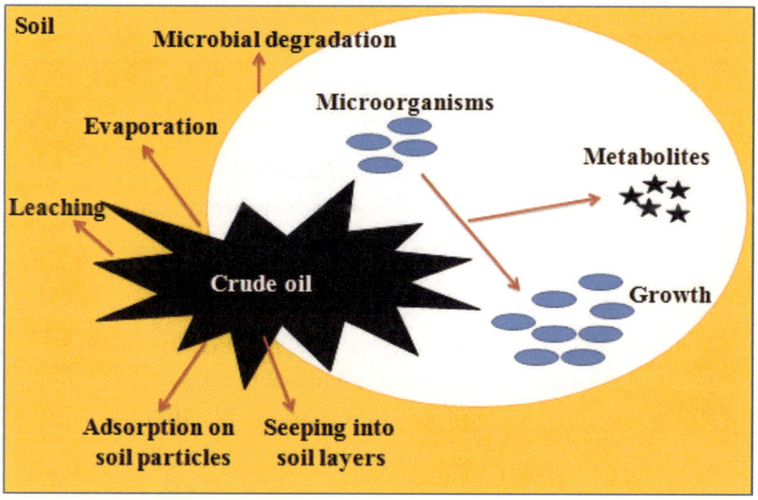

Figure 1. Fate of crude oil during land spillage.

Nevertheless, human health effects from environmental exposure to low levels of PAHs are unknown. Additionally, environmental impacts of PHs are numerable. There are many reports on the contamination of drinking water supplies by oil spillage [5]. Contamination can have an economic impact on tourism and marine resource extraction industries. More importantly, marine animals and birds exposed to oil spills are severely affected. Because of the health implications, less than 1% of oil-soaked birds survive, even after cleaning [6]. Additionally, fluctuations in body temperature, hypothermia, blindness, dehydration, impaired digestive process, and disorder of lungs and liver were observed in many heavily furred marine mammals exposed to oil spills.

By considering all the above facts, remediation and reclamation of PH-contaminated sites are essentially important to protect the health of ecosystem. There are two main remediation technologies, namely *ex situ* and *in situ* methods. *Ex situ* methods involve excavation of affected soils or extraction of contaminated groundwater and subsequent treatment at the surface. These methods consist of soil excavation and disposal to landfill and groundwater "pump and treat." In contrast, in *in situ* methods, the contaminated soils or groundwater are treated at the spill sites. These methods include but are not limited to solidification and stabilization, soil vapor extraction, permeable reactive barriers, monitored natural attenuation, bioremediation-phytoremediation, chemical oxidation, and steam-enhanced extraction and *in situ* thermal desorption. However, the further sections of this chapter give detailed information about microbial remediation of crude oil contaminated soil.

2. Enrichment and isolation of crude oil-degrading microorganisms

An enrichment culture is a medium with specific and known qualities that favors the growth of a particular microorganism. Enrichment cultures are used to increase the small number of desired microorganism to the detectable level. Enrichment cultures are often used for soil sample. This type of technique is very useful for the detection and isolation of PH-degrading microorganisms from PH-contaminated soil. Brief description of enrichment method to be used for the isolation of crude oil-degrading bacteria is presented below.

Two 100-mL erlenmeyer flasks containing 25 mL of mineral salt medium (MSM) are prepared separately, one is for the isolation of crude oil-degrading bacteria and another is used for the isolation of crude oil-degrading fungi. Since bacteria and fungi have been reported as principal microorganisms of PH degradation, the information provided in this chapter is related to these two organisms only, unless otherwise it is stated. The composition of MSM is as follows (g L^{-1}): NaCl, 5.0; KH_2PO_4, 5.0; $(NH_4)_2SO_4$, 1.0; $K_2HPO_4 \cdot 3H_2O$, 1.0; $MgSO_4 \cdot 7H_2O$, 0.25; $NaNO_3$, 2.0; $FeCl_2 \cdot 4H_2O$, 0.02; and $CaCl_2$, 0.02. The pH of the medium in first and second flaks is adjusted to 7.2 and 5.5 for bacteria and fungi, respectively. After sterilizing the medium at 121°C for 30 min [7], 1.0% of crude oil contaminated soil is used as an inoculum to inoculate the medium separately. Then the medium is enriched for 7 days at respective temperatures (25°C for fungi and 30°C for bacteria) and 180 rpm on the rotary shaker. The culture is enriched by four consecutive inoculations of 1.5-mL inoculum to 100-mL Erlenmeyer flasks containing 25 mL of fresh MSM medium. Following enrichment, parts of the medium are plated onto the MSM medium containing 2.0% of agar and 1.0% of crude oil and incubated for 3–7 days separately for bacteria and fungi [8]. Finally, different pure colonies obtained from the plates are stored in the Luria-Bertani medium (bacteria) or Czapek Dox medium (fungi) with 15% of glycerol at −80°C until further use. Schematic representation of enrichment and isolation of crude oil-degrading bacteria is shown in **Figure 2**. However, microorganisms have also been enriched and isolated by using methods of various modifications. For instance, oil-degrading bacteria were isolated using sterile crude oil as the medium [9].

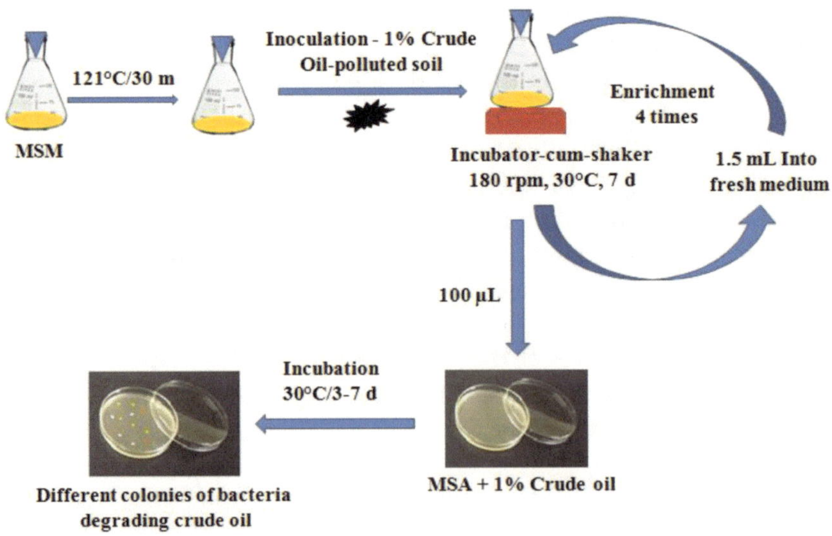

Figure 2. Enrichment and isolation of crude oil-degrading bacteria.

3. Microbial ecology of oil fields

Knowing which organisms are present in a particular habitat is critical to research in microbiology. Sequencing DNA enables researchers to determine which types of microbes may be present at the site of sample collection. Nowadays, DNA-based technologies playing a major role in the analysis of microbial communities at the PHs or organic compounds contaminated soils, water, and sediments [10]. A study of genetic material recovered directly from environmental samples is called "metagenomics." This field of science may also be referred to as "environmental genomics," "ecogenomics," or "community genomics." In early days, cultivated clonal cultures, early environmental gene sequencing cloned specific genes (16S or 18S rRNA genes) were used to produce a profile of diversity in a natural sample. A vast majority of microbial biodiversity had been missed by cultivation-based methods [11]. It is being frequently reported that the environmental samples contain more number of noncultivable microorganisms than cultivable one. Thus, recent studies focusing on either "shotgun" or Polymerase chain reaction (PCR) directed sequencing to get largely unbiased samples of all genes from all the members of the sampled communities [12]. Due to the ability of metagenomics to reveal the previously hidden diversity of microbial life and as the price of DNA sequencing continues to fall, metagenomics now allows microbial ecology to be investigated at a much greater scale and detail than before.

In the analysis of microbial community structure of PH-contaminated soil or sediment, the total chromosomal DNA is to be extracted by using one of several available commercial DNA kits. The extracted DNA is stored at −20°C until further use. In order to analyze the bacterial community structure, 16S rRNA genes are PCR amplified from the bulk DNA by using PCR reaction mixture. The volume of the PCR mixture is usually 20–50 μL, which contains template DNA, universal primers (e.g., 27F/1492R), each of four dNTPs, polymerase enzyme buffer, and polymerase enzyme (Taq or pfu polymerases). Likewise, the 16S rRNA region is amplified by PCR using the forward primer 27F (5-AGA GTT TGA TCC TGG CTC AG-3) and reverse primer 1492R (5-CGG CTA CCT TGT TAC GAC TT-3) [13]. Generally, DNA amplification is performed under the specified cycling conditions: 1 cycle of 2 min at 94°C, then 25 cycles of 30 s at 94°C, 30 s at 55°C, and 1 min at 72°C, followed by a final cycle of 10 min at 72°C. After amplification, PCR products, also called "amplicons," are tested by 2% agarose gel to confirm the specific length of DNA amplicons. Furthermore, these amplicons are purified for sequencing purpose. Nowadays, gel extraction and purifications combo kits are available commercially. These kits have ability to perform both a gel extraction and a purification of amplicons in a single step. In the case of fungal community structure analysis, ITS (internal transcribed spacer) regions of their whole DNA are amplified by using ITS1 and ITS2 primers. Finally, purified PCR products are used for sequencing purpose.

There are several high-throughput methods (formerly "next-generation") available for sequencing the genome. These methods include massively parallel signature sequencing (MPSS), polony sequencing, 454-Pyrosequencing, Illumina (Solexa) sequencing, SOLiD sequencing, Ion Torrent semiconductor sequencing, DNA nanoball sequencing, Heliscope single-molecule sequencing, single-molecule real-time (SMRT) sequencing, and nanopore DNA sequencing. Comparison of selective high-throughput sequencing methods [14, 15] is shown in **Table 1**.

Method	Read length	Accuracy	Time per run	Cost per 1 million base pairs in US$	Advantages	Disadvantages
Chain termination (Sanger)	400–900 bp	99.9%	20 min–3 h	2400	Long individual reads, useful for many applications	More expensive, time-consuming step of PCR
Pyro-sequencing	700 bp	99.9%	24 h	10	Long read size, fast	Runs are expensive, homopolymer errors
Ion semiconductor (Ion Torrent)	400 bp	98%	2 h	1	Less expensive equipment, fast	Homopolymer errors
Sequencing by synthesis (Illumina)	HiSeq 2500: 50–500 bp	99.9%	1–11 d	0.05–0.15	Potential for high sequence yield	Very expensive equipment requires high concentrations of DNA

Table 1. Comparison of selective high-throughput sequencing methods.

Once sequences are ready, they are analyzed using different bioinformatics tools (**Figure 3**). For instance, QUIIME software package (Quantitative insights into microbial ecology) [16] is used to analyze the sequences. More often, Basic Local Alignment Search Tool (BLAST) program of the National Centre for Biotechnology Information (NCBI, https://www.ncbi.nlm.nih.gov/) is used to search and identify the closest species. In-house Perl Scripts are used to analyze alpha- and beta-diversities within and among the samples, respectively. In addition to these, QUIIME software package (http://qiime.org/) and UPARSE pipeline (http://drive5.com-/uparse/) are used to analyze the reads and pick operational taxonomic units (OTUs). Then, sequences are assigned to OTUs at particular percent similarity. UPARSE is a method for generating clusters (OTUs) from next-generation sequencing reads of marker genes such as 16S rRNA, the fungal ITS region and the COI gene. Finally, representative sequence for each OUT is picked, and RDP (Ribosomal database project) classifier [17] is used to assign taxonomic data to each representative sequence. RDP provides quality-controlled, aligned, and annotated bacterial and archaeal 16S rRNA gene sequences and fungal 28S rRNA gene sequences and a suite of analysis tools to the scientific community. Finally, Simpsons Index of Diversity [18, 19] is used to define the community structure. Simpson's Index of Diversity = 1 − D:

$$D = \sum n(n-1)/N(N-1) \tag{1}$$

where n = total number of organisms of a particular species and N = total number of organisms of all species. More recently, in Ref. [20] researchers have studied and reported about the microbial community structure in crude oil-contaminated seawaters by using bioinformatics tools such as QUIIME, UPRASE, and RDP.

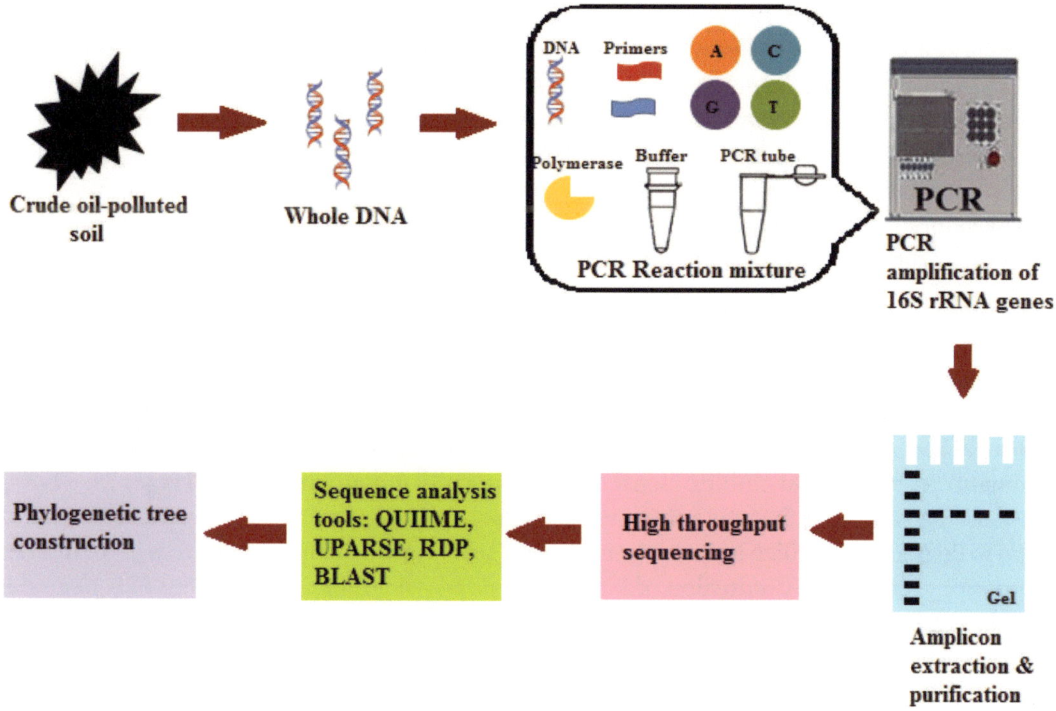

Figure 3. Schematic representation of microbial community analysis of crude oil-polluted soil by using high throughput sequencing methods.

4. Microbial degradation of PHs

Biodegradation is the disintegration of materials by bacteria, fungi, or other biological means. Substances to be degraded by microorganisms are generally organic materials. Materials in organic nature are degraded aerobically with oxygen or anaerobically without oxygen. More often, organic materials are the good nutrient sources for microorganisms. Since there is a large diversity in the microorganisms, a huge range of compounds are biodegraded, including hydrocarbons (e.g., oil), polychlorinated biphenyls (PCBs), PAHs, pharmaceutical substances, etc.

In most of the studies, microbial groups such as bacteria, yeast, and fungi have been identified as principal agents in the degradation of PHs, even though their degradation efficiencies are varying. However, bacteria are the most active and primary degraders of spilled oil in the environment [21], and some of them are known to grow exclusively on PHs as their sole carbon and energy source. In one of our recent investigations, we isolated two bacterial strains (*Bacillus thuringiensis* strain B3, *B. cereus* strain B6) from Ecuadorian oil fields, they grew exclusively in MSM containing 1% diesel as their carbon source [22]. So far, several bacterial genera, namely, *Acinetobacter, Aeromicrobium, Alcaligenes, Bacillus, Brevibacterium, Burkholderia, Corynebacterium, Dietzia, Flavobacterium, Gordonia, Micrococcus, Mycobacterium, Pseudomonas, Sphingomonas*, etc., isolated from petroleum contaminated soil proved to be the potential organisms for PHs' degradation [23, 24]. Similarly, several fungal genera isolated from PH-contaminated soils, and reported as PH degraders. They include *Amorphoteca, Aspergillus, Candida, Cochliobolus, Fusarium, Graphium, Neosartorya, Penicillium, Phaenerochaete, Pichia, Pseudallescheria, Talaromyces*, and *Yarrowia* [24–27]. Recently, we found two indigenous fungal strains in Ecuadorian oil fields, they were belonging to the genus *Geomyces*, which could remove 77–80% of crude oil in medium and soil experiments, respectively [28]. In Ref. [29], researchers found 30% removal of crude oil by immobilized bacterial cells. In a more recent laboratory-based study, *Pseudomonas* sp. has removed 74% of PHs from the crude oil sludge in 7 days [30].

In the natural environment, biodegradation of crude oil involves a succession of species within the consortia of the present microbes. Impact of a microbial consortium on a contaminant is always much higher than by an individual organism. A single species can metabolize only a limited range of hydrocarbon substrates. Instead, a consortium of many different bacterial and/or fungal species, with broad enzymatic capacities, can degrade the maximum amount of contaminant. So far, several studies focused on the microbial degradation of PHs [8, 22, 31–34]; these studies reported that the microorganisms possess specific enzyme systems that enable them to degrade and utilize hydrocarbons as their sole carbon and energy sources [35]. Another important aspect is the production of biosurfactants by microorganisms during PHs degradation. Biosurfactants are the extracellular surfactants of the microorganisms, play major role in enhancing the bioavailability of contaminant to the microorganisms.

There are many environmental factors come into action during the degradation of PHs by microorganisms either *in vitro* or *in vivo*. Considering the physical factors, temperature plays an important role in biodegradation of hydrocarbons. It acts directly by affecting the chemistry of the pollutants, physiology, and diversity of the microbial flora. The highest degradation rates can be seen in the range of 30–40°C, 20–30°C, 15–20°C in soil, freshwater, and marine environments, respectively [36, 37]. Similar to above findings, members of *Geomyces* have shown optimum sporulation rates at 25°C on the medium containing either diesel or crude oil [28]. Another considerable factor that influences the microbial degradation of PHs is nutrients. Nitrogen and phosphorous are very important elements; they influence the rate of degradation greatly. Since PHs mainly contain carbon and hydrogen, microorganisms need additional elements for their growth on PHs. Additionally, pH, concentration, type and age of the contaminants also play major role in influencing the degradation of PHs by microorganisms.

With respect to the aerobic and anaerobic environments, nevertheless, the most rapid and complete degradation of the majority of organic pollutants is principally achieved under aerobic conditions. The susceptibility of hydrocarbons to microbial degradation can be generally ranked as follows: linear alkanes > branched alkanes > small aromatics > cyclic alkanes [38]. Some compounds, such as the high molecular weight polycyclic aromatic hydrocarbons (PAHs), may not be degraded at all [39]. Several microbial enzymes have been identified as important agents in the degradation of PHs. For instance, oxygenases [40], monooxygenases [41], dioxygenases [42], and hydrolases [43] were among them.

The most widely used technique for the detection of residual PHs during microbial degradation is gas chromatography-flame ionizing detection (GC-FID). Helium or hydrogen or nitrogen is used as inert carrier gas in gas chromatography. Carrier gas carries the gaseous mixture or aqueous liquids with boiling points < 400°C, which are to be analyzed. Analysis takes place in a capillary column onto a detector. This allows better resolution of components in complex mixtures. This method determines the content of TPHs in the range C_{10}–C_{40} (*n*-alkanes), from solids including soils and wastes. GC-FID is used for both quantitative and qualitative applications with detection limits of 10 mg TPHs per kg soil. There is another method in which GC is coupled with mass spectroscopy called GC-MS. MS is described as a universal detector because of its versatility in the measurement of TPHs and PAHs. Another analytical method is available for the characterization of PHs, called infrared spectroscopy (IR). In this method, a spectrum is produced with stretching and bending vibrations associated with a molecule when it absorbs energy in the IR region of the electromagnetic spectrum. The spectra of hydrocarbon derivatives originate mainly from combinations or overtones of the C–H stretching modes of saturated CH_2 and terminal –CH_3 or aromatic C–H functional groups. Thus, IR-based detection is very helpful the elucidation of functional groups of residual and parent PHs during microbial degradation. More recently, in Ref. [44], TPHs in a biopile system of crude oil-contaminated dessert soil were measured by using "in-house" gravimetric and Fourier transform infrared spectroscopy (FTIR) methods.

5. Bioremediation

Removal or neutralization of pollutants from a contaminated site by using organisms is called bioremediation. It is one type of waste management technique. Principally, hazardous substances are broken down to less toxic or nontoxic substances by organisms. Bioremediation technologies have different approaches. In one kind of approach, the bioremediation process can be either *in situ* or *ex situ*. The *in situ* approach involves treating the contaminated soil or water at the site of contamination, whereas the *ex situ* approach involves the removal of contaminated materials to be treated elsewhere. There is another kind of approach in the bioremediation process in which bioremediation can be achieved either by biostimulation or bioaugmentation. Biostimulation is a widely used approach, which involves stimulating naturally occurring microbial communities, either by nutrients or other needs (such as pH, moisture, aeration, electron donors, electron acceptors, etc.), to break down a contaminant. In bioaugmentation, organisms selected for high degradation efficiencies are used to inoculate the contaminated site. Most widely used bioremediation methods are shown in **Figure 4**.

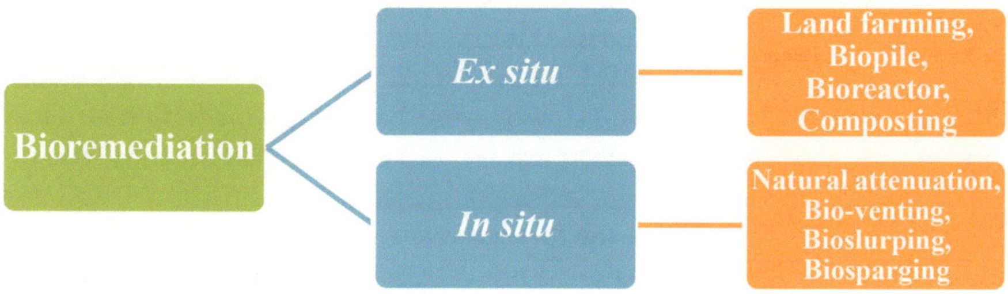

Figure 4. Types of bioremediation techniques.

5.1. Land farming

Land farming is the simplest method, which is inexpensive and requires less equipment (**Figure 5a**). It can be used either in the form of *ex situ* or *in situ* mode. The *in situ* or *ex situ* type of land farming method is applied when pollutant lies <1 or >1.7 m below the ground level, respectively [45]. Land farming consists of careful application of excavated polluted soil on a fixed layer of support above the ground surface. This allows the aerobic biodegradation of pollutant by autochthonous (indigenous) microorganisms. The major activities of land farming are soil tillage, which brings about aeration, addition of N-P-K fertilizers, and irrigation. Expectedly, all such operations greatly stimulate the indigenous microorganisms to enhance bioremediation during land farming. Practically, at the field level, this method is giving encouraging results. For instance, a land farming-based field trial experiment conducted in Canada for 3 years, where there was 80% removal of diesel contaminant from the soil [46].

5.2. Biopile

In this approach, there is above-ground piling of excavated polluted soil followed by amendments (nutrients and aeration) (**Figure 5b**). The remediated soil is placed in a liner to prevent further contamination of the soil, they may also be covered with plastic to control runoff, evaporation, and volatilization. This technique is widely used in nowadays due to easy controlling of nutrients, aeration, and temperature [47]. When the biopile system was combined with bioaugmentation and biostimulation approaches, >90% of TPHs were reduced in PH-contaminated soil in 94 days [48]. Nevertheless, the biopile system has its own disadvantages, such as conserve much space, robust engineering, cost or maintenance, and operation, lack of power supply at remote areas, heat generation resulted in the decreased microbial activities. Periodic turning (to enhance the aeration and subsequent hike in the biodegradation activities) of piled polluted soil is the principle of another bioremediation method called "windrows."

5.3. Bioreactor

A bioreactor is a vessel in which contaminated materials are converted to specific product(s) following series of biological reactions. There are different operating modes of bioreactor, such as batch, fed-batch, sequencing batch, continuous, and multistage. Polluted samples can be fed into a bioreactor either in the form of solid or slurry. One of the major advantages of bioreactor-based bioremediation is excellent control of bioprocess parameters such as temperature, pH, agitation and aeration rates, and substrate and inoculum concentrations. Another advantage of bioreactor is that it can be used for the treatment of either polluted water or soil. In a practical application of stirred tank bioreactor (2.5 L), 82–97% of TPHs were removed from crude oil-polluted sediment [49]. Yet, bioreactor-based bioremediation is not a full-scale practice due to several reasons. This approach is cost ineffective, because volume of polluted sample to be treated may be too large, requiring more manpower, capital, and safety measures for transporting the samples to the treatment site. Another disadvantage is due to several bioprocess parameters or variables of a bioreactor, if any parameter that is not properly controlled at optimum, this in turn will reduce microbial activities and will make process less effective. In addition to these, pollutants are likely to respond differently to different bioreactors. Thus, it is difficult to design a specific reactor for every pollutant.

http://dx.doi.org/10.5772/intechopen.70109

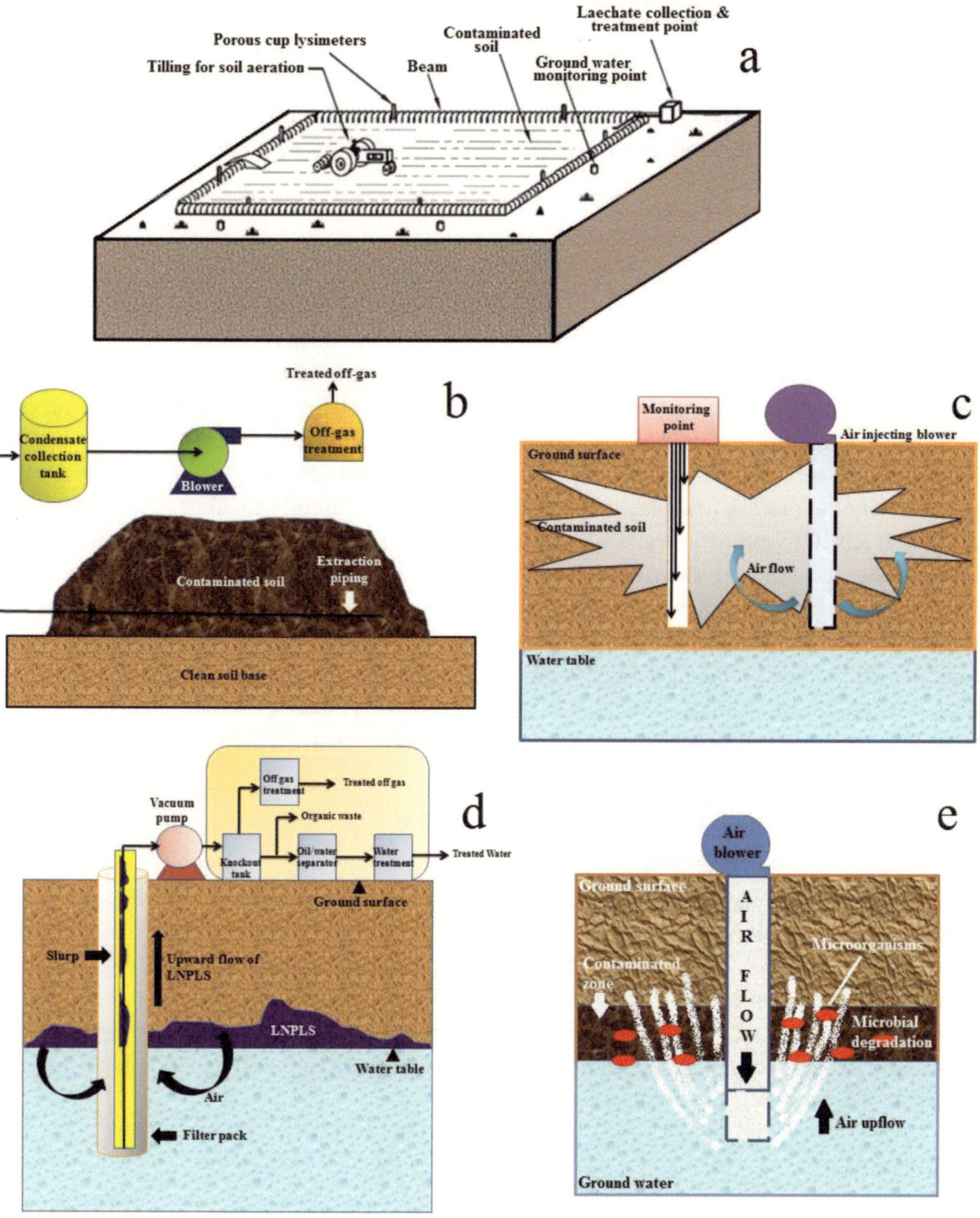

Figure 5. Bioremediation methods—(a) landfarming, (b) biopile, (c) bioventing, (d) bioslurping, and (e) biosparging.

5.4. Composting

Composting is a process of piling contaminated-soil along with organic substances such as manure, yard waste, or food-processing wastes. These are often added to supplement the amount of nutrients and readily degradable organic matter in soil. Stimulation of microbial growth by added nutrients results in effective biodegradation in a relatively short period of time. Efficiency of compositing in the removal of PHs from soil has been tested practically by using several lab- and field-scale studies. For instance, 85% reduction in diesel content was reached when a soil spiked with diesel oil was mixed with biowaste (vegetable, fruit, and garden waste) at a 1:10 ratio (fresh weight) and composted in a monitored composting bin system for 12 weeks [50].

5.5. Natural attenuation

Reduction of concentration and amount of pollutants at contaminated sites by natural process is called "natural attenuation." It can also be termed as intrinsic remediation, bioattenuation, and intrinsic bioremediation. In the process of natural attenuation, contaminants are left on the site and the naturally occurring processes are left to clean up the site. Several processes are come into action during natural attenuation. For example, biological degradation, volatilization, dilution, dispersion, dilution of the contaminant and sorption of the contaminant onto the organic matter, and clay minerals in the soil. It is mainly used to remediate the contaminated aquifer when the contamination source has been removed. In particular, it is used for benzene, toluene, ethylbenzene, and xylene (BTEX) and more recently for chlorinated hydrocarbons. Other contaminants that could potentially be remediated by natural attenuation include pesticides and inorganic compounds. The success of natural attenuation greatly depends on the subsurface geology, hydrology, and microbiology. Major disadvantages of natural attenuation are as follows: (i) it is relatively very slow process, since it is nonengineered biodegradation process; (ii) long-term monitoring is an absolute necessity since there must be no risk to the environment and to humans.

5.6. Bioventing

Bioventing is an *in situ* remediation technology, it is used to treat the contaminated groundwater system (**Figure 5c**). However, recently, this technique has also been used to remediate contaminated soil. Bioventing enhances the activity of indigenous microorganisms and stimulates the natural *in situ* biodegradation of hydrocarbons by inducing air or oxygen flow (by direct air injection), and nutrients into the unsaturated zone. In a field-level application of bioventing process for cleaning the phenanthrene-contaminated soil, in Ref. [51], researchers observed 93% contaminant removal after 7 months.

5.7. Bioslurping

Bioslurping is a unique *in situ* technique, is a combination of bioventing and vacuum-enhanced pumping, and is used to bioremediate soils and water (**Figure 5d**). Principle of this method is pumping or separation of free-product that is lighter than water (light nonaqueous phase liquid or LNAPL) to recover free product from the groundwater and soil. The bioslurping system uses a "slurp" tube that extends into the free-product layer; the pump draws liquid (including free-product) and soil gas up the tube in the same process stream. Thus, slurp is much similar to a straw in a glass draws liquid. The pumping mechanism brings about upward movement of LNAPLs to the surface, where it becomes separated from water and air. Once free products (contaminants) are separated, there is final treatment of contaminants by a conventional bioventing system to complete remediation process. This technique is cost effective because only a small amount of groundwater and soil vapor are pumped at a time, therefore the treatment plant used to treat the vapor and free product can be small.

5.8. Biosparging

It is also another *in situ* remediation technique. In biosparging, like bioventing, there is injection of air into soil subsurface to stimulate microbial activities in order to promote pollutant removal from polluted sites (**Figure 5e**). However, unlike bioventing, air is injected at the saturated zone. This causes upward movement of volatile organic compounds to the unsaturated zone to promote biodegradation. Biosparging has been widely used in treating aquifers contaminated with petroleum products, especially diesel and kerosene. This technique has shown effective results when applied to contaminated ground water. Practically, biosparging was used to clean benzene, toluene, ethylbenzene, and xylene (BTEX)-contaminated ground water, where they observed >70% reduction in BTEX [52].

5.9. Ecotoxicology

Ecotoxicology is the study of the effects of toxic chemicals on biological organisms, especially at the population, community, ecosystem levels. With regard to the present contest, ecotoxicity tests are conducted after the completion of bioremediation experiments. Toxicity of residual PHs and/or their products of microbial degradation present in the soil samples are tested though the survival, growth, behavior, and reproductions of organisms. Hence, bioassays can serve as a complementary tool in environmental risk assessment of bioremediated places, which help to determine whether the contaminant concentration at remediated sites is high enough to cause adverse effects on organisms. Frequently used toxicity tests are shown in **Figure 6**.

5.10. Earthworm survival tests

The common earthworm species, *Eisenia fetida,* is used to determine acute toxicity of the PH-contaminated soils before, during and after bioremediation. In this method, animals (~10) are placed into soil (~200 g) in 1-L wide-mouth jars with loose fitting lids. Lethal concentration-50

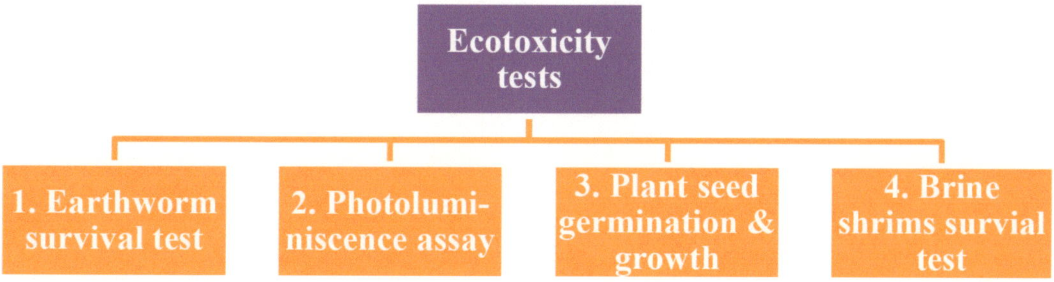

Figure 6. Types of ecotoxicity tests.

(LC50) for each soil is estimated using five concentrations of bioremediated soil (100, 50, 25, 12.5, 6.5, and 0%) prepared with control (contaminant free) soil. The soil water content is adjusted as per the requirement. Surviving earthworms are counted after 14 days of incubation at room temperature under constant fluorescent lighting conditions. Survival percent is inversely proportional to the toxicity of PHs. In a bioremediation experiment, it was found that the earthworm survival percentages were 28 and 100 after 4 and 12 weeks of treatment of heavy oil-contaminated soil [53].

5.11. Photo luminescence assay

In this method, there is a response of luminescent bacteria (*Photobacterium phosphoreum*) to residual PHs present in the treated contaminated-soils. However, this process needs special equipment and reagents such as Microtox analyzers and solid-phase test kits. Initially, soil dilutions are prepared (with Microtox diluent) and incubated for 20 min with reconstituted lyophilized bacteria. During the incubation, photoluminescence activity is induced in the bacteria (by kit reagents). Finally, activity of photoluminescence is detected by Microtox analyzer. Higher toxicity of PHs results in the lesser luminescence activity and vice versa. Photo luminescence assay is widely used in the bioremediation experiments. For instance, in Ref. [53], there was an observation of the loss of Microtox inhibiting activities by bioremediated soils, which were treated for 3 months.

5.12. Plant seed germination and growth

Plants depend on soil for germination and growth. Therefore, any alterations in the seed development may reflect the presence of toxic substances in the soil. Seed germination tests in ecotoxicological assays are considered short-term and evaluate acute toxicity effects. The effects of untreated and bioremediated oil soils are determined by using different plant species such as corn, wheat, oat, grass, cowpea, garden cress, etc. In this method, oily and oil-free soils are dispensed into wood or plastic containers having sufficient number of cells. Each cell should accommodate approximately 100 g soil. Then, 5–10 seeds are placed 1–1.5 cm below the soil surface. Generally, seed cultures are exposed to 12-h light/dark cycles at a soil surface light intensity of 310–350 lm with fluorescent lamps. Room temperature is maintained at 20–23°C and around 30% soil moisture capacity is maintained by spraying the soil surface with water. Time and germination percentages of seeds, plant growth (mg dry weight/plant) are determined before and after bioremediation. In one of our most recent investigations [54], we observed substantial improvement in germination time and percent germination of cowpea seeds in bioremediated soil over control soil.

6. Summary

Taken together, details provided in this chapter would seem to suggest that microbial processes are favorable tools for remediation of oil-contaminated sites. In this area, genome-based global studies are attracting widespread interest due to better understanding of metabolic and regulatory network, new information on the evolution of microbial degradation pathways and molecular adaptability to environmental changes. Methods that we described in this chapter are essentially the same as we used previously in lab- and at field-based experiments in Ecuador. Our research underlined the importance of native microflora (*Bacillus cereus, Bacillus subtilis, Geomyces* sp., *Geomyces pannorum*) of Ecuadorian amazon rainforest in degrading petroleum hydrocarbons and metal biosorption. The most important limitation lies in "bioaugmentation," where adaptability of microorganisms to new environment is limited by multiple existing local environmental conditions. The findings of this study indicate that "biostimulation" is practically and economically more feasible than "bioaugmentation" for cleaning the oil-polluted sites. Future investigation focusing on "How to improve porosity and aeration of the contaminated soil?" is considerably important for biostimulation-based remediation techniques. Mixing of soils with rice hulls causes increased porosity and aeration. Additionally, soil treatment with hydrogen peroxide increases the oxygen content in the soil. Future studies on the current topic are therefore recommended in order to validate applicability of biostimulation for cleaning the petroleum hydrocarbons-contaminated soils on a large scale.

Author details

Maddela Naga Raju[1]* and Laura Scalvenzi[2]

*Address all correspondence to: raju@mail.sysu.edu.cn

1 Department of Life Sciences, Universidad Estatal Amazónica, Puyo, Pastaza, Ecuador

2 Department of Earth Sciences, Universidad Estatal Amazónica, Puyo, Pastaza, Ecuador

References

[1] Oil spill. 2016. Available from: https://en.wikipedia.org/wiki/Oil_spill#cite_note-34 [Accessed: December 20, 2016]

[2] CDC. Polycyclic Aromatic Hydrocarbons (PAHs). 2009. Available from: https://www.epa.gov/sites/production/files/2014-03/documents/pahs_factsheet-_cdc_2013.pdf [Accessed: December 15, 2016]

[3] Bostrom CE, Gerde P, Hanberg A, Jernstrom B, Johansson C, Kyrklund T, Rannug A, Tornqvist M, Victorin K, Westerholm R. Cancer risk assessment, indicators, and guidelines for polycyclic aromatic hydrocarbons in the ambient air. Environmental Health Perspectives. 2002;**110**(Suppl 3):451-488

[4] Korashy HM, El-Kadi AOS. The role of aryl hydrocarbon receptor in the pathogenesis of cardiovascular diseases. Drug Metabolism Reviews. 2006;**38**(3):411-450

[5] Oil Spill Disrupts Water Supply—Nation—The Star Online. 2012. Available from: http://www.thestar.com.my/news/nation/2012/03/05/oil-spill-disrupts-water-supply/ [Accessed: 20 April, 2015]

[6] Expert Recommends Killing Oil-Soaked Birds. Spiegel Online. 2010. Available from: http://www.spiegel.de/international/world/gulf-of-mexico-spill-expert-recommends-killing-oil-soaked-birds-a-693359.html [Accessed: 10 January, 2017]

[7] Liu YC, Li LZ, Wu Y, Tian W, Zhang LP, Xu L, Shen QR, Shen B. Isolation of an alkane-degrading *Alcanivorax* sp. strain 2B5 and cloning of the alkB gene. Bioresource Technology. 2010;**101**:310-316

[8] Wang XB, Chi CQ, Nie Y, Tang YQ, Tan Y, Wu G, Wu XL. Degradation of petroleum hydrocarbons (C6-C40) and crude oil by a novel *Dietzia* strain. Bioresource Technology. 2011;**102**(17):7755-7761

[9] Chai LJ, Jiang XW, Zhang F, Zheng BW, Shu FC, Wang ZL, Cui QF, Dong HP, Zhang ZZ, Hou DJ, She YH. Isolation and characterization of a crude oil degrading bacteria from formation water: Comparative genomic analysis of environmental *Ochrobactrum intermedium* isolate versus clinical strains. Journal of Zhejiang University Science B. 2015;**16**(10):865-874

[10] Paissé S, Coulon F, Goñi-Urriza M, Peperzak L, McGenity TJ, Duran R. Structure of bacterial communities along a hydrocarbon contamination gradient in a coastal sediment. FEMS Microbiology Ecology. 2008;**66**:295-305

[11] Hugenholtz P, Goebel BM, Pace NR. Impact of culture-independent studies on the emerging phylogenetic view of bacterial diversity. Journal of Bacteriology. 1998;**180**(18): 4765-4774

[12] Eisen JA. Environmental shotgun sequencing: Its potential and challenges for studying the hidden world of microbes. Public Library of Science Biology. 2007;**5**(3):e82

[13] Liu H, Yao J, Yuan Z, Shang Y, Chen H, Wang F, Masakorala K, Yu C, Cai M, Blake RE, Choi MF. Isolation and characterization of crude-oil-degrading bacteria from oil-water mixture in Dagang oilfield, China. International Biodeterioration & Biodegradation. 2014;**87**:52-59

[14] Liu L, Li Y, Li S, Hu N, He Y, Pong R, Lin D, Lu L, Law M. Comparison of next-generation sequencing systems. Journal of Biomedicine and Biotechnology. 2012;**2012**:1-11

[15] Quail MA, Smith M, Coupland P, Otto TD, Harris SR, Connor TR, Bertoni A, Swerdlow HP, Gu Y. A tale of three next generation sequencing platforms: Comparison of Ion Torrent, Pacific biosciences and illuminaMiSeq sequencers. BMC Genomics. 2012;**13**(1):341

[16] Caporaso JG, Kuczynski J, Stombaugh J, Bittinger K, Bushman FD, Costello EK. QIIME allows analysis of high-throughput community sequencing data. Nature Methods. 2010;**7**(5):335-336

[17] Edgar RC. Search and clustering orders of magnitude faster than BLAST. Bioinformatics. 2010;**26**(19):2460-2461

[18] Simpson EH. Measurement of diversity. Nature. 1949;**163**:688

[19] Bowman JS, Rasmussen S, Blom N, Deming JW, Rysgaard S, Sicheritz-Ponten T. Microbial community structure of Arctic multiyear sea ice and surface seawater by 454 sequencing of the 16S RNA gene. ISME Journal. 2012;**6**(1):11-20

[20] Meng L, Liu H, Bao M, Sunc P. Microbial community structure shifts are associated with temperature, dispersants and nutrients in crude oil-contaminated seawaters. Marine Pollution Bulletin. 2016;**111**(1-2):203-212

[21] Brooijmans RJ, Pastink MI, Siezen RJ. Hydrocarbon-degrading bacteria: The oil-spill clean-up crew. Microbial Biotechnology. 2009;**2**(6):587-594

[22] Maddela NR, Masabanda M, Leiva-Mora M. Novel diesel-oil degrading bacteria and fungi from Ecuadorian Amazon rainforest. Water Science and Technology. 2015;**71**(10): 1554-1561

[23] Daugulis AJ, McCracken CM. Microbial degradation of high and low molecular weight polyaromatic hydrocarbons in a two-phase partitioning bioreactor by two strains of *Sphingomonas* sp. Biotechnology Letters. 2003;**25**(17):1441-1444

[24] Chaillan F, Le Flèche A, Bury E, Phantavong YH, Grimont P, Saliot A, Oudot J. Identification and biodegradation potential of tropical aerobic hydrocarbon-degrading microorganisms. Research in Microbiology. 2004;**155**(7):587-595

[25] Kiran B, Rajender K, Narsi RB. Biodegradation of polycyclic aromatic hydrocarbons by white rot fungi *Phanerochaete chrysosporium* in sterile and unsterile soil. Journal of Scientific & Industrial Research. 2008;**67**:538-542

[26] Al-Nasrawi H. Biodegradation of crude oil by fungi isolated from Gulf of Mexico. Journal of Bioremediation & Biodegradation. 2012;**3**:147. DOI: 10.4172/2155-6199.1000147

[27] Covino S, D'Annibale A, Stazi SR, Cajthaml T, Cvancarová M, Stella T, Petruccioli M. Assessment of degradation potential of aliphatic hydrocarbons by autochthonous filamentous fungi from a historically polluted clay soil. Science of the Total Environment. 2015;**505**:545-554

[28] Maddela NR, Scalvenzi L, Pérez M, Montero C, Gooty JM. Efficiency of indigenous filamentous fungi for biodegradation of petroleum hydrocarbons in medium and soil: Laboratory study from Ecuador. Bulletin of Environmental Contamination and Toxicology. 2015;**95**(3):385-394

[29] Zhen-Yu W, Ying XU, Hao-Yun W, Jian Z, Dong-Mei G, Li FM, Xing B. Biodegradation of crude oil in contaminated soils by free and immobilized microorganisms. Pedosphere. 2012;**22**(5):717-725

[30] Obi LU, Atagana HI, Adeleke RA. Isolation and characterisation of crude oil sludge degrading bacteria. SpringerPlus. 2016;**5**(1):1946

[31] Maiti A, Das S, Bhattacharyya N. Isolation and characterization of a new bacterial strain from petroleum oil contaminated soil, India. Journal of Science. 2012;**2**(2):103-108

[32] Ahirwar S, Dehariya K. Isolation and characterization of hydrocarbon degrading microorganisms from petroleum oil contaminated soil sites. Bulletin of Environmental and Science Research. 2013;**2**(4):5-10

[33] El-Din SMB, Moussa TA, Moawad H, Sharaf OA. Isolation and characterization of polyaromatic hydrocarbons degrading bacteria from compost leachate. Journal of Advances in Biology. 2014;**5**(2):651-660

[34] Maddela NR, Scalvenzi L, Venkateswarlu K. (In Press) Microbial degradation of total petroleum hydrocarbons in crude oil: A field-scale study at the low-land rainforest of Ecuador. Environmental Technology. 2016. DOI: 10.1080/09593330.2016.1270356

[35] Panda S, Kar R, Panda C. Isolation and identification of petroleum hydrocarbon degrading microorganisms from oil contaminated environment. International Journal of Environmental Science Technology. 2013;**3**(5):1314-1321

[36] Bartha R, Bossert I. The treatment and disposal of petroleum wastes. In: Atlas RM, editor. Petroleum Microbiology. New York, NY, USA: MacMillan; 1984. pp. 553-578

[37] Cooney JJ. The fate of petroleum pollutants in fresh water ecosystems. In: Atlas RM, editor. Petroleum Microbiology. New York, NY, USA: MacMillan; 1984. pp. 399-434

[38] Ulrici W. Contaminant soil areas, different countries and contaminant monitoring of contaminants. In: Rehm H, Reed G, editors. Environmental Process II. Soil Decontamination Biotechnology. 2nd ed. Vol. 11. Chichester: Wiley; 2000. pp. 5-42

[39] Atlas R, Bragg J. Bioremediation of marine oil spills: When and when not—the Exxon Valdez experience. Microbial Biotechnology. 2009;**2**(2):213-221

[40] Van Beilen JB, Funhoff EG, Van Loon A, Just A, Kaysser L, Bouza M, Holtackers R, Rothlisberger M, Li Z, Witholt B. Cytochrome P450 alkane hydroxylases of the CYP153 family are common in alkane-degrading eubacteria lacking integral membrane alkane hydroxylases. Applied and Environmental Microbiology. 2006;**72**(1):59-65

[41] Mc Donald IR, Miguez CB, Rogge G, Bourque D, Wendlandt KD, Groleau D, Murrell J. Diversity of soluble methane monooxygenase-containing methanotrophs isolated from polluted environments. FEMS Microbiology Letters. 2006;**255**(2):225-232

[42] Maeng JHO, Sakai Y, Tani Y, Kato N. Isolation and characterization of a novel oxygenase that catalyzes the first step of n-alkane oxidation in *Acinetobacter* sp. strain M-1. Journal of Bacteriology. 1996;**178**(13):3695-3700

[43] Jan B, Beilen V, Neuenschwunder M, Suits THM, Roth C, Balada SB, Witholt B. Rubredoxins involved in alkane degradation. The Journal of Bacteriology. 2003;**184**(6):1722-1732

[44] Benyahia F, Embaby AS. Bioremediation of crude oil contaminated desert soil: Effect of biostimulation, bioaugmentation and bioavailability in biopile treatment systems. International Journal of Environmental Research and Public Health. 2016;**13**:1-11

[45] Nikolopoulou M, Pasadakis N, Norf H, Kalogerakis N. Enhanced ex situ bioremediation of crude oil contaminated beach sand by supplementation with nutrients and rhamnolipids. Marine Pollution Bulletin. 2013;**77**:37-44

[46] Paudyn K, Rutter A, Rowe RK, Poland JS. Remediation of hydrocarbon contaminated soils in the Canadian Arctic by landfarming. Cold Regions Science Technology. 2008;**53**:102-114

[47] Whelan MJ, Coulon F, Hince G, Rayner J, McWatters R, Spedding T, Snape I. Fate and transport of petroleum hydrocarbons in engineered biopiles in polar regions. Chemosphere. 2015;**131**:232-240

[48] Gomez F, Sartaj M. Optimization of field scale biopiles for bioremediation of petroleum hydrocarbon contaminated soil at low temperature conditions by response surface methodology (RSM). International Biodeterioration & Biodegradation. 2014;**89**:103-109

[49] Chikere CB, Okoye AU, Okpokwasili GC. Microbial community profiling of active oleophilic bacteria involved in bioreactor based crude-oil polluted sediment treatment. Journal of Applied & Environmental Microbiology. 2016;**4**:1-20

[50] Van Gestel K, Mergaert J, Swings J, Coosemans J, Ryckeboer J. Bioremediation of diesel oil-contaminated soil by composting with biowaste. Environmental Pollution. 2003;**125**:361-368

[51] Frutos FJG, Escolano O, Garcıa S, Mar Babın M, Fernandez MD. Bioventing remediation and ecotoxicity evaluation of phenanthrene-contaminated soil. Journal of Hazardous Materials. 2010;**183**:806-813

[52] Kao CM, Chen CY, Chen SC, Chien HY, Chen YL. Application of in situ biosparging to remediate a petroleum hydrocarbon spill site: Field and microbial evaluation. Chemosphere. 2008;**70**:1492-1499

[53] Salanitro JP, Dorn PB, Huesemann MH, Moore KO, Rhodes IA, Rice Jackson LM, Vipond TE, Western MM, Wisniewski HL. Crude oil hydrocarbon bioremediation and soil ecotoxicity assessment. Environmental Science & Technology. 1997;**31**:1769-1776

[54] Maddela NR, Burgos R, Kadiyala V, Banganegiri M, Carrión AR. Removal of crude oil from soil by using novel microorganisms of Ecuador soils: Solid and slurry phase methods. International Biodeterioration & Biodegradation. 2016;**108**:85-90

Chapter 18

Organic Contaminants in Refinery Wastewater: Characterization and Novel Approaches for Biotreatment

Taghreed Al-Khalid and Muftah H. El-Naas

Additional information is available at the end of the chapter

http://dx.doi.org/10.5772/intechopen.72206

Abstract

Addressing major environmental issues, such as water pollution, is essential nowadays in realizing sustainable development. The ever-increasing world population and industrial development have led to the introduction of different types of chemicals to the environment, leading to considerable deterioration in environmental quality. A major class of these chemicals is phenolic compounds, which are hazardous pollutants and highly toxic even at low concentrations. In recent years, researchers have realized the importance of extracting new bacterial strains that are effective in treating different types of highly contaminated wastewaters at different severe conditions. They also focused considerable amount of research on developing new types of reactors that would provide efficient mixing and reduce mass transfer limitations. The aim is to develop and evaluate effective reactor systems and biocatalysts for the biodegradation of major contaminants in petroleum refinery wastewater. This chapter examines the different available options for the treatment of refinery wastewater with more focus on novel biotreatment options.

Keywords: biodegradation, wastewater, biotreatment, immobilization, phenols

1. Introduction

1.1. Pollution problem

Effluents from the chemical and petroleum industries contain many hazardous chemicals, which have resulted in the accumulation of severe environmental impacts. These effluents are rich in aromatic organic compounds such as polyaromatic hydrocarbons (PAHs) and phenolic substances that are barely degradable by nature; so they remain as a serious threat to the environment. They may accumulate in human and animal tissue upon long distance transportation. Some of these compounds like phenols are highly soluble in water and can be detected in wastewater in a wide range of concentration from a few milligrams per liter to as high as 7000 mg/l [1]. They may remain in water, and under favorable conditions, they can go through various reactions, such as chlorination and methylation, to produce even more harmful or more recalcitrant toxic materials like chlorophenols and cresols [1]. These organic compounds are among the most common forms of contaminants in industrial wastewater, and many of them show carcinogenic, teratogenic or mutagenic properties and were considered by the US Environmental Protection Agency EPA (USA) as priority pollutants. Thus, a great concern has been raised worldwide to remove these contaminants from industrial effluents before discharge into aqueous ecosystems, and it has been obligatory for industries to treat their wastewater effluents to ensure safe disposal to the environment.

Although collected efforts have been directed towards the replacement of fossil fuels, crude oil still keeps its place as a major source of energy and is expected to account for 32% of the world's energy supply by 2030 [2]. Refinery processes consume large amounts of water, and this makes them the main source of organic contaminants in wastewaters [3–5]. It was estimated that approximately 0.4–1.6 times the volume of the crude oil processed is generated as refinery wastewater (RWW) [2]. On the average, processing a barrel of crude oil consumes 65–90 gallons (246–341 l) of water [6]. Therefore, the oil industry will continue to discharge toxic waste into the marine environment. A decreased productivity of algae (a very important link in the food chain) was observed for water bodies receiving these effluents [2].

Because of a rising social and political concern on the environment, this water has to be properly treated to comply with the disposal limits imposed by the environmental legislations or to be reused [7]. The management of wastewater represents major economical and environmental challenges to most industries. Only a limited amount of work was directed to this topic for RWW. This chapter addresses the main concerns associated with the presence and treatment of organic contaminants in RWW with a focus on novel biotreatment approaches.

1.2. Sources of wastewater in petroleum refinery

Maximized benefits of crude oil are derived by processing crude oil in a refinery into a wide variety of products. More than 2500 refined products are produced, including liquefied petroleum gas, gasoline, kerosene, jet fuel, diesel fuel, lubricants, waxes and bitumen. Meanwhile, petrochemical industry derives its feedstock from refined products, which are transformed into valuable products such as plastics, synthetic materials and agro chemicals [4]. Large amounts of water are consumed in oil refineries for cooling systems, crude desalting, distillation, hydrotreating and during maintenance and shut down [8].

After initial fractionation, the crude passes through several treatment and conversion processes to reach the final blending stocks. Conversion processes include thermal and catalytic cracking, steam reforming, isomerization, alkylation and lube oil units, whereas treatment processes include naphtha and gas oil desulfurization, sour water strippers and catalyst regeneration units [8]. The composition of RWW is highly dependent on the complexity of the process. Al Zarooni and Elshorbagy [8] classified refineries as either hydroskimming or complex. A hydroskimming refinery comprises three sub units: a distillation unit in which

Unit	Wastewater major pollutants
Crude desalting	Free oil, ammonia, sulfides and suspended solids
Crude oil distillation	Sulfides, ammonia, phenols, oil, chloride, mercaptans
Thermal cracking	H_2S, ammonia, phenols
Catalytic cracking	Oil, sulfides, phenol, cyanide, ammonia
Hydrocracking	High in sulfides
Polymerization	Sulfides, mercaptans, ammonia
Alkylation	Spent caustic, oil, sulfides
Isomerization	Low level of phenols
Reforming	Sulfide
Hydrotreating	Ammonia, sulfides, phenol

Table 1. Major water sources in petroleum refining processes (adapted from [6]).

Parameter	Value range
pH	8.3–8.9
Conductivity (ms/cm)	5.2–6.8
Total suspended solid (mg/l)	30–40
Total dissolved solid (mg/l)	3800–6200
SO_4 (mg/l)	14.5–16
COD (mg/l)	3600–5300
Total phenol (mg/l)	160–185
Phenol (mg/l)	11–14
o-cresol (mg/l)	14–16.5
m, p-cresol (mg/l)	72–75
N-hexane (mg/l)	1.8–1.85
2,4- and 2,5-DCP (mg/l)	28–32

Table 2. Main characteristics of petroleum refinery wastewater (adapted from [6]).

crude oil is fractionated into various components, a reforming unit for reformate production and a desulfurization unit for reducing the sulfur content of some fractions such as kerosene and naphtha. A complex refinery incorporates a catalytic cracking unit additional to the hydroskimming refinery. Regardless of configuration, the waste effluent is the overall contribution of the units involved in crude oil processing. Al Zarooni and Elshorbagy [8] carried out an extensive program for the identification of major process and utilities wastewater streams and quantification of these streams relative to the total wastewater generated from all refinery processes.

1.3. Characteristics of refinery wastewater

RWW is characterized by a high chemical oxygen demand (COD) [5, 6], which results from the overall contribution of several aliphatic and aromatic hydrocarbons, emulsified oil and grease and inorganic substances, including ammonia, sulfides and cyanides [6, 9, 10]. Typical reported levels are 300–600 mg/l for COD, 20–200 mg/l for phenol, up to 3000 mg/l for oil and suspended solids of more than 100 mg/l [4, 8]. However, the quantity and characteristics of wastewater depend on the process configuration and complexity [4]. COD levels in the range of 3600–5300 mg/l were also reported [6]. **Table 1** summarizes the main pollutants in different petroleum refining units.

Typical analyses of effluents from a petroleum refinery are presented in **Table 2** as ranges of values. More data about the characteristics of RWW can be found elsewhere [2, 5, 8, 10–13].

2. Current treatment options

In line with the quest for a cleaner environment and bylaws of environmental compliance, removal of organic contaminants from RWW is a challenging task for achieving a sustainable development.

Al Zarooni and Elshorbagy [8] reported that dilution of the wastewater with process cooling water serves as the main approach applied to the RWW before disposal into the sea. It was strongly recommended to include primary as well as secondary treatment utilities to reduce the pollutant concentrations below the allowable standards for marine discharge.

Traditionally, there are two basic treatment stages. The first stage is a sequence of pretreatment steps, which employ mechanical and physicochemical techniques, whereas the second stage is an advanced treatment, which involves mainly a biological technique in the integrated activated sludge unit [2, 6]. The primary treatment is crucial for the efficient and prolonged performance of the secondary treatment unit. It targets the reduction of suspended matter like oil and grease to be achieved mechanically by gravity in separation tanks and then by a physiochemical step. In the advanced treatment, contaminants are decreased to specific acceptable discharge limits. In this regard, various solutions are proposed, including electrocoagulation, photocatalytic oxidation, wet oxidation, photodegradation, catalytic vacuum distillation, coagulation-flocculation, fenton oxidation, adsorption, microbial degradation, membrane bioreactor, ultrasonic degradation and chemical precipitation [5, 6].

Diya'uddeen et al. [2] and Rasalingam et al. [14] presented detailed reviews on the different treatment technologies for RWW, with a focus on photocatalytic degradation as an advanced oxidation process (AOP). AOPs have gained extensive attention due to the possibility of destroying a wide variety of organic substances by chemical oxidation, resulting in complete mineralization [2, 14]. In particular, heterogeneous photocatalysis has been demonstrated to be a promising efficient and cost-effective technique for RWW treatment at an advanced stage. However, the industrial application of this technique is greatly limited by the scarcity of available information in the literature [2]. Remya and Lin [15] reviewed the current status of microwave application in wastewater treatment. The following sections will present a brief summary of the main options and highlight the main benefits and drawbacks of each option.

2.1. Electrocoagulation

Electrochemical technology has attracted great attention in recent years for wastewater treatment for the many featured advantages such as environmental compatibility, versatility, energy efficiency, safety, easy automation, selectivity and cost effectiveness [3, 5]. Traditional electrochemical methods include electrocoagulation, electroflotation, electroflocculation, electrochemical reduction and electrochlorination. Electrocoagulation (EC) is based on utilizing "sacrificed" anode, which corrodes by an applied electrical current, for in situ formation of a coagulant. Small dispersed particles combine into larger agglomerates which can be removed by precipitation, floatation or filtration. EC is efficient for removing suspended solids, oil and greases. The reactions occurring in an electrochemical cell with aluminum electrodes are as follows [9, 13]:

At the anode:

$$Al_{(s)} \rightarrow Al^{3+}{}_{(aq)} + 3e^{-} \quad (1)$$

At the cathode:

$$3H_2O + 3e^{-} \rightarrow 3/2\, H_{2(g)} + 3OH^{-}{}_{(aq)} \quad (2)$$

In the solution:

$$Al^{3+}{}_{(aq)} + 3H_2O \rightarrow Al\,(OH)_{3(s)} + 3H^{+}{}_{(aq)} \quad (3)$$

To explain the theory, EC occurs through a sequence of steps as (i) electrolytic reactions at electrode surfaces, where Al ions form at the anode and hydroxyl ions are generated at the cathode, (ii) in situ oxidation of Al ions followed by precipitation of aluminum hydroxide in aqueous phase and (iii) adsorption of soluble or colloidal contaminants on coagulants which are removed by sedimentation or flotation. Electrode material has a crucial role in the mechanism of electrocoagulation, which is also highly influenced by the chemistry of the aqueous medium, especially conductivity [4]. The performance of the electrochemical system is highly affected by the initial composition of the wastewater and the current density [9]. Treatment efficiency, in terms of energy and electrode consumption, is affected by the operating conditions [13].

In spite of the attraction of electrocoagulation, its application for real RWW is rather scarce and limited in the literature. Nevertheless, real application and related performance and design considerations have been addressed by a few studies [4, 5, 9, 12, 13].

2.2. Activated carbon adsorption

Adsorption using activated carbon has frequently been considered the most efficient technique for removing nondegradable waste pollutants, with a high adsorption capacity, flexibility and simplicity of design, low operational cost and the advantage that activated carbon can be regenerated and reused. During the adsorption, phenol is not degraded but rather transferred from the RWW to another phase, which results in the formation of serious by-products (secondary pollution). As such, regeneration is essential for reusing the activated carbon and for the collection and reuse of the contaminant substance [14, 16]. Activated carbon is characterized by a wide variety of pores. The adsorption process proceeds through a sequence of diffusion steps from the bulk phase into the micropores, which are thought to be the major adsorption sites on activated carbon. Also, the amount adsorbed was reported to be positively influenced by increased temperatures and decreased particle size of the adsorbent [14]. In adsorption, adsorbate-adsorbent interaction plays a key role. The lack of such effective interaction results in only moderate adsorption capacity [14]. Adsorption isotherms are established by estimating uptake (or equilibrium adsorption capacity), q, calculated from the difference between the initial and the final phenol concentrations as follows:

$$q = \frac{(C_i - C_f)V}{m} \tag{4}$$

where q is the uptake (mg/g), C_i and C_f (mg/l) are the pollutant initial and final concentrations, respectively, m is the adsorbent dosage (g) and V is the solution volume.

A wide variety of equilibrium isotherm models have been developed. However, the Langmuir isotherm and the Freundlich isotherm are the most commonly used. The former has been applied to a variety of pollutant processes, which involve homogeneous surfaces and negligible interaction between the adsorbed molecules. A main assumption is an arrangement of monolayer adsorption on the adsorptive sites. On the other hand, the Freundlich isotherm is applied for nonideal and multilayer adsorption processes [14].

The high cost of commercial activated carbon has drawn attention to other alternatives. El-Naas et al. [16] evaluated the effectiveness of activated carbon locally prepared from date-pits (DP-AC) for the removal of phenol from RWW. The results proved DP-AC to be a promising low-cost alternative to commercial activated carbon. Furthermore, different approaches were tested for the regeneration of saturated activated carbon and 86% regeneration efficiency could be achieved after four regeneration cycles. Effective utilization of DP-AC was also proved for the reduction of COD in refinery wastewater [17].

2.3. Biological processes

The conventional activated sludge process has been widely used for the removal of organic contaminants from RWW, in an integrated system that includes mechanical and physio-chemical pretreatment [6]. Biological means are incapable of completely removing recalcitrant organic material usually encountered in RWW. These obstacles may be tackled by bioaugmentation, which involves the introduction of robust indigenous or genetically modified organisms. However, a major drawback of bioaugmentation is the uncertainty of reproducibility when the process is transferred to a full scale due to the effects of many variables [2]. To enhance biodegradability, biofilm reactors have been proposed as they prove to be more efficient than conventional biological systems. Biological processes have been identified as suitable and cost-effective method for wastewater treatment. However, conventional processes suffer from many different operational problems such as inhibition at high concentration of toxic substances, long retention time and/or start-up periods, weak tolerance to shock loads and excessive sludge formation [10]. Bacterial activity can be enhanced in immobilized cell reactors, which offer many advantages over suspended cell reactors as will be explained later [18].

Substitution of the conventional activated sludge process continues to be an attractive topic of research. In this regard, Viero et al. [10] treated RWW in a submerged membrane bioreactor as a robust process that allows operation under shock loading rates and hydraulic fluctuations. This is mainly a combined process that utilizes a bioreactor and a granulated activated carbon filter. Also great attention has been directed to anaerobic-aerobic treatments. Chan et al. [19] provided a detailed review of the various types of anaerobic-aerobic wastewater treatment techniques including the new technologies aimed at developing high rate bioreactors and integrated anaerobic-aerobic bioreactors. However, most of these integrated bioreactors lack information on industrial implementation.

2.4. Combined or integrated methods

Complete degradation of persistent organics like chlorinated phenols through biological means proves to be difficult due to the biorefractory nature of these compounds. So, there still exists a need for advanced schemes and devised combinations of treatments for the complete removal of such contaminants. Pretreatment technologies are very effective in decreasing the priority pollutants concentration before the biodegradation step. Several solutions are proposed including the use of coagulants and electrochemical oxidation, Fenton oxidation and ozonation [6]. Combined photochemical or electrochemical pretreatment and biological processes are well documented [6, 20]. In addition, the effective combination of adsorption and biodegradation processes has long been approached by many researchers [6, 21]. For example, ozonation was combined as a pretreatment step with biodegradation for the decomposition of chlorinated phenols. Ozonation is a chemical oxidation method and its combination with a bioprocess results in less toxic compounds, thus enhancing the overall process efficiency and reducing the treatment time and cost. A removal efficiency of 85% could be achieved for 4 chlorophenol (4 CP) that initially could not be degraded in a pure biological step, whereas the degradation of 2,4-dichlorophenol (2,4-DCP) was improved from 40 to 87%. However, several issues were raised as to the high cost and lack of information on technical aspects of the process [22].

The combination of AOP as chemical pretreatment and biological processes was found particularly useful in enhancing the biodegradability and was recommended as a successful technology for industrial wastewater treatment. In this context, Fenton oxidation and/or reductive dehalogenation were used as a pretreatment of chlorinated aromatic compounds before the start of the biological process [23]. The combination of microwave irradiation with AOPs has also been discussed in detail [15]. The choice of a pretreatment method cannot be generalized as it depends upon several factors, including the type of contaminants, real conditions and process costs [23].

3. Advantages and drawbacks of current options

Many of the aforementioned methods have been recognized as efficient techniques for the treatment of RWW and to offer a lot of featured advantages such as energy efficiency, safety and environmental compatibility [3, 13]. However, most of these physiochemical methods suffer from noticeable drawbacks such as high capital and operating costs. Ultimately, most of them do not destroy the contaminant, but rather transfer it to another phase, which results in the formation of harmful by-products [13, 16]. Formation of chlorinated organic compounds has been reported during some electrochemical applications, and activated carbon adsorption was recommended as a further polishing treatment to remove them [5, 12]. Also formation of intermediate by-products like catechol and hydroquinone was detected after ozonation of phenolic compounds [22]. Moreover, most of these methods are incapable of treating heavily contaminated water with COD levels above 4000 mg/l [5].

Although AOPs have emerged as effective methods, which offer a chance for the mineralization of various biorefractory organics [14], the running cost of many AOPs is still relatively high [15]. Also, they were reported for the formation of by-products and to be limited to treating wastewater with relatively low COD concentration [24]. Many other methods which are relatively cheap and easy to operate are characterized by strict technical limitations, in terms of operating conditions and effluent hydraulic rates (e.g. Fenton and membrane applications), low efficiencies and excessive sludge generation (e.g. membrane applications) or in terms of high energy consumption like microwave effects, which cannot be utilized by conventional heating [2, 15]. On the other hand, some applications are limited by the hazards associated with them like in ozone utilization, being an unstable gas [2]. Others are difficult to be commercialized for real-time RWW treatment, and large-scale industrial applications seem to be lacking.

In view of the above discussion, it is essential to search for more viable alternatives that can be utilized in novel biological treatment systems. As for the several mixed processes that have been proposed recently, a lot of these treatment schemes not only have noticeable advantages but also have important drawbacks. These problems can be sorted in two main areas: the first relates to all economic aspects including the high cost needed for the implementation of these techniques; the second includes all technical issues related to the resources needed for the transformation from very toxic compounds to environmentally compatible ones [22].

4. Why is biodegradation favorable?

Biodegradation is the decomposition of organic substances by microorganisms into metabolic by-products with lower toxicity. Enzymes play a catalytic role in this process, where a chemical is converted stepwise into end products through various intermediates. This transformation is called mineralization [25]. Biodegradation is a cost effective and environmentally compatible option that is often preferred, thanks to the possibility of complete mineralization [26, 27]. Because of the aromatic structure of many organic compounds (e.g. phenols), they are highly stable due to the difficulty of cleaving the benzene ring. However, several microorganisms have the capability to utilize these compounds for their metabolic activities as carbon and energy sources. Biological transformation has been recognized as one of the key solutions to deal with environmental pollution caused by many problematic organic contaminants. In this regard, the use of pure and mixed cultures of organisms is considered a favorable and most promising approach [28]. Many strains of bacteria, fungi and algae have the ability to degrade toxic organic substances. Bacterial cultures of *Pseudomonas* genus are the most commonly utilized biomass for the biodegradation of organic contaminants, with special interest paid to *Pseudomonas putida* due to its high removal efficiency [29]. However, a main drawback in bioprocesses is the inhibition of the enzymatic activity at high substrate concentrations. Under certain conditions, organic material can be decomposed aerobically or anaerobically [30]. Conventionally, aerobic processes are preferred. Aerobic microorganisms grow faster and are more efficient because they can achieve complete mineralization of toxic organic substances to inorganic constituents (CO_2, H_2O) [31]. This is in addition to low associated costs [16]. On the other hand, the end products of biochemical reactions in anaerobic processes often produce esthetically displeasing colors and fouling odors in water [2]. Therefore, there is a limited interest in the utilization of anaerobic microorganism for the degradation of organic waste. However, there have been several studies in this regard [32–34]. Since most biological treatment studies have used aerobic biomasses, discussion in the following sections of this chapter will focus on aerobic biodegradation. Detailed reviews on the biodegradation of some organic compounds can be found in the literature [27, 35–37].

4.1. Mechanisms of biodegradation

Biodegradation is a multivariable process, which is affected by a combination of many biotic and abiotic factors, including pH, temperature, oxygen content and availability, microbial abundance and substrate concentration [26, 27]. The chemical structure of aromatic compounds plays a key role as reflected by the number, type and position of substituents on the aromatic ring and degree of branching. The greater the number of substituents in the structure, the more toxic and less degradable it is.

Metabolic processes are dominated by the catalysis of enzymes, which are particular to each type of biomass and reaction. A metabolic reaction is ultimately a process of energy conversion. Little is known about the biodegradation mechanism by fungi and algae; so the following is a brief discussion of this mechanism by aerobic bacteria, as typically represented by the biodegradation of aromatics.

In aerobic biodegradation, enzymatic attack on the aromatic ring is initiated by oxygen. A typical pathway for metabolizing phenols (phenol is a basic structural unit for a variety of synthetic organic compounds) is to hydroxylate the ring by the enzyme phenol hydroxylase, form catechol and then open the ring through *ortho*- or *meta*-oxidation. Thus, phenol hydroxylase is the first enzyme and catechol is a basic intermediate in the degradation pathways of many aromatic substances. In the *ortho*-pathway, the aromatic ring is cleaved by the enzyme catechol 1,2-dioxygenase (C12O). In the *meta*-pathway, the ring is cleaved by the enzyme catechol 2,3-dioxygenase (C23O). The ring is thus opened and then degraded [27].

(C12O) and (C23O) designate two different orientations as to how the ring cleavage can occur. However, the biodegradation of many aromatics proceeds through the *ortho*-cleavage pathway after the formation of catechol because the *meta*-cleavage results in the formation of dead end metabolites from catechol; the enzyme gets inactivated by the accumulation of a toxic intermediate [38]. As a rule, conversion of catechol does not follow the *meta*-cleavage pathway, and generally, the *ortho*-cleavage pathway is required for the complete degradation of many aromatic organics [39]. Discussion of biodegradation pathways and mechanisms can be found in [27, 35, 36, 40].

4.2. Biodegradation kinetics and modeling

Kinetic studies indicate how effectively a bioprocess is functioning. This is essential to improve process control and removal efficiency [31]. The main step in modeling a biodegradation process is to relate the specific growth rate of the biomass to the consumption rate of the substrate [26]. Different kinetic models have been used to describe the dynamics of microbial growth on phenols [35].

Based on material balance, the rate of biomass growth and the rate of substrate utilization (both in mg/l h) can be represented by Eqs. (5) and (6), respectively:

$$\frac{\mathrm{dX}}{\mathrm{dt}} = \mu \mathrm{X} - k_{\mathrm{d}}\mathrm{X} = \mu_{\mathrm{net}}\mathrm{X}\ (\text{or } \frac{\mathrm{dlnX}}{\mathrm{dt}} = \mu_{\mathrm{net}}) \tag{5}$$

$$\frac{\mathrm{d}S}{\mathrm{dt}} = -\frac{\mu \mathrm{X}}{\mathrm{Y}} \tag{6}$$

where Y is the cell mass yield (g/g) = dX/dS; X is the biomass concentration (mg/l); S is the substrate concentration (mg/l); k_d is the decay coefficient (h^{-1}) (often ignored) and μ is the specific growth rate (h^{-1}).

There are two most common models for cell growth during the biodegradation: the Monod model and the Haldane model, represented by Eqs. (7) and (8), respectively:

$$\mu = \frac{\mu_{max} S}{K_s + S} \tag{7}$$

$$\mu = \frac{\mu_{max} S}{K_s + S + \left(\frac{S^2}{K_i}\right)} \tag{8}$$

where K_s (mg/l) is the half saturation coefficient (related to microorganism's affinity to the substrate) and K_i is the substrate inhibition constant (mg/l). The first model neglects the substance inhibition, whereas the second model is the most widely used since it accounts for the inhibitory effect of toxic material and it is mathematically simple [26, 41]. It is noteworthy that when K_i is very large, the Haldane equation reduces to the Monod model.

Although cell immobilization by entrapment offers significant advantages, it is problematic in terms of diffusion limitation owing to the resistance imposed by the protective structure [42]. Also, in the design and modeling of bioreactors, it is equally important to assess external mass transfer coefficients for the transfer of substrate from the bulk phase to the surface of the biofilm. Diffusion limitations considerably affect the intrinsic reaction kinetics. Therefore, kinetic models for degradation in biofilm reactors are complex and difficult to develop. Several studies addressed this topic and proposed comprehensive diffusion-reaction models, which account for mass transfer limitations [42, 43].

5. Novel approaches in biotreatment

In light of the fact that biodegradation of aromatic organic pollutants has a unique environmental and ecological impact, there is a rising interest in this area of biotechnology and a strong motivation to develop efficient strategies and novel cost-effective methods for RWW management. This section will briefly review some of the achievements on this track.

5.1. Immobilization techniques

Immobilization is an effective technique that is usually used for many purposes, mainly the protection of the biomass from toxicity effects of the contaminant, in addition to the ease of separation and reutilization of the biomass [35]. Other advantages were highlighted in the previous section. Bacterial activity can be increased by various immobilization methods like bead entrapment, carrier binding, encapsulation, cell coating and film attachment [44].

Polyvinyl alcohol (PVA) is a synthetic polymer that shows attractive properties to be used as a carrier in bacterial immobilization; it is inexpensive, nontoxic, water soluble and has excellent electrical insulation [45]. El-Naas et al. [46] evaluated the characteristics of PVA gel matrices with immobilized *P. putida*, prepared by crosslinking in the gel structure by repeated cycles of freezing-thawing, resulting in a highly porous fibril structure with high mechanical strength and elastic rubbery nature. The effectiveness of these gel pellets with immobilized *P. putida* was investigated for the biodegradation of phenol in different reactor configurations, in batch and continuous modes, and proved to be effective even at high phenol concentrations [26, 29, 47, 48]. Emphasis was placed on the contaminant uptake per mass of PVA, which is of particular economical importance in the design of RWW treatment processes. Compared to natural biodegradable Ca-alginate, which has long been used in the area of biocatalysts, PVA gel is more durable with lower resistance to mass transfer due to its porous structure.

In view of that fact that biodegradation is in essence a series of enzyme-catalyzed reactions, there has been an increasing attention to the direct application of immobilized robust enzymes obtained from degrading microorganisms for the treatment of contaminated wastewater. Enzymatic treatment of organic pollutants is documented in a few recent studies, which addressed novel, practical and inexpensive immobilization methods. Extensive discussion on this topic can be found in a review by Demarche et al. [49].

Attention in recent years has focused on developing aerobic granules in sequencing batch reactors (SBRs) [50]. This is a novel biotechnique, characterized by self-immobilization of microorganisms through cell-to-cell adhesion, without any carrier material. Aerobic granules are reported for a very high cell concentration (up to 15,000 mg/l) and to be able to decontaminate high-strength wastewater (up to 15 kg COD/m^3/day) [51]. However, this technology is still in a development stage. Detailed information on aerobic granulation can be found in a review by Khan et al. [50].

5.2. Utilization of spouted bed bioreactor (SBBR)

The SBBR is superior to the conventional bubble column bioreactor with its characteristic advantage of efficient intense mixing that is induced by cyclic motion of particles within the bed. The cyclic motion of particles results from a single air jet injected through an orifice in the bottom of the reactor [29, 46–48]. This reactor design offers better contact between substrate and cells as well as faster oxygen and nutrient transfer rate, which leads to higher removal rates of the contaminant. Details about the SBBR can be found elsewhere [32]. El-Naas et al. [29] proved that the efficient mixing in the SBBR has a significant role in overcoming the external mass resistance for the transfer of substrate from the bulk phase to the surface of the biofilm, and thus uniform concentration across the height of the reactor is a justified assumption. Effective utilization of the SBBR for the biodegradation of phenol and its derivatives and for the treatment of RWW is reported in several studies as will be illustrated in the next section.

5.3. Different applications

Example applications of novel biotreatment approaches for RWW treatment are briefly discussed in the following sections.

5.3.1. Biodegradation of phenol

Owing to the diversity of RWW composition, phenols have been adopted as a suitable measure for the performance of biodegradation [10]. *P. putida*, immobilized in PVA gel particles, has been successfully utilized for the removal of phenol from simulated wastewater, and the immobilized bacteria proved to be resilient in sustaining deprivation of nutrients and sudden exposure to high concentrations of phenol [48]. Batch and continuous biodegradation of phenol by PVA-immobilized *P. putida*, in bubble column reactor and a specially designed SBBR, was investigated in several studies and assessed for the effects of various variables. SBBR was shown to have superior performance. The biodegradation rate, being highly dependent on temperature, pH and initial phenol concentration, was optimized at 30°C, 7 and 75 mg/l, respectively. The experimental data fitted better to the Haldane inhibitory model (Eq. (8)) than the Monod model (Eq. (7)). However, the inhibition effect was not encountered in the continuous operation up to an initial phenol concentration of 150 mg/l, thanks to the continuous dilution effect. A mathematical dynamic model that incorporates the effect of internal mass transfer resistance and growth kinetics in the SBBR as represented by the Haldane model was proposed. The experimental data fitted fairly well to the model; the dynamics of the system is mainly controlled by the mass transfer [47]. The biodegradation rate also depends on biomass abundance as reflected by a linear increase in the biodegradation rate with the amount of PVA gel, the phenol uptake per mass of PVA reached a maximum at a PVA volume of 10 ml within a total volume of 1 liter; more PVA particles in the bioreactor is expected to have adverse effect by hindering particle movement and mixing [47]. The importance of the uptake value in the design of industrial RWW treatment processes was emphasized in Section 5.1.

It is a matter of fact that mass transfer limitations have significant effect on the biodegradation rate. Two other important parameters, which are directly related to the mass transfer limitations, are the PVA particle size and the air flow rate [26, 29]. Mass transfer is directly related to the accessibility of the biomass to phenol; it was enhanced by decreasing the PVA particle size and increasing the air flow rate. This effect was more pronounced for high air flow rates and low phenol concentrations. This could be due to improvements in mass transfer induced by the combined effect of good mixing and reduced particle size. The effect of the air flow rate was closely examined; the biodegradation rate increased linearly as the air flow rate was increased from 1 to 3 l/min; then, it became almost constant. In this regard, two main factors are believed to be involved: mixing and aeration. The effects of these two factors were assessed separately, and it was shown that at air flow rates higher than 1 l/min, the main one of the two factors is the availability of enough oxygen to complete the biodegradation reaction. However, the effect of mixing may be more significant for smaller PVA particle size or very low initial phenol concentrations. Therefore, the external diffusion plays a dominant role as a controlling mechanism at low bulk concentrations. The intense mixing that is characteristic of the SBBR proves to be effective in overcoming these mass transfer limitations, which makes it an efficient reactor for the biodegradation of phenol.

5.3.2. *Biodegradation of cresols*

The biodegradation of *p*-cresol in simulated wastewater was carried out using *P. putida* immobilized in PVA gel, in batch and continuous modes in SBBR to evaluate the effects of several operating conditions. The effects of initial substrate concentration, temperature, solution pH and PVA volume fraction on the biodegradation of *p*-cresol were evaluated in a batch SBBR. The same trends for the effects of these conditions were observed as for phenol indicating a high dependence of the biodegradation capabilities of *P. putida* on temperature, pH and biomass abundance, which is represented by PVA volume fraction. The process was optimized at 35°C, 8 and 40%, respectively. As for the effect of the initial substrate concentration, there was no sign of inhibition up to a substrate concentration of 200 mg/l, which justifies the Monod non-inhibitory model (Eq. (7)). Continuous operation is essential when considering the industrial applicability of biodegradation process. Thus, the effects of air flow rate and residence time (as affected by liquid flow rate) were investigated in continuous operation. Continuous biodegradation results indicated that *P. putida* proved to be effective for the biodegradation of *p*-cresol up to 200 mg/l, with a removal efficiency of more than 85% and with the steady state achieved within one residence time. Optimal air flow rate was 2 l/min, with the same observation as for phenol that the biodegradation rate could not be improved beyond a certain air flow rate, which could be due to slugging and bubble coalescence. The increase in the liquid flow rate reduced the residence time inside the reactor and consequently the biodegradation rate was reduced. Assuming *p*-cresol is the limiting reactant and perfect mixing in the bioreactor, which is the case in SBBR, the global biodegradation rate was calculated based on the specific consumption rate and the mass balance of the continuous flow reactor is expressed as:

$$\frac{dM_A}{dt} = F(S_{Ao} - S_A) - r_A V \tag{9}$$

where M_A is the mass (mg) of substrate A in the reactor, t is the time (h), F is the volumetric flow rate (l/h), S_A is the substrate concentration at a certain time, $-r_A$ is the rate of removal of substrate A (mg/l h) and V is the reactor volume (l). Dividing the equation by V:

$$\frac{dS}{dt} = \frac{F}{V}(S_o - S) - r_S \tag{10}$$

The substrate uptake is given by:

$$-r_s = q \tag{11}$$

Assuming constant yield Y is constant, which is valid if the substrate concentration is much higher than K_s (i.e. $S \gg K_s$) [26], the Monod model can then be expressed in terms of the degradation rate q (mg/l h) as:

$$q = \frac{q_{max} S}{K_s + S} \tag{12}$$

Combining Eqs. (10) and (12), the change of substrate concentration in the reactor can be evaluated as:

$$\frac{dS}{dt} = \frac{F}{V}(S_o - S) - \frac{q_{max} S}{K_s + S} \tag{13}$$

The experimental results from the continuous biodegradation process showed very good fit with the model proposed in Eq. (13). The effects of temperature, pH and initial substrate concentration were also evaluated for the other two isomers, *o*- and *m*-cresol. The optimum ranges were 30–35°C for temperature and 6–8 for pH, with different optimum values for each isomer. The immobilized *P. putida* could sustain cresols concentration up to 200 mg/l, but substrate inhibition was observed in the biodegradation of *o*-cresol at 150 mg/l, which could be attributed to the formation of toxic intermediates.

A RWW treatment plant generally receives influents with mixtures of recalcitrant organic chemicals. This is a complex situation where biodegradation of a compound could be strongly affected by the presence of other components in the mixture. This is particularly related the biological transformation called co-metabolism, which involves the transformation of a non-growth substrate by cells growing on a growth substrate [35]. It could also be affected by the accumulation of metabolic intermediates. For a treatment scheme to be effective, all the interactions among the different substrates need to be considered. Scant information is available in the literature on the biodegradation of mixtures of cresols. The potential of *P. putida* in degrading binary (*o*- and *p*-cresol) and ternary (*o*-, *p*- and *m*-cresol) mixtures of cresols was also investigated in the continuous mode. It was obviously shown that the biodegradation of *o*-cresol was strongly inhibited by the presence of *p*-cresol and *m*-cresol.

Reference is made to the studies by Surkatti and El-Naas [45, 52] for details on the biodegradation of cresols.

5.3.3. Biodegradation of chlorophenols

The work presented in this section aimed at developing an integrated system for biodegradation of chlorophenols, considering 2,4-dichlorophenol (2,4-DCP) as a model contaminant. The system utilizes the advanced technique for the immobilization of *P. putida* in PVA gel matrix, in a specially designed SBBR that promises to be effective in industrial applications for the treatment of real RWW. Details on this work can be found elsewhere [53, 54].

Batch experiments were performed to confirm that the removal efficiency was due to the biodegradation effect and not to other abiotic effects such as stripping or adsorption on the carrier. It was confirmed that the contribution by these factors to the reduction in DCP concentration ranged from 2.5 to 7%, which confirms the biodegradation effect by immobilized *P. putida.*

Process optimization is critical for any large-scale industrial application. A preliminary screening led to approaching the optimum range for each of the main factors, namely temperature, pH and initial substrate concentration. The full potential of response surface methodology (RSM) with Box-Behnken design was consequently utilized to determine the effects of significant parameters, and their interactions, in the removal of 2,4-DCP and to specify the optimal conditions for its degradation based on the degradation rates achieved in the SBBR batch experiments. A quadratic regression model that is capable of predicting the rate in terms of the main independent variables was developed. The maximum biodegradation rate within the experimentation region was determined as: temperature 32.6°C, pH 5.0 and the initial DCP concentration was 70.5 mg/l. Under these conditions, the predicted global biodegradation rate was 41.8 mg/l h. By analysis of variance, the model showed a good fit with R^2 and adjusted R^2 94.03 and 83.28%, respectively, which justifies the adequacy of the model. It was further experimentally validated. In the batch process, 100% removal of 2,4-DCP was achieved for all DCP initial concentrations, up to 200 mg/l. This is a compound that is naturally difficult for biodegradation, and for which removal efficiencies in the range of 40–60% have often been reported. *P. putida* showed good adaptation to the contaminant with repeated use and showed continued improvement; its degradation capability almost doubled after 4 months of repeated use from 38 to 70 mg/l h. For the sake of comparison, the degradation rates from other reported batch studies are shown in **Table 3**.

Although the biodegradation of 2,4-DCP followed a Haldane inhibition model, the inhibition was not significantly observed at any initial DCP concentration, which could be mainly due to the advantage of the immobilization of bacteria in the PVA gel pellets.

The process was thereafter operated in the continuous mode to investigate the hydrodynamics and the performance of the reactor. The various issues investigated were the hydraulic residence time (*HRT*), the initial concentration of DCP in the feed and the degradation capacity (*DC*), which is a characteristic design factor for continuous reactors. It has an economic significance as a measure of the hydraulic throughput and is defined as:

$$DC = (S_i - S_e)D \tag{14}$$

Initial DCP concentration (mg/l)	Degradation rate (mg/l h)	% Removal	Ref.
1000 (fungi), free	8.3	100	[55]
500 (fungi), free	5.21	100	[56]
113	28.75	100	[57]
120, free	0.48	40	[22]
80, free	2.67	100	[31]
75	70	100	[54]

Table 3. Degradation rates of 2,4-DCP from batch studies in the literature [54].

where S_i and S_e are influent and effluent substrate concentrations, respectively, and D is the dilution rate (1/*HRT*).

Another criterion often used in continuous operations is the organic loading rate (*OLR*) of the feed that is related to the amount of contaminant that can be processed over a given period of time, defined as:

$$OLR = \frac{S_i F}{V} = \frac{S_i}{HRT} = S_i D \tag{15}$$

ORL is directly affected by changes in the influent flow rate and its initial concentration.

OLR and *DC* values from the combined sets of experiments (those for the effect liquid flow rate and those for the effect of initial DCP concentration) were plotted in **Figure 1**, with

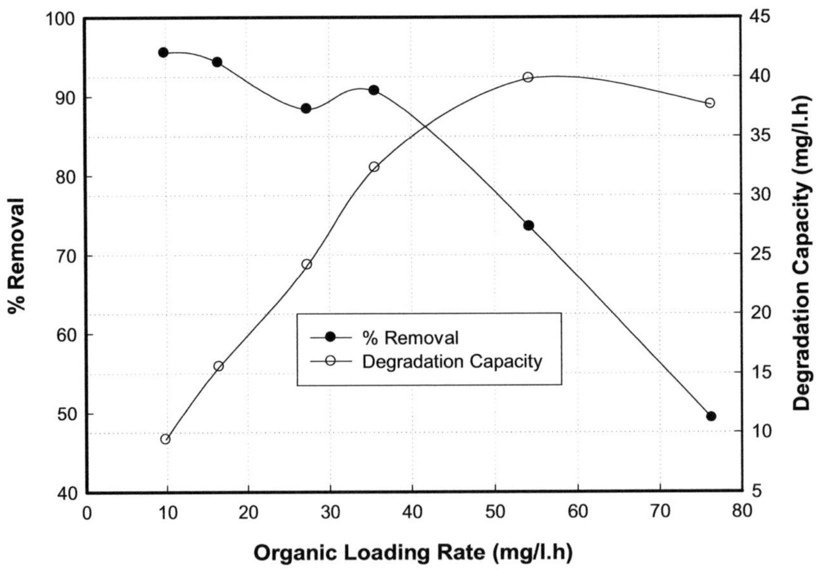

Figure 1. Effect of organic loading rate on degradation capacity and percent removal [54].

corresponding percent removal efficiencies. The results demonstrate that operating conditions are selected with a compromise based on either the desired degradation efficiency (percent removal) or the desired degradation capacity. This is different for a batch operation, in which a bioreactor is operated with a predetermined percent of substrate removal of substrate percent. A combined 80% removal at a throughput of 1400 g/m^3 day was obtained during continuous operation with a *HRT* of 1 h and an initial DCP concentration of 75 mg/l.

Dynamic modeling of DCP degradation in continuous operation was based on the same discussion that applied for cresols and led to the derivation of Eq. (13). However, the Monod expression was replaced with the Haldane one in the equation to give Eq. (16):

$$\frac{dS}{dt} = \frac{F}{V}(S_o - S) - \frac{q_{max}\ S}{K_s + S + \left(\frac{S^2}{K_i}\right)} \quad (16)$$

The experimental results from the continuous biodegradation process showed very good fit with the model proposed in Eq. (16). The performance of the SBBR was further tested in the continuous operation for tolerance to organic and hydraulic shock loads, resulting from concentration and liquid flow rate fluctuations. The SBBR showed stability and sustainability, as it could withstand a 100% increase in organic loading rate and recover quickly to its normal operation once the shock load had been terminated. The SBBR performance was also compared to that of a packed bed reactor and proved to be more stable and achieved higher removal efficiencies (77% versus 53%) with a HRT of 1 h and an initial DCP concentration of 75 mg/l. Such a difference between the two reactors is expected as explained by the better mixing properties of the SBBR and the mass and heat transfer limitations of the packed bed reactor.

The process was tested for the treatment of real RWW with characteristics similar to those given in **Table 2**. The immobilized bacteria were acclimatized to the RWW and then used for the batch treatment in the SBBR. The process was shown to be effective, with 100% removal efficiencies achieved for all cresols, whereas the removal efficiencies for phenol and DCP were approximately 87% and 63%, respectively. Significant overall reduction in total phenols of almost 90% was observed and a maximum COD reduction of 59%. These results are merely due to the biodegradation effect without any pre- or post-treatment. With this in mind, they may be considered quite satisfactory.

6. Conclusions

The presence of organic contaminants in RWW represents a major environmental concern for the oil industry. The current options for RWW management are rather economically or technically limited in application. A new approach that utilizes a state of the art technique for the immobilization of an effective organics-degrading bacterial strain (*P. putida*) in a specially designed spouted bed bioreactor has proven to be effective for the treatment of RWW.

Author details

Taghreed Al-Khalid[1] and Muftah H. El-Naas[2*]

*Address all correspondence to: muftah@qu.edu.qa

1 Chemical and Petroleum Engineering Department, UAE University, Al-Ain, UAE

2 Gas Processing Center, Qatar University, Doha, Qatar

References

[1] Praveen P, Loh K-C. Simultaneous extraction and biodegradation of phenol in a hollow fiber supported liquid membrane bioreactor. Journal of Membrane Science. 2013;**430**:242-251

[2] Diya'uddeen BH, Daud WMAW, Abdul Aziz AR. Treatment technologies for petroleum refinery effluents: A review. Process Safety and Environmental Protection. 2011; **89**(2):95-105

[3] Yavuz Y, Koparal AS. Electrochemical oxidation of phenol in a parallel plate reactor using ruthenium mixed metal oxide electrode. Journal of Hazardous Materials. 2006;**136**(2):296-302

[4] Yavuz Y, Koparal AS, Öğütveren ÜB. Treatment of petroleum refinery wastewater by electrochemical methods. Desalination. 2010;**258**(1-3):201-205

[5] Yan L, Wang Y, Li J, Ma H, Liu H, Li T, et al. Comparative study of different electrochemical methods for petroleum refinery wastewater treatment. Desalination. 2014;**341**:87-93

[6] El-Naas MH, Alhaija MA, Al-Zuhair S. Evaluation of a three-step process for the treatment of petroleum refinery wastewater. Journal of Environmental Chemical Engineering. 2014;**2**(1):56-62

[7] Santos ID, Afonso JC, Dutra AJB. Behavior of a Ti/RuO_2 anode in concentrated chloride medium for phenol and their chlorinated intermediates electrooxidation. Separation and Purification Technology. 2010;**76**(2):151-157

[8] Al Zarooni M, Elshorbagy W. Characterization and assessment of Al Ruwais refinery wastewater. Journal of Hazardous Materials. 2006;**136**(3):398-405

[9] El-Naas MH, Al-Zuhair S, Al-Lobaney A, Makhlouf S. Assessment of electrocoagulation for the treatment of petroleum refinery wastewater. Journal of Environmental Management. 2009;**91**(1):180-185

[10] Viero AF, de Melo TM, Torres APR, Ferreira NR, Sant'Anna GL Jr, Borges CP, et al. The effects of long-term feeding of high organic loading in a submerged membrane bioreactor treating oil refinery wastewater. Journal of Membrane Science. 2008;**319**(1-2):223-230

[11] Jou C-JG, Huang G-C. A pilot study for oil refinery wastewater treatment using a fixed-film bioreactor. Advances in Environmental Research. 2003;**7**(2):463-469

[12] Rajkumar D, Palanivelu K. Electrochemical treatment of industrial wastewater. Journal of Hazardous Materials. 2004;**113**(1-3):123-129

[13] Abdelwahab O, Amin NK, El-Ashtoukhy ESZ. Electrochemical removal of phenol from oil refinery wastewater. Journal of Hazardous Materials. 2009;**163**(2-3):711-716

[14] Rasalingam S, Peng R, Koodali R. Removal of hazardous pollutants from wastewaters: Applications of TiO_2-SiO_2 mixed oxide materials. Journal of Nanomaterials. 2014;**2014** http://dx.doi.org/10.1155/2014/617405

[15] Remya N, Lin J-G. Current status of microwave application in wastewater treatment—A review. Chemical Engineering Journal. 2011;**166**:797-813

[16] El-Naas MH, Al-Zuhair S, Alhaija MA. Removal of phenol from petroleum refinery wastewater through adsorption on date-pit activated carbon. Chemical Engineering Journal. 2010;**162**(3):997-1005

[17] El-Naas M, A-Zuhair S, Alhaija M. Reduction of COD in refinery wastewater through adsorption on date-pit activated carbon. Journal of Hazardous Materials. 2009;**173**:750-757

[18] Sá CSA, Boaventura RAR. Biodegradation of phenol by pseudomonas putida DSM 548 in a trickling bed reactor. Biochemical Engineering Journal. 2001;**9**(3):211-219

[19] Chan YJ, Chong MF, Law C, Hassell DG. A review on anaerobic–aerobic treatment of industrial and municipal wastewater. Chemical Engineering Journal. 2009;**155**:1-18

[20] Suryaman D, Hasegawa K. Biological and photocatalytic treatment integrated with separation and reuse of titanium dioxide on the removal of chlorophenols in tap water. Journal of Hazardous Materials. 2010;**183**(1-3):490-496

[21] Ma X, Li N, Jiang J, Xu Q, Li H, Wang L, et al. Adsorption–synergic biodegradation of high-concentrated phenolic water by *Pseudomonas putida* immobilized on activated carbon fiber. Journal of Environmental Chemical Engineering. 2013;**1**(3):466-472

[22] García-Peña EI, Zarate-Segura P, Guerra-Blanco P, Poznyak T, Chairez I. Enhanced phenol and chlorinated phenols removal by combining ozonation and biodegradation. Water, Air & Soil Pollution. 2012;**223**(7):4047-4064

[23] Kastanek F, Maleterova Y. Combination of advanced oxidation and/or reductive dehalogenation and biodegradation for the decontamination of waters contaminated with chlorinated organic compounds. Separation Science and Technology. 2007;**42**:1613-1625

[24] Mustafa Y, Shihab A. Removal of 4-chlorophenol from wastewater using a pilot-scale advanced oxidation process. Desalination and Water Treatment. 2013;**51**:6663-6675

[25] Crawford RL, Moo-Young M. 6.02—Biodegradation: Principles, Scope, and Technologies. Comprehensive Biotechnology. 2nd ed. Burlington: Academic Press; 2011. pp. 3-13

[26] El-Naas MH, Al-Muhtaseb SA, Makhlouf S. Biodegradation of phenol by *Pseudomonas putida* immobilized in polyvinyl alcohol (PVA) gel. Journal of Hazardous Materials. 2009;**164**(2-3):720-725

[27] Nair CI, Jayachandran K, Shashidhar S. Biodegradation of phenol. African Journal of Biotechnology. 2008;**7**(25):4951-4958

[28] Shourian M, Noghabi KA, Zahiri HS, Bagheri T, Karballaei G, Mollaei M, et al. Efficient phenol degradation by a newly characterized *Pseudomonas* sp. SA01 isolated from pharmaceutical wastewaters. Desalination. 2009;**246**(1-3):577-594

[29] El-Naas MH, Al-Zuhair S, Makhlouf S. Continuous biodegradation of phenol in a spouted bed bioreactor (SBBR). Chemical Engineering Journal. 2010;**160**(2):565-570

[30] Wang S-G, Liu X-W, Zhang H-Y, Gong W-X, Sun X-F, Gao B-Y. Aerobic granulation for 2,4-dichlorophenol biodegradation in a sequencing batch reactor. Chemosphere. 2007;**69**(5):769-775

[31] Sahinkaya E, Dilek FB. Biodegradation kinetics of 2,4-dichlorophenol by acclimated mixed cultures. Journal of Biotechnology. 2007;**127**(4):716-726

[32] Bajaj M, Gallert C, Winter J. Treatment of phenolic wastewater in an anaerobic fixed bed reactor (AFBR)—Recovery after shock loading. Journal of Hazardous Materials. 2009; **162**(2-3):1330-1339

[33] Hussain A, Kumar P, Mehrotra I. Nitrogen biotransformation in anaerobic treatment of phenolic wastewater. Desalination. 2010;**250**(1):35-41

[34] Puyol D, Mohedano AF, Rodriguez JJ, Sanz JL. Effect of 2,4,6-trichlorophenol on the microbial activity of adapted anaerobic granular sludge bioaugmented with Desulfitobacterium strains. New Biotechnology. 2011;**29**(1):79-89

[35] Al-Khalid TT, El-Naas MH. Aerobic biodegradation of phenols: A comprehensive review. Critical Reviews in Environmental Science and Technology. 2012;**42**(16):1631-1690

[36] Peng R-H, Xiong A-S, Xue Y, X-Y F, Gao F, Zhao W, et al. Microbial biodegradation of polyaromatic hydrocarbons. FEMS Microbiology Reviews. 2008;**32**(6):927-955

[37] Bhatt P, Kumar MS, Mudliar S, Chakrabarti T. Biodegradation of chlorinated compounds—A review. Critical Reviews in Environmental Science and Technology. 2006; **37**(2):165-198

[38] Solyanikova IP, Golovleva LA. Bacterial degradation of Chlorophenols: pathways, biochemica, and genetic aspects. Journal of Environmental Science and Health. 2004; **B39**(3):333-351

[39] Farrell A, Quilty B. The enhancement of 2-chlorophenol degradation by a mixed microbial community when augmented with *Pseudomonas putida* CP1. Water Research. 2002; **36**(10):2443-2450

[40] Krastanov A, Alexieva Z, Yemendzhiev H. Microbial degradation of phenol and phenolic derivatives. Engineering in Life Sciences. 2013;**13**:76-87

[41] Kumar A, Kumar S, Kumar S. Biodegradation kinetics of phenol and catechol using *Pseudomonas putida* MTCC 1194. Biochemical Engineering Journal. 2005;**22**(2):151-159

[42] Tepe O, Dursun AY. Combined effects of external mass transfer and biodegradation rates on removal of phenol by immobilized *Ralstonia eutropha* in a packed bed reactor. Journal of Hazardous Materials. 2008;**151**(1):9-16

[43] Banerjee I, Modak JM, Bandopadhyay K, Das D, Maiti BR. Mathematical model for evaluation of mass transfer limitations in phenol biodegradation by immobilized *Pseudomonas putida*. Journal of Biotechnology. 2001;**87**(3):211-223

[44] Quan X, Shi H, Zhang Y, Wang J, Qian Y. Biodegradation of 2,4-dichlorophenol and phenol in an airlift inner-loop bioreactor immobilized with Achromobacter sp. Separation and Purification Technology. 2004;**34**(1-3, 103):97

[45] Surkatti R, El-Naas M. Competitive interference during the biodegradation of cresols. International Journal of Environmental Science and Technology. 2017. DOI: 10.1007/s13762-017-1383-2

[46] El-Naas MH, Mourad A-HI, Surkatti R. Evaluation of the characteristics of polyvinyl alcohol (PVA) as matrices for the immobilization of *Pseudomonas putida*. International Biodeterioration & Biodegradation. 2013;**85**:413-420

[47] El-Naas MH, Al-Zuhair S, Makhlouf S. Batch degradation of phenol in a spouted bed bioreactor system. Journal of Industrial and Engineering Chemistry. 2010;**16**(2):267-272

[48] Al-Zuhair S, El-Naas M. Immobilization of *Pseudomonas putida* in PVA gel particles for the biodegradation of phenol at high concentrations. Biochemical Engineering Journal. 2011;**56**(1-2):46-50

[49] Demarche P, Junghanns C, Nair RR, Agathos SN. Harnessing the power of enzymes for environmental stewardship. Biotechnology Advances. 2012;**30**(5):933-953

[50] Khan MZ, Mondal PK, Sabir S. Aerobic granulation for wastewater bioremediation: A review. The Canadian Journal of Chemical Engineering. 2013;**91**(6):1045-1058

[51] Adav SS, Chen MY, Lee DJ, Ren NQ. Degradation of phenol by aerobic granules and isolated yeast *Candida tropicalis*. Biotechnology and Bioengineering. 2007;**96**:844-852

[52] Surkatti R, El-Naas M. Biological treatment of wastewater contaminated with p-cresol using *Pseudomonas putida* immobilized in polyvinyl alcohol (PVA) gel. Journal of Water Process Engineering. 2014;**1**:84-90

[53] Al-Khalid T, El-Naas M. Biodegradation of 2, 4 dichlorophenol. American Journal of Engineering and Applied Sciences. 2017;**10**:175-191

[54] Al-Khalid T. Aerobic biodegradation of 2,4 dichlorophenol in a spouted bed bio-reactor (SBBR) [PhD thesis]. UAE University; 2014

[55] Stoilova I, Krastanov A, Stanchev V, Daniel D, Gerginova M, Alexieva Z. Biodegradation of high amounts of phenol, catechol, 2,4-dichlorophenol and 2,6-dimethoxyphenol by *Aspergillus awamori* cells. Enzyme and Microbial Technology. 2006;**39**(5):1036-1041

[56] Stoilova I, Krastanov A, Yanakieva I, Kratchanova M, Yemendjiev H. Biodegradation of mixed phenolic compounds by *Aspergillus awamori* NRRL 3112. International Biodeterioration & Biodegradation. 2007;**60**(4):342-346

[57] Ma J-Y, Quan X-C, Yang Z-F, Li A-J. Biodegradation of a mixture of 2,4-dichlorophenoxyacetic acid and multiple chlorophenols by aerobic granules cultivated through plasmid pJP4 mediated bioaugmentation. Chemical Engineering Journal. 2012;**181-182**:144-151

Chapter 19

Analytical Methods for Polycyclic Aromatic Hydrocarbons and their Global Trend of Distribution in Water and Sediment: A Review

Abiodun Olagoke Adeniji,
Omobola Oluranti Okoh and Anthony Ifeanyi Okoh

Additional information is available at the end of the chapter

http://dx.doi.org/10.5772/intechopen.71163

Abstract

Polycyclic aromatic hydrocarbons (PAHs) are major organic pollutants in the environment, which are toxic to humans and biota, given their carcinogenic, mutagenic and teratogenic nature. In this chapter, we carried out an overview of the sources and toxicity of PAHs, their common analytical methods of determination in the water and sediment samples, and also their global trend of distribution, with a view to provide baseline guidance for relevant control authorities. The choice methods for determining these contaminants are high-performance liquid chromatography (HPLC) with UV/fluorescence detectors and GC/MS. Mass spectrometer coupled with GC is preferred because it offers robust identification of the analyte compounds both by retention time and mass spectrum, with additional structural information. Results collated revealed an extensive distribution of PAHs with total mean concentrations ranging from 0.0003 to 42,350 μg/L in water and 0 to 1.266×10^9 μg/kg (dw) in the sediment. PAHs in the two environmental matrices were much higher in the regions with intense oil exploration, shipping and industrial activities. It is therefore necessary to regularly monitor their levels in the aquatic environment, so as to provide mitigation options that will prevent risk to humans and aquatic animals.

Keywords: polycyclic aromatic hydrocarbons, carcinogenicity, endocrine system disruption, aquatic environment, bioaccumulation

1. Introduction

Water is the most abundant compound on the surface of the earth and needed in all aspects of life in adequate quantity and quality. However, water in nature picks up some amounts of chemical impurities, most of which are anthropogenic and that consequentially impact greatly on its quality [1]. There are three major groups of chemical pollutants: stable trace elements, organic materials, and radionuclides. The organic contaminants include persistent organic pollutants (POPs) and some new compounds such as pharmaceuticals, veterinary medicines and hormones [2].

POPs are synthetic chemicals which are either deliberately or unintentionally produced. They are mostly lipophilic, acutely toxic and persistent in the environment with long-range of transport, thereby leading to global pollution. They accumulate in food chain and are found at highest levels in marine mammals [3]. Highly significant and most commonly determined POPs include organochlorine compounds (OCs), e.g. polychlorinated biphenyls (PCBs), pesticides like dichlorodiphenyltrichloroethane (DDT) and its allied metabolic products such as

PAHs	Chemical formula	Molecular weight (g/mol)	CAS number	Ring number	Melting point (°C)	Boiling point (°C)	IARC group	ERL (μg/kg)	ERM μg/kg)
Naphthalene	$C_{10}H_8$	128	91-20-3	2	80.2	218	2B	160	2100
Acenaphthylene	$C_{12}H_8$	152	208-96-8	3	92.5	280	NA	44	640
Acenaphthene	$C_{12}H_{10}$	152	83-32-9	3	93.4	279	3	16	500
Fluorene	$C_{13}H_{10}$	166	86-73-7	3	115	295	3	19	540
Phenanthrene	$C_{14}H_{10}$	178	85-01-8	3	99.2	340	3	240	1500
Anthracene	$C_{14}H_{10}$	178	120-12-7	3	215	340	3	843	1100
Fluoranthene	$C_{16}H_{10}$	202	206-44-0	4	108	384	3	600	5100
Pyrene	$C_{16}H_{10}$	202	129-00-0	4	151	404	3	665	2600
Benzo[a]anthracene	$C_{18}H_{12}$	228	56-55-3	4	167	435	2B	261	1600
Chrysene	$C_{18}H_{12}$	228	218-01-9	4	258	448	2B	384	2800
Benzo[b] fluoranthene	$C_{20}H_{12}$	252	205-99-2	5	168	481	2B	NA	NA
Benzo[k] fluoranthene	$C_{20}H_{12}$	252	207-08-9	5	217	480	2B	NA	NA
Benzo[a]pyrene	$C_{20}H_{12}$	252	50-32-8	5	177	495	1	430	1600
Dibenzo[a,h] anthracene	$C_{22}H_{14}$	278	53-70-3	5	270	524	2A	63.4	260
Indeno[1,2,3-cd] pyrene	$C_{22}H_{12}$	276	193-39-5	6	164	536	2B	NA	NA
Benzo[g,h,i]perylene	$C_{22}H_{12}$	276	191-24-2	6	278	550	3	NA	NA
ΣPAHs								**4000**	**44,792**

NA: not applicable; ERL: effects range low; ERM: effects range median; IARC: International Agency for Research on Cancer; and CAS: chemical abstract services [21–23].

Table 1. Physicochemical properties of the 16 priority PAHs, their toxicity classification and standard pollution criteria concentrations for sediment.

p,p′-DDT, p,p′-DDE, and p,p′-DDD [2]. Others are polycyclic aromatic hydrocarbons (PAHs), phthalate esters, polybrominated diphenyl ethers (PBDEs), polychlorinated naphthalenes (PCNs), BPA (bisphenol A) and alkyl phenols [4, 5].

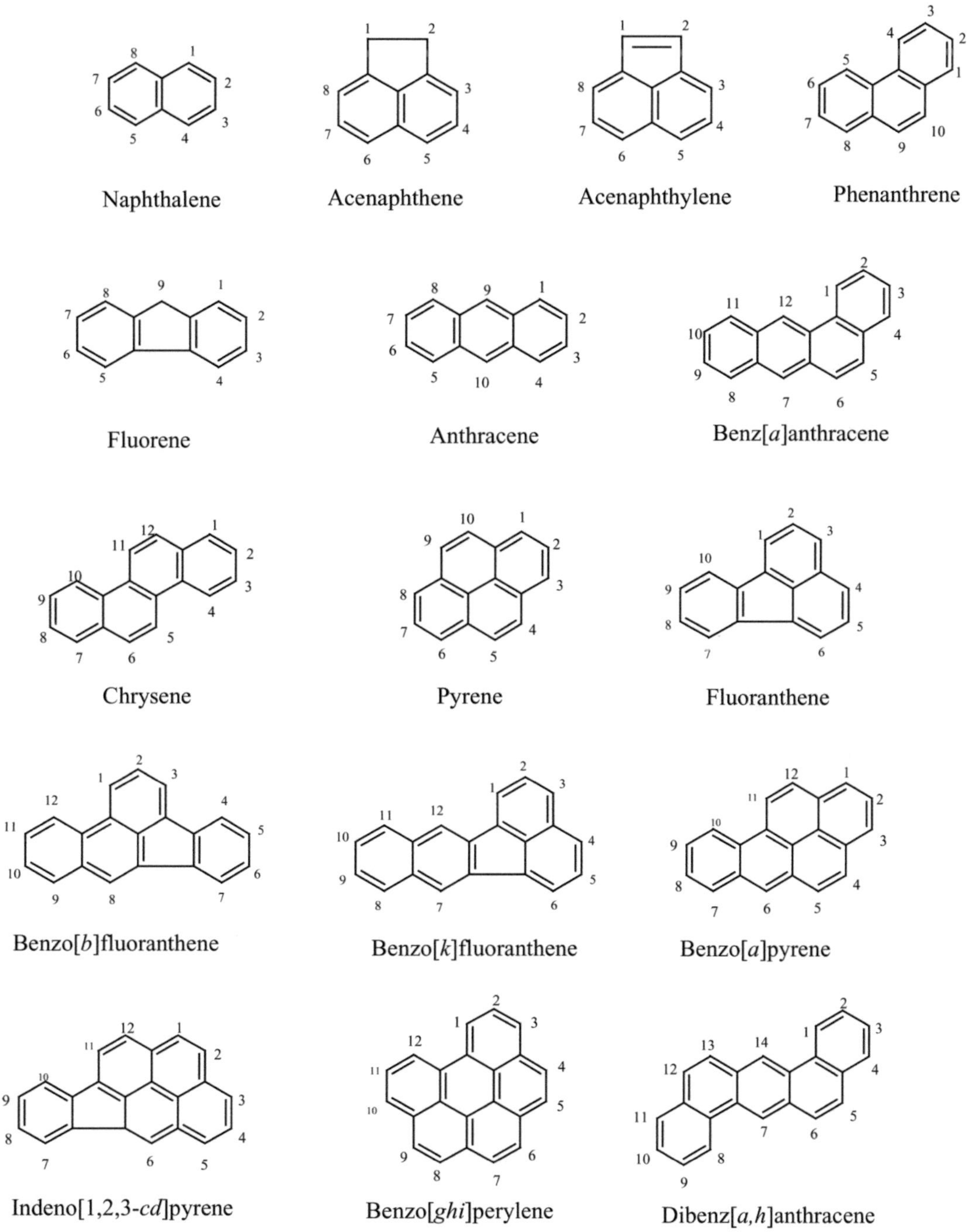

Figure 1. Structures of the 16 priority PAHs [24].

Polycyclic aromatic hydrocarbons (PAHs) are a class of hazardous organic chemicals which can be in linear, angular, and cluster in arrangements. They have two to seven fused carbon rings that can have substituted groups attached, and range from naphthalene ($C_{10}H_8$, two rings) to coronene ($C_{24}H_{12}$, seven rings) with molecular masses ranging from 128 to 278 Da (**Table 1**). They generally have low vapour pressure and are globally distributed in atmospheric, terrestrial and aquatic systems [6–8].

PAHs are basically classified into two: low molecular weight PAHs (LPAHs) and high molecular weight PAHs (HPAHs). LPAHs (e.g. naphthalene, acenaphthene, acenaphthylene, fluorene, anthracene, phenanthrene) tend to have a core structure of two to three benzenoid rings (six-sided aromatic rings of carbon) while HPAHs have molecular structures of four or more benzenoid rings (e.g. fluoranthene, pyrene, benzo[a]pyrene, and benzofluoranthenes) (**Figure 1**). The hydrophobicity, bioaccumulation tendency, resistance to biodegradation, and overall environmental persistence of the compounds generally increase with increasing molecular weight [9, 10]. The alkylated PAHs which are formed through diagenetic processes are more toxic than the parent compounds. They usually co-exist with their parent compounds in the environmental matrices [11, 12].

The toxicity of these alkyl substituted PAHs has generated great concerns in the recent times owing to their huge contribution to the total level of PAHs in the crude oils, mineral oils and diesel, compared to their non-alkylated counterparts which are typically from combustion sources. They bioaccumulate to a greater degree and degrade more slowly than their parent homologues. Alkyl substitution may be inducing phototoxicity in exposed organisms after the parent compounds have been broken down. The alkylated PAHs are also more persistent in the environment than the parent PAHs. Although polycyclic aromatic hydrocarbons in the fuels or crude oils spilled into aquatic environments represent a small percentage of the total mass of the volume, the proportion usually increases greatly after a few months of the incidence, thereby constituting a serious health risk to the aquatic biota [13].

Here, we review the analytical techniques for PAHs in water and sediment, as well as the existing data on their sources, toxicity and distribution in aquatic systems around the world.

2. Physicochemical properties of PAHs

2.1. Solubility, structures, melting and boiling points of PAHs

Solubility of PAHs in water generally decreases as the molecular weight increases while their boiling and melting point increases correspondingly [14]. Four-ring and five-ring aromatic hydrocarbons such as chrysene and benzo[a]pyrene are virtually insoluble in water [15]. Solubility also decreases as ring structure increases, as degree of substitution increases, vapour pressure decreases and molecular weight increases [6, 16]. Molecules with a linear arrangement are more likely to be less soluble than the angular or perifused molecules. For example, anthracene is less soluble compared to phenanthrene. Alkyl (i.e. CH_2— group) substitution of the aromatic ring also results in an overall decrease in the solubility of PAHs, although there are some exceptions, e.g. benz[a]anthracene is less soluble than either methyl- or ethylbenz[a] anthracene. The solubility of PAHs in water is enhanced three- to fourfold by a rise in temperature from 5 to 30°C. Dissolved and colloidal organic fractions also enhance the solubility of PAHs which are incorporated into micelles [13, 14].

PAHs in aquatic environments rapidly tend to become associated with the particulate matter or organic substances such as biopolymers, humic substances and black carbon ending in sedimentation due to their low solubility in water [8, 16]. Once adsorbed to the sediments, they are much more stable than pure compounds and are resistant to oxidation and nitration reactions to which they would otherwise be quite sensitive due to photochemical processes. Most PAHs are classified as semi-volatile organic compounds because of their low volatility [8]. Some processes such as photo-oxidation, hydrolysis, biotransformation, biodegradation and mineralization in the aquatic system lead to transformation of PAHs to other substances. Likewise, other activities such as bioaccumulation, adsorption, desorption and re-suspension are accountable for the cycling of the compounds all the way through the aquatic environment. High molecular weight PAHs in the aquatic systems are mostly degraded by photo-degradation [17].

2.2. Sorption of PAHs to organic carbon

Total organic carbon (TOC) and black carbon (BC) play an important role in the sequestration of PAHs in soils and sediments despite their relatively low level in the environmental media [14]. Sorption of PAHs to soil and sediments increases with increasing organic carbon content and with increasing surface area of the sorbent particles [10]. Hence, the concentration of PAHs varies largely among different soil size fractions. A particular study reported that the highest total PAH concentration was found in the 250–500 μm size fraction, while the maximum level of individual PAHs was detected in the 250–500 μm or 500–2000 μm size fraction [18].

The organic carbon content depends on geographical location, pollutants entering rivers, pH and layer depth of the tested sediments. Sediment samples with lowest pH recorded the least amount of organic content [19, 20]. The association of PAHs with the solid phase depends on their molecular weight and octanol-water partitioning coefficient (Kow).

Among the EPA 16 priority PAHs, phenanthrene, fluoranthene, benzo[a]anthracene, chrysene, benzo[b&k]fluoranthene, benzo[a]pyrene and indeno[1,2,3-cd]pyrene are the most important contaminants in soil. Furthermore, the high partitioning to organic carbon is the root cause of the high rate of bioconcentration for these compounds and the ease with which they enter the food web [14].

3. Sources of PAHs in the environment

PAHs enter the shallow coastal, estuarine, lake and riverine environments from petroleum spills, treated industrial and municipal waste water discharges, urban and suburban stormwater runoff, chemical refineries, recreational and commercial boats, volcanoes and atmospheric fallout of vehicle exhaust [2, 25]. They are generated from both natural and anthropogenic sources, being ubiquitous.

3.1. Natural sources

These include natural petroleum seeps, forest fires, prairie fires, agricultural burning and post-depositional transformation of biogenic precursors [26]. The actual amount of PAHs and particulates emitted from these sources varies with the type of organic material burned, type of fire, nature of the blaze and intensity of the fire. PAHs from fires tend to sorb to suspended particulates and eventually enter the terrestrial and aquatic environments as atmospheric fall-out. Treated wood has also been recognized as a source of PAHs in water and sediments [14]. Other natural sources include volcanoes, chlorophyllous plants, fungi and bacteria [10].

3.2. Anthropogenic sources

Anthropogenic sources are basically grouped as pyrolytic and petrogenic. Pyrolytic sources include combustion processes (e.g. fossil fuel combustion, electric power generation, refuse incineration, home heating and industrial emissions), while the petrogenic input is closely related to releases from petroleum products (e.g. oil spills, road construction materials such as production of coke, carbon black, coal tar, and asphalt) [7, 9, 10, 27–30].

Some PAH contaminant sources exhibit co-dominance of both pyrogenic and petrogenic types owing to the complex processes involved in the PAH source type. Crankcase and other lubricating oils used in internal combustion engines, for example, contain a variety of petrogenic PAHs. The PAH composition changes, however, as the oil is used, since various combustion-derived PAHs accumulate with the increased use. Coal tar residues from former manufactured gas plant (MGP) sites tend to have a highly complex and highly variable PAH composition, since the coal tar was produced by the heating of PAH containing coal or coke to produce coal gas as a heating and light source prior to the wide-spread availability of natural gas in the mid- to late-twentieth century [9]. Polycyclic aromatic hydrocarbons (PAHs) in water bodies are subject to distribution and accumulation among the water column, suspended particulate matter, bottom sediments, and biota [31], because of their low vapour pressure, non-polarity, lipophilicity and high hydrophobicity [8].

3.3. The use of molecular diagnostic ratios for source identification

Sources of PAHs in the environment are usually identified using several ratios and indices of some particular PAHs on the basis of their composition and distribution pattern [32–34]. An ideal source ratio would be distinctive to a particular source [35, 36]. The molecular ratios are mostly used to distinguish between PAHs from pyrogenic and petrogenic origins [37]. In general, two- to three-ringed and some alkyl-substituted PAHs are good for distinguishing petrogenic contamination [33, 38–40], whereas four- to six-ringed PAHs which are more toxic and thermodynamically stable than those from petrogenic sources are appropriate for identifying those from pyrogenic origins [16, 33, 34, 41], given the high level of non-alkylation in their own composition [32, 33, 42, 43].

The ratio of low molecular weight to high molecular weight PAHs (LMW/HMW) is a measure of weathering process in the aquatic environment. A lower value of this ratio indicates high level of resistance to microbial degradation in the high molecular weight PAHs or a higher solubility of the low molecular weight PAHs in the water column [33]. However, many diagnostic ratios involving the low molecular weight PAHs (e.g. LMW/HMW, anthracene/Phen, etc.) are very unstable because they are usually influenced by some environmental factors like photodegradation and volatilization; hence, they are very much affected by mobility in the aquatic environment. On the other hand, the higher molecular PAH ratios are relatively more stable and are less affected by such factors [44, 45].

Statistical analysis revealed that the following molecular ratios: Anth/178, BaA/228, Chry/BaA, Flt/Flt + Pyr, Flt/Pyr and Inp/Inp + BghiP are strongly correlated with pyrogenic PAHs. Similarly, Chry/BaA, Inp/BghiP, BaP/BghiP and some other similar ones have shown tendency to provide more useful information needed to differentiate PAHs from several pyrogenic sources such as vehicle emission, coal burning etc. [41, 46]. For instance, BaA/(BaA + Chry) in the range of 0.2–0.35 indicates combustion of coal, grass or wood burning [47] and BaP/BghiP > 0.6 suggests vehicular emission [48]. Although, all the ratios have been found useful in the identification of sources of pollution in the aquatic environment, the information they provide are not very definite but are only rough idea of the origin. The ratios are sometimes inconsistent, as they occasionally offer contradictory information, pointing to different sources of pollution at the same time owing to their instability. Researchers must therefore exercise caution in using them. Cross plot and more ratios can therefore be employed to achieve a robust interpretation of the information provided by these isomeric ratios [33, 39, 41].

The ratios of the alkyl substituted PAHs to their parent compounds have been found appropriate to provide more reliable information for source distinction because the alkylated PAHs decrease in abundance as temperature increases [49]. The formation of these compounds is very distinctive at low temperatures (~100–150°C) than at higher temperatures (~2000°C) because of their thermal instability. Most commonly used of these compounds are the methylated species of phenanthrene, fluoranthene, pyrene, chrysene and benz[a]anthracene (**Table 2**) [39].

3.4. Toxicity of PAHs and risk of exposure

More than 80% of the total PAH contribution to environmental and health concerns could be attributed to the 16 EPA priority PAHs (**Figure 1** and **Table 1**). Carcinogenic nature of these compounds is a major worry, although not all of them are affected [48]. In water, the toxicity of individual PAHs to plants and animals decreases beyond molecular weight of 202 because solubility drops rapidly afterwards. However, sub-lethal effects can result from exposure to very low concentrations of the high molecular weight (HMW) compounds. In most cases, environmental concentrations of PAHs in water are enormously below levels that are acutely toxic to aquatic organisms except in the vicinity of chemical or petroleum spills. However, concentration of PAHs in the sediment can be much higher, even though their limited bioavailability often reduces their toxic potential to a large extent [6].

PAHs are generally classified into five groups based on their carcinogenic tendencies (**Table 1**). Group 1 consists of substances with carcinogenic potential for humans and benzo[a]pyrene with sufficient toxicological data is the only member [34, 51]. Group 2A PAHs are those that are probably carcinogenic to humans and a major example in this group is dibenzo(a,h)

Ratio	Petrogenic	Pyrolytic	References
LMW/HMW	>1.0	<1.0	[38]
Phen/anthracene	>15	<10	[32]
Anth/178	<0.1	≥0.1	[41]
BaA/228	<0.2	0.2–0.35	[50]
Chry/BaA	<0.4	>0.9	[44]
Anth/Anth + Phen	<0.1	>0.1	[33]
Flt/Flt + Pyr	<0.4	≥0.5	[37]
Flt/Pyr	<1.0	>1.0	[32]
Inp/Inp + BghiP	<0.2	>0.5	[37]
BaA/BaA + Chry	<0.2	>0.35	[50]
MPhen/Phen	=0.4	<0.4	[39]
MPF/PF	=0.5	<0.5	[39]
MCB/CB	=1.0	<1.0	[39]
MPsp/Psp	=0.8	<0.8	[39]

Phen: phenanthrene; Anth: anthracene; Chry: chrysene; BaA: benzo[a]anthracene; LMW: low molecular weight; HMW: high molecular weight; Flt: fluoranthene; Pyr: pyrene; Inp: indeno(123,cd)pyrene; BghiP: benzo[g,h,i] perylene; MPsp: sum of all the methyl PAH species of phenanthrene, fluoranthene, pyrene, chrysene and benz[a] anthracene; Psp: sum of the parent compounds of the methyl PAH species (i.e. phenanthrene, fluoranthene, pyrene, chrysene and benz[a]anthracene); MPhen/Phen: sum of 3-methylphenanthrene, 2-methylphenanthrene, 9-methylphenanthrene and 1-methylphenanthrene; MPF: sum of three peaks of methylpyrenes/methylfluoranthenes; PF: sum of pyrene and fluoranthene; MCB: sum of five peaks of methylchrysenes/methylbenz[a]anthracenes; and CB: sum of chrysene and benz[a]anthracene.

Table 2. Molecular diagnostic ratios and possible sources of PAHs in the environment.

anthracene. Although, the list contains more PAHs in this category, however, they are not among the EPA priority contaminants. Therefore, the sentence is very correct as far as the 16 priority PAHs are concerned. They are the substances that are possibly carcinogenic to humans. Group 3 compounds are those not classifiable as being carcinogenic to humans and its members include acenaphthene, fluorene, phenanthrene, anthracene, fluoranthene, pyrene and benzo[g,h,i]perylene. There is another group for those which are probably not carcinogenic to humans but none of the 16 EPA PAHs falls into the category. However, acenaphthylene was not classified into any of the 5 existing groups, probably because there is no sufficient data to decide on where it should belong [22, 24, 37, 52].

Seven of the pyrogenic PAHs (4–7 rings) in the group(s) 1, 2A and/or 2B have been identified by the United States Environmental Protection Agency (USEPA) and the International Agency for Research on Cancer (IARC) to be of high risk to humans and are therefore used to assess the level of pollution in the environment. They are benzo[a]pyrene, dibenzo[a,h]anthracene, benz[a]anthracene, benzo[b]fluoranthene, benzo[k]fluoranthene, indeno[1,2,3-cd]pyrene and chrysene [34, 37, 53, 54]. Aside their confirmed carcinogenicity, all the seven PAHs were also found genotoxic alongside benzo[g,h,i]perylene that was not classified as being carcinogenic to humans [24]. The USEPA, therefore, established toxicity equivalency factors (TEFs) for the

PAH congeners	TEF
Benzo[a]anthracene	0.1
Chrysene	0.01
Benzo[b]fluoranthene	0.1
Benzo[k]fluoranthene	0.1
Benzo[a]pyrene	1
Dibenzo[a,h]anthracene	0.1
Indeno[1,2,3-cd]pyrene	0.1

Table 3. Toxicity equivalent factors (TEFs) of the carcinogenic PAHs.

quantification of their level of toxicity. The highest TEF of 1 was assigned to benzo[a]pyrene while lower values as shown in **Table 3** were assigned to other PAHs. Toxicity equivalent concentration (TEC) of individual PAH can be calculated by multiplying the concentration of the congener in the environmental sample with the respective TEF, as shown in Eq. (1). TECs for the seven carcinogenic PAHs (cPAHs) are summed up to obtain the total toxicity equivalent concentration (TTEC) in the chemical mixture or the sample under investigation [37, 55].

$$\text{Total toxicity equivalent concentration (TTEC)} = \sum C_n \cdot TEF_n \quad (1)$$

where C_n is the concentration of each congener n in the PAH mixture and TEF_n is the toxicity equivalency factor for the specific congener n. Compliance is determined by comparing the TTEC for the sample with the appropriate cleanup level (Method B or C) for benzo[a]pyrene. The Method B and C cleanup levels for the reference substance are 0.137 and 18 mg/kg, respectively, although the estimates are only considered applicable for exposures through the oral route [55, 56].

Some of PAHs can affect health when exposure is at levels higher than the maximum concentration limit (MCL) for relatively short periods of time. The damage can include suppressed immune systems, or red blood cell damage leading to anaemia. Long-term exposure is believed to lead to potential developmental and reproductive effects and some forms of cancer [57]. Occupational exposures to high levels of mixtures containing PAHs have resulted in symptoms such as eye irritation, nausea, vomiting, diarrhoea and confusion. Reactive metabolites of some PAHs, such as epoxides and dihydrodiols, have the ability to bind with cellular proteins and DNA. The resulting biochemical disruptions and cell damage lead to mutations, development of malformations, tumours and cancer. The most common PAH that causes cancer in laboratory animals is benzo[a]pyrene. Ingestion of high levels of these compounds during pregnancy could result in birth defects and decreased body weight of the offspring. Most of the PAHs are not genotoxin by themselves; they need to be metabolized to the diol-epoxides which react with DNA, thus inducing genotoxic damage [10].

Bioconcentration and bioaccumulation of PAHs in organisms occurs through various routes including ingestion, inhalation or dermal contact pathways. PAHs are toxic and their carcinogenicity is initiated by their metabolic conversion to peroxides that bind covalently to cellular macromolecules, including DNA, causing an increase of elevated levels of DNA adducts and developing errors in DNA replication which cause carcinogenesis in both humans and other organisms [48]. Although unmetabolized PAHs can have toxic effects, the primary concern for animals is the ability of reactive metabolites, such as epoxides and dihydrodiols, of some PAHs to bind to cellular proteins and DNA. The resulting biochemical disruptions and cell damage lead to mutations, mental malformations, tumours, and cancer [6, 9].

3.4.1. Sediment quality criteria

The contamination of aquatic sediment by PAHs is evaluated by comparing the PAH concentrations in such sediment sample with the effects-based guideline values like the effects range low (ERL) and effects range median (ERM) established by the United States National Oceanic and Atmospheric Administration (USNOAA) [58, 59]. **Table 1** presents the ERL and ERM values for the 16 priority PAHs. The ERL values suggest a likelihood of serious biological effects on aquatic creatures, while ERM values indicates a great risk of posing damaging biological effects on the aquatic organisms [59]. A lesser PAH concentration than the ERL would guarantee no adverse effects on the benthic invertebrates. However, any concentration between the ERL and ERM suggests the occurrence of such effects occasionally in the aquatic milieu. Meanwhile, higher PAHs concentrations above the ERM value would impact adverse effects on the organisms from time to time [37, 58, 60].

3.5. Extraction methods for PAHs in water and sediment

3.5.1. Water extraction

PAHs are usually present in water samples in trace quantity, owing to their low solubility [61]. Therefore, the analytical method to use should be of high enrichment factors for the target compounds and should allow concentrating and increasing the low levels to values that can be detected using analytical instrument [20, 62]. Method validation is very necessary for the recovery of PAHs to show that the extraction procedures for the samples are very effective [63, 64].

Recent techniques for the extraction and concentration of PAHs from environmental water samples include solid-phase extraction (SPE), liquid-liquid extraction (LLE), continuous liquid-liquid extraction (CLLE), solid-phase microextraction (SPME), hollow fibre liquid-phase microextraction (HF-LPME) and stirring bar sorptive extraction (SBSE) [20, 65–68]. The two most commonly used, which were also recommended by the USEPA for pre-concentration of PAHs in drinking water samples are LLE and SPE [63, 69–71].

3.5.2. Sediment extraction

Analysis of PAHs in the solid matrices is not simple and therefore requires careful sample pre-treatment such as isolation or extraction, prior to instrumental determination. The extraction procedure may be followed by cleanup if necessary [72].

Extraction methods for PAHs in soils and sediments includes Soxhlet extraction (SE), mechanical shaking or agitation [30, 73], automated Soxhlet extraction, pressurized fluid extraction (PFE), microwave extraction (ME) [74], surfactant promoted extraction, accelerated solvent extraction (ASE), supercritical fluid extraction (SFE), sub-critical fluid extraction, solvent washing, extraction by vegetable oils and extraction by cyclodextrins [72], ultrasonic extraction [75], microwave dissolution, pressurized liquid extraction (PLE) [76–79], solid phase micro-extraction (SPME) and micellar solid-phase microextraction (MSPME) [80, 81].

An efficient extraction technique should be able to produce good results within a short time with minimum operator involvement. It should also be cheap, and safe for both the analyst and the environment [82]. The choice of extraction technique depends on several factors including capital cost, operating cost, sample matrix, simplicity of operation, sample throughput and the availability of a standardized method [30, 83].

3.6. Analytical methods for polycyclic aromatic hydrocarbon in water and sediment

Polycyclic aromatic hydrocarbons with their derivatives are globally determined after extraction from food, environmental or biological samples using analytical methods approved by certain agencies and/or organizations such as the United States Environmental Protection Agency (USEPA), International Organization for Standardization (ISO) and National Institute for Occupational Safety and Health (NIOSH). The methods are basically grouped into three: immunoassay, spectrometric and chromatographic methods [57, 84].

Immunoassay methods (EPA 4030 and 4035, Update III) which exist mostly as kits are not very popular because of their tendency to introduce strong biases in the final results. Besides, the precision, accuracy and affinity of the methods for many aromatic compounds are significantly lower than what could be obtained from other standard methods. Hence, they are mostly used for field screening in the analysis of soil and water [84, 85]. Among the spectrometries, ultraviolet (UV) and infrared (IR) methods are the most prominent. However, UV methods (absorption and fluorescence) which are considered being sensitive and selective to aromatic compounds like PAHs are more often affected by interference due to the presence of some other compounds like lipids in the sample matrix. Similarly, the IR spectrometric method which although is fast and cheap, requires the sample to undergo a mandatory cleanup step after extraction before analytical determination and is also poorly selective [57].

Liquid and gas chromatographic methods are the most frequently used for the analysis of PAHs, in spite of being more expensive and time consuming [8, 9, 36]. Available liquid chromatographic (LC) methods include high-performance liquid chromatography (HPLC) with UV and fluorescence (FL) detectors in series (EPA 550, 610 and 8310; NIOSH 5506 and 5800) [74, 86], mass spectrometer [87], photo-diode array detector (PDA) [88], PDA and fluorescence detectors combined [89], or fluorescence detector alone (ISO 17993) [22]. Whereas gas chromatographic (GC) methods could be used after the EPA recommended organic extraction procedures with flame ionization detector (EPA 610 and 8100, NIOSH 5515), Fourier transform-infrared detector (EPA 8410), mass spectrometer (EPA 525, 625 and 8270; ISO 18287; ISO/DTS 28581; ISO/FDIS 28540 and EN 15527), or mass spectrometer with thermal extraction (EPA 8275) [22, 74, 90].

3.6.1. Liquid chromatographic methods

HPLC methods are more patronized because of the ease of determining thermally labile, semi-volatile, less volatile or non-volatile compounds. They are also more sensitive, specific and reproducible than some GC based methods [8, 63, 91–93]. EPA methods 550, 610 and 8310 require the aqueous samples to be treated prior the instrumental analysis with liquid-liquid or solid phase extraction procedure for isolation of the compounds of interest. Solid samples are extracted using either Soxhlet or sonication devices with suitable solvent(s) [94–96]. The sample extracts are required to be dissolved in any solvent miscible with the one used for extraction before injection into the instrument [84].

However, NIOSH 5506 is limited in that, it can only respond to samples extracted by acetonitrile only and not any other extraction solvent. The performance of the method is also affected negatively by huge presence of extremely adsorptive particulate matter such as fly ash, asphalt fumes, or diesel soot in the sample. But it is sensitive to PAHs and capable to determine them in the mixture of aliphatic compounds, even in trace level. NIOSH 5800 is also different from others because it uses a flow-injection method for the determination of total polycyclic aromatic compounds in the sample at two different sets of fluorescent wavelengths [21]. The HPLC methods (EPA 550, 610, 8310 and NIOSH 5506) determine PAHs in the sample extracts using UV and fluorescence (FL) detectors connected in series. The first four PAHs (naphthalene, acenaphthylene, acenaphthene, and fluorene) in the priority list of USEPA which are not very fluorescent are analysed with the less sensitive ultraviolet detector while the remaining ones are determined using fluorescence detector with better sensitivity [87].

EPA 8310 provides adequate sensitivity at part per billion levels for the detection of PAHs in water samples. The method is also applicable to the analysis of PAHs in the effluent samples [84]. UV and fluorescence detectors are the most extensively used in the liquid chromatographic measurement of PAHs, especially those with high molecular weights. UV/Visible detector can detect almost all the PAHs in the UV range of 190 to 360 nm. The sensitivity and selectivity of fluorescence detector in the quantitation of PAHs in the complex environmental mixture is much higher, particularly when appropriate excitation and emission wavelengths are selected [63, 69, 87, 88]. Photodiode array (PDA) detector which is much more sensitive than UV/Visible detector is sometimes used as an alternative, either alone [8, 70, 88] or with fluorescence detector [89, 97]. Notwithstanding, the latter remains the most sensitive among the modern LC detectors, which is highly recommended for trace determination of PAHs [98].

Another valued detector in liquid chromatography is mass spectrometer (MS). It is very useful for identification and characterization of trace polar components, especially if many compounds of interest are involved [92, 99]. An all-inclusive use of LC/MS for the analysis of aromatic compounds has been reported. It involves the introduction of sample into the chromatograph for separation, after which it passes through a sophisticated interface (such as thermospray, electrospray, moving belt and particle beam) where separation of target compounds from aqueous mobile phase takes place, then finally into the mass spectrometer for characterization [92, 100]. However, there is no particular interface suitable for the separation of all the polycyclic aromatic hydrocarbons in the sample [87].

3.6.2. Gas chromatographic methods

Gas chromatographic methods are used for the separation and detection of non-polar organic compounds that are volatile and thermally stable. It is also used for the analysis of certain semi-volatile compounds including PAHs [92, 99]. Applicable detectors for the analytical determination of aromatic compounds including PAHs in samples using GC include photoionization detector (PID), flame ionization detector (FID), Fourier transform-infrared (FT-IR) and mass spectrometer (MS) [75, 84, 101, 102]. The technique can offer high resolving power with the use of capillary column. PAHs with molecular weight (MW) above 300 atomic mass unit (amu) are always difficult to analyse with GC because of their low volatilities, tendency to decompose when subjected to high temperature, and possibility of adsorbing to the GC inlet and column [87].

EPA 8021B is a GC method that is majorly used with PID to determine the level of some volatile aromatics such as benzene, toluene, ethylbenzene, and xylenes (BTEX), as well as certain oxygenates (e.g. methyl-t-butyl ether (MTBE)), a number of olefins, cycloalkanes, branched alkanes and some halogenated compounds in so many sample types, including ground water, aqueous sludges, waste solvents, oily wastes, soils and sediments. It has the advantages of being selective and sensitive, although it is easily desensitized and contaminated by compounds with heavy molecular weights [84, 103].

EPA 610 was specifically designed for PAHs determination in wastewater after liquid-liquid extraction (LLE) with methylene chloride. The advantage of the method is that it is available for use as a liquid chromatographic technique with UV and fluorescence detection. It can also be used as a gas chromatographic method with either packed or capillary column coupled with flame ionization detector (GC/FID) [87, 94]. EPA 8100 is another GC/FID method that can be used to determine the target compounds in both water and solid samples. It is a packed column gas chromatographic method. The method allows the water samples to be adjusted to neutral pH and extracted prior to analytical determination using either LLE or continuous LLE, while the extraction of solid samples could be achieved through the use of Soxhlet or ultrasonic extraction set-up. The packed column in the procedures conversely present a major challenge of adequate resolution of four pairs of PAHs, which are anthracene and phenanthrene; chrysene and benzo[a]anthracene; benzo[b]fluoranthene and benzo[k]fluoranthene; and dibenzo[a,h]anthracene and indeno[1,2,3-cd]pyrene, although the listed pairs of compounds can be well resolved if capillary column is used in place of the packed column. Otherwise, the silica gel cleanup is suggested as a necessary component of the methods, except if the sample matrix is relatively clean [87, 90].

NIOSH 5515 which is also a GC/FID method was designed to analyse air sample with capillary column (30 m × 0.32-mm ID, fused silica capillary, 1-μm DB-5) coupled with FID, after being subjected to filter extraction [86, 104]. Even though FID is a sensitive detector for PAHs, the use is limited by its sensitivity to background interferences from some common environmental pollutants like phthalates and other co-extracted non-target compounds from carbonaceous sources. Therefore, cleanup step is very mandatory for appropriate quantification and identification of analytes using this detector [84, 87, 105, 106].

EPA 525, 625, and 8270 are GC/MS methods for identification and quantitation of PAHs and some other contaminants in various environmental media [87]. EPA 525 was designed for the determination of several groups of semi-volatile compounds including PAHs, phthalates, PCBs and adipates in drinking or river water. The pH of the water sample is to be adjusted to <2 using 6 N HCl before a solid phase extraction (SPE) procedure [107]. EPA 625 was, however, intended for the analysis of the listed classes of contaminants in the wastewater samples. The municipal and/or industrial wastewater samples are recommended for extraction at two different pHs using LLE, followed by GC/MS analysis. The updated version of the method (EPA 625.1) has included the use of SPE as an alternative extraction procedure [108].

Moreover, EPA 8270 can be employed for the quantitative determination of the levels of semi-volatile organic pollutants in water, air, soils, and other solid waste matrices. Recommended sample preparation techniques for aqueous samples in this method include LLE and CLLE. Solid samples are to be prepared by Soxhlet, automated Soxhlet, ultrasonic, supercritical fluid, or solvent extraction device while non-aqueous solvent-soluble wastes are to be made ready for instrumental analysis by solvent dilution. In addition to the listed extraction methods, EPA 8270 also made provision for the isolation of semi-volatile compounds in the leachates obtained from toxicity characteristic leaching procedure (TCLP) by SPE. Aside PAHs, the method can be applied to several other classes of organic compounds, such as chlorinated hydrocarbons, pesticides, aromatic nitro compounds, phenols, phthalate esters, organophosphate esters, nitrosamines, ethers, ketones, anilines, haloethers, aldehydes, pyridines, and quinolones [109].

EPA 8272, another GC/MS technique is used for the analysis of ten parent PAHs and two alkylated PAHs (naphthalene, acenaphthylene, acenaphthene, fluorene, phenanthrene, anthracene, fluoranthene, pyrene, benz[a]anthracene, chrysene, 2-methylnaphthalene and 1-methylnaphthalene) in the pore water recovered from sediments, groundwater, and any other water sample (e.g. seawater, tap water) [110, 111]. The target compounds determinable with the method are those that are very soluble in the environmental waters. The interstitial water is usually removed from the sediment by centrifugation, followed by supernatant collection, while other sample types are prepared for analysis by removal of colloid with flocculation using aluminium potassium sulphate (alum) and sodium hydroxide. The flocculation step is repeated, followed by centrifugation, supernatant collection and solid phase micro-extraction (SPME). The PAHs in the extract are thereafter determined using GC/MS operated in the selected ion monitoring (SIM) mode [112].

GC/MS is the most frequently used technique for the analytical determination of many organic compounds, including PAHs because it identifies analytes not only by retention time but also by mass spectrum, providing structural information and high sensitivity needed for quantification in selected ion monitoring (SIM) mode [8, 66]. It is more expensive than the non-selective procedures like GC/FID and GC/PID and is often employed to determine the concentrations of target volatile and semi-volatile constituents of petroleum. However, the identification of unknown compounds with MS is limited using electron ionization (EI) because mass spectra only are not sufficient for suitable identification of substances. Chemical ionization (CI) or high resolution mass spectrometry if available can provide additional information needed for exhaustive elucidation [57, 84, 99].

EPA 8410 and 8275 are other GC based methods for PAHs. EPA 8410 uses Fourier transform-infrared spectrometer (FTIR) for detection and quantitation of the compounds of interest in the sample extract. The method is considered a valuable complement to EPA 8270 because it enables proper identification of certain isomers that are ordinarily difficult to be differentiated using GC/MS. The capillary GC procedure is applicable to the analysis of many semi-volatile organic compounds in wastewater, soils, sediments, and solid wastes. It requires compulsory cleanup of extract before injection into the gas chromatograph [113]. EPA 8275 is another capillary GC/MS procedure with an online thermal extraction for quantitative analysis of specific PCBs and the 16 EPA priority PAHs in soils, sludges, and solid wastes [87]. It involves the extraction of the analyte compounds from the sample matrix by heating to 340°C in the extraction chamber, where the extract is held for about three minutes before being swept into the column by split injection in the ratio ~35:1 or ~400:1, depending on the concentration of the sample. The injected extract is concentrated on the head of the GC column. The analytes are appropriately separated in the column and are qualitatively and quantitatively determined using mass spectrometer [114].

Of all the methods discussed, the most prevalent for the determination of PAHs in environmental media are the GC/MS and HPLC with UV/fluorescence detectors. The choice of the two is based on their major advantages of high sensitivity and selectivity, in addition to the ability of mass spectrometer to provide additional structural information when used as detector [115].

3.7. Levels of PAHs in some waters and sediments across the world

3.7.1. Levels in water

The levels of PAHs in some environmental waters across the globe are presented in **Table 4**. The reported concentrations varied from 0.0003 to 42,350 μg/L. The minimum and maximum values were both recorded in the Asian continent, precisely from the Kor River water, Iran [16] and major rivers of Southern Thailand, respectively. Foremost rivers in the southernmost part of Thailand were categorized as natural inland water Class II, implying that the water must be made to pass through a number of treatment steps before it can be considered for reuse, given their level of pollution. The most abundant PAHs in the water samples from these rivers were the 4 and 5 rings congeners, which suggest discharge of anthropogenic wastes as major source of contaminants in the water bodies [116].

Other higher levels published emanated from African continent. For instance, Edokpayi et al. [117] reported PAH concentration in the range of 126–7510 μg/L for Mvudi River water and BDL—7805 μg/L for Nzhelele River water, both in South Africa. The elevated concentrations of the organic contaminants in the two rivers were linked to contamination by the wastewater treatment facilities in the study area. Similarly, rivers from Mutshundudi and Nzhelele areas of Limpopo, South Africa were reportedly polluted up to 137 μg/L by high degree of refuse dumping, organic waste burning, activities of the roadside mechanics and traffic density in the area, with HPAHs predominating in the rainy season [64]. Gorleku et al. [102] also accounted for total PAHs as high as 84.50 μg/L in the water column of Tema Harbour, Ghana. The levels determined at the inner fishing harbour (e.g. canoe landing site and some locations in the main harbour) were higher than at the outer fishing harbour. Canoe landing site recorded significantly high values because of possible leakage of petroleum products from the two-stroke engines of the fishing canoes, industrial effluent discharge into the sea through Chemu Lagoon [118], smoking of fish and other anthropogenic activities in the vicinity.

Sample source	Obtained concentration (µg/L)	% LPAHs	Reference
Europe			
Evoikos Gulf Water, Greece (Shipyards)	0.0009–0.14	68.6	[70]
Malliakos Gulf Water, Greece (Ag. Marina)	0.002–0.05	75.0	[70]
Malliakos River Water, Greece (Sperchios)	0.001–0.07	96.3	[70]
Surface Waters of Northern Greece	0.001–0.68	85.2	[127]
Asia			
Kor River Water, Iran	0.0003–0.24	81.6	[16]
Surface waters in Hangzhou, China	0.99–9.66	–	[128]
Gao-ping River, Taiwan	0.01–9.40	–	[129]
Xiamen Harbour, China	0.11–0.95	–	[130]
Water Resources in Mahasarakham University, Thailand	0.001–0.009	13.4	[123]
Major Rivers of Southern Thailand	2250–42,350	–	[116]
Well Water and River Water, Xuanwei and Fuyuan, China	0.0009–0.06	–	[131]
Mumbai Harbour Line, India	0.009–0.047	–	[132]
Reclaimed and Surface Water of Tianjin, China	1.80–35	95.9	[121]
Africa			
Suez Canal, Egypt	0.01–0.5	16.2	[124]
Shallow Lake, Zeekoevlei, South Africa	0.00001–0.01 (individual)	95.0	[133]
Mvudi River water, South Africa	126–7510	25.3	[117]
Nzhelele River water, South Africa	BDL – 7805	23.1	[117]
Surface Water in Nigerian Coastal Communities—Eleme, Nigeria (highly industrialized)	22.10 (total)	–	[97]
Surface Water in Nigerian Coastal Communities—Eleme, Nigeria (less industrialized)	8.39 (total)	–	[97]
Lakes and Rivers around Johannesburg/Pretoria in South Africa	0.021–0.62	86.6	[20]
Tema Harbour Water, Ghana	33.20–84.50	–	[102]
Rivers Water, Limpopo, South Africa	0.10–137	–	[64]
Hartbeespoort Dam, South Africa	0.030–0.05	–	[134]
America			
Paraíba do Sul River, Brazil	0.255 (benzo[a]pyrene only)	–	[135]
Mississippi river, USA	0.06–0.15	–	[136]
Chesapeake Bay, USA	0.02–0.09	–	[137]

Sample source	Obtained concentration (µg/L)	% LPAHs	Reference
York river, USA	0.002–0.12	–	[138]
Petroleum Produced Water, Sergipe, Brazil	3.50–44.30	52.4	[67]
San Francisco Estuary, USA	0.007–0.12	–	[139]

BDL: below detection limit.

Table 4. Levels of PAHs in waters around the world.

The EU Directive 98/83/EC gave a permissible level of 0.1 µg/L for PAHs in water intended for consumption by humans [119, 120]. Unfortunately, the target limit was found exceeded in many regions of the world as shown in **Table 4**; this may be attributed to the increased urbanization, industrialization and incessant discharge of untreated or partially treated effluents into the water bodies. Other possible sources of high PAHs in the water phase of the aquatic systems include vehicular exhaust emission, intense shipping activities [121], and run-off from petroleum filling stations [97]. The results compiled revealed that pollution due to PAHs in water is more pronounced in the Asian and African continents than other parts of the globe. Hence, there is need for strict compliance.

Table 4 indicated that LPAHs are more soluble in the water phase than the sediments, as they recorded large percentages in most of the water samples analysed [16, 63], although their solubility is usually reduced significantly in the marine water [122]. Two- and three-ring PAHs which are referred to as low molecular weight PAHs account for between 52.4 and 96.3% in the water column, except in few studies conducted in Thailand, South Africa and Egypt [117, 123, 124].

3.7.2. Levels in sediment

The global distribution of PAHs in the sediments as shown in **Table 5** revealed that the total concentrations of the contaminants ranged between 0 and 1.266×10^9 µg/kg. The highest levels of these organic pollutants were reported from Taylor Creek, Bayelsa State in the Southern part of Nigeria, an environment known for oil exploration. The extremely high level of PAHs in the area can be associated with the activities of the Etelebou flow station that discharges liquid effluents into the tributary of Taylor Creek. The contaminant levels were particularly high at Ogboloma, Koroama and Okolobiri of the community. This could be very injurious to the health of humans in the locality and also affect the metabolic and behavioural systems of the aquatic organisms adversely [125].

Similarly, PAHs in the range of 310–528,000 µg/kg dry weight was recorded in the surficial marine sediments obtained from Kitimat Harbour, Canada. The pollution reported in the harbour sediments could principally come from an aluminium smelter at the head of Kitimat Arm through atmospheric deposition, run-off water and effluent discharge. Remarkably high PAH levels (1×10^7 µg/kg) were even obtained from one of the settling ponds at Alcan aluminium smelter site. The outfall from the site reportedly flows directly into the harbour while

Sample source	Obtained concentration (μg/kg dry weight)	% HPAHs	Reference
Europe			
West Mediterranean Sea (French Riviera, Corsica, Sardinia)	1.5–20,440	–	[141, 142]
The Czech Republic (Industrial)	3500–61,700	–	[143, 144]
Inshore coastal areas around Malta	0–14,990	–	[145]
Asia			
Gulf and the Gulf of Oman, UAE	0.6–9.4	–	[146]
Coastal of Bushehr, Persian Gulf	41.7–227.5	–	[25]
Industrial Port Area of Southern Kaohsiung Harbour	4425–51,261	56.3	[75]
Mumbai Harbour Line, India	17–134,134	–	[132]
Gulf and the Gulf of Oman, Qatar	0.55–92	–	[146]
Khure-Musa Estuarine, Persian Gulf	703–3302 (total)	80.9	[140]
Gulf and the Gulf of Oman, Bahrain	13–6600	–	[146]
Marine Environment, Korea	8.80–18,500	–	[147]
Coastal and Estuarine Areas of the Northern Bohai and Yellow Seas, China	52.3–1871	78.3	[148]
Gulf and the Gulf of Oman, Oman	1.6–30	–	[146]
Delhi, India	920–19,321 (total)	62.4	[8]
Kor River, Iran	167.4–530.3	82.5	[16]
Kyeonggi Bay, Korea	10–1400	–	[58]
Middle of Muggah Creek Estuary, Sydney (Australia)	142 (total)	–	[149, 150]
America			
South San Francisco Bay	~120 to 9560 (total)	–	[151]
Gulf of Naples and nearby Coastal Areas, US	0.34–31.77	–	[2, 152, 153]
Casco Bay, Maine, Texas	16–20,798 (total)	93.6	[154]
Kitimat Harbour, Canada	310–528,000	–	[126]
Fore River and Portland Harbor, Maine	2953–278,300	69	[122]
San Francisco Bay, California	36–6273 (total)	87.5	[155]
Todos Santos Bay, Mexico	7.6–813	–	[156]
Bagnoli Surface Sediment (industrial side of the Gulf of Naples)	0.1–2947	–	[2, 157, 158]
Tabasco state, Mexico	454–3120	–	[159]
Northwest Coast Mediterranean Sea	86.5–48,090	–	[160]
Africa			
Suez Canal, Egypt	103.41–238.76	76.2	[124]

Sample source	Obtained concentration (μg/kg dry weight)	% HPAHs	Reference
Central South Africa (Industrial, Residential and Agricultural)	44–39,000 (total)	–	[144, 161]
Lakes and Rivers around Johannesburg/Pretoria in South Africa	61–45,281	52.3	[20]
Rivers in Thohoyandou, Limpopo Province, South Africa	111.6–61,764 (total)	–	[64]
Taylor Creek, Southern Nigeria (non-tidal freshwater-Bayelsa)	1.781×10^8–1.266×10^9	–	[125]
Mvudi River water, South Africa	266–21,600	55.3	[117]
Nzhelele River water, South Africa	206–13,710	87.4	[117]
Tema Harbour, Ghana	28,600–190,300	–	[102]

Table 5. Levels of PAHs in sediments around the world.

surface run-off from the northern part of the site and effluent from the anode plant wet scrubbers in the area enter through the lagoon, raising the concentration of PAHs in the harbour sediments [126]. Another elevated PAH levels from American sediments were reported from Fore River and Portland Harbour, Maine. The total PAHs at most of the sampling locations were high. For instance, levels detected at the Gas Works/China Clay Docks, Maine State Pier and Casco Bay Ferry Terminal were 278,300, 161,990 and 63,533 μg/kg, respectively. High concentrations at these sites were possibly from combustion-related sources because the ratio of LPAHs to the total PAHs in all the study sites were very low, signifying they are less likely to have come from petrogenic origins [122].

The study conducted in 2011 at the Suez Canal, Port Said in Egypt also revealed the effects of heavy shipment, sewage outfalls, highest traffic density of oil tanker, and discharge of industrial effluents on the health status of the marine sediment at the sites. The total PAHs at Port Said and Suez Harbours were up to 239 μg/kg [124]. Likewise, Gorleku et al. [102] and Dhananjayan et al. [132] reported concentration ranges of 28,600–190,300 μg/kg and 17–134,134 μg/kg in the Tema Harbour, Ghana and Mumbai Harbour Line, India, respectively. Reasons for these levels of PAHs include oil spillage, coastal and river run-off, industrial and domestic wastes discharges. Diagnostic analysis suggested that the PAH assemblages in the Mumbai Harbour Line could have originated from combustion of fossil materials, while the results from Tema Harbour indicated a possible contamination from both petrogenic and pyrogenic sources [102, 132].

Generally, the extremely high levels of PAHs in the freshwater and marine sediments globally were mostly from oil exploration areas, harbours/shipping yards and industrial locations. The compiled results showed that the effects of pollution due to industrial growth were not as high as those from the two other sources. PAHs occurred well above the ERL (4000 μg/kg) in most of the areas where the studies were carried out, and not less than 25% of the sites exceeded the ERM value of 44,792 μg/kg. This implies there is possibility of high biological risk to the aquatic creatures in the freshwater and marine environments discussed if not curtailed within a reasonable period of time [37, 58, 59].

Table 5 revealed that the ratio of high molecular weight PAHs to the total PAHs (expressed in percentage) in the sediments ranges generally from 56.3 to 93.6%, corroborating previously reported works [16, 63, 102, 140] and suggesting sources associated with run-off, air deposition and combustion of fossil materials [122].

4. Conclusion

PAHs are among the major organic pollutants found in the aquatic systems. They are mostly generated from both natural and anthropogenic sources. Many of them are highly carcinogenic in nature and are also linked with endocrine system disruption at levels higher than the maximum concentration limit within a short period of time. Hence, they are frequently determined in the environmental samples. Liquid and gas chromatography remain the most commonly used analytical methods for the qualitative and quantitative determination of the organic contaminants, owing to their several advantages, which include better resolution, high sensitivity and selectivity. HPLC with UV/fluorescence detectors in series and GC/MS have proven to be the best techniques amongst others. GC/MS is the most frequently used because of the advantages of identification using both retention time and mass spectrum, providing added information on the chemical structures of the analyte compounds. It also offers high sensitivity required for quantification in selected ion monitoring (SIM) mode. Although, LC/MS can offer similar advantages, it is limited in that, there is no specific interface suitable for the separation of all the PAHs at the same time.

The results collated revealed that fresh and marine waters in many parts of the world are polluted with PAHs, possibly due to increased urbanization, industrialization, incessant discharge of untreated or partially treated effluents into the water bodies and intense shipping activities, amongst many other inputs. The pollution was more pronounced in the Asian and African continents than other parts of the globe. Similarly, the levels reported for the sediments were in most cases above the ERL, signifying the likelihood of serious biological effects on aquatic organisms worldwide. Particularly high concentrations of PAHs in the sediment compartment were generally from areas characterized with intense oil exploration, shipping and industrial activities. LPAHs which are more soluble in the water phase recorded high percentage in most of the water samples and conversely, HPAHs were found generally higher in the sediment samples with percentage ranging from 56.3 to 93.6%. Therefore, there is need for strict compliance with environmental laws, in order to achieve a very safe environment for humans and biota.

Acknowledgements

Special thanks to South African Medical Research Council and Govan Mbeki Research Development Centre, University of Fort Hare, South Africa for providing funds and the enabling environment to write this article.

Conflict of interest

We declare that there is no conflict of interest concerning the publication of this paper.

Author details

Abiodun Olagoke Adeniji[1,2]*, Omobola Oluranti Okoh[1,2] and Anthony Ifeanyi Okoh[1,3]

Address all correspondence to: adenijigoke@gmail.com

1 SAMRC Microbial Water Quality Monitoring Centre, University of Fort Hare, Alice, South Africa

2 Department of Chemistry, University of Fort Hare, Alice, South Africa

3 Applied and Environmental Microbiology Research Group, Department of Biochemistry and Microbiology, University of Fort Hare, Alice, South Africa

References

[1] WHO (World Health Organization). Guidelines for Drinking-water Quality. 3rd ed. Incorporating the First and Second Addenda. Volume 1 Recommendations [Internet]. 2008. Available from: http://www.who.int/water_sanitation_health/dwq/fulltext.pdf [Accessed: 2017-08-25]

[2] Tornero V, d'Alcalà MR. Contamination by hazardous substances in the gulf of Naples and nearby coastal areas: A review of sources, environmental levels and potential impacts in the MSFD perspective. Science of the Total Environment. 2014;**466-467**:820-840. DOI: 10.1016/j.scitotenv.2013.06.106

[3] WHO (World Health Organization). Persistent Organic Pollutants (POPs): Children's Health and the Environment [Internet]. 2008. Available from: http://www.who.int/ceh/capacity/POPs.pdf [Accessed: 2017-06-16]

[4] Wilson NK, Chuang JC, Lyu C. Levels of persistent organic pollutants in several child day care centers. Journal of Exposure Science and Environmental Epidemiology. 2001; **11**(6):449-458. DOI: 10.1038/sj.jea.7500190

[5] Mai B, Qi S, Zeng EY, Yang Q, Zhang G, Fu J, Sheng G, Peng P, Wang Z. Distribution of polycyclic aromatic hydrocarbons in the coastal region off Macao, China: Assessment of input sources and transport pathways using compositional analysis. Environmental Science & Technology. 2003;**37**:4855. DOI: 10.1021/es034514k

[6] Albers PH. Sources, fate, and effects of PAHs in shallow water environments: A review with special reference to small watercraft. In: Michael Kennish, editor. Impacts of

Motorized watercraft on Shallow Estuarine and Coastal Marine Environments. Journal of Coastal Research (special issue). 2002;**37**:143-150 [Internet]. Available from: https://www.jstor.org/stable/pdf/25736349.pdf?refreqid=excelsior:6ca468674d04aa5a13deb051bdf96f68 [Accessed: 17-05-2015]

[7] Scally K. The use of forensic polycyclic aromatic hydrocarbon signatures and compound ratio analysis techniques (CORAT) for the source characterisation of petrogenic/pyrogenic environmental releases [thesis]. Galway Mayo Institute of Technology (GMIT); 2005

[8] Kumar B, Verma VK, Gaur R, Kumar S, Sharma CS, Akolkar AB. Validation of HPLC method for determination of priority polycyclic aromatic hydrocarbons (PAHs) in waste water and sediments. Advances in Applied Science Research. 2014;**5**(1):201-209

[9] CCME (Canadian Council of Ministers of the Environment). Canadian Soil Quality Guidelines for Carcinogenic and Other Polycyclic Aromatic Hydrocarbons (PAHs) (Environmental and Human Health Effects). Scientific Criteria Document (Revised), Publication No. 1445 [Internet]. 2010. Available from: http://www.ccme.ca/files/Resources/supporting_scientific_documents/pah_soqg_scd_1445.pdf [Accessed: 2016-06-13]

[10] Brazkova M, Krastanov A. Polycyclic aromatic hydrocarbons: Sources, effects and biodegradation. In: Proceedings of the International Scientific Conference of University of Ruse, Razgrad, Bulgaria; 2013;**52**(10.2):1-5

[11] Amdany R. Passive Samplers: Development and Application in Monitoring Organic Micropollutants in South African Water Bodies and Wastewater [thesis]. South Africa: University of the Witwatersrand, Johannesburg; 2013

[12] Alegbeleye OO. Bioremediation of Polycyclic Aromatic Hydrocarbons (PAHs) in water using indigenous microbes of Diep- and Plankenburg Rivers, Western Cape, South Africa [thesis]. South Africa: Cape Peninsula University of Technology; 2015

[13] Irwin RJ, van Mouwerik M, Stevens L, Seese MD, Basham W. Environmental Contaminants Encyclopedia: PAHs Entry [Internet]. 1997. Available from: http://fliphtml5.com/cbzk/bilj/basic [Accessed: 2017-05-11]

[14] Prabhukumar G, Pagilla K. Polycyclic Aromatic Hydrocarbons in Urban Runoff—Sources, Sinks and Treatment: A Review. DuPage River Salt Creek Workgroup [Internet]. 2010. Available from: http://drscw.org/dissolvedoxygen/DPAH1.pdf [Accessed: 2015-05-03]

[15] Whittle KJ, Hardy R, Mackie PR, McGill AS, Straughan D, Crisp DJ, Baker JM, Bonner WN. A quantitative assessment of the sources and fate of petroleum compounds in the marine environment. Philosophical Transactions of the Royal Society of London B. 1982;**297**:193-218. DOI: 10.1098/rstb.1982.0038

[16] Kafilzadeh F, Shiva AH, Malekpour R. Determination of polycyclic aromatic hydrocarbons (PAHs) in water and sediments of the Kor River, Iran. Middle-East Journal of Scientific Research. 2011;**10**(1):01-07

[17] CCME (Canadian Council Of Ministers of the Environment). Canadian Water Quality Guidelines for the Protection of Aquatic Life: Polycyclic Aromatic Hydrocarbons (PAHs). Publication No. 1299 [Internet]. 1999. Available from: http://ceqg-rcqe.ccme.ca/download/en/201/ [Accessed: 2015-05-03]

[18] Li H, Chen J, Wu W, Piao X. Distribution of polycyclic aromatic hydrocarbons in different size fractions of oil from a coke oven plant and its relationship to organic carbon content. Journal of Hazardous Materials. 2010;**176**:729-234. DOI: 10.1016/j.jhazmat.2009.11.095

[19] Niemirycz E, Gozdek J, Maron D. Variability of organic carbon in water and sediments of Odra River and its tributaries. Polish Journal of Environmental Studies. 2006;**15**:557-553

[20] Sibiya PN. Modification, development and application of extraction methods for PAHs in sediments and water [thesis]. South Africa: University of the Witwatersrand, Johannesburg; 2012

[21] NIOSH (National Institute for Occupational Safety and Health). Method 5506: Polynuclear Aromatic Hydrocarbons by HPLC. Fourth Edition, Issue 3 [Internet]. 1998. Available from: https://www.cdc.gov/niosh/docs/2003-154/pdfs/5506.pdf [Accessed: 2017-08-04]

[22] Lerda. Polycyclic Aromatic Hydrocarbons (PAHs) Factsheet, 4th Edition. European Union Joint Research Centre. JRC Technical Notes, 66955-2011 [Internet]. 2011. Available from: https://ec.europa.eu/jrc/sites/jrcsh/files/Factsheet%20PAH_0.pdf [Accessed: 2017-07-18]

[23] Stogiannidis E, Laane R. Source characterization of polycyclic aromatic hydrocarbons by using their molecular indices: An overview of possibilities. Reviews of Environmental Contamination and Toxicology. 2015;**234**:49-133. DOI: 10.1007/978-3-319-10638-0_2

[24] Yan J, Wang L, Fu PP, Yu H. Photomutagenicity of 16 polycyclic aromatic hydrocarbons from the US EPA priority pollutant list. Mutation Research. 2004;**557**(1):99-108. DOI: 10.1016/j.mrgentox.2003.10.004

[25] Mirza R, Mohammady M, Dadoloahi A, Safahieh AR, Savari A, Hajeb P. Polycyclic aromatic hydrocarbons in seawater, sediment and oyster (Saccostrea cucullata) from the northern part of the Persian Gulf (Bushehr Province). Water, Air and Soil Pollution. 2011;**233**(6):189-198. DOI: 10.1007/s11270-011-0850-5

[26] Lopez-Avila V, Young R, Benedicto J, Ho P, Kim R, Beckert WF. Extraction of organic pollutants from solid samples using microwave energy. Analytical Chemistry. 1995;**67**:2096-2102. DOI: 10.1021/ac00109a031

[27] Abbas AO, Brack W. Polycyclic aromatic hydrocarbons in Niger Delta soil: Contamination sources and profiles. International Journal of Environmental Science and Technology. 2005;**2**(4):343-352. DOI: 10.1007/BF03325895

[28] Elordui-Zapatarietxe S, Moore B, Bowkett M. Development of a Standard Industry Method for Total Petroleum Hydrocarbon (TPH) Monitoring which Fully Meets Regulatory

Requirements [Internet]. 2008. Available from: https://www.qub.ac.uk/research-centres/ATWARM/DisseminationTemplates/PostersMay2011/Filetoupload,243776,en.pdf [Accessed: 2016-12-17]

[29] Sabin LD, Maruya KA, Lao W, Diehl D, Tsukada D, Stolzenbach KD, Schiff KC. Exchange of polycyclic aromatic hydrocarbons among the atmosphere, water, and sediment in coastal embayments of Southern California, USA. Environmental Toxicology and Chemistry. 2010;**29**(2):265-274. DOI: 10.1002/etc.54

[30] Oluseyi T, Olayinka K, Alo B, Smith RM. Comparison of extraction and clean-up techniques for the determination of polycyclic aromatic hydrocarbons in contaminated soil samples. African Journal of Environmental Science and Technology. 2011;**5**(7):482-493. DOI: 10.5897/AJEST10.307

[31] Qi W, Liu H, Qu J, Hu C, Lan H, Berg M, Ren H, Xu W. Polycyclic aromatic hydrocarbons in effluents from wastewater treatment plants and receiving streams in Tianjin, China. Environmental Monitoring and Assessment. 2011;**177**:467-480. DOI: 10.1007/s10661-010-1648-4

[32] Kannan K, Kober JL, Khim JS, Szymczyk K, Falandysz J, Giesy JP. Polychlorinated biphenyls, polycyclic aromatic hydrocarbons and alkylphenols in sediments from the Odra River and its tributaries, Poland. Environmental Toxicology and Chemistry. 2003;**85**(4-6):51-60. DOI: 10.1080/0277221042000

[33] Sany SBT, Rezayi M, Hashim R, Salleh A, Mehdinia A, Safari O. Polycyclic aromatic hydrocarbons in coastal sediment of Klang Strait, Malaysia: Distribution pattern, risk assessment and sources. PLOS ONE. 2014;**9**(4):1-14. https://doi.org/10.1371/journal.pone.0094907

[34] Dahle S, Savinov V, Petrova V, Klungsøyr J, Savinova T, Batova G, Kursheva A. Polycyclic aromatic hydrocarbons (PAHs) in Norwegian and Russian Arctic marine sediments: Concentrations, geographical distribution and sources. Norwegian Journal of Geology. 2006;**86**:41-50

[35] Havenga WJ, Rohwer ER. The use of SPME and GC/MS for the chemical characterization and assessment of PAH pollution in aqueous environmental samples. International Journal of Environmental Analytical Chemistry. 2000;**78**:3-4, 205-221. DOI: 10.1080/03067310008041342

[36] Pampanin DM, Sydnes MO. Polycyclic aromatic hydrocarbons a constituent of petroleum: Presence and influence in the aquatic environment. In: Kutcherov V, Kolesnikov A, editors. Physical and Theoretical Chemistry: Hydrocarbon. Croatia: InTech; 2013. DOI:10.5772/48176

[37] Hussein RA, Al-Ghanim KA, Abd-El-Atty MM, Mohamed LA. Contamination of Red Sea Shrimp (*Palaemon serratus*) with polycyclic aromatic hydrocarbons: A health risk assessment study. Polish Journal of Environmental Studies. 2016;**25**(2):615-620. DOI: 10.15244/pjoes/60767

[38] Zhang W, Zhang S, Wan C, Yue D, Ye Y, Wang X. Source diagnostics of polycyclic aromatic hydrocarbons in urban road runoff, dust, rain and canopy through fall. Environmental Pollution. 2008;**153**:594-601. http://dx.doi.org/10.1016/j.envpol.2007.09.004

[39] Saha M, Togo A, Mizukawa K, Murakami M, Takada H, Zakaria MP, Chiem NH, Tuyen BC, Prudente M, Boonyatumanond R, Sarkar SK, Bhattacharya B, Mishra P, Tana TS. Sources of sedimentary PAHs in tropical Asian waters: Differentiation between pyrogenic and petrogenic sources by alkyl homolog abundance. Marine Pollution Bulletin. 2009;**58**:189-200. DOI: 10.1016/j.marpolbul.2008.04.049

[40] Riccardi C, Di Filippo P, Pomata D, Di Basilio M, Spicaglia S, Buiarelli F. Identification of hydrocarbon sources in contaminated soils of three industrial areas. Science of the Total Environment. 2013;**450**:13-21. DOI: 10.1016/j.scitotenv.2013.01.082

[41] Jiang J, Lee C, Fang M, Liu JT. Polycyclic aromatic hydrocarbons in coastal sediments of southwest Taiwan: An appraisal of diagnostic ratios in source recognition. Marine Pollution Bulletin. 2009;**58**:752-760. DOI: 10.1016/j.marpolbul.2008.12.017

[42] Zakaria MP, Takada H, Tsutsumi S, Ohno K, Yamada J, Kouno E, Kumata H. Distribution of polycyclic aromatic hydrocarbons (PAHs) in rivers and estuaries in Malaysia: A widespread input of petrogenic PAHs. Environmental Science and Technology. 2002;**36**:1907-1918. DOI: 10.1021/es011278+

[43] Khairy MA, Kolb M, Mostafa AR, El-Fiky A, Bahadir M. Risk assessment of polycyclic aromatic hydrocarbons in a Mediterranean semi enclosed basin affected by human activities (Abu Qir Bay, Egypt). Journal of Hazardous Materials. 2009;**170**:389-397. DOI: 10.1016/j.jhazmat.2009.04.084

[44] Zhang XL, Tao S, Liu WX, Yang Y, Zuo Q, Liu SZ. Source diagnostics of polycyclic aromatic hydrocarbons based on species ratios: A multimedia approach. Environmental Science and Technology. 2005;**39**:9109-9114. DOI: 10.1021/es0513741

[45] Tobiszewski M, Namiesnik J. PAH diagnostic ratios for the identification of pollution emission sources. Environmental Pollution. 2012;**162**:110-119. DOI: 10.1016/j.envpol.2011.10.025

[46] Dickhut RM, Canuel EA, Gustafson KE, Liu K, Arzayus KM, Walker SE, Gaylor MO, MacDonlad EH. Automotive sources of carcinogenic polycyclic aromatic hydrocarbons associated with particulate matter in the Chesapeake Bay Region. Environmental Science & Technology. 2000;**34**:4635-4640. DOI: 10.1021/es000971e

[47] Tobiszewski M. Application of diagnostic ratios of PAHs to characterize the pollution emission sources. In: Proceedings of the 5th International Conference on Environmental Science and Technology. IPCBEE. 2014;**69**:41-44. DOI: 10.7763/IPCBEE

[48] Kumar B, Tyagi J, Verma VK, Gaur R, Sharma CS. Concentrations, source identification and health risk of selected priority polycyclic aromatic hydrocarbons in residential street soils. Advances in Applied Science Research. 2014;**5**(3):130-139

[49] Garrigues P, Budzinski H, Manitz MP, Wise SA. Pyrolytic and petorgenic inputs in recent sediments: A definitive signature through phenanthrene and chrysene compounds distribution. Polycyclic Aromatic Compounds. 1995;**7**:275-284. DOI: 10.1080/10406639508009630

[50] Yunker MB, Macdonald RW, Vingarzan R, Mitchell RH, Goyette D, Sylvestre S. PAHs in the Fraser River basin: A critical appraisal of PAH ratios as indicators of PAH source and composition. Organic Geochemistry. 2002;**33**:489-515. DOI: 10.1016/S0146-6380(02)00002-5

[51] Peters CA, Knightes CD, Brown DG. Long-term composition dynamics of PAH-containing NAPLs and implications for risk assessment. Environmental Science and Technology. 1999;**33**:4499-4507. DOI: 10.1021/es981203e

[52] CNRS (Centre national de la recherche scientifique). CLP Criteria (Classification, Labelling and Packaging for Substances and Mixtures) [Internet]. 2011. Available from: www.prc.cnrs-gif.fr [Accessed: 2017-07-25]

[53] UNEP (United Nations Environment Programme). Evaluation of Urban Pollution of Surficial and Groundwater Aquifers in Africa Project [Internet]. 2002. Available from: https://wedocs.unep.org/bitstream/handle/20.500.11822/8306/-Evaluation%20of%20Urban%20Pollution%20of%20Surficial%20and%20Groundwater%20Aquifers%20in%20Africa-20023725.pdf?sequence=3&isAllowed=y [Accessed: 2017-08-21]

[54] IARC (International Agency for Research on Cancer). Monographs on the Evaluation of Carcinogenic Risks to Humans, Volume 92 [Internet]. 2010. Available from: http://monographs.iarc.fr/ENG/Monographs/vol92/index.php. [Accessed: 2017-04-05]

[55] Van den Berg M, Birnbaum LS, Denison M, De Vito M, Farland W, Feeley M, Fiedler H, Hakansson H, Hanberg A, Haws L, Rose M, Safe S, Schrenk D, Tohyama C, Tritscher A, Tuomisto J, Tysklind M, Walker N, Peterson RE. The 2005 World Health Organization re-evaluation of human and Mammalian toxic equivalency factors for dioxins and dioxin-like compounds. Toxicological Sciences. 2006;**93**(2):223-241. https://doi.org/10.1093/toxsci/kfl055

[56] Schoeny R, Poirier K. Provisional Guidance for Quantitative Risk Assessment of Polycyclic Aromatic Hydrocarbons. EPA/600/R-93/089 [Internet]. 1993. Available from: https://cfpub.epa.gov/ncea/risk/recordisplay.cfm?deid=49732 [Accessed: 2016-12-23]

[57] Pavlova A, Ivanova R. Determination of petroleum hydrocarbons and polycyclic aromatic hydrocarbons in sludge from wastewater treatment basins. Journal of Environmental Monitoring. 2003;**25**:319-323. DOI: 10.1039/B208157C

[58] Kim GB, Maruya KA, Lee RF, Lee JH, Koh CH, Tanabe S. Distribution and sources of polycyclic aromatic hydrocarbons in sediments from Kyeonggi Bay, Korea. Marine Pollution Bulletin. 1999;**38**(1):7-15. DOI: 10.1016/S0025-326X(99)80006-X

[59] MacDonald DD, Ingersoll CG, Berger TA. Development and evaluation of consensus-based sediment quality guidelines for freshwater ecosystems. Archives of Environmental Contamination and Toxicology. 2000;**39**:20-31. DOI: 10.1007/BF00118995

[60] EPAI (Environmental Protection Agency, Ireland). Parameters of Water quAlity: Interpretation and Standards [Internet]. 2001. Available from: https://www.epa.ie/pubs/advice/water/quality/Water_Quality.pdf [Accessed: 2016-09-18]

[61] Kouzayha A, Iskandarani M, Mokh S, Rabaa A, Budzinski H, Jaber F. Optimization of a solid phase extraction method using centrifugation for the determination of 16 polycyclic aromatic hydrocarbons in water. Journal of Agricultural and Food Chemistry. 2011;**14**:7592-7600. DOI: 10.1021/jf200123v

[62] Kiss G, Varga-Puchony Z, Hlavay J. Determination of polycyclic aromatic hydrocarbons in precipitation using solid-phase extraction and column liquid chromatography. Journal of Chromatography A. 1996;**725**:261-272. DOI: 10.1016/0021-9673(95)00940-X

[63] Manoli E, Samara C. Polycyclic aromatic hydrocarbons in natural waters: Sources, occurrence and analysis. Trends in Anaytical Chemistry. 1999;**18**:417-428. DOI: 10.1016/S0165-9936(99)00111-9

[64] Nekhavhambe TJ, van Ree T, Fatoki OS. Determination and distribution of polycyclic aromatic hydrocarbons in rivers, surface runoff, and sediments in and around Thohoyandou, Limpopo Province, South Africa. Water SA. 2014;**40**(3):415-425. http://dx.doi.org/10.4314/wsa.v40i3.4

[65] Wells MJM. Principles of extraction and the extraction of semivolatile organics from liquids. In: Somenath Mitra, Editor. Sample Preparation Techniques in Analytical Chemistry. John Wiley & Sons; 2003. p. 37-225. DOI: 10.1002/0471457817

[66] Ma J, Xiao R, Li J, Yu J, Zhang Y, Chen L. Determination of 16 polycyclic aromatic hydrocarbons in environmental water samples by solid-phase extraction using multi walled carbon nanotubes as adsorbent coupled with gas chromatography-mass spectrometry. Journal of Chromatography A. 2010;**1217**:5462-5469. http://dx.doi.org/10.1016/j.chroma.2010.06.060

[67] Bispo JRL, Navickiene S, Dórea HS. Method validation for SPE applied to determination of PAH in petroliferous industry effluent water. American Journal of Analytical Chemistry. 2011;**2**:971-978. DOI: 10.1016/j.microc.2006.06.002

[68] Johnson B. Determining Trace Amounts of Contaminants in Water [Internet]. 2011. Available from: http://www.horizontechinc.com/PDF/WCP_article_2011.pdf [Accessed: 2015-07-14]

[69] Filipkowska A, Lubecki L, Kowalewska G. Polycyclic aromatic hydrocarbon analysis in different matrices of the marine environment. Analytica Chimica Acta. 2005;**547**(2):243-254. DOI: 10.1016/j.aca.2005.05.023

[70] Triantafyllaki S, Dassenakis M. Determination of polycyclic aromatic hydrocarbons in seawater by high performance liquid chromatography with photodiode array and fluorescence detection. In: Proceedings of the 9th international conference on environmental science and technology; 1-3 September, 2005; Rhodes Island, Greece. p. 939-944

[71] Hassan J, Izadi M, Homayonnejad S. Application of low density homogeneous liquid-liquid extraction combined with GC for TPH and PAH determination in semi-micro solid samples. Journal of the Brazilian Chemical Society. 2013;**24**(4):639-644. DOI: 10.5935/0103-5053.20130080

[72] Pakpahan EN, Isa MH, Kutty SRM. Polycyclic aromatic hydrocarbons in petroleum sludge cake: Extraction and origin—a case study. International Journal of Applied Science and Technology. 2011;**1**(5):201-207. DOI: 10.1080/09593330.2012.698648

[73] Lau EV, Gan S, Ng HK. Extraction techniques for polycyclic aromatic hydrocarbons in soils. International Journal of Analytical Chemistry. 2010;**2010** Article ID 398381, 9 p. http://dx.doi.org/10.1155/2010/398381

[74] EPA (Environmental Protection Agency). Method 3500C: Organic Extraction and Sample Preparation; Revision 3 [Internet]. 2007. Available from: https://www.epa.gov/sites/production/files/2015-12/documents/3500c.pdf [Accessed: 2017-08-12]

[75] Dong C, Chen C, Chen C. Determination of polycyclic aromatic hydrocarbons in industrial harbor sediments by GC-MS. International Journal of Environmental Research and Public Health. 2012;**9**:2175-2188. DOI: 10.3390/ijerph9062175

[76] Hawthorne SM, Grabanski CB, Martin E, Miller DJ. Comparisons of soxhlet extraction, pressurized liquid extraction, supercritical fluid extraction and subcritical water extraction for environmental solids: Recovery, selectivity and effects on sample matrix. Journal of Chromatograph A. 2000;**892**:421-433. DOI: 10.1016/S0021-9673(00)00091-1

[77] Richter BE. Extraction of hydrocarbon contamination from soils using accelerated solvent extraction. Journal of Chromatography A. 2000;**874**(2):217-224. DOI: 10.1016/S0021-9673(00)00073-X

[78] Wennrich L, Popp P, Möder M. Determination of chlorophenols in soils using accelerated solvent extraction combined with solid-phase micro-extraction. Analytical Chemistry. 2000;**72**:546-551. DOI: 10.1021/ac990463r

[79] Wong PK, Wang J. The accumulation of polycyclic aromatic hydrocarbons in lubricating oil over time—A comparison of supercritical fluid and liquid–liquid extraction methods. Environmental Pollution. 2001;**112**:407-415. DOI: 10.1016/S0269-7491(00)00142-1

[80] Pino V, Ayala JH, Afonso AM, Gonzalez V. Micellar microwave-assisted extraction combined with solid-phase microextraction for the determination of polycyclic aromatic hydrocarbons in a certified marine sediment. Analytica Chimica Acta. 2003;**477**:81-91. DOI: 10.1016/S0003-2670(02)01410-1

[81] Nikolaou A, Kostopoulou M, Lofrano G, Meric S. Determination of PAHs in marine sediments: Analytical methods and environmental concerns. Global NEST Journal. 2009; **11**(4):391-405

[82] Dean JR. Extraction Methods for Environmental Analysis [Internet]. Chichester: John Wiley and Sons Ltd, England; 1998. Available from: http://www.ssu.ac.ir/fileadmin/

templates/fa/daneshkadaha/daneshkadah_pezeshki/goroha/farmacology/Upload_DP_farmacology/book/book_air_analysis.pdf [Accessed: 2017-05-21]

[83] Banjoo DR, Nelson PK. Improved ultrasonic extraction procedure for the determination polycyclic aromatic hydrocarbons in sediments. Journal of Chromatography. 2005; **1006**:9-18. DOI: 10.1016/j.chroma.2005.01.033

[84] Weisman W, editor. Analysis of Petroleum Hydrocarbons in Environmental Media, Total Petroleum hydrocarbon Criteria Working Group Series [internet]. Vol. 1. Amherst Scientific Publishers; 1998. Available from: http://www.hawaiidoh.org/references/TPHCWG%201998.pdf [Accessed: 2017-07-09]

[85] Adeniji AO, Okoh OO, Okoh AI. Analytical methods for the determination of the distribution of total petroleum hydrocarbons in the water and sediment of aquatic systems: A review. Journal of Chemistry. 2017; Article ID 5178937: 13 pages. DOI: 10.1155/2017/5178937

[86] SKC. Polynuclear Aromatic Hydrocarbons—NIOSH 5506, 5515. SKC Chemical Fact File, Publication 1464, Rev 1304 [Internet]. 2013. Available from: https://www.skcinc.com/catalog/pdf/cff/1464.pdf [Accessed: 2017-07-07]

[87] Cai S, Syage JA, Hanold KA, Balogh MP. Ultra-performance liquid chromatography atmospheric pressure photo ionization-tandem mass spectrometry for high-sensitivity and high-throughput analysis of U.S. Environmental Protection Agency 16 priority pollutants polynuclear aromatic hydrocarbons. Analytical Chemistry. 2009;**81**:2123-2128. DOI: 10.1021/ac802275e

[88] Nemcik M. Low-level PAH Analysis Using the Finnigan Surveyor HPLC System with PDA deteCtion: Application Note 341 [Internet]. 2004. Available from: https://tools.thermofisher.com/content/sfs/brochures/App-Note-341-Low-Level-PAH-Analysis-using-the-Surveyor-HPLC-System-with-PDA-Detection.pdf [Accessed: 2017-07-02]

[89] Knauer. Application Note: Rapid Analysis of 17 Polycyclic Aromatic Hydrocarbons with UV- and FL-Detection According to DIN EN 17993:2002 [Internet]. 2011. Available from: http://www.knauer.net/fileadmin/user_upload/produkte/files/Dokumente/application_notes/vev0054n_17_pah_en_iso_17993_uhplc_fld.pdf [Accessed: 2017-08-17]

[90] EPA (Environmental Protection Agency). Method 8100: Polynuclear Aromatic Hydrocarbons [Internet]. 1986. Available from: http://www.epadatadump.com/pdf-files/8100.pdf [Accessed: 2017-06-08]

[91] Emmenegger C, Kalberer M, Morrical B, Zenobi R. Quantitative analysis of polycyclic aromatic hydrocarbons in water in the low-nanogram per liter range with two-step laser mass spectrometry. Analytical Chemistry. 2003;**75**:4508-4513. DOI: 10.1021/ac0340197

[92] EPA (Environmental Protection Agency). Method 8000D Determinative Chromatographic Separations [Internet]. 2003. Available from: https://www.epa.gov/sites/production/files/2015-12/documents/8000d.pdf [Accessed: 2017-05-16]

[93] Sakuma T, Leigh D, Seto C, Schreiber A, Wittrig R. Analysis of Polycyclic Aromatic Hydrocarbons (PAH), Alkylated Derivatives, and Photo-degradation Products in Environmental and Food Samples Using LC-FLD-MS/MS with Q TRAP® Technology [Internet]. 2011. Available from: https://sciex.com/Documents/brochures/PAH_seafood_water_QTRAP4k_4520411.pdf [Accessed: 2017-05-13]

[94] EPA (Environmental Protection Agency). Method 610: Polynuclear Aromatic Hydrocarbons (Methods for Organic Chemical Analysis of Municipal and Industrial Wastewater) [Internet]. 1984. Available from: https://www.epa.gov/sites/production/files/2015-10/documents/method_610_1984.pdf [Accessed: 2017-06-08]

[95] EPA (Environmental Protection Agency). Method 8310: Polynuclear Aromatic Hydrocarbons [Internet]. 1986. Available from: https://www.epa.gov/sites/production/files/2015-12/documents/8310.pdf [Accessed: 2017-06-08]

[96] EPA (Environmental Protection Agency). Method 550.1: Determination of Polycyclic Aromatic Hydrocarbons in Drinking Water by Liquid-Solid Extraction and HPLC with Coupled Ultraviolet and Fluorescence Detection [Internet]. 1990. Available from: https://www.o2si.com/docs/epa-method-550.1.pdf [Accessed: 2017-06-08]

[97] Ana GREE, Sridhar MKC, Emerole GO. Contamination of surface waters by polycyclic aromatic hydrocarbons in two Nigerian coastal communities. Journal of Environmental Health Research. 2011;**11**(2):77-86

[98] Fu R, Zou Y. Analysis of Polynuclear Aromatic Hydrocarbons (PAHs) in Water with ZORBAX Eclipse PAH Column [Internet]. 2008. Available from: https://www.agilent.com/cs/library/applications/5989-7953EN.pdf [Accessed: 2016-06-04]

[99] Cheng C, Lai J, Huang M, Oung J, Shiea J. Chapter 6: Analysis of polar components in crude oil by ambient mass spectrometry. In: Abdel-Raouf ME-S, editor. Crude Oil Emulsions—Composition Stability and Characterization. 2012. p. 107-121. DOI: 10.5772/36261

[100] Hsu CS, Qiant K. High-boiling aromatic hydrocarbons characterized by liquid chromatography-thermospray-mass spectrometry. Energy & Fuels. 1993;**7**:268-272. DOI: 10.1021/ef00038a017

[101] Shibamoto T. Chromatographic Analysis of Environmental and Food Toxicants, Chromatographic Science Series. Vol. 77. CRC Press; 1998. 344 p. Available from: https://drive.google.com/file/d/0B6Vu2Q2dNVv6bFQ4YWJ1ZVJILWc/edit [Accessed: 2017-02-17]

[102] Gorleku MA, Carboo D, Palm LMN, Quasie WJ, Armah AK. Polycyclic aromatic hydrocarbons (PAHs) pollution in marine waters and sediments at the Tema Harbour, Ghana. Academia Journal of Environmental Sciences. 2014;**2**(7):108-115. DOI: 10.15413/ajes.2014.0112

[103] EPA (Environmental Protection Agency). Method 8021B: Aromatic and Halogenated Volatiles by Gas Chromatography Using Photo ionization and/or Electrolytic Conductivity Detectors [Internet]. 2014. Available from: http://www.epadatadump.com/pdf-files/8021b.pdf [Accessed: 2017-08-12]

[104] NIOSH (National Institute for Occupational Safety and Health). Method 5515: Polynuclear Aromatic Hydrocarbons by GC. Fourth Edition, Issue 2 [Internet]. 1994. Available from: https://www.cdc.gov/niosh/docs/2003-154/pdfs/5515.pdf [Accessed: 2017-08-04]

[105] Dionex Corporation. Application Note 95: Polycyclic Aromatic Hydrocarbon Determination by Reversed-phase High-performance Liquid Chromatography [Internet]. 1994. Available from: http://tools.thermofisher.com/content/sfs/brochures/AN-95-Polycyclic-Aromatic-Hydrocarbon-Determination-LPN0570.pdf [Accessed: 2017-08-12]

[106] Buddhadasa S. Methodologies for the Analysis of Petroleum Hydrocarbons Extracted from Contaminated Soils [Thesis]. Australia: Victoria University of Technology; 2002

[107] EPA (Environmental Protection Agency). Method 525.2: Determination of Organic Compounds in Drinking Water by Liquid-Solid Extraction and Capillary Column Gas Chromatography/Mass Spectrometry [Internet]. 1995. Available from: https://www.ssi.shimadzu.com/industry/methods/m_525_2.pdf [Accessed: 2017-06-08]

[108] EPA (Environmental Protection Agency). Method 625: Base/Neutrals and Acids (Methods for Organic Chemical Analysis of Municipal and Industrial Wastewater) [Internet]. 1984. Available from: https://www.epa.gov/sites/production/files/2015-10/documents/method_625_1984.pdf [Accessed: 2017-07-17]

[109] EPA (Environmental Protection Agency). Method 8270D: Semi-volatile Organic Compounds by Gas Chromatography/Mass Spectrometry (GC/MS), Revision 5 [Internet]. 2014. Available from: https://www.epa.gov/sites/production/files/2015-12/documents/8270d.pdf [Accessed: 2017-08-09]

[110] Cam D, Gagni S, Meldolesi L, Galletti G. Determination of polycyclic aromatic hydrocarbons in sediment using solid-phase microextraction with gas chromatography–mass spectrometry. Journal of Chromatographic Science. 2000;**38**:55-60. https://doi.org/10.1093/chromsci/38.2.55

[111] Qingling L, Xiaoqin X, Sen-Chun LF, Xiaoru W. Determination of trace PAHs in seawater and sediment pore-water by solid-phase microextraction (SPME) coupled with GC/MS. Science in China Series B: Chemistry. 2006;**49**(6):481-491. https://doi.org/10.1007/s11426-006-2026-5

[112] EPA (Environmental Protection Agency). Method 8272: Parent and Alkyl Polycyclic Aromatics in Sediment Pore Water by Solid-Phase Micro-extraction and Gas Chromatography/Mass Spectrometry in Selected Ion Monitoring Mode, Revision 0. 2007. Available from: https://www.epa.gov/sites/production/files/2015-12/documents/8272.pdf [Accessed: 2017-08-08]

[113] EPA (Environmental Protection Agency). Method 8410: Gas Chromatography/Fourier Transform Infrared Spectrometry for Semi-volatile Organics: Capillary Column. Revision 1, Update V. 2014. Available from: http://www.epadatadump.com/pdf-files/8410.pdf [Accessed: 2017-08-12]

[114] EPA (Environmental Protection Agency). Method 8275A: Semivolatile Organic Compounds (PAHs and PCBs) in Soils/Sludges and Solid Wastes Using Thermal Extraction/Gas Chromatography/Mass Spectrometry (TE/GC/MS); Revision 1 [Internet]. 1996. Available from: https://www.epa.gov/sites/production/files/2015-12/documents/8275a.pdf [Accessed: 2017-06-08]

[115] ATSDR (Agency for Toxic Substances and Disease Registry). Toxicological Profile for Total Petroleum Hydrocarbon. Department of Health and Human Services; 1999. 315 p. DOI: 10.1201/9781420061888_ch155

[116] Hajisamoh A. Pollution levels of 16 priority PAHs in the major rivers of Southern Thailand. Research and Reviews. Journal of Chemistry. 2013;**2**(1):7-11

[117] Edokpayi JN, Odiyo JO, Popoola OE, Msagati TAM. Determination and distribution of polycyclic aromatic hydrocarbons in rivers, sediments and wastewater effluents in Vhembe District, South Africa. International Journal of Environmental Research and Public Health. 2016;**13**:387. DOI: 10.3390/ijerph13040387

[118] Nukpezah D. Corporate environmental governance in Ghana: studies on industrial level environmental performance in manufacturing and mining [thesis]. Germany: Brandenburg University of Technology, Cottbus; 2010

[119] EU (European Union). EU's Drinking Water Standards: Council Directive 98/83/EC on the Quality of Water Intended for Human Consumption. Adopted by the Council, on 3 November 1998 [Internet]. 1998. Available from: http://www.lenntech.com/applications/drinking/standards/eu-s-drinking-water-standards.htm. [Accessed: 2017-08-12]

[120] EC (European Communities). European Communities (Drinking Water) (No. 2) Regulations 2007 (S.I. 278 of 2007). Handbook on implementation for Water Services Authorities for private water supplies [internet]. 2007. Available from: https://www.epa.ie/pubs/advice/drinkingwater/privatewatersupplieshandbook/Section%202.pdf [Accessed: 2017-08-12]

[121] Cao Z, Wang Y, Mab Y, Xua Z, Shi G, Zhuang Y, Zhua T. Occurrence and distribution of polycyclic aromatic hydrocarbons in reclaimed water and surface water of Tianjin, China. Journal of Hazardous Materials A. 2005;**122**:51-59. DOI: 10.1155/2012/403615

[122] Doan M. Concentrations of Polycyclic Aromatic Hydrocarbons in Surficial Sediments of the Fore River and Portland Harbor, Maine. A Report to the Natural Resource Damage Trustees' Agreement Number: 604195 [Internet]. 2005. Available from: http://www.cascobayestuary.org/wp-content/uploads/2014/07/2005_focb_fore_river_pahs_report6.pdf [Accessed: 2016-11-15]

[123] Kanchanamayoon W, Tatrahun N. Determination of polycyclic aromatic hydrocarbons in water samples by solid phase extraction and gas chromatography. World Journal of Chemistry. 2008;**3**(2):51-54

[124] Al-Agroudy N, Soliman YA, Hamed MA, Zaghloul GY. Distribution of PAHs in water, sediments samples of Suez Canal during 2011. Journal of Aquatic Pollution and Toxicology. 2017;**1**(1):1-10

[125] Okafor EC, Opuene K. Preliminary assessment of trace metals and polycyclic aromatic hydrocarbons in the sediments of Taylor Creek, Southern Nigeria. International Journal of Environmental Science and Technology. 2007;**4**(2):233-240. https://doi.org/10.1007/BF03326279

[126] Simpson CD, Mosi AA, Cullen WR, Reimer KJ. Composition and distribution of polycyclic hydrocarbons in surficial marine sediments from Kitimat Harbour, Canada. Science of the Total Environment. 1996;**181**:265-278. DOI: 10.1016/0048-9697(95)05026-4

[127] Manoli E, Samara C, Konstantinou I, Albanis T. Polycyclic aromatic hydrocarbons in the bulk precipitation and surface waters of Northern Greece. Chemosphere. 2000;**41**:1845-1855. https://doi.org/10.1016/S0045-6535(00)00134-X

[128] Chen BL, Xuan XD, Zhu LZ, Wang J, Gao YZ, Yang K, Shen XY, Lou BF. Distribution of polycyclic aromatic hydrocarbon in surface waters, sediment and soils of Hangzhou City, China. Water Research. 2004;**38**:3558-3568. DOI: 10.1016/j.watres.2004.05.013

[129] Doong RA, Lin YT. Characterization and distribution of polycyclic aromatic hydrocarbon contaminations in surface sediments and water from Gao-ping River, Taiwan. Water Research. 2004;**38**:1733-1744. DOI: 10.1016/j.watres.2003.12.042

[130] Zhou JL, Hong H, Zhang Z, Maskaoui K, Chen W. Multi-phase distribution of Organic micro pollutants in Xiamen Harbour, China. Water Research. 2000;**34**:2132-2150. DOI: 10.1016/S0043-1354(99)00360-7

[131] JunGang L, RenJi X, QingHua Z, JiYan L, ChunYang L, FuSheng W. Primary investigation of the pollution status of polycyclic aromatic hydrocarbons (PAHs) in water and soil of Xuanwei and Fuyuan, Yunnan Province, China. Chinese Science Bulletin. 2009;**54**:3528-3535. https://doi.org/10.1007/s11434-009-0499-2

[132] Dhananjayan V, Muralidharan S, Peter VR. Occurrence and distribution of polycyclic aromatic hydrocarbons in water and sediment collected along the harbour line, Mumbai, India. International Journal of Oceanography. 2012 Article ID 403615: 7 p. http://dx.doi.org/10.1155/2012/403615

[133] Das SK, Routh J, Roychoudhury AN. Sources and historic changes in polycyclic aromatic hydrocarbon input in a shallow lake, Zeekoevlei, South Africa. Organic Geochemistry. 2008;**39**:1109-1112. http://dx.doi.org/10.1016/j.orggeochem.2008.04.005

[134] Amdany R, Chimuka L, Cukrowska E, Kukučka P, Kohoutek J, Vrana B. Investigating the temporal trends in PAH, PCB and OCP concentrations in Hartbeespoort Dam, South Africa, using semipermeable membrane devices (SPMDs). Water SA. 2014;**40**(3):425-436. http://dx.doi.org/10.4314/wsa.v40i3.5

[135] Azevedo D, Gerchon E, Reis E. Monitoring of pesticides and polycyclic aromatic hydrocarbons in water from Paraíba do Sul River Brazil. Journal of Brazilian Chemical Society. 2004;**15**:292-299. http://www.scielo.br/pdf/jbchs/v15n2/19948.pdf [Accessed: 29-08-2017]

[136] Zhang S, Zhang Q, Darisaw S, Ehie O, Wang G. Simultaneous quantification of polycyclic aromatic hydrocarbons (PAHs), polychlorinated biphenyls (PCBs), and pharmaceuticals

and personal care products (PPCPs) in Mississippi river water, in New Orleans, Louisiana, USA. Chemosphere. 2007;**66**:1057-1069. DOI: 10.1016/j.chemosphere.2006.06.067

[137] Gustrafson KE, Dickhut RM. Distribution of polycyclic aromatic hydrocarbons in southern Chesapeake Bay surface water: Evaluation of three methods for determining freely dissolved water concentrations. Environmental Toxicology and Chemistry. 1997;**16**:452-461. DOI: 10.1002/etc.5620160310

[138] Countway RE, Dickhut RM, Canuel EA. Polycyclic aromatic hydrocarbon (PAH) distributions and associations with organic matter in surface waters of the York River, VA Estuary. Organic Geochemistry. 2003;**34**:209-224. DOI: 10.1016/S0146-6380(02)00162-6

[139] Ross JR, Oros DR. Polycyclic aromatic hydrocarbons in the San Francisco Estuary water column: Sources, spatial distributions, and temporal trends (1993-2001). Chemosphere. 2004;**57**:909-920. DOI: 10.1016/j.chemosphere.2004.08.010

[140] Mirza R, Faghiri I, Abed E. Contamination of polycyclic aromatic hydrocarbons in surface sediments of Khure-Musa Estuarine, Persian Gulf. World Journal of Fish and Marine Sciences. 2012;**4**(2):136-141

[141] Baumard P, Budzinski H, Garrigues P, Sorbe JC, Bourgeot T, Bellocq J. Concentration of PAHs (polycyclic aromatic hydrocarbons) in various marine organisms in relation to those in sediments and to tropic level. Marine Pollution Bulletin. 1998;**30**:951-960. DOI: 10.1016/S0025-326X(98)00088-5

[142] Baumard P, Budzinski H, Garrigues P. PAHs in Arachon Bay, France: Origin and biomonitoring with caged organisms. Marine Pollution Bulletin. 1998;**36**:577-586. http://dx.doi.org/10.1016/S0025-326X(98)00014-9

[143] Hilscherova K, Kannan K, Kang YS, Holoubeck I, Machala M, Masunaga S, Nakanishi J, Giesy JP. Characterization of dioxin-like activity of sediments from a Czech River Basin. Environmental Toxicology and Chemistry. 2001;**20**(12):2768-2777. DOI: 10.1002/etc.5620201216

[144] Quinn L, Pieters R, Nieuwoudt C, Borgen AR, Kylinc H, Bouwmana H. Distribution profiles of selected organic pollutants in soils and sediments of industrial, residential and agricultural areas of South Africa. Journal of Environmental Monitoring. 2009;**11**:1647-1657. DOI: 10.1039/B905585A

[145] Axiak V. Monitoring pollution by petroleum hydrocarbons in inshore coastal areas around Malta: 2000-2002. Malta Environment and Planning Authority [internet]. p. 2003. https://era.org.mt/en/Documents/MonitoringPollution_PHC_FinalJan03.pdf [Accessed: 2017-08-12]

[146] Tolosa I, de Mora S, Fowler SW, Villeneuve JP, Bartocci J, Cattini C. Aliphatic and aromatic hydrocarbons in marine biota and coastal sediments from the Gulf and the Gulf of Oman. Marine Pollution Bulletin. 2005;**50**:1619-1633. DOI: 10.1016/j.marpolbul.2005.06.029

[147] Yim UH, Hong SH, Shim WJ. Distribution and characteristics of PAHs in sediments from the marine environment of Korea. Chemosphere. 2007;**68**:85-92. DOI: 10.1016/j.chemosphere.2006.12.032

[148] Jiao W, Wang T, Khim JS, Luo W, Hu W, Naile JE, Giesy JP, Lu Y. PAHs in surface sediments from coastal and estuarine areas of the northern Bohai and Yellow Seas, China. Environmental Geochemistry and Health. 2011;**34**:445-456. DOI: 10.1007/s10653-011-9445-8

[149] Hutcheson MS, Popham DJ, Odense R, Boyle D, Wangersky PJ. Evaluation of Micro-organisms for Assessing the Toxicity of and Mutagenicity of Contaminated Sediments. Seakem Oceanography, Final Report to Supply and Services Canada. Contract No. 03SB.KE603-3-0806. 128 pp + 2 Appendices [Internet]. 1986. Available from: http://publications.gc.ca/collections/collection_2015/mpo-dfo/Fs97-18-108-eng.pdf [Accessed: 2016-12-09]

[150] Stewart PL, White L. A Review of Contaminants on the Scotian Shelf and in Adjacent Coastal Waters: 1970 to 1995 [Internet]. 2001. Available from: http://www.dfo-mpo.gc.ca/Library/261398.pdf [Accessed: 2017-08-12]

[151] Nilsen EB, Rosenbauer RJ, Fuller CC, Jaffe BJ. Sedimentary organic biomarkers suggest detrimental effects of PAHs on estuarine microbial biomass during the 20th century in San Francisco Bay, CA, USA. Chemosphere. 2015;**119**:961-970. https://doi.org/10.1016/j.chemosphere.2014.08.053

[152] Feo ML, Sprovieri M, Gherardi S, Sammartino S, Marsella E. Polycyclic aromatic hydrocarbons and polychlorinated biphenyls in the Harbour of Naples (Southern Italy): Time and spatial distribution patterns. Environmental Monitoring and Assessment. 2011;**174**:445-459. DOI: 10.1007/s10661-010-1469-5

[153] Bergamin L, Romano E, Finoia MG, Bianchi J, Colasanti A, Ausili A. Benthic foraminifera from the coastal zone of Baia (Naples Italy): Assemblage distribution and modification as tools for environmental characterisation. Marine Pollution Bulletin. 2009;**59**:234-244. DOI: 10.1016/j.marpolbul.2009.09.015

[154] Kennicutt MC II, Wade TL, Presley BJ, Requejo AG, Brooks JM, Denoux GJ. Sediment contaminants in Casco Bay, Maine: Inventories, sources, and potential for biological impact. Environmental Science & Technology. 1994;**28**(1):1-15. DOI: 10.1021/es00050a003

[155] Pereira E, Hosttler FD, Luona SN, van Geen A, Fuller CC, Anima RJ. Sedimentary record of anthropogenic and biogenic polycyclic aromatic hydrocarbons in San Fransisco Bay, California. Marine Chemistry. 1999;**64**:99-113. https://doi.org/10.1016/S0304-4203(98)00087-5

[156] Benlahcen KT, Chaoui A, Budzinski H, Bellocq J, Garrigues PH. Distribution and sources of polycyclic aromatic hydrocarbons in some mediterranean coastal sediments. Marine Pollution Bulletin. 1997;**34**:98-305. DOI: 10.1016/S0025-326X(96)00098-7

[157] Romano E, Bergamin L, Ausili A, Pierfranceschi G, Maggi C, Sesta G. The impact of the Bagnoli industrial site (Naples, Italy) on sea-bottom environment. Chemical and textural features of sediments and the related response of benthic foraminifera. Marine Pollution Bulletin. 2009;**59**:245-256. DOI: 10.1016/j.marpolbul.2009.09.017. 616

[158] Albanese S, De Vivo B, Lima A, Cicchela D, Civitillo D, Cosenza A. Geochemical baselines and risk assessment of the Bagnoli brownfield site coastal sea sediments (Naples, Italy). Journal of Geochemical Exploration. 2010;**105**:19-33

[159] Botello AV, Gonzalez C, Diaz G. Pollution by petroleum-hydrocarbons in sediments from continental-shelf of Tabasco State, Mexico. Bulletin of Environmental Contamination and Toxicology. 1991;**47**:565-571. DOI: 10.1007/BF01700947

[160] Macias-Zamora JV, Mendoza-Vega E, Villaescusa-Celaya JA. PAHs composition of surface marine sediments: A comparison to potential local sources in Todos Santos Bay, B.C., Mexico. Chemosphere. 2002;**46**:459-468. https://doi.org/10.1016/S0045-6535(01)00069-8

[161] Nieuwoudt C, Pieters R, Quinn LP, Kylin H, Borgen AR, Bouwman H. Polycylic aromatic hydrocarbons (PAHs) in soil and sediment from industrial, residential and agricultural areas in Central South Africa: An initial assessment. Soil and sediment contamination. 2011;**20**:188-204. DOI: 10.1080/15320383.2011.546443